Water Pollution: Effects and Mitigation Strategi

Water Pollution: Effects and Mitigation Strategies

Editor: Bonnie Wakeling

Callisto Reference
www.callistoreference.com

Callisto Reference,
118-35 Queens Blvd., Suite 400,
Forest Hills, NY 11375, USA

Visit us on the World Wide Web at:
www.callistoreference.com

ISBN: 978-1-64116-121-3 (Hardback)

Printed in China.

Cataloging-in-Publication Data

Water pollution : effects and mitigation strategies / edited by Bonnie Wakeling.
p. cm.
Includes bibliographical references and index.
ISBN 978-1-64116-121-3
1. Water--Pollution. 2. Pollution prevention. I. Wakeling, Bonnie.
TD420 .W38 2019
628.168--dc23

Table of Contents

Preface

Water pollution is caused by a variety of organic, inorganic and macroscopic contaminants from natural and anthropogenic sources. Some significant factors responsible for this contamination are industrial waste, oil spills, sewage discharge, chemicals and acid mine drainage. Water pollution is analyzed for the nature of contamination by a variety of sampling and testing techniques, which can be divided into physical, chemical and biological testing methods. A lot of mitigation strategies, like instituting green infrastructure, municipal wastewater treatment, industrial pollution prevention, etc. are being investigated and implemented with varying degrees of success. This book contains some path-breaking studies in the field of water pollution, its effects and mitigation strategies. It unravels the recent advancements in the field of water pollution mitigation. This book will be immensely beneficial to students and researchers alike.

After months of intensive research and writing, this book is the end result of all who devoted their time and efforts in the initiation and progress of this book. It will surely be a source of reference in enhancing the required knowledge of the new developments in the area. During the course of developing this book, certain measures such as accuracy, authenticity and research focused analytical studies were given preference in order to produce a comprehensive book in the area of study.

This book would not have been possible without the efforts of the authors and the publisher. I extend my sincere thanks to them. Secondly, I express my gratitude to my family and well-wishers. And most importantly, I thank my students for constantly expressing their willingness and curiosity in enhancing their knowledge in the field, which encourages me to take up further research projects for the advancement of the area.

Editor

Ecotoxicity Evaluation of Industrial Discharge Waters and Metallic Solutions using Two Organisms (*Lactuca sativa and Daphnia magna*)

Anne Priac, Amandine Poupeney, Coline Druart and Grégorio Crini*
Chrono-Environment, UMR 6249 usc INRA, University of Franche-Comté/CNRS, 16 route de Gray, 25000 Besançon, France

Abstract

Surface treatment industrial discharge water is a complex anthropogenic source of pollutants, including organic pollutants (PAHs, VOCs...) and numerous metal ions. We attempted to identify the main toxicants comparing impact assessment of real polycontaminated effluents and reconstituted polymetallic solutions via ecotoxicological bioassays performed with *Daphnia magna* immobilization test (24 h) and *Lactuca sativa* germination test (168 h). We focused first on 2 (Ni and Zn) then on 5 metals (Ni, Zn, Co, Cr, Al). Our results showed differences between metal toxicity order: Zn>Al>Ni>Cr~Co, for daphnids and Ni>Zn>Al~Co>Cr for lettuce. However, discharge waters remained more toxic than synthetic solutions: those 5 metals were not entirely responsible for the discharge water ecotoxicity. We also found *D. magna* to be more sensitive than *L. sativa*. This last assessment should be interpreted with care, knowing that immobilization and germination tests are respectively acute and chronic toxicity bioassays. Thus, battery tests are appropriate to evaluate industrial discharge water samples, and should be increasingly used as eco toxicological standards.

Keywords: Bioassay; Heavy metals; Reconstituted solutions; Waste water

Introduction

Industrial discharge waters, especially those from the surface treatment (ST) industry, released into the aquatic ecosystems have their own set of various environmental and sanitary issues, due to the fact that various loads of hazardous substances including: metallic trace elements (MTE; mostly Zn, Ni, Cu, Cr, Sn and Al), organic matter (oils, solvents, etc.) and diverse organics such as polycyclic aromatic hydrocarbons (PAHs) and volatile organic compounds (VOCs) [1]. ST industries are, like other industrial sectors, subject to specific release regulations, notably for metals. Although the discharge waters usually respect the regulatory standards, the present metals could be assimilated by fauna and flora and thus lead to long term toxic effects on the environment [2,3]. Nowadays, while pollutant mixtures present in discharge water after treatment are relatively easy to characterize chemically, assessing their impact on the environment is usually difficult and has rarely been reported [4]. Finally, the toxicity of treated ST waste remains poorly defined.

To assess the biological and chemical quality of water, 4 main kinds of approach can be used: (1) Chemical analysis to characterize the water mass studied qualitatively and quantitatively, (2) Comparing the analytical data to ecotoxicological information available in the literature to reach an *a priori* assessment of the hazard of substances (as in Draft Assessment Reports for pesticides), (3) Laboratory bioassays to assess the toxicity of substances and (4) *in situ* studies using native organisms or via active bio indication to assess the risk of natural populations exposed to substances released in the environment. Laboratory bioassays for water quality assessment are numerous and offer a large choice of indicators [5-7]. Three different types of standardized bioassays are the most commonly used, notably for the regulatory framework for chemicals management. They represent 3 trophic levels: primary producers with algae, primary consumers with crustaceans and secondary consumers with fish. Among them, the short-term bioassay based on the immobilization of a freshwater crustacean, *Daphnia magna*, is a test also used in the ecotoxicological assessment of industrial discharge waters. Nevertheless, it was pointed out that toxicity strongly relies on the choice of bio indicators and the endpoints used in the bioassays since sensitivity varies among taxonomic groups and species [8-10]. Consequently, it may be very useful to assess discharge water thanks to various bio-indicators in order to increase the ecological representativeness, to include a panel of sensitivity and to avoid a major risk of environmental effects and toxicity underestimation [11,12]. Recently, phytotoxicity tests using plants such as *Lactuca sativa* have been also proposed to assess the impact of industrial effluents by our group for the first time [3]. Our results demonstrated that these tests were simple, quick and reliable. Moreover, the use of these bioassays also presented the advantage of being inexpensive and not requiring major equipment as also reported in other works [8,10,13]. However, these tests were mainly used to assess the toxicity of single substances, such as metals (Table 1) [10,14-28] and there is a lack of studies concerning the impact of complex matrices such as discharge waters [29] or synthetic solutions of several metals.

The aim of this work was to assess the environmental impact of industrial discharge waters poly-contaminated with metals and to determine which metal(s) is (are) most responsible for the toxicity through the use of 2 bio-indicators *Daphnia magna* and *Lactuca sativa* via reconstituted solutions.

Materials and methods

Toxicity bioassays

Standardized germination tests [30] were performed following the method previously described in detail by Charles et al. [3]. The test assessed the germination of 30 plump lettuce seeds (*Lactuca sativa*

***Corresponding author:** Grégorio Crini, Chrono-Environment, UMR 6249 usc INRA, University of Franche-Comté/CNRS, 16 route de Gray, 25000 Besançon, France, E-mail: gregorio.crini@univ-fcomte.fr

Element	Bioassay indicator	Index	Endpoint	Concentration [mg L^{-1}]	Reference
Zn	*L.sativa (var n.r.)*	96h EC_{50}	Root elongation	1	[14]
	P.subcapitata	72h EC_{50}	Growth rate	0.042	[15]
	D.magna	48h LC_{50}	Death	0.970	[16]
	G.pulex	48h LC_{50}	Death	4.920	[17]
Ni	*L.sativa (var Tro.)*	NOEC	Growth rate	1.8	[10]
	P.subcapitata	96h EC_{50}	Population	0.233	[18]
	D.magna	48h LC_{50}	*n.r.*	6.9	[19]
	G.sp	96h LC_{50}	*n.r.*	13	[20]
Cr	*L.sativa (var Rav.)*	72h EC_{50}	Growth rate	5.9	[21]
	P.subcapitata	72h EC_{50}	Population	0.030	[22]
	D.magna	48h LC_{50}	Immobilization	0.290	[23]
	G.pulex	48h LC_{50}	Death	0.809	[24]
Co	*D.magna*	48h LC_{50}	Death	4.4	[25]
Al	*D.magna*	48h LC_{50}	Immobilization	3.9	[26]
Cu	*L.sativa (var n.r.)*	96h EC_{50}	Root elongation	3	[14]
	P.subcapitata	72h EC_{50}	Growth rate	0.020	[15]
	D.magna	48h LC_{50}	Death	0.0111	[27]
	G.pulex	48h LC_{50}	Death	0.047	[28]

Specific lettuce varieties: *n.r.*, non reported; *Tro.*, Trocadero; *Rav.*, Ravel
End-points: LC_{50}, lethal concentration for 50% of the individuals tested; NOEC, no observed effect concentration.

Table 1: Toxicities of metallic trace elements (published data) on different organisms (*Lactuca sativa*, *Daphnia magna*, *Pseudoskirchneriella subcapitata* and *Gammarus pulex* or *sp.*).

Sample	Concentrations [mg L^{-1}]		EC_{50} [% of DW]	Germination (%)
	Ni	Zn	*D. magna*	*L. sativa*
DW1	0.34	1.91	17	54
DW2	0.28	1.51	5.2	58
DW3	0.54	2.46	21	42
DW4	0.35	1.84	11.3	48
DW5	0.51	2.06	32	44

Every control GR was higher than the required 90% of seed germination
All concentrations were above the quantification limits

Table 2: Ecotoxicity (EC_{50} and GR) of different discharge water samples (DW1 to DW5) on *Daphnia magna* and *Lactuca sativa* respectively in relation to Zn and Ni concentrations.

(L.) *var* Batavia Dorée de Printemps) watered with Reverse Osmosis Water ROW (controls; pH=6 ± 0.2), Discharge Water DW or Synthetic Solution SS, in triplicates, for 7 days, in the dark at 24 ± 1°C, on a filter paper substrate. As recommended by the standard, DW or SS pH must be between 5.5 and 9. To validate the test, germination rates (GR) of controls must be higher than or equal to 90%.

Bioassays using *Daphnia magna* were carried out by an accredited analysis laboratory (Carso, Lyon, France). The test was performed according to the "Inhibition Protocol Mobility" described in the standardized biomonitoring test ISO [31].

Industrial discharge waters & synthetic solutions

Five DW (denoted DW1 to DW5) were firstly collected in a ST company in Franche-Comté over a one-year period. Effluents were average sample characteristic of that day's activity. As the 2 main issues to be dealt with in DWs were Ni and Zn (specific company threshold emission values for these 2 MTE were 3.5 mg L^{-1}), DW5 Ni and Zn concentrations were mimicked in single and binary solutions S: S1 (Zn=2 mg L^{-1}), S2 (Ni=0.5 mg L^{-1}) and S3 (Zn=2 mg L^{-1} and Ni=0.5 mg L^{-1}). Each solution ecotoxicity was evaluated with germination test. We also determined Ni and Zn EC_{50} (concentrations range: 0 to 300 mg L^{-1}) for *L.sativa* and *D.magna*.

Four others DWs (DW6 to DW9) were then collected in the same company. Ni, Zn, Al, Cr and Co concentrations of these DWs were mimicked in mixture SS denoted SS6 to SS9. Each solution ecotoxicity was evaluated with germination and immobilization test (Table 2). We also determined Al, Co and Cr EC_{50} (concentrations range: 0 to 1000 mg L^{-1}) for *L.sativa* and *D.magna*.

For each of these 9 DWs, EC_{50} (expressed in percentage of DW) was determined through lettuce germination and daphnids immobilization tests. DWs samples were diluted with ROW. Every metallic synthetic solutions were prepared in ROW from sulfate salts of Al, Co, Cr, Ni and Zn (purchased from Fisher Scientific, France).

Chemical analyses

For each DW sample and synthetic solution, pH was determined (pH meter, model 3110, WTW, Alès, France). Metal concentrations were measured by spectrophotometry (cuvette test and/or reagent tests; portable Spectroflex 6100, WTW, Alès, France) or by ICP-AES (ThermoFisher, iCAP 6500 radial model, Courtaboeuf, France) after acid digestion for DWs, following a previously reported method [1]. All results are expressed in mg L^{-1}.

Statistical analysis

Germination rates of control, DW5, S1, S2 and S3 were compared using the Kruskal-Wallis test, with a significance threshold of $p<0.05$. All statistical analyses were performed with R (2.15.1) (R Development Core Team, 2013). Dose-dependent curves and EC_{50} values were calculated with Hill's model using the macro Excel Regtox free version EV 7.0.6.

Results and discussion

The toxicity of the first 5 DWs was studied through 2 bio-indicators (Table 2). The results showed deleterious effects on both bio-indicators since EC_{50} were low for daphnids (below 32%) and lettuce seed germination rates were significantly lower than those of controls (>90%). Due to activities of the industry focused on in our study, investigations of toxicity were firstly led on Ni and Zn. Concentrations

Bio-indicator	EC_{50} [mg L^{-1}]				
	Al	Co	Cr	Ni	Zn
D.magna	8.45	11.69	10.36	9.8	6.35
L.sativa	237	247.7	265	58.3	154.3

Table 3: EC_{50} values for daphnia and lettuce for 5 MTE detected in DW samples ICP-AES analysis (Al, Co, Cr, Ni and Zn).

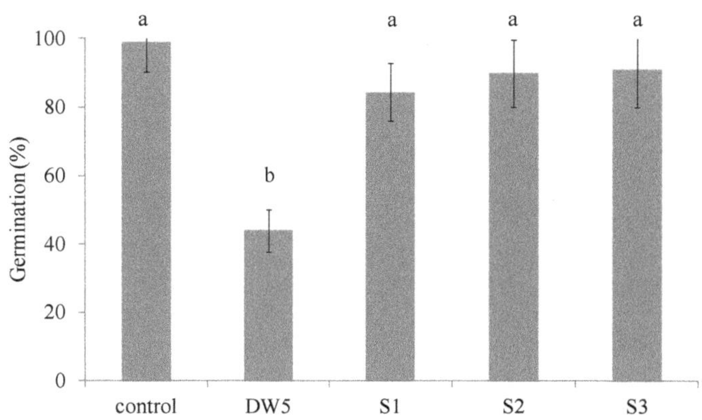

Figure 1: Germination (%) of *L. sativa* seeds for the DW5 sample and synthetic solutions S1, S2 and S3. Different letters indicated significant differences (p-value <0.01).

Sample	Metal ion [mg L^{-1}]					EC_{50} [% of DW or SS]	
	Al	Co	Cr	Ni	Zn	*D. magna*	*L. sativa*
DW6	5.09	2.75	0.15	0.62	2.67	6.1	45
SS6	5.06	2.72	0.15	0.63	2.69	56.8	86
DW7	5.70	4.05	0.35	0.74	1.97	5	68
SS7	5.58	4.06	0.24	0.70	1.96	52.1	84
DW8	2.66	1.69	0.24	0.31	1.45	ND	66
SS8	3.64	2.28	0.29	0.32	1.49	72.2	84
DW9	5.36	3.58	0.25	0.40	2.05	18.4	68
SS9	5.26	3.71	0.26	0.41	2.79	47.9	86

ND, not determined
All concentrations were above the quantification limits

Table 4: Concentrations of 5 metals (mg L^{-1}) in 4 discharge waters (DW6 to 9) and synthetic solutions (SS6 to 9) in relation to toxicity on *D. magna* and *L. sativa* (EC_{50} in % of DW or SS).

of both these suspected toxicants are presented in Table 2 and showed daily variability (as previously reported by Charles et al. [3]). From an analytical point of view, the chemical composition in Ni and Zn can be ranked as follows: DW3 > DW5 > DW1 ~ DW 4 > DW2. For the 2 bioassays, the sample toxicity range (decreasing order) was:

- DW2 > DW4 > DW1 > DW3 > DW5 for *D. magna* and
- DW3 > DW5 > DW4 > DW1 > DW2 for *L. sativa*.

The more toxic DWs for *D. magna* were the less toxic for *L. sativa*. This was also confirmed by Castillo et al. [32] studying the impact of final tannery industrial effluent (daphnids EC_{50} 24 h=77.9%) and lettuce ($EC_{50\text{-root growth}}$ 120 h>90%). Despite its low coefficient of variation CV (14%), it appeared that lettuce DW toxicity could be linked to Ni and Zn concentrations (except for DW4 and DW1 which were inverted). For daphnids, no correlation was shown (CV=59%). This variability was explained by the production activity, as suggested by Hitchcock et al. [33] who calculated a CV reaching 133.7% for the mortality of nematodes exposed to industrial effluents (pulp and paper industries).

To verify the hypothesis that Ni and Zn concentrations in the DW can be linked to lettuce ecotoxicological response, we ran germination tests on synthetic solutions S1, S2 and S3 containing Ni and Zn, alone or in a mixture, in the same concentrations as those found in DW5. We also performed ecotoxicological tests on both *D. magna* and *L. sativa* (Table 3), to assess individual EC_{50} of nickel and zinc. The results showed that GR of single (S1 and S2) and binary (S3) synthetic solutions were not significantly different from the control (Figure 1). This result was not surprising in regard to the values of EC_{50} determined in *L. sativa* for Ni and Zn (Table 3) which were far above the concentrations found in DWs. However, the EC_{50} results were not expected considering those found in the literature (Table 1). Indeed, toxicity values for other endpoints were much lower, of the order of 1 mg L^{-1}, both for Zn and Ni [10]. This major disparity could be explained by the variety of lettuce used for the assay, as criticized by Priac et al. (unpublished work) who demonstrated that among 4 varieties, Batavia (used in the present paper) was the least sensitive. Significant differences were found between the germination rates of lettuce exposed to synthetic solutions of Zn and Ni and those exposed to the DWs at same concentrations (Figure 1). Similar experiments and interpretations were reported by Yoo et al. [34] with Cu, Ag and cyanides, to reproduce an effluent from a lead frame manufacturing factory. Unlike our results, those of Yoo et al. [34] demonstrated that these 3 substances were responsible for the toxicity of the effluent on daphnids since they observed a similarity in the toxicity of the real and the synthetic effluents. In the present study, the DW toxicity observed on *L. sativa* was not explained only by the presence and the concentrations in Zn and Ni.

Investigations were conducted on a larger number of metals potentially responsible for the toxicity of DWs. Among 23 elements measured, 15 were present at quantifiable levels at least once, and 5 of them (Al, Co, Cr, Ni and Zn) were selected for the following experiments owing to their concentrations in DW6 to DW9 (higher than 1 mg L^{-1}; Table 4) and/or their known effects on the environment (Table 1). DWs 6, 7 and 9 appeared to be much more toxic (EC_{50} 6.1, 5, 18.4% of the sample) than their respective SS (56.8, 52.1, 47.9% of the solution tested) on *D. magna*. Results showed the same tendency for the GR of *L. sativa*, but not as dramatic: for instance daphnid EC_{50} values were 5 and 52.1% for DW7 and SS7, respectively, whereas lettuce EC_{50} values were 68 and 84%. Like for Ni and Zn, the presence of Al, Co and Cr did not explain all the DW toxicity on both *D. magna* and *L. sativa*, even though daphnid EC_{50} values showed these 5 metals to be toxic (Table 3).

To our knowledge, few studies have assessed the environmental impact of discharge water or synthetic solutions on more than one bio-indicator [7,12,14,35,36]. Bioassay batteries have already been shown to be a relevant way to evaluate toxicity, irrespective of the ecosystem studied [7,36,37]. Sensitivity differences observed between daphnids and lettuce (Tables 2-4) also occurred on comparison with data from the literature (Table 1). General differences can be explained by bioassay endpoint (acute or chronic toxicities) or protocol variability (bioindicator subspecies or cultivars, animal gender, lapse of exposure, number of individuals per Petri dish or tube, etc.; [38]). Yet, it appears that differences between bioindicator sensitivity remain in bibliographic data. For 3 metals for which we found comparative results (Zn, Ni, Cr), toxicity ranges were different for lettuce (Zn<Ni<<Cr) compared to algae, daphnids and gammarids (Cr<Zn<<Ni) as described in Table 1. Table 4 also shows single EC_{50} differences between indicators: toxicity range for daphnids being (from less to more toxic) Cr, Ni and Zn while the lettuce toxicity range was Cr, Zn and Ni. Another difference

between these 2 bioindicators was related to the order of magnitude of the EC_{50} (e.g. lettuce nickel EC_{50}: 58.3 mg L^{-1} vs daphnids: 9.8 mg L^{-1}).

Conclusions

In this study, the 2 bioindicators *Lactuca sativa* and *Daphnia magna* were proved to be pertinent to assess the ecotoxicity of polycontaminated discharge water from the surface treatment industry. The results showed that metal-based synthetic single and mixed solutions were less toxic than the discharge water, meaning that the ecotoxicity of these effluents could not be explained only by the 5 metals chosen in this work. Consequently, it would be interesting to lead future investigations not only towards a more exhaustive determination of the chemical composition of discharge water but also possible interactions (e.g. additivity, antagonism, synergy) between metals and/or trace organics and/or other minerals. Results also demonstrated that lettuce was more resistant than daphnids to the discharge waters and synthetic solutions. Ecotoxicological assessments complete chemical analyses as they integrate all chemical interactions. As reported in this study the use of a battery of tests was a relevant tool to include the whole variability of toxicity.

Acknowledgment

The authors are grateful to *Ville de Besançon* which funded Anne Priac's PhD, to Sophie Gavoille and Céline Lagarrigue from the *Agence de l'Eau Rhône Méditerrannée Corse*, the *Conseil Régional de Franche-Comté*, and the FEDER (*Fonds Européens de Développement Régional*) for financial support (NIRHOFEX Program 2013-2016).

Conflict of Interests

The authors declare that they have no conflict of interest.

References

1. Morin-Crini N, Druart C, Gavoille S, Lagarrigue C, Crini G (2013) Analytical monitoring of the chemicals present in the discharge water generated by the surface treatment Industry. J Environ Protect 4: 53-60.
2. Komjarova I, Blust R (2009) Effect of Na, Ca and pH on simultaneous uptake of Cd, Cu, Ni, Pb, and Zn in the water flea *Daphnia magna* measured using stable isotopes. Aquatic Toxicol 94: 81-86.
3. Charles J, Sancey B, Morin-Crini N, Badot PM, Degiorgi F, et al. (2011) Evaluation of the ecotoxicity of polycontaminated industrial effluents using the lettuce plant (*Lactuca sativa*) as a bioindicator. Ecotox Environ Saf 74: 2057-2064.
4. Kadirova ZC, Hojamberdiev M, Katsumata KI, Isobe T, Matsushita N, et al. (2014) Photodegradation of gaseous acetaldehyde and methylene blue in aqueous solution with titanium dioxide-loaded activated carbon fiber polymer materials and aquatic plant ecotoxicity tests. Environ Sci Pollut R 21: 4309-4319.
5. Cairns JJ, Pratt JR (1989) The scientific basis of bioassays. Hydrobiologia 188: 5-20.
6. Farré M, Barceló D (2003) Toxicity testing of wastewater and sewage sludge by biosensors, bioassays and chemical analysis. Trac-Trend Anal Chem 22: 299-310.
7. Pandard P, Devillers J, Charissou AM, Poulsen V, Jourdain MJ, et al. (2006) Selecting a battery of bioassays for ecotoxicological characterization of wastes. Sci Total Environ 363: 114-125.
8. Wang W, Freemark K (1995) The use of plants for environmental monitoring and assessment. Ecotoxicol Environ Saf 30: 289-301.
9. Wangberg SA, Bergström B, Blanck H, Svanberg O (1995) The relative sensitivity and sensitivity patterns of short-term toxicity tests applied to industrial waste water. Environ Toxic Water 10: 81-90.
10. Di Salvatore M, Carafa AM, Carratu G (2008) Assessment of heavy metals phytotoxicity using seed germination and root elongation tests: A comparison of two growth substrates. Chemosphere 73: 1461-1464.
11. Chapman PM, Fairbrother A, Brown D (1998) A critical evaluation of safety (uncertainty) factors for ecological risk assessment. Environ Toxicol Chem 17: 99-108
12. Hernando MD, Fernandez-Alba AR, Tauler R, Barcelo D (2005) Toxicity assays applied to wastewater treatment. Talanta 65: 358-366.
13. Dutka BJ (1989) Methods for microbiological and toxicological analysis of waters, wastewaters and sediments. National Water Research Institute NWRI, Environment Canada.
14. Fjällborg B, Li B, Nilsson E, Dave G (2006) Toxicity identification evaluation of five metals performed with two organisms (*Daphnia magna* and *Lactuca sativa*). Arch Environ Contam Toxicol 50: 196-204.
15. Aruoja V, Dubourguier HC, Kasemets K, Kahru A (2009) Toxicity of nanoparticles of CuO, ZnO and TiO_2 to microalgae *Pseudokirchneriella subcapitata*. Sci Total Environ 407: 1461-1468.
16. Gale NL, Wixson BG, Erten M (1992) An evaluation of the acute toxicity of lead, zinc and cadmium in Missouri Ozark groundwater. Trace Subst Environ Health 25: 169-183.
17. McLoughlin N, Yin D, Maltby L, Wood RM, Yu H (2000) Evaluation of sensitivity and specificity of two crustacean biochemical biomarkers. Environ Toxicol Chem 19: 2085-2092.
18. Chen CY, Lin KC, Yang DT (1997) Comparison of relative toxicity relationships based on batch and continuous algal toxicity tests. Chemosphere 35: 1959-1965.
19. Cabejzek I, Stasiak M (1960) Investigation on the influence of some metals on the biocoenosis of water with the use of *Daphnia magna* as an indicator (Part I). Roczyn Zabl Hig Warsaw 11: 303-312.
20. Rehwoldt R, Lasko L, Shaw C, Wirhowski E (1973) The acute toxicity of some heavy metals ions toward benthic organisms. Bull Environ Contam Toxicol 10: 291-294.
21. Adema DMM, Henzen L (1989) A comparison of plant toxicities of some industrial chemicals in soil culture and soilless cultures. Ecotoxicol Environ Saf 18: 219-229.
22. Nyholm N (1990) Expression of results from growth-inhibition toxicity tests with algae. Arch Environ Contam Toxicol 19: 518-522.
23. Diamantino TC, Guilhermino L, Almeida MF, Soares AMVM (2000) Toxicity of sodium mobylate and sodium dichromate to *Daphnia magna* Straus evaluated in acute, chronic, and acetylcholinesterase inhibition tests. Ecotoxicol Environ Saf 45: 253-259.
24. EPA (1993) Methods for measuring the acute toxicity of effluents and receiving waters to freshwater and marine organisms (EPA/600/4-90/02TF), US Environmental Protection Agency; Office of Toxic Substances: Washington, D.C.
25. EPA (1975) Methods for acute toxicity test with fish, macroinvertebrates and amphibians (EPA/660/3-75-009), US Environmental Protection Agency; Office of Toxic Substances: Washington, D.C.
26. Biesinger KE, Christensen GM (1972) Effects of various metals on the survival, growth, reproduction, and metabolism of *Daphnia magna*. J Fish Res Board Can 29: 1691-1700.
27. Long KE, Van Genderen EJ, Klaine SJ (2004) The effects of low hardness and pH on copper toxicity to *Daphnia magna*. Environ Toxicol Chem 23: 72-75.
28. Taylor EJ, Maund SJ, Pascoe D (1991) Toxicity of four common pollutants to the freshwater macroinvertebrates *Chironomus riparius* Meigen (Insecta: diptera) and *Gammarus pulex* (L.) (Crustacea: amphipoda). Arch Environ Contam Toxicol 21: 371-376.
29. Backhaus T, Altenburger R, Arrhenius Å, Blanck H, Faust M, et al. (2003) The BEAM-project: prediction and assessment of mixture toxicities in the aquatic environment. Cont Shelf Res 23: 1757-1769.
30. ISO (2005) Soil Quality – Determination of the effects of pollutants on soil flora – Screening test for emergence of lettuce seedlings (*Lactuca sativa* L.). International Organisation for Standardization ISO 17126. Geneve, Switzerland.
31. ISO (1989) Water Quality – Determination of the inhibition of the mobility of *Daphnia magna* Straus (Cladocera, Crustacea). Acute Toxicity Assay. International Organisation for Standardization ISO 6341. Geneve, Switzerland.
32. Hitchcock DR, Black HC, Williams PL (1997) Investigations into using the nematode *Caenorhabditis elegans* for municipal and industrial wastewater toxicity testing. Arch Environ Contam Toxicol 33: 252-260.
33. Castillo GC, Vila IC, Neild E (2000) Ecotoxicity assessment of metals and wastewater using multitrophic assays. Environ Toxicol 15: 370-375.

34. Yoo J, Kim SB, Kim WK, Jung J (2014) Toxicity identification of effluent from a semiconductor lead frame manufacturing factory. J In Eng Chem 20: 494-498.

35. Bohorquez-Echeverry P, Campos-Pinilla C (2007) Assessment of *Lactuca sativa* and *Selenastrum capricornotum* like indicators of water toxicity. Universitas Scientiarum 12: 83-98.

36. Wilke BM, Riepert F, Koch C, Kühne T (2008) Ecotoxicological characterization of hazardous wastes. Ecotox Environ Saf 70: 283-293.

37. Ewell WS, Gorsuch JW, Kringle RO, Robillard KA, Spiegel RC (1986) Simultaneous evaluation of the acute effects of chemicals on seven aquatic species. Environ Toxicol Chem 5: 831-840.

38. Alexander PD, Alloway BJ, Dourado AM (2006) Genotypic variations in the accumulation of Cd, Cu, Pb and Zn exhibited by six commonly grown vegetables. Environ Pollut 144: 736-745.

2

Heavy Metals in an Important Section of the São Francisco River (Northeast Brazil): Distribution Profile, Accumulation Mechanisms, and Risks of Dissemination through the Food Chain

Pereira MDG[1*], Souza CLM[2], Sachdev RDL[1], Santos AVD[1], Pinto PADC[2], Souza LA[1], Ribeiro JN[3], Oliveira JPD[3], Guimarães MCC[3] and Ribeiro AVFN[4]

[1]State University of Bahia, Department of Exact and Earth Sciences, R. Silveira Martins, Salvador-BA, Brazil
[2]State University of Bahia, Department of Technology and Social Sciences Av. Dr. Chastinet Guimarães, S/N, Juazeiro-BA, Brazil
[3]Federal University of Espírito Santo, Center for Health Sciences, Maruípe Avenue, S/N, Vitória-ES, Brazil
[4]Federal Institute of Espírito Santo, Undergraduate Coordination, Av. Vitória, Vitória-ES, Brazil

Abstract

Background: This work assessed the distribution of Cd, Cr, Cu, Mn, Ni, Pb, and Zn in an important Brazilian fluvial section, which has never been evaluated before this study

Procedures: Along six samplings, total and bioavailable concentrations of these metals were quantified in water and sediments, respectively, by atomic spectrometry. Bioavailable metal concentrations were extracted from sediments by 0.1 $molL^{-1}$ HCl, while total metal concentrations were released in water after acidic decomposition with 14 $molL^{-1}$ HNO_3. Sediments were submitted to analyses of X-ray diffractometry, thermogravimetry, infrared spectrophotometry, electronic scanning microscopy, granulometry, and density. Subsequently, adsorption tests were performed with those elements that tend to exist as cations in river water (Cd^{2+}, Cu^{2+}, Ni^{2+}, Pb^{2+}, and Zn^{2+}).

Results: Water pH (6.9 to 8.3) favored precipitation of Cr and Mn, explaining their undetectable concentrations in water. Although cadmium, copper, lead, nickel, and zinc are stabilized as cations at water pH and chemical oxygen demand (<0.5 mgL^{-1}), total concentrations of these metals in water were undetectable. Sediments presented prevalence of quartz, sandy granulometry, maximum density of 2.8 gcm^{-3}, and remarkable bioavailable concentrations of Cr (up to 778.4 ± 37.4 mg kg^{-1}) and Mn (up to 230.9 ± 6.2 mg kg^{-1}, which can be readily assimilated by benthic organisms. Bioavailable concentrations of Cd, Cu, Ni, Pb, and Zn in sediments were considerable small (≤ 43.4 ± 2.5 $mgkg^{-1}$). Although sediments are limited for adsorbing Cd^{2+}, Cu^{2+}, Ni^{2+}, Pb^{2+}, and Zn^{2+} (≤ 1.49 mgg^{-1}), these cations are adsorbed with low reversibility, being progressively accumulated and disseminated through the food chain over a long period of time.

Conclusions: This work improved the knowledge about the distribution of heavy metals in fluvial environments, as well as evaluated the risks of an eventual contamination of humans.

Keywords: Heavy metals; São Francisco river; Sediments; Adsorption; Mathematical model of Langmuir; Bioavailability

Abbreviations: FAAS: Flame Atomic Absorption Spectrometry; ICPOES: Inductively Coupled Plasma Optical Emission Spectrometry

Introduction

The São Francisco is one the most important rivers in Brazil and South America, with a length of 2,860 km and a watershed of 630,000 km² [1]. It drains the areas of Minas Gerais (MG), Bahia (BA), Goiás (GO), Pernambuco (PE), Alagoas (AL), and Sergipe (SE), as well as part of the Federal District. The São Francisco River is of huge relevance for the Brazilian semiarid region (969,582 km²) that hosts 22 million of people and the Caatinga (84,453 km²), which is a biome exclusively occurring in Brazil [2].

In the lower São Francisco basin, this river separates Juazeiro, BA and Petrolina, PE, which are two cities with approximately 600,000 inhabitants and an economy based on cultivation of fruits (mostly grape and mango) and leather tanning. Besides its essential importance for the maintenance of all regional agricultural and industrial activities, the São Francisco is largely responsible for the supply of water for human consumption. Despite this, studies concerning the distribution and accumulation of heavy metals in sediments and water of this important Brazilian fluvial section have never been performed before this study. This lack of information continues, since no work has been published since February 2013, when the present work was finished. Ultimately, it means that the potential risks of human contamination by heavy metals through the consumption of contaminated aquatic organisms and water are still completely unknown in the assessed area.

In this work, the following heavy metals were selected for evaluation: trivalent Cr, Cd, Pb, Mn, Ni, and Zn. Trivalent Cr was chosen due to its use in the leather tanning industries, which are widely distributed in the cities of Juazeiro, BA and Petrolina, PE. Moreover, when in large excesses in a food chain, trivalent Cr can reach humans and cause stomach cancer [3]. Cd and Pb were evaluated because these elements are very toxic to humans even if small amounts are considered. Of the many human diseases that can be caused by Cd and Pb, infertility,

***Corresponding author:** Madson de Godoi Pereira, State University of Bahia, Department of Exact and Earth Sciences, R Silveira Martins, 2555, 41.150-000, Salvador-BA, Brazil, E-mail: madson.pereira444@gmail.com

leukemia, bone weakness, and psychiatric disorders should be focused on [4]. Although Cu, Mn, Ni, and Zn are essential elements, they can also exhibit toxic effects to humans if certain limits are exceeded. Excesses of Cu can promote blindness, while large ingestions of Mn manifest neurological damages. On the other hand, high exposure to Ni and Zn can be responsible for carcinogenicity [5,6]. Besides the potential risks concerning human health, Cu, Mn, Ni and Zn have widespread use in diverse types of industries and agricultural activities [4].

Thus, this work aimed at assessing the concentrations of Cd, Cr, Cu, Pb, Mn, Ni, and Zn in fluvial sediments between Juazeiro, BA and Petrolina, PE, as well as to correlate these concentrations with water chemical conditions (pH, chemical oxygen demand, and total concentrations of the cited metals) and sediment structural characteristics. This integrated approach is intended to evaluate the risks of dissemination of Cd, Cr, Cu, Mn, Ni, Pb, and Zn through the food chain with the consequent possibility of reaching humans.

Materials and Methods

Equipment and reactants

The following instruments were used: Gehaka Master P&D (Brazil) water purification unit, Quimis Q225M (Brazil) horizontal table, IKA basic A11 blade mill (Germany), Tecnal Tec-3MP potentiometer (Brazil), GBC SensAA flame atomic absorption spectrophotometer (Malaysia), Varian Vista-PRO inductively coupled plasma optical emission spectrometer (Australia), Rigaku X'Pert Pro PW 3040/60 X-ray diffractometer (Japan), TA Instruments TGA 2950 thermogravimetric analyser (USA), Perkin-Elmer Spectrum 100 infrared spectrophotometer (USA), and Jeol JSM-6610LV electron scanning microscope (Japan).

Single stock standard solutions of metals at 1,000 mgL^{-1} (SCP Science, USA) were used to calibrate the spectrometers and prepare all solutions used in the retention tests. 14 $molL^{-1}$ HNO_3 (Merck, Germany) and 0.1 $molL^{-1}$ HCl (Vetec, Brazil) were used to perform total decomposition of water and extraction of bioavailable metals from sediments, respectively. All solutions were prepared with ultrapure water (18.2Ω cm^{-1}).

Experimental part

Sampling points and procedures

Geographical locations of Juazeiro (BA) and Petrolina (PE) are indicated in Figure 1.

During six samplings, sediments and water were periodically taken at the six points indicated in Figure 2 in order to determine bioavailable and total contents of metals, respectively. Thus in all, 36 sediment samples and 36 water samples were analysed. These periodic samplings of sediments and water were made on 10/17/2011 (first), 12/05/2011 (second), 04/24/2012 (third), 05/29/2012 (fourth), 09/20/2012 (fifth), and 02/05/2013 (sixth). Chemical oxygen demand and pH were also measured in the 36 water samples. The sampling points indicated in Figure 2 are located near the following places: the campus of Bahia State University (point 1), urban riverbanks (points 2 and 3), a fluvial island (point 4), and leather tanneries (points 5 and 6).

Sediments, which were submitted to structural analyses, were collected at the points 1 to 9, indicated in Figure 3. For this purpose, sediments were collected only once, so that structural analyses were performed with nine sediment samples. The sediments collected at point 1 were also used to perform the retention tests with metallic cations. These sampling points are located near the following places: urban riverbanks (points 1, 2, 6 and 9), a fluvial island (point 3), leather tanneries (points 4 and 7), a rural property (point 5), and a sailing club (point 8).

Table 1 lists geographical coordinates of the sampling points shown in Figures 2 and 3.

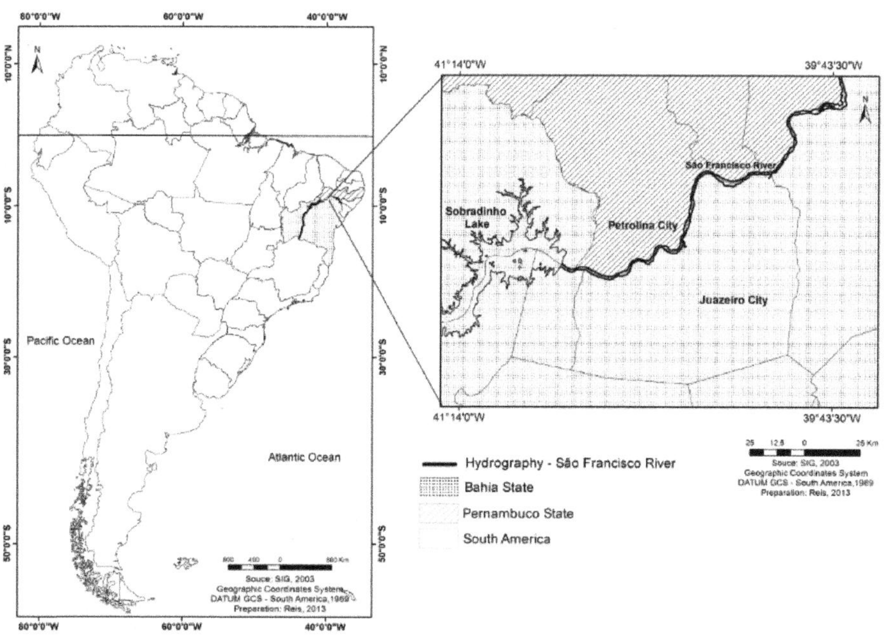

Figure 1: General geographic location of Juazeiro, BA and Petrolina, PE.

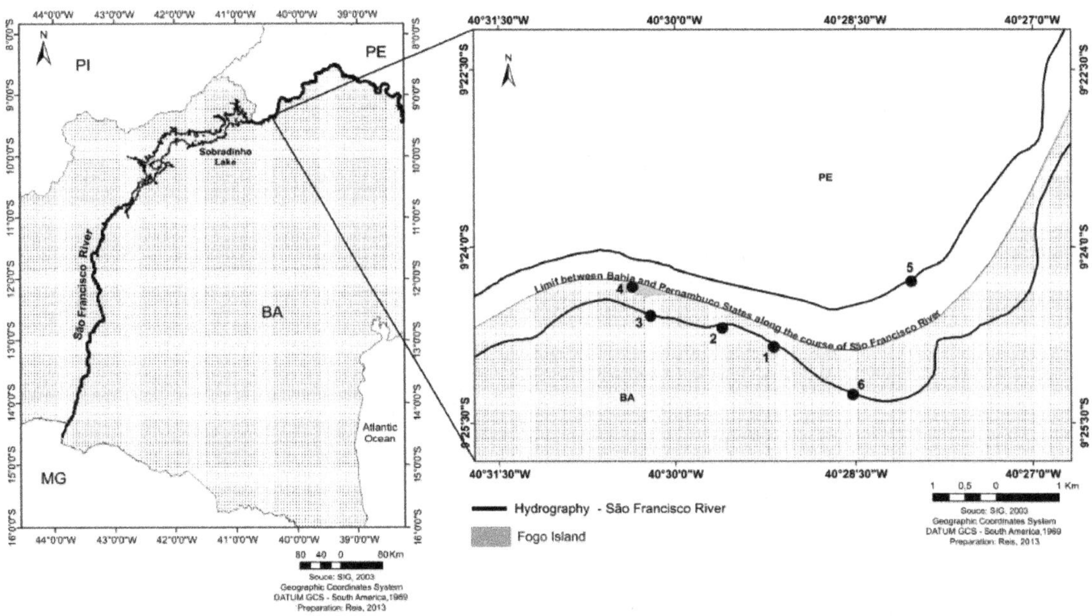

Figure 2: Sampling points of water and sediments for periodic analyses.

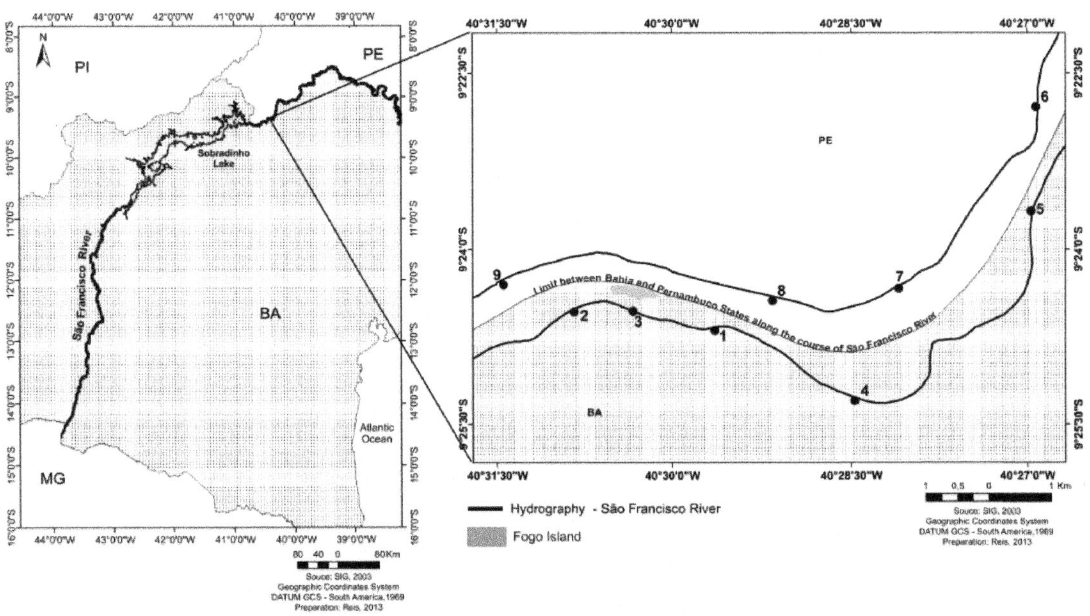

Figure 3: Sampling points of sediments for structural characterization.

All marginal and superficial samples of sediments (depth of 5 cm) and water (depth of 10 cm) were collected and preserved by following well-established procedures [7].

Analytical procedures

Periodic analyses of sediment and water samples

For periodic quantifications of bioavailable metals in sediments, the samples were dried in an oven at 100° C for 72 h, passed through a 10-mesh sieve, crushed in a blade mill, and sieved again at 270-mesh (nominal sieve opening of 0.053 mm). This procedure was performed to ensure a high sample homogenization, as well as precise and accurate results [8].

Portions of one gram of sediments were stirred at 200 rpm and 25°C for 2 h with 25.00 ml of 0.1 $molL^{-1}$ HCl [9]. The supernatants were filtered and analyzed by flame atomic absorption spectrometry (FAAS) to quantify the bioavailable concentrations of Cd, Cu, Mn, Ni, Pb, and Zn, and by inductively coupled plasma optical emission spectrometry (ICP OES) to determine the bioavailable concentrations of Cr.

All water samples were pre-concentrated 10X by evaporation and submitted to acidic decomposition with 14 $molL^{-1}$ HNO_3 [10] to

Sampling points	Coordinates	
	Figure 2	Figure 3
1	S 09° 24' 51.5" WO 40° 29' 10.8"	S 09° 24' 41.6" W 040° 29' 38.4"
2	S 09° 24' 41.5" WO 40° 29' 36.7"	S 09° 24' 32.0" W 040° 30' 50.2"
3	S 09° 24' 35.3" WO 40° 30' 12.6"	S 09° 24' 22.9" W 040° 30' 18.6"
4	S 09° 24' 19.3" WO 40° 30' 18.2"	S 09° 25' 17.2" W 040° 28' 27.7"
5	S 09° 24' 17.6" WO 40° 28' 01.7"	S 09° 23' 40.5" W 040° 26' 58.0"
6	S 09° 25' 15.4" WO 40° 28' 31.6"	S 09° 22' 47.2"W 040° 26' 55.8"
7	-	S 09° 24' 20.1" W 040° 28' 05.7"
8	-	S 09° 24' 26.5" W 040° 29' 09.5"
9	-	S 09° 24' 18.2" W 040° 31' 26.6"

Table 1: Geographical coordinates of the sampling points indicated in Figure 2 and Figure 3.

subsequent quantifications of total concentrations of Cd, Cu, Pb, Mn, Ni, and Zn by FAAS. Total Cr concentrations were quantified by ICP OES. Other water samples were also collected (Figure 2) to measure pH by means of a potentiometer. Water samples were also submitted to analyses of chemical oxygen demand [10] to estimate their contents of dissolved organic matter.

All analyses by FAAS were made by using air-acetylene flame (fluxes of air and acetylene of 8.0 and 2.0 L min^{-1}, respectively) and background correction with a deuterium lamp. Hallow cathode lamps were operated in the following wavelengths (nm): 228.8 (Cd), 324.8 (Cu), 279.5 (Mn), 232.0 (Ni), 283.3 (Pb), and 213.9 (Zn). For quantification of Cr by ICP OES at 426.2 nm, the equipment was operated with a power of 1300 W and 40 MHz radiofrequency. All spectrometers were calibrated by means of daily calibration curves, which were built from successive dilutions of stock solutions of each metal at 1,000 mg L^{-1}.

Structural analyses of the sediments

Before analyses of thermogravimetry, X-ray diffractrometry, scanning electron microscopy, granulometric composition, and density, sediments were dried in an oven at 100°C for 72 hours, and passed through a 10-mesh sieve (nominal sieve opening of 2.0 mm) to remove large fragments of plants and rocks. In this way, the natural granulometric distribution of sand (diameters from 2.0 to 0.5 mm), silt (diameters from 0.5 to 0.02 mm) and clay (diameters<0.02 mm) of these sediments was kept in the analyses described below.

The analyses of thermogravimetry were performed by heating samples from room temperature to 1,000°C (at 10°C min^{-1}) in the presence of synthetic air. For the X-ray analysis, the samples were exposed to radiation (Co-Kαλ, =1.79026 Å) with 2θ angles varying from 5 to 50°. The applied current and voltage were 30 mA and 40 kV, respectively. Analyses by infrared spectrophotometry were performed using Kbr discs to prepare the sediment samples. The spectral range varied from 4,000 to 400 cm^{-1}. For electron scanning microscopy, sediments were covered with a thin layer of gold and an acceleration voltage of 20 kV was applied. The peaks were associated to specific minerals following recommended procedures [11].

The granulometric compositions of sediments were determined by considering the vertical displacement time of clays (Stokes's law) after addition of chemical dispersants [12].

To determine the density of sediments, 20 g of them were added to 50.00 ml volumetric flasks, whose volumes were subsequently filled with 95% (v/v) ethanol. The volumes of 95% (v/v) ethanol were recorded with a burette and the values of density were calculated by dividing the sediment mass (20 g) by the difference between the volume of volumetric flask (50.00 ml) and the spent volume of 95% (v/v) ethanol [12].

Adsorption curves for Cd^{2+}, Cu^{2+}, Ni^{2+}, Pb^{2+}, and Zn^{2+}

To elucidate the accumulation mechanisms of Cd^{2+}, Cu^{2+}, Ni^{2+}, Pb^{2+}, and Zn^{2+} in sediments, adsorption curves were built for these cations. For this purpose, sediments were dried and sieved as described in the previous section. Afterwards, adsorption curves were obtained by following a procedure developed in our laboratories, in which 50 mg of sediments were stirred at 200 rpm for 60 minutes with 25.00 ml of different solutions containing Cd^{2+}, Cu^{2+}, Ni^{2+}, Pb^{2+}, or Zn^{2+} (3 to 10 mgL^{-1}) at pH 7 and 25°C. These solutions were prepared with water of the São Francisco River, which was collected at point 1 of Figure 3 and was previously monitored for total concentrations of these metals. For each concentration used to plot these curves, one analytical blank and three replicates were analyzed. The remaining Cd^{2+}, Cu^{2+}, Ni^{2+}, Pb^{2+}, and Zn^{2+} concentrations in the supernatants were quantified by FAAS. Subsequently, all data that originated adsorption curves were linearized by using the mathematical model of Langmuir, which allows estimation of the following thermodynamic parameters for each one of the mentioned cations: the maximum adsorptive capacity of the sediments, Langmuir constant related to adsorption energy, and separation factor, which predicts the adsorption reversibility.

Results and Discussion

The analytical quality of all spectrometric quantifications of metals (by FAAS or ICP OES) was checked and confirmed by analyses of standard solutions with traceability for cadmium, chromium, copper, lead, manganese, nickel, and zinc.

Periodic analyses of metals in sediments and water

Table 2 lists the bioavailable concentrations of metals after extraction with 0.1 molL^{-1} HCl, which is an extractor that simulates the capacity of benthic organisms to assimilate nutrients/pollutants from sediments [9]. This extractor can release adsorbed metals from aluminosilicates and silicates, as well as dissolve alkaline and amphoteric oxides, thus releasing metals that were occluded into or adsorbed onto these oxides.

Bioavailable Cr concentrations were much higher than 90 mgkg^{-1}, which is the guiding value for fluvial sediments [13]. At points 5 and 6 (Figure 2), the high concentrations of Cr in sediments can be attributed to the continuous discharges of untreated leather tanning effluents, whose composition is rich in Cr^{3+}. At both sampling points, which are close to this type of industry, Cr was precipitated as $Cr(OH)_3$ due to the water pH, which had an average of 7.2 (6.9 to 8.3) for 36 samples. At pH 7.2, precipitations of $Cr(OH)_3$ are expected at $[Cr^{3+}]$ concentrations as small as 1.2×10^{-7} molL^{-1}. Because of the high tendency of Cr to form insoluble $Cr(OH)_3$, total concentration of this element in water was smaller than its analytical detection limit (0.05 mg L^{-1}).

Another source of Cr in the sediments at points 5 and 6 is the erosion of soils naturally enriched with chromite (Cr_2O_3) [14], whose presence also explains the concentrations of Cr at points 1, 2, 3 and 4 which are located upstream of the tanneries. For comparison, fluvial sediments of another Brazilian area heavily polluted by tanneries had total concentrations of Cr as elevated as 2,878 ± 78 mgkg^{-1} [15]. Sediments from the Dunajec River (Poland) also showed alarming concentrations of chromium, which exceeded 1,000 mg kg^{-1} at some

Samplings	Sampling points	Cd	Cr	Cu	Mn	Ni	Pb	Zn
First	1	<0.5*	121.4 ± 7.6	3.8 ± 0.2	28.6 ± 1.1	<2.5*	<5.0*	9.3 ± 0.06
	2	<0.5	377.0 ± 8.9	<0.5*	26.3 ± 0.1	5.9 ± 0.2	<5.0	13.9 ± 0.07
	3	<0.5	408.7 ± 9.2	<0.5	20.5 ± 0.7	5.7 ± 0.1	<5.0	9.2 ± 0.2
	4	<0.5	324.4 ± 8.1	<0.5	20.5 ± 0.3	<2.5	<5.0	4.6 ± 0.09
	5	<0.5	505.7 ± 19.4	7.8 ± 0.2	26.3 ± 0.1	<2.5	<5.0	9.2 ± 0.3
	6	<0.5	444.3 ± 26.7	9.6 ± 0.1	37.7 ± 0.5	<2.5	<5.0	13.2 ± 0.2
Second	1	<0.5	86.6 ± 2.4	3.1 ± 0.2	16.8 ± 1.5	<2.5	<5.0	7.2 ± 0.3
	2	<0.5	274.4 ± 15.2	3.3 ± 0.1	87.8 ± 1.1	<2.5	13.2 ± 0.8	<0.1*
	3	<0.5	501.2 ± 46.4	2.8 ± 0.1	166.2 ± 1.6	7.1 ± 0.6	<5.0	9.4 ± 0.6
	4	<0.5	456.0 ± 37.4	3.3 ± 0.2	19.0 ± 0.2	<2.5	<5.0	5.7 ± 0.2
	5	<0.5	328.9 ± 11.7	1.6 ± 0.2	30.8 ± 1.8	<2.5	<5.0	7.9 ± 0.1
	6	<0.5	497.7 ± 46.8	2.9 ± 0.1	171.8 ± 1.8	<2.5	<5.0	6.1 ± 0.06
Third	1	<0.5	174.9 ± 13.6	<0,5	38.9 ± 1.4	<2.5	<5.0	28.8 ± 0.4
	2	<0.5	404.2 ± 25.5	<0,5	221.9 ± 7.3	<2.5	<5.0	32.3 ± 0.6
	3	<0.5	602.7 ± 32.6	<0,5	219.9 ± 2.3	4.8 ± 0.4	<5.0	9.4 ± 0.6
	4	<0.5	517.6 ± 24.8	<0,5	30.9 ± 0.6	<2.5	<5.0	<0.5
	5	<0.5	577.6 ± 48.9	<0,5	142.2 ± 2.4	<2.5	<5.0	10.9 ± 0.9
	6	<0.5	416.6 ± 21.4	1.9 ± 0.1	129.8 ± 1.4	<2.5	<5.0	7.7 ± 0.3
Fourth	1	<0.5*	111.4 ± 5.8	<0,5*	18.3 ± 0.9	28.6 ± 1.1	<5.0*	<0.5
	2	<0.5	208.7 ± 21.4	2.0 ± 0.1	165.9 ± 7.5	26.3 ± 0.1	<5.0	2.0 ± 0.1
	3	<0.5	302.9 ± 24.7	1.7 ± 0.01	230.0 ± 2.6	20.5 ± 0.7	<5.0	1.7 ± 0.01
	4	<0.5	517.9 ± 38.4	<0.5	32.9 ± 0.4	20.5 ± 0.3	<5.0	<0.5
	5	<0.5	274.1 ± 11.3	<0.5	111.1 ± 0.7	26.3 ± 0.08	<5.0	<0.5
	6	<0.5	413.7 ± 10.7	<0.5	75.4 ± 0.5	37.7 ± 0.5	<5.0	<0.5
Fifth	1	<0.5	398.5 ± 25.3	<0.5*	67.2 ± 1.1	<2.5*	<5.0*	7.2 ± 0.5
	2	<0.5	778.4 ± 37.4	<0.5	127.3 ± 8.5	<2.5	<5.0	14.3 ± 0.9
	3	<0.5	587.3 ± 46.6	2.2 ± 0.1	230.9 ± 6.2	<2.5	<5.0	10.1 ± 0.8
	4	<0.5	422.2 ± 26.4	5.8 ± 0.9	198.2 ± 9.1	<2.5	<5.0	38.0 ± 2.0
	5	<0.5	314.7 ± 9.9	9.3 ± 0.8	80.2 ± 6.2	<2.5	<5.0	23.3 ± 1.8
	6	<0.5	187.6 ± 17.3	6.2 ± 0.9	274.3 ± 6.2	<2.5	< 5.0	29.4 ± 2.1
Sixth	1	<0.5	329.4 ± 28.8	<0.5*	43.7 ± 2.4	<2.5*	<5.0*	11.2 ± 0.7
	2	<0.5	429.3 ± 17.7	<0.5	99.4 ± 4.2	<2.5	<5.0	19.7 ± 0.6
	3	<0.5	217.1 ± 7.1	<0.5	335.7 ± 5.2	<2.5	<5.0	7.7 ± 0.4
	4	<0.5	660.4 ± 34.4	<0.5	176.1 ± 3.0	<2.5	<5.0	23.1 ± 1.1
	5	<0.5	523.6 ± 24.0	<0.5	128.1 ± 5.1	<2.5	<5.0	43.4 ± 2.5
	6	<0.5	404.5 ± 31.9	<0,5	204.2 ± 4.8	<2.5	< 5.0	37.9 ± 1.7

*detection of the analytical method

Table 2: Bioavailable contents of metals (mg kg^{-1}) in sediments from points indicated in Figure 2 (N=3).

sampling points close to tanneries [16]. In turn, average chromium concentrations of 36.67 and 74 mg kg^{-1} were found in sediments of non-polluted (by chromium) Iranian rivers [17,18].

The compounds $Cr(OH)_3$ and Cr_2O_3 have trivalent Cr whose prolonged ingestion can cause stomach cancer in humans [3]. This excessive ingestion of trivalent Cr can come from the eating of benthic organisms (mainly fishes) that live directly in contact with the contaminated sediments. The alarming bioavailable concentrations of trivalent Cr in sediments between Juazeiro, BA and Petrolina, PE can be readily disseminated through the food chain, eventually reaching humans. It is important to note that local cuisine is based on the consumption of native fishes of the São Francisco. Moreover, for many needy families of both cities, the ingesting of São Francisco fishes is the main nutritional option. Therefore, there is an imminent risk related to public health.

Total concentrations of Mn were not detectable (<0.03 mgL^{-1}) in water samples. On the other hand, bioavailable Mn concentrations in sediments were also much higher than those found for Cd, Cu, Pb, Ni, and Zn, although the highest Mn concentration (230.9 ± 6.2 $mgkg^{-1}$) was lower than 300 $mgkg^{-1}$, which is considered an environmental limit for sediments [19].

Mn is commonly found in sediments as insoluble pyrolusite (MnO_2), which is formed from modifications of other minerals such as rhodochrosite ($MnCO_3$). Once in the sediments, pyrolusite can be reduced to Mn^{2+} by a reaction that is represented in Equation 1. In this case, electrons would come from organic matter or reduced inorganic compounds.

$$MnO_{2(s)} + 4\,H^+_{(aq)} + 2\,e^-\Omega \rightarrow Mn^{2+}_{(aq)} + 2\,H_2O_{(l)} \quad (1)$$

However, at the average pH of the São Francisco water (7.2), the transfer of electrons to pyrolusite is limited by the small concentrations of protons. Moreover, this limited concentration of protons also ensures that Mn^{2+} ions directly released into water column, from some natural or anthropogenic source, are converted to MnO_2. The chemical conditions of this water favor the accumulation/maintenance of Mn as MnO_2 in sediments, thus explaining the undetectable concentrations of this element in water.

Mn dioxide (MnO_2) has considerable toxicity to the human gastrointestinal system [4] if high doses of this compound are ingested. Similarly to the insoluble Cr compounds, the risks of excessive assimilation of MnO_2 by humans can be readily realized by the consumption of contaminated benthic organisms.

As discussed above, water pH of the evaluated fluvial section favors accumulation and/or maintenance of insoluble compounds of Cr and Mn in sediments. Nevertheless, the same reasoning cannot be extended to Cd, Cu, Ni, Pb, and Zn, whose hydroxides require much higher water pH to be formed. At the average water pH (7.2) of the São Francisco, precipitations of $Cd(OH)_2$, $Cu(OH)_2$, $Ni(OH)_2$, $Pb(OH)_2$, and $Zn(OH)_2$ would occur at high Cd^{2+}, Cu^{2+}, Ni^{2+}, Pb^{2+}, and Zn^{2+} concentrations [20], which are highly unlikely to be found in a river as voluminous as the São Francisco, even considering the worst cases of aquatic pollution. Thus, if Cd, Cu, Pb, Ni, and Zn were introduced into the São Francisco water, they would be stabilized as Cd^{2+}, Cu^{2+}, Ni^{2+}, Pb^{2+}, and Zn^{2+}. Between Juazeiro, BA, and Petrolina, PE, possible sources of these cations include untreated industrial and/or agricultural wastes, as well as natural weathering of minerals and rocks.

Once in water, Cd^{2+}, Cu^{2+}, Ni^{2+}, Pb^{2+}, and Zn^{2+} can be electrostatically adsorbed on sediment particles, whose superficial chemical groups (hydroxyl, carboxyl, and silanol, among others) are deprotonated at the São Francisco's water pH, thus generating negative charges. In fluvial waters, dissolved organic matter can form stable complexes with Cd^{2+}, Cu^{2+}, Ni^{2+}, Pb^{2+}, and Zn^{2+} and if these complexes have positive charges or are neutral, they could also be adsorbed on sediment particles. Nevertheless, all water samples had chemical oxygen demand lower than analytical detection limits (0.5 mgL^{-1}), indicating small concentrations of dissolved organic matter, independent of its biodegradable or recalcitrant nature. The small values of chemical oxygen demand minimize the occurrence of complexation reactions involving Cd^{2+}, Cu^{2+}, Ni^{2+}, Pb^{2+}, and Zn^{2+}.

All water samples had total concentrations of Cd, Cu, Pb, Ni, and Zn smaller than the analytical detection limits (in mgL^{-1}): 0.001 (Cd), 0.02 (Cu), 0.03 (Ni), 0.05 (Pb), and 0.03 (Zn). This outcome indicates that, at the sampling points, there were not active emissions sources of these elements. Furthermore, as listed in Table 2, bioavailable levels of Cd, Cu, Ni, Pb, and Zn in sediments were much smaller than their average concentrations in the upper Earth's crust ($mgkg^{-1}$): 0.102 (Cd), 14.3 (Cu), 18.6 (Ni), 20 (Pb), and 71 (Zn) [21]. Together with undetectable concentrations in water, the low concentrations of Cd, Cu, Pb, Ni, and Zn in sediments reinforce the finding of limited distribution and accumulation of these metals in the evaluated fluvial section.

For comparative purposes, maximum bioavailable concentrations of Cd, Cu, Ni, Pb, and Zn in fluvial sediments of an industrialized areas of Brazil were, respectively: 4.5 ± 0.0, 17.5 ± 0.2, 4.4 ± 0.4, 5,281 ± 119, and 5,247 ± 452 $mgkg^{-1}$ [9] respectively. In turn, sediments of another Brazilian river exhibited the following maximum bioavailable concentrations of cadmium, copper, lead and zinc: 1.3 ± 0.03, 44.6 ± 0.4, 33.4 ± 4.0, and 43.7 ± 0.2 mg kg^{-1} [22]. In the present work, the bioavailable concentrations of nickel in sediments (fourth sampling) (Table 2) were comparable to those found in Odra River (Poland) [23].

Over a long period of time, low contents of Cd, Cu, Pb, Ni, and Zn in sediments do not necessarily imply low risks of human contamination, since the extension of this contamination depends strongly on adsorption reversibility. Thus, if adsorptions of these metals have small reversibility, they can be progressively accumulated in sediments and disseminated through the food chain, eventually reaching humans. To evaluate the reversibility of Cd^{2+}, Cu^{2+}, Ni^{2+}, Pb^{2+} and Zn^{2+} adsorptions on sediments, it was necessary to carry out structural analyses and adsorption tests.

Structural analyses of the sediments

To evaluate the greatest possible fluvial extension, sediments were taken at a greater number of sampling points (Figure 3). However, the analyses revealed very similar results and because of this, the discussions concerning the thermogravimetry, infrared spectrophotometry, X-ray diffractometry, and electron scanning microcopy were performed considering only for the sediments collected at point 1.

The sediment thermogram highlights small contents of volatile or thermally unstable compounds (Figure 4), including organic matter. It is possible to observe that only 3% (approximately 0.3 mg) of the initial sediment mass were lost between 250 and 600°C, which is the temperature range in which organic matter and many carbonates typically decompose.

Diffractogram peaks (Figure 5) revealed that quartz is responsible for the most intense peak, while kaolinite, calcite, valerite, hemathite, and illite created the other peaks. This prevalence of quartz indicates sediments with a homogeneous distribution of superficial chemical groups able to adsorb dissolved ions such as Cd^{2+}, Cu^{2+}, Ni^{2+}, Pb^{2+}, and Zn^{2+}. In Table 3, specific mineral structures are associated with their respective 2θ values (Figure 3).

The infrared spectrum (Figure 6) revealed an intense peak between 1,100 and 1,000 cm^{-1}, which is attributed to the stretch of Si-O-Si bonds of minerals as quartz. No peak associated to common organic functional groups was identified in Figure 6, thus corroborating the small amounts of organic matter previously highlighted by the thermogram (Figure 4). Other Brazilian fluvial sediments with total organic matter of up to 7.8 ± 0.4% (m/m) originated a very distinct infrared spectrum, which exhibited characteristic peaks associated to hydroxyls and carboxyls [22].

The analyses of sediments by electron scanning microscopy revealed particles with heterogeneity in sizes and shapes (Figure 7a and Figure 7b).

Table 4 lists the data concerning the sediment granulometry and density. All sediments had high percentages of sand, which can be

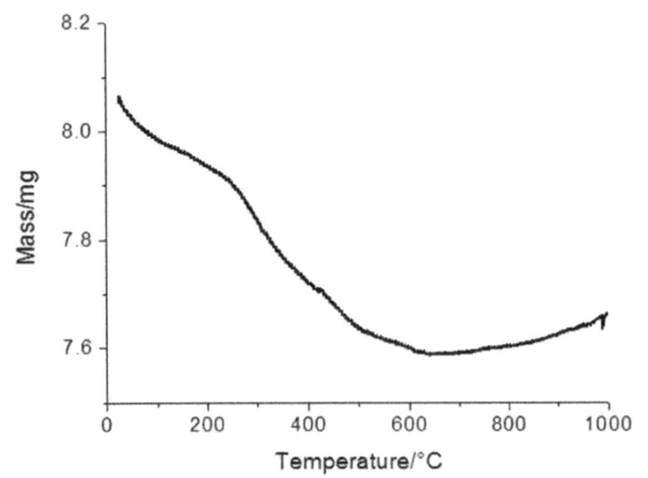

Figure 4: Sediment thermogram.

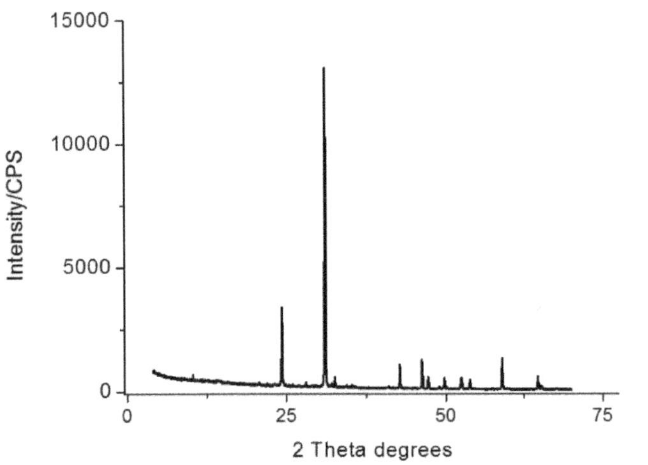

Figure 5: Sediment X-ray diffractogram.

explained by erosion of sandy marginal soils [24]. As can be seen in an aerial photo taken between Juazeiro, BA and Petrolina, PE, the fluvial bed was quite silty during the samplings (Figure 8). Due to the accelerated mineral weathering rates, areas with accentuated temperatures tend to have soils enriched with minerals with high chemical resistance (quartz). This is also the case of Cerrado (another Brazilian region subjected to hot weather), where many soils had quartz prevalence in their mineralogical composition [25].

A maximum difference of 2.9% was observed between the smallest and the highest density (average value of 2.72 g cm^{-3}), thus revealing great uniformity in the sediment mineral composition. Pure quartz has a density (2.65 g cm^{-3}) close to 2.72 g cm^{-3}, reinforcing the wide distribution of this mineral in the evaluated sediments.

This prevalence of quartz in all sediments suggests great similarity concerning the mechanisms of Cd^{2+}, Cu^{2+}, Ni^{2+}, Pb^{2+} and Zn^{2+} adsorptions and values of maximum adsorptive capacities. Moreover, this consistency in the mineralogical composition points to similarities in the adsorption reversibilities for these cations. As discussed below, all these expectations were confirmed.

Adsorption curves for Cd^{2+}, Cu^{2+}, Ni^{2+}, Pb^{2+}, and Zn^{2+}

As previously stated, structural analyses showed the existence of very similar sediments along the points indicated in Figure 3. For this reason, adsorption curves were built with only sediments from point 1. Amounts of Cd^{2+}, Cu^{2+}, Ni^{2+}, Pb^{2+}, and Zn^{2+} that were adsorbed on a fixed sediment mass (q_e, mgg^{-1}) and cation equilibrium concentrations in the supernatants (C_e, mgL^{-1}) are listed in Table 5, while the correspondent adsorption curves are indicated in Figures 9a-9e.

The Langmuir mathematical model uses the Equation 2 to achieve a linear relationship among the values of q_e and C_e. Thus, the closer to 1 the correspondent linear correlation coefficient (R) is, the more the Langmuir mathematical model will fit the treated data derived from an adsorption curve.

In Equation 2, besides the parameters q_e and C_e that were already defined, Q_0 (mgg^{-1}) is the maximum Cd^{2+}, Cu^{2+}, Ni^{2+}, Pb^{2+}, and Zn^{2+} amounts that can be adsorbed in a fixed sediment mass, and b (Lmg^{-1}) is the constant of Langmuir related to the adsorption energy [26].

$$\left(\frac{Ce}{Qe}\right)=\left(\frac{1}{Qo.b}\right)+\left(\frac{1}{Qo}\right).Ce \qquad (2)$$

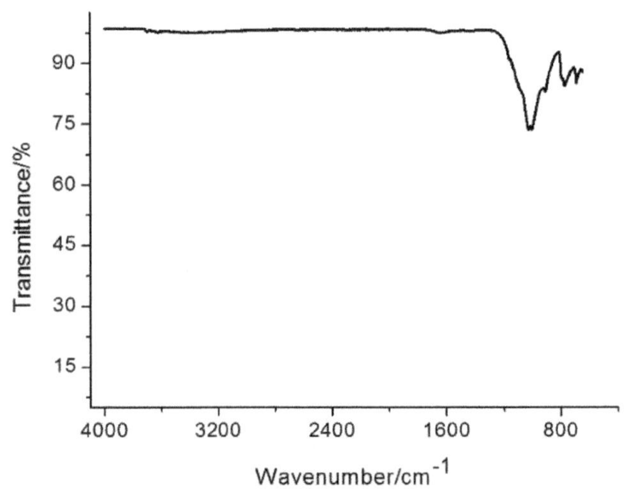

Figure 6: Sediment infrared spectrum.

Figure 7a: Sediment scanning electron microscopy: Magnification of 30X.

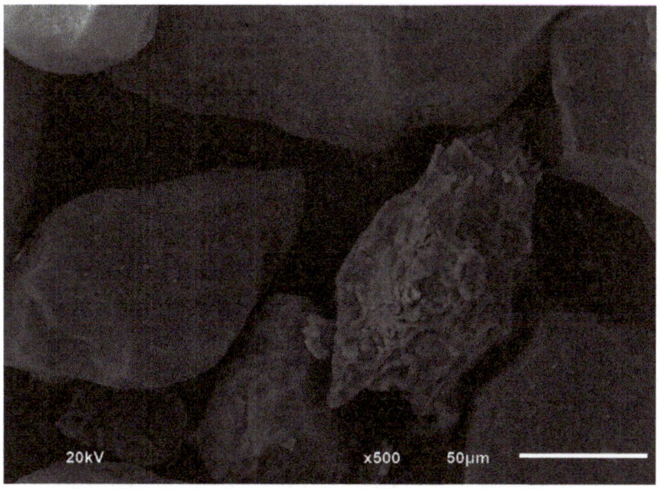

Figure 7b: Sediment scanning electron microscopy: Magnification of 500X.

Figure 8: Aerial photo of the São Francisco between Juazeiro, BA and Petrolina, PE.

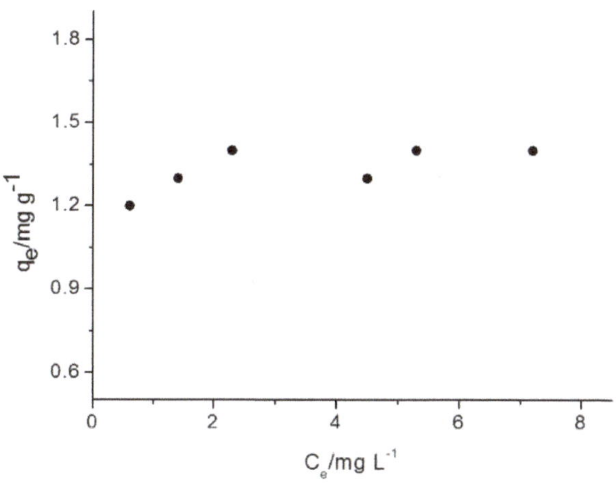

Figure 9a: Adsorption curves of Cd^{2+}.

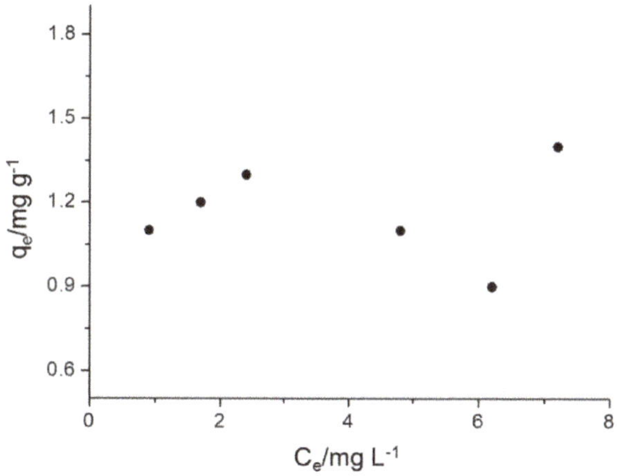

Figure 9b: Adsorption curves of Cu^{2+}.

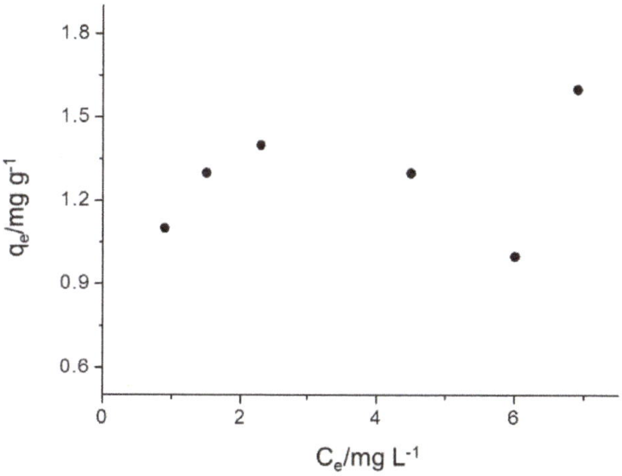

Figure 9c: Adsorption curves of Ni^{2+}.

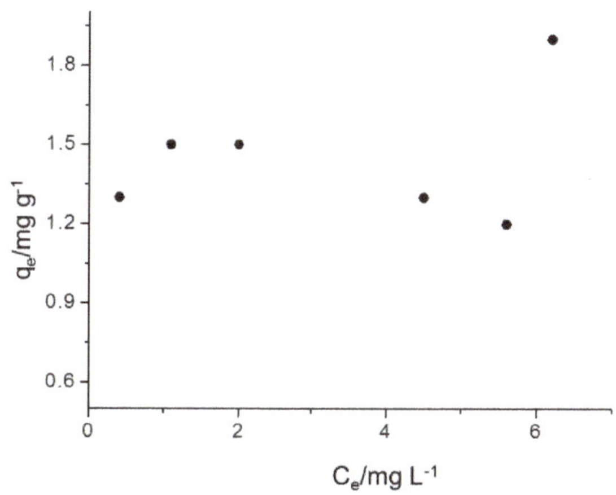

Figure 9d: Adsorption curves of Pb^{2+}.

It is possible to calculate Q_o as the inverse of the angular coefficient ($1/Q_o$) of Equation 2, while the constant of Langmuir (b) can be calculated with the values of Q_o and linear coefficient ($1/Q_ob$) of this equation.

After applying Equation 2 to the values of q_e and C_e (Table 5), line equations and their respective linear correlation coefficients (R) were calculated (Table 6). Table 6 also lists the thermodynamic parameters Q_o and b.

The linear correlation coefficients (R) were ≥ 0.93, thus showing good adjustments of the Langmuir mathematical model to the data

Mineral	Chemical formula	2θ value from X-ray diffractogram
Quartz	SiO_2	31.062
Hemathite	Fe_2O_3	24.279 e 64.705
Calcite	$CaCO_3$	46.192 e 49.728
Valerite	μ-$CaCO_3$	52.346 e 58.976
Illite	$[K_{0.8}(Al_{1.8}Mg_{0.2})(Si_{3.4}A_{l0.6})O_{10}(OH)_2$	10.339 e 20.692
Kaolinite	$[Al_2Si_2O_5(OH)_4]$	42.724

Table 3: Major mineral composition of sediments collected at points of Figure 3.

Sampling point	Sand (%, m/m)	Silt (%, m/m)	Clay (%, m/m)	Density (g cm^{-3})
1	88.3	9.4	2.3	2.66
2	91.7	7.6	0.7	2.74
3	98.7	0.5	0.8	2.74
4	93.8	5.2	1	2.71
5	98.8	0.4	0.8	2.78
6	98.2	1.8	0	2.74
7	98.2	1.7	0.2	2.7
8	98.9	0.9	0.2	2.66
9	98.9	0.8	0.3	2.74

Table 4: Granulometric fractioning and density of sediments from points indicated in Figure 3.

Metallic cation														
Cd^{2+}			Cu^{2+}			Ni^{2+}			Pb^{2+}			Zn^{2+}		
C_0	C_e	Q_e	C_0	C_e	Q_e	C_0	C_e	Q_e	C_0	C_e	Q_e	C_0	C_e	Q_e
3	0.6	1.2	3	0.9	1.1	3	0.9	1.1	3	0.4	1.3	3	1.4	0.8
4	1.4	1.3	4	1.7	1.2	4	1.5	1.3	4	1.1	1.5	4	2.3	0.9
5	2.3	1.4	5	2.4	1.3	5	2.3	1.4	5	2	1.5	5	2.9	1.1
7	4.5	1.3	7	4.8	1.1	7	4.5	1.3	7	4.5	1.3	7	5.4	0.8
8	5.3	1.4	8	6.2	0.9	8	6	1	8	5.6	1.2	8	6.2	0.9
10	7.2	1.4	10	7.2	1.4	10	6.9	1.6	10	6.2	1.9	10	7.7	1.2

Table 5: Values of C_0 (Initial metallic cation concentration, mg L^{-1}), C_e (Equilibrium metallic cation concentration, mg L^{-1}) and Q_e (Adsorbed amount of metallic cation in the sediments, mg g^{-1}) for Cd^{2+}, Cu^{2+}, Ni^{2+}, Pb^{2+} and Zn^{2+} adsorption curves.

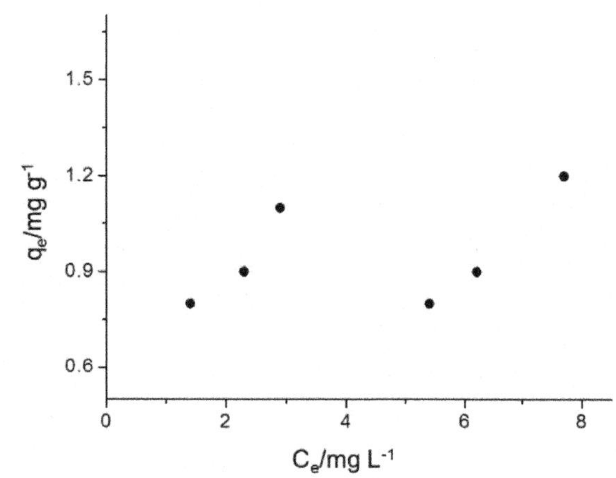

Figure 9e: Adsorption curves of Zn^{2+}.

that originated the adsorption curves (Figures 9a-9e). This adequacy can be attributed to the large mineralogical similarity among the sediments, thus revealing the existence of adsorptive chemical groups with energetic homogeneity, which is one the most important premises of the Langmuir mathematical model [26].

Although interactions of Cd^{2+}, Cu^{2+}, Ni^{2+}, Pb^{2+} and Zn^{2+} with sediments can be efficiently described by adsorptive mechanisms, the extensions of these adsorptions were not appreciable, as indicated by the small values of Q_o. The evaluated sediments had maximum adsorptive capacities that were comparatively much smaller than those of other natural materials, including argillaceous minerals [27] and humic organic matter [28]. These limited Q_o values agree with the predominance of sandy particles in sediments, which have a small superficial area, few adsorptive chemical groups, and low porosity.

Another important parameter derived from the Langmuir model is the RL (equilibrium or separation factor), which is defined by Equation 3 [29], where b is the Langmuir constant and C_o is the highest initial equilibrium cation concentration (10 mgL^{-1}) (Table 5).

$$RL = 1/(1 + b.Co) \qquad (3)$$

Values of RL between 0 and 1 indicate favorable and reversible adsorptions, while R_L=0 is associated to irreversible adsorptions [30,31]. Table 7 lists the R_L values that were calculated from b and C_o listed in Table 5.

Considering the R_L values listed in Table 7, Cd, Cu, Ni, Pb and Zn adsorptions can be classified as favorable and reversible. However, the reversibility's associated with these adsorptions were small, since RL values were closer to zero than one. It is important to note that R_L=0 means total irreversibility [30]. The decreasing reversibility order was: $Zn^{2+}>Cd^{2+}>Ni^{2+}>Pb^{2+}>Cu^{2+}$. Thus, over a long period of time, alarming amounts of cadmium, copper, lead, nickel and zinc tend to be retained on sediment particles, and accumulated in benthic organisms, eventually reaching humans.

Metallic cation	Line equation	Linear correlation coefficient (R)	Maximum adsorptive capacity Q_o mgg^{-1})	Constant of Langmuir b (mgL^{-1})
Cd^{2+}	Y=0.71X+0.09	0.99	1.41	7.88
Cu^{2+}	Y=0.88X+0.02	0.94	1.14	43.86
Ni^{2+}	Y=0.76X+0.09	0.93	1.31	8.48
Pb^{2+}	Y=0.67X+0.09	0.94	1.49	7.46
Zn^{2+}	Y=0,91X+0.57	0.93	1.1	1.59

Table 6: Data obtained after adsorption curves linearization.

Metallic cation	R_L
Cd^{2+}	0.013
Cu^{2+}	0.0023
Ni^{2+}	0.012
Pb^{2+}	0.013
Zn^{2+}	0.059

Table 7: Values of separation factor (R_L).

Conclusions

By using an integrated approach, this work assessed the distribution, accumulation mechanisms, and bioavailability of Cd, Cr, Cu, Mn, Ni, Pb and Zn in an important fluvial section of the Brazilian semiarid region, whose environmental conditions had not been studied before.

Considering the average river water pH (7.2), Cr and Mn are settled as insoluble compounds, while Cd, Cu, Ni, Pb and Zn are preferentially retained in sediments by adsorptive processes from their respective hydrated cations. No alarming bioavailable concentrations of Cd, Cu, Ni, Pb and Zn were found in the sediments, but considering their small adsorption reversibility's, it is possible to expect potential risks related to contamination of benthic organisms. Over long periods of time, this scenario could cause problems concerning human contamination and public health. On the other hand, Cr and Mn, whose bioavailable concentrations in sediments had alarming values, can be readily accumulated in benthic organisms and disseminated through the food chain.

Acknowledgements

We would like to thank the National Council for Scientific and Technological Development (CNPq), the Coordination for the Improvement of Higher Education Personnel (CAPES), the Research Support Foundation of the State of Bahia (FAPESB) and the State University of Bahia (UNEB) for fellowships and financial support (CNPq: 555522/2006-7 and 620041/2006-4 and FAPEB: APP0076/2009).

Conflict of Interests

The authors declare that there is no conflict of interests regarding the publication of this manuscript.

References

1. Maneta MP, Torres M, Wallender WW, Vosti S, Kirby M, et al. (2009) Water demand and flows in the São Francisco river basin (Brazil) with increased irrigation. Agr Water Manage 96: 1191-1200.
2. Almeida-Cortez JS, Cortez PHM, Franco JMV, Uzunian A (2007) Caatinga. Editora Harbra, São Paulo.
3. Washington DC (1998) US Environmental Protection Agency.
4. Nordberg GF, Fowler BA, Nordberg M, Friberg LT (2007) Handbook on the toxicology of metals. (3rd edn), Academic press, San Diego.
5. Denkhaus E, Salnikow K (2002) Nickel essentiality, toxicity, and carcinogenicity. Crit Rev Oncol Hemat 42: 35-56.
6. Plum LM, Rink L, Haase H (2010) The essential toxin: Impact of zinc on human health. Int J Environ Res Public Health 7:1342-1365.
7. Jordão CP, Pereira MG, Pereira JL (2002) Metal contamination of river waters and sediments from effluents of kaolin processing in Brazil. Water Air Soil Poll 140: 119-138.
8. Villaescusa-Celaya JA, Gutiérrez-Galindo EA, Flores-Munõz G (2000) Heavy metals in the fine fraction of coastal sediments from Baja California (Mexico) and California (USA). Environ Pollut 108: 453-462.
9. Cotta JAO, Rezende MOO, Piovani MR (2006) Evaluation of the content of metals in Betari River sediment in the upper Ribeira tourist state park-PETAR, São Paulo, Brasil. Quim Nova 29: 40-45.
10. McCrady MH (1996) Standard methods for examination of water and wastewater. (12th edn), Am J Public Health Nations Health 56: 684.
11. Resende M, Curi N, Ker JC, de Rezende SB (2005) Mineralogy of Brazilian soils-interpretation and applications. Publisher of the Federal University of Lavras, Lavras.
12. Embrapa (1997) Manual of chemical analyzes of soils, plants and fertilizers. Embrapa-CNPS, Rio de Janeiro.
13. CCME (1999) Canadian environmental quality guidelines for the protection of aquatic life.
14. MME (2009) Ministry of mines and energy.
15. Jordão CP, Pereira JL, Jham GN (1997) Chromium contamination in sediment, vegetation and fish caused by tanneries in the state of Minas Gerais, Brazil. Total Sci Environ 207: 1-11.
16. Pawlikowski M, Szalińska E, Wardas M, Dominik J (2006) Chromium originating from tanneries in river sediments: a preliminary investigation from the upper Dunajec river (Poland). Polish J Environ Stud 15: 885-894.
17. Malvandi, H (2017) Preliminary evaluation of heavy metal contamination in the Zarrin-Gol river sediments, Iran. Mar Pollut Bull 117: 547-553.
18. Shanbehzadeh S, Dastjerdi MV, Hassanzadeh A, Kiyanizadeh T (2014) Heavy metals in water and sediment: a case study of Tembi river. Scien World J 2014: 1-5.
19. Sampaio ACS (2003) Heavy metals in water and sediments of the rivers of the upper Paraguay basin (dissertation). Mato Grosso do Sul Federal University, Campo Grande.
20. Skoog DA, West DM, Holler FJ (1992) Fundamentals of analytical chemistry. (6th edn), Saunders College Publishing, Orlando 69.
21. Wedepohl KH (1995) The composition of the continental crust. Geochim Cosmochim Ac 59: 1217-1232.
22. Pereira MGD, Andrade MVD, Ornelas VC, Almeida RAD, Fontes MP, et al. (2012) Assessment of physical-chemical characteristics of water and sediments from a Brazilian tropical estuary: status and environmental implications. Scien World J 2012: 1-14.
23. Ibragimow A, Glosinska G, Siepak M, Walna B (2010) Heavy metals in fluvial sediments of the Odra River flood plains-introductory research. Quaestiones Geographicae, 29: 37-47.
24. Embrapa Solos do Nordeste.
25. Silva MB, dos Anjos LHC, Pereira MG, Cavassani LR (2017) Soils in the karst landscape of Bodoquena plateau in cerrado region of Brazil. Catena 154: 107-117.
26. Foo KY, Hameed BH (2010) Insights into the modeling of adsorption isotherm systems. Chem Eng J 156: 2-10.
27. Sparks DL (2003) Soil environmental chemistry. (2nd edn), Academic press, Amsterdam.

28. Matos GD, Arruda MAZ (2003) Vermicompost as natural adsorbent for removing metal ions from laboratory effluents. Process Biochem 39: 81-88.

29. Tagliaferro GV, Pereira PHF, Rodrigues LA, da Silva MLCP (2011) Adsorption of lead, cadmium and silver in hydrated niobium (V) oxide prepared by the precipitation method in homogeneous solution. Quim Nova 34: 101-105.

30. Meroufel B, Benali O, Benyahia M, Benmoussa Y, Zenasni MA (2013) Adsorptive removal of anionic dye from aqueous solutions by Algerian kaolin: Characteristics, isotherm, kinetic and thermodynamic studies. J Mater Environ Sci 4: 482-491.

31. Krishna RH, Swamy AVVS (2012) Physico-chemical key parameters, Langmuir and Freundlich isotherm and lagergren rate constant studies on the removal of divalent nickel from the aqueous solutions onto powder of calcined brick. Int J Eng Res Dev 4: 29-38.

3

Water Quality Assessment of Mahmoudia Canal in Northern West of Egypt

Ali M Abdullah[1]* and Salah El-Dien Hussona[2]

[1]Holding Company for Water and Wastewater and IGSR, Alexandria University, Cairo, Egypt
[2]Department of Environmental Studies, IGSR, Alexandria University, Cairo, Egypt

Abstract

The main water source for Alexandria and Behiera governorate which locate at northern west of Egypt are Mahmoudia canal (receive about 15 Mm^3/day), which receive water from Rosetta branch at Mahmoudia city. The canal receives domestic and agriculture wastes from Zarcon Drain and other non-point sources. Drinking water in Alexandria is supplied from surface water sources of Mahmoudia canal. The water is transferred to the eight water treatment plants for treatment operations. The primary objective of conventional drinking water treatment and distribution is to deliver to the consumer water that satisfy the requirements and does not constitute human health risk. To achieve this, water utilities employ a range of physical and chemical barriers in order to reduce risks. The present study offered comprehensive water quality information of Mahmoudia canal. Cluster analysis grouped 10 sampling sites into three groups, less polluted (LP), medium polluted (MP) and highly polluted (HP). The dissolved oxygen concentrations, as indicated by the results of Mahmoudia canal during the period of study, ranged from 4.9 mg O_2/l at the beginning of canal after mixing with discharge of Zarcon drain to 5.9 mg O_2/l at the Manshia water treatment plant intake while DO values in Khandak canal, (at 15.2 km from starting of the canal) varied from 6 to 6.2 mg O_2/l. Nutrient concentrations did not comply with the permissible limits of ammonia, as well as the values of COD and BOD were exceeded the Egypt standards. Ammonia, 20 samples out of 36 samples were not complying with Egypt standard. All COD and BOD data obtained were exceeding of Egypt standard.

Keywords: Water quality; Mahmoudia canal

Introduction

Degradation of Water quality is a major issue in Egypt. The severity of present water quality problems in Egypt varies among different water bodies depending on: flow, use pattern, population density, extent of industrialization, availability of sanitation systems and the social and economic conditions existing in the area of the water source. Discharge of untreated or partially treated industrial and domestic wastewater, leaching of pesticides and residues of fertilizers; and navigation are often factors that affect the quality of water [1].

The industrial sector is an important user of natural resources and a contributor to pollution of water and soil. There are estimated to be some 24,000 industrial enterprises in Egypt, about 700 of which are major industrial facilities. The spatial distribution of industry in Egypt is influenced by the size of the employment pool, availability of services, access to transportation networks, and proximity to principal markets. The manufacturing facilities are therefore often located within the boundaries of major cities, in areas with readily available utilities and supporting services. In general, the majority of heavy industry is concentrated in Greater Cairo and Alexandria [2]. Industrial demand for water in the year 2000 has been reported to be 3.6 BCM/year. By the year 2017, the industrial demand for water is expected to reach 5.5 BCM/year. Consequently, a corresponding increase in the volume of industrial wastewater is expected [1].

Based on the population studies and rates of water consumption, the total wastewater flows generated by all governorates, assuming full coverage by wastewater facilities is estimated to be 3.5 BCM/year. Approximately 1.6 BCM/year receives treatment. By the year 2017, an additional capacity of treatment plants equivalent to 1.7 BCM is targeted. Although the capacity increase is significant, it will not be sufficient to cope with the future increase in wastewater production from municipal sources and therefore, the untreated loads that will reach water bodies are not expected to decline in the coming years [1].

The constituents of concern in domestic and municipal wastewater are: pathogens, parasites, nutrients, oxygen demanding compounds and suspended solids. In Greater Cairo and other cities, the sewerage systems also serve industrial and commercial activities. Therefore, instances of high levels of toxic substances in wastewater have been reported. As these toxic substances (heavy metals & organic micro-pollutants) are mainly attached to suspended material, most of it accumulates in the sludge. Improper sludge disposal and/or reuse may lead to contamination of surface and ground water [1].

In general, the bulk of treated and untreated domestic wastewater is discharged into agricultural drains. Total coliform bacteria reach 106 CFU/100 ml as recorded in some drains of Eastern Delta. It is important to mention that all drains of Upper Egypt flow back into the Nile. Moreover, it has become a national policy to maximize the reuse of drainage water by mixing it with canal water. Many irrigation canals may be contaminated with pollutants from domestic sources as a result [1].

Apart from being the largest consumer of water, agriculture is also a contributor to water pollution. Drainage water seeping from agriculture fields are considered non-point sources of pollution. These non-point sources are, however, collected and concentrated in agricultural drains and become point sources of pollution for the River Nile, the Northern Lakes, and irrigation canals in case of mixing water for reuse. Moreover, these non-point sources of pollution may

***Corresponding author:** Ali Mahmoud Ahmed Abdullah, Ph. D, Technical Manager at Holding Company for Water and Wastewater, Cairo, Egypt
E-mail: dr2252000@dr.com

also influence the groundwater quality. Major pollutants in agricultural drains are salts; nutrients (phosphorus & nitrogen); pesticide residues (from irrigated fields), pathogens (from domestic wastewater), and toxic organic and inorganic pollution (from domestic and industrial sources) [2].

Few studies were published about Mahmoudia canal, Nadia [3] reported that Lindane, DDE and DDT were found in Mahmoudia canal and the concentrations were ranged from 0.91 to 2.32 μg/l. Nadia [3] also cited that the heavy metals (Cu, Pb, Zn, Mn and Fe) were found in the canal at levels less than the maximum permissible limits, the same researcher found that the total dissolved solids (TDS), total alkalinity, total hardness (TH) and sulphates were less than the maximum permissible limits [3].

This study has been performed to follow the effect of agricultural wastewater drains and seasonal variation effects on Mahmoudia canal, which may help in giving information about variability in Mahmoudia canal quality that it is the main water sources for Alexandria residents for human uses.

Material and Methods

One hundred and twenty water samples were collected from ten stations. Ten samples were collected monthly, starting from June 2011 to May 2012 as shown in Figure 1. From each station two samples were collected using Van Dorn water sampler, the experimental procedure of the studied parameters were carried out according to APHA [4] as follows:

pH was measured on the spot by using pH meter model (CG-840 SCOTT, Germany) after buffer calibration at pH 4 and 10. Dissolved Oxygen (DO): DO was determined by using DO meter model (HACH SENION 8.) and expressed as (mg/l). The Electrical Conductivity (EC) was measured using an analytical Electrical Conductivity-Jenway 4320 after calibrated by calibrated standard 413, 1413 and 2776 uS/cm. The temperature was measured using an analytical thermometer embedded in (Electrical Conductivity-Jenway 4320).

Heavy metals: heavy metals were measured in water samples using (coupled plasma 400 emission spectrometer Perkin Elemer Emission Spectrometer). The Preservation and digestion were carried out as follows:

- Preservation: Samples were preserved immediately after collection by acidifying with concentrated HNO_3 to pH <2 by adding 5 ml nitric acid to 1 liter water samples. After acidification store in refrigerator.

- Digestion: Heavy metals were extracted in water samples using digestion by nitric acid method as follow: Mix well 500 ml of sample in a beaker at hot plate, add 10 ml nitric acid bring slow boil, evaporate on a hot plate to reduce the volume 200-100 ml, continue heating add another 10 ml of nitric acid until the volume reached to 100-80 ml until digestion is complete. The samples do not dry during digestion, mix well and complete to 100 ml by distilled water, and take portion of their solution for required determination [4].

Chemical oxygen demand (COD), biological oxygen demand (BOD) and total organic carbon (TOC) were determined according to standard methods for water and wastewater (5220 A, 5210 A and 5310

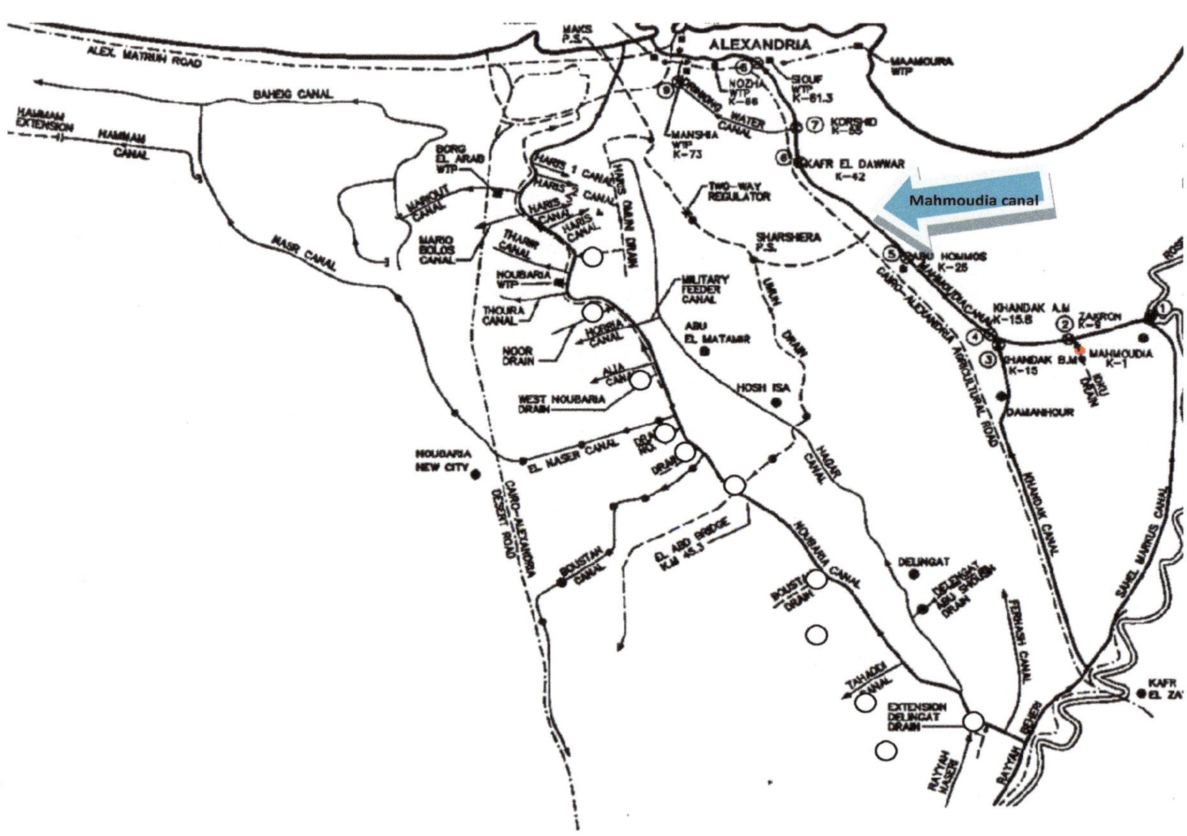

Figure 1: Sampling sites a long of Mahmoudia canal (northern west of Egypt).

A, respectively) [4]. Ammonia (NH_3), Nitrates (NO_3), Nitrites (NO_2) and Phosphates (PO_4) were determined according to standard methods for water and wastewater (4500-NH_3 A, 4500-NO_3 A, and 4500-NO_2 and 4500-P A, respectively) [4].

Total hardness (TH), chloride (Cl), UV_{254} absorption, total dissolved solids at 105°C (TDS) and sulphates (SO_4) were determined according to standard methods for water and wastewater [4].

Chlorinated pesticides, n-methyl Carbamates and PCBs were determined according to EPA 508.1, 537.1 and 508.1 methods [5] by using GC/ECD and HPLC/FLD instruments.

Total algal count (TGC), filter clogging algae, taste and odour algae and polluted water algae were investigating according to standard methods for water and wastewater [4].

Total plate count (TPC), total coliform (TC) and fecal coliform (FC) were investigating according to standard methods for water and wastewater [4].

All analytical test methods used in the present study were verified and validated according to the test method validation for chemistry methods and biological examination [6].

The MS Excel 2007 software and NordTest TR 537 were used for calculations of validation parameters for analytical test method in water: uncertainty, selectivity, linearity, accuracy, precision, range, detection and quantitaition limit, system suitability and solution stability [7].

SPSS 18 software was used for calculation of the principal component, factor analysis and cluster analysis (LEAD Technologies, Inc).

The Canadian Council of Ministries of the Environment (CCME) Water quality Index (WQI) was used to evaluate surface water qualities. Calculations of the index are based on scope (F_1)-number of parameters that are not complaint with water quality guidelines; frequency (F_2)-number of times that the guide lines are not respected and the amplitude (F_3)-the difference between non-complaint measurement and the corresponding guidelines [8].

WQI was calculated according to the following equation:

$$WQI = 100 - \left(\sqrt{f_1^2 + f_2^2 + f_3^2}\right)/1.732$$

$$\text{Scope } F_1 = \frac{\text{Number of failed variables}}{\text{Total no. of variables}} \times 100$$

$$\text{Frequency } F_2 = \frac{\text{Number of failed tests}}{\text{Total no. of tests}} \times 100$$

F_3 represents Amplitude: The extent (excursion) to which the failed test exceeds the guideline.

This is calculated in three stages. First, the excursion is calculated

$$Excursion = \left(\frac{\text{failed test value}}{\text{guideline value}}\right) - 1$$

Second, the normalized sum of excursions (nse) is calculated as follows:

$$nse = \left(\frac{\sum excursion}{\text{total of tests}}\right)$$

Then, F_3 calculated as the following:

$$F_3 = \left(\frac{nse}{0.01 \times nse + 0.01}\right)$$

Results

Results presented in Tables 1 and 2 and Figures 2-5 showed the seasonal and spatial variations of Mahmoudia canal parameters (physical, chemical and biological) from June 2011 to May 2012 and also showed the correlation matrix and principle component analysis of the same canal. The bolded values were exceeding the Egypt maximum permissible limit [9].

The Atrazine, Chlordane-alpha, Chlordane-gamma, Aldrin, 4,4'-DDD, 4,4'-DDE, 4,4'-DDT, Endosulfan I, Endosulfan II, Endrin Aldehyde, HCH-delta, Lindane, Heptachlor Epoxide, Methoxychlor, 3-hydroxy carbofuran, carbofuran, chloropyrofos, Aroclor 1016, Aroclor 1232, Aroclor 1242, Aroclor 1248 and Aroclor 1260 were not

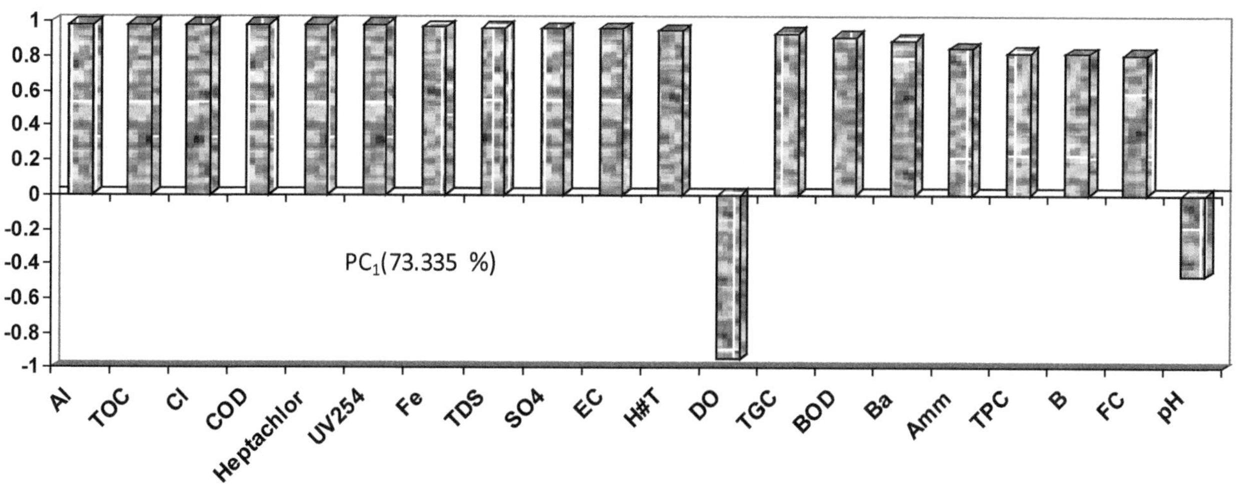

Figure 2: Component loading for first component (PC1).

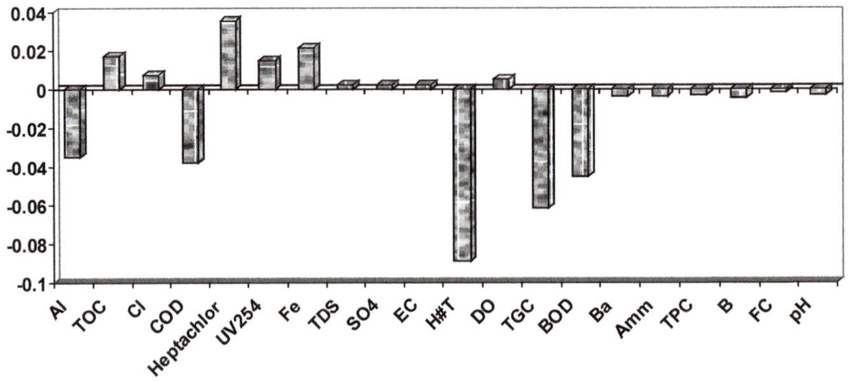

Figure 3: Component loading for first component (PC2).

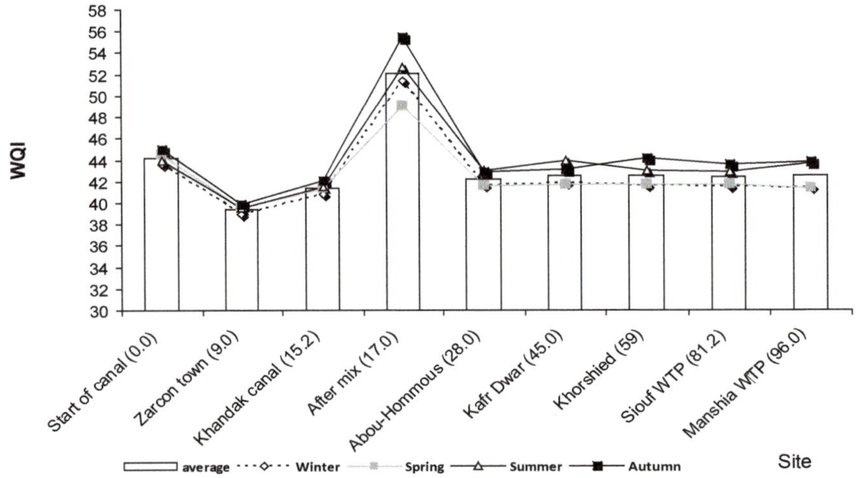

Figure 4: WQI of Mahmoudia canal.

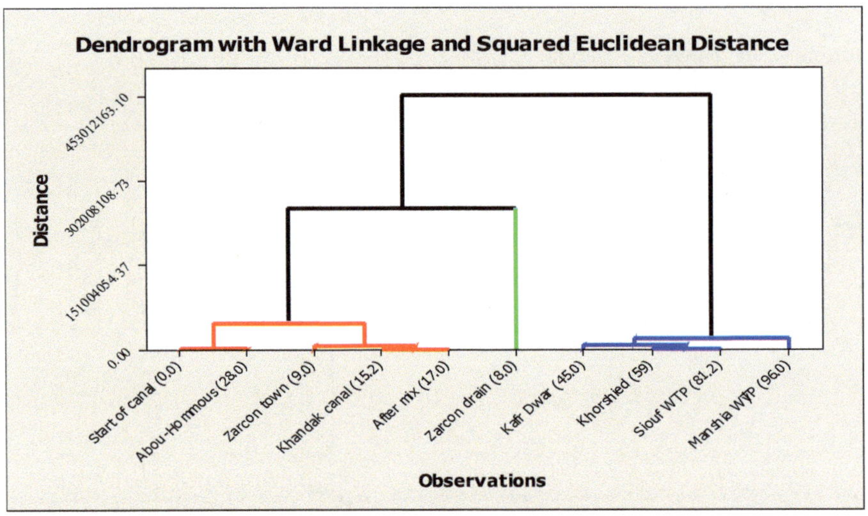

Figure 5: Dendrogram showing the relationship among sampling sites in Mahmoudia canal.

detected during the present study period were not detected during the present study period.

The beryllium, copper, nickel, silver, cadmium, lead and zinc were not detected during the present study period were not detected during the present study period.

The seasonal variation are shown in the values of standard deviation

(RSD), EC, UV254 absorption values, NH3, chloride, TOC, COD, BOD, sulphate, nitrite, nitrates, phosphates, heavy metals, a-BHC, b-BHC, heptachlor and the biological parameters (Table 1).

The values of COD and BOD exceed the Egyptian maximum permissible limit (10 and 5 mg/l, respectively) [10].

All analytical test methods used in the present study were verified and validated according to the test method validation for chemistry methods and biological examination as shown in Table 2.

The test method validation parameters were calculated for analytical test method in water: uncertainty, linearity, accuracy, precision, range, detection and quantitaition limit, system suitability and solution stability as shown in Table 2.

Data analysis for Mahmoudia canal water parameters (correlation matrix, principle component and cluster analysis) were carried out by using SPSS 18 as shown in Tables 3-7 and Figures 2, 3 and 5.

Cluster analysis

To verify data of the most influential factor among the stations of Mahmoudia canals; cluster analysis was done using Word's method (linkage between groups), with Euclidian distance as a similarity measure. Data were then amalgamated into Dendrogram plots (Figure 5).

	Start of canal (0.0)	**Zarcon drain (8.0)**	**Zarcon town (9.0)**	**Khandak canal (15.2)**	**After mix (17.0)**	**Abou-Hommous (28.0)**	**Kafr Dwar (45.0)**	**Khorshied (59)**	**Siouf WTP (81.2)**	**Manshia W TP (96.0)**	**RSD**
Temp. (°C)	21.8	20.4	20.6	20.7	20.7	20.7	20.9	21.0	21.1	20.7	12.7
pH	7.9	7.8	7.9	7.9	7.9	7.8	7.8	8.0	7.9	7.8	2.92
EC (μmosh/cm)	534	956	475	347	467	466	480	485	478	480	31.3
UV_{254}	0.082	0.248	0.105	0.076	0.086	0.08	0.078	0.08	0.079	0.073	55.4
DO (mg/l)	5.5	**4.3**	5.1	6.1	5.5	5.4	5.4	5.5	5.7	5.7	8.9
NH_3 (mg/l)	**1.7**	**2.6**	**2.1**	0.3	1.3	**0.8**	**0.8**	**0.6**	**0.7**	**0.7**	83.2
Chloride (mg/l)	48.4	106.5	54.0	27.0	38.1	39.0	40.5	41.5	39.5	41.8	50.12
H. total (mg/l)	175	218	176	137	161	160	161	162	160	161	19.2
TOC(mg/l)	5.6	17.0	7.2	5.1	5.9	5.5	5.4	5.5	5.4	5.0	55.66
COD (mg/l)	**41.0**	**66.5**	**46.3**	**28.8**	**38.8**	**40.5**	**39.3**	**36.0**	**36.0**	**36.5**	25.8
BOD (mg/l)	**17.3**	**25.8**	**19.5**	**12.3**	**18.0**	**19.5**	**16.5**	**15.0**	**13.0**	**14.0**	26.07
TDS (mg/l)	320	574	285	208	280	279	288	291	287	288	31.3
Sulphate (mg/l)	85.4	153.0	76.0	55.6	74.8	74.6	76.9	77.6	76.6	76.9	31.3
Nitrite (mg/l)	**0.2**	**0.3**	0.1	**0.2**	**0.2**	**0.2**	**0.3**	**0.2**	**0.2**	**0.2**	48.8
Nitrate (mg/l)	3.5	4.0	1.3	2.8	3.6	3.6	4.0	4.2	3.2	3.5	94.9
PO_4 (μg/l)	496	712	105	340	481	602	554	506	269	496	58.37
Al (mg/l)	0.0228	0.125	0.0445	0.018	0.0215	0.0233	0.0223	0.0233	0.019	0.0185	104.3
Fe (mg/l)	0.0095	0.081	0.0118	0.0048	0.0023	0.0013	ND	ND	0.0035	0.0008	155.4
Ba (mg/l)	0.0058	0.0205	0.012	0.0063	0.0023	0.003	0.0028	ND	ND	ND	58.4
B (mg/l)	0.009	0.0145	0.0115	ND	0.0065	0.0013	0.001	ND	ND	ND	49.2
a-BHC (μg/l)	0.0023	0.0395	0.0058	0.0003	0.002	0.002	0.0008	ND	ND	ND	364
b-BHC (μg/l)	ND	0.0395	0.0043	0.003	ND	ND	ND	ND	ND	ND	434
Heptachlor (μg/l)	0.0025	0.393	0.0065	0.0023	0.004	0.0038	0.0018	ND	ND	ND	317
Aldrin (μg/l)	ND	ND	ND	ND	ND	ND	ND	ND	ND	ND	0.0
TGC (u/ml)	351	2720	963	906	642	592	586	532	379	343	116
F. clog (u/ml)	328	1528	810	825	628	490	535	492	384	221	86.8
T & O (u/ml)	75	366	173	160	129	116	153	118	108	53	93
P. W (u/ml)	31	148	65	89	60	81	31	29	33	30	105
TPC (CFU/ml)	15200	29300	20450	17800	18700	14000	11950	9975	9325	6950	76.4
TC (CFU/ml)	5100	9750	7300	7800	7450	6850	5875	6425	4500	3475	64
FC (CFU/ml)	**175**	**1600**	**1100**	**475**	**595**	**670**	**895**	**650**	**175**	**150**	107

- The value between two bracket meant that "the distance by km from the beginning of canal"
- The Atrazine, Chlordane-alpha, Chlordane-gamma, Aldrin, 4,4'-DDD, 4,4'-DDE, 4,4'-DDT, Endosulfan I, Endosulfan II, Endrin Aldehyde, HCH-delta, Lindane, Heptachlor Epoxide, Methoxychlor, 3-hydroxy carbofuran, carbofuran, chloropyrofos, Aroclor 1016, Aroclor 1232, Aroclor 1242, Aroclor 1248 and Aroclor 1260 were not detected during the present study period.
- . The beryllium, copper, nickel, silver, cadmium, lead and zinc were not detected during the present study period.
- RSD is the standard deviation of the result over the period of study

Table 1: Average and standard deviation values of physical, chemical and biological analysis of Mahmoudia water characteristics (2011/2012).

	MDL (mg/l)	LOQ (mg/l)	Repeatability limit (r) (mg/l)	Reproducibility limit (R) (mg/l)	Recovery %	Linearity (R^2)	Measurement uncertainty %	Notes
Temp. (°C)	NA	NA	NA	NA	NA	NA	NA	
pH	0.1	0.2	0.2	0.3	102.6	0.992	8	
EC (µmosh/cm)	2	4	5	7	98	0.994	7	
UV_{254}	NA	NA	NA	NA	NA	NA	NA	
DO (mg/l)	NA	NA	NA	NA	NA	NA	NA	
NH_3 (mg/l)	0.01	0.02	0.04	0.05	98	0.991	12	
Chloride (mg/l)	0.5	1.0	1.0	1.0	99	0.994	10	
H. total (mg/l)	0.5	1.0	1.0	1.0	98	0.993	10	
TOC(mg/l)	0.2	0.4	0.4	0.4	97.5	0.991	15	
COD (mg/l)	0.2	0.4	0.3	0.4	94.8	0.996	12	
BOD (mg/l)	NA	NA	NA	NA	NA	NA	NA	
TDS (mg/l)	1.0	2.0	2.0	3.0	96.8	0.991	11	
Sulphate (mg/l)	1.0	2.0	2.0	3.0	97.4	0.992	12	
Nitrite (mg/l)	0.01	0.02	0.04	0.04	93.1	0.992	20	
Nitrate (mg/l)	0.01	0.02	0.04	0.04	93.1	0.992	20	
Heavy metals	0.0005	0.001	0.002	0.004	91.8	0.995	25	
Chlorinated pesticides (µg/l)	0.01	0.02	0.02	0.025	94.8	0.996	25	
PCBs (µg/l)	0.01	0.02	0.02	0.025	98.1	0.997	25	
N-methyl Carbamates (µg/l)	1.0	2.0	2.0	3.0	103.0	0.998	30	
TGC (u/ml)	2	4	5	6	95.8	0.99	20	
F. clogging (u/ml)	2	4	5	6	95.8	0.99	20	
T & O (u/ml)	2	4	5	6	95.8	0.99	20	
P. W (u/ml)	2	4	5	6	95.8	0.99	20	
TPC (CFU/ml)	1	2	2	4	96.8	0.992	25	
TC (CFU/ml)	1	2	2	4	96.8	0.992	25	
FC (CFU/ml)	1	2	2	4	96.8	0.992	25	

Table 2: Performance data and quality control for the analytical test methods.

The physico-chemical water parameters pH, EC, $UV_{254,}$ DO, ammonia, Cl, TDS, TH, SO_4, NO_3, NO_2, PO_4, TOC, Al, Fe, Ba, B, a-BHC,b-BHC, Heptachlor, Aldrin, TGC, TPC, FC, BOD and COD were used as variables to show the spatial heterogeneity among the stations (Figure 5).

Water Quality Index" (WQI):

Estimation of WQI of Mahmoudia canals: Results presented in Table 7 and Figure 4 showed the seasonal and spatial variations of WQI in Mahmoudia canal from June 2011 to May 2012.

WQI as judged according to The Canadian Council of Ministries of the Environment (CCME) Water quality Index as follows: Values of WQI of Mahmoudia canal ranged from 38.98 (poor water quality) to 55.39 (marginal water quality). The lowest value was observed at Zarcon town (Km9.0) during winter, and the highest value was observed at after mix (Km17.0) during autumn. WQI values of Khandak canal ranged from 40.84 (poor water quality) during winter to 42.04 (poor water quality) during autumn as shown in Figure 4.

Discussion

In Egypt, the used primary and secondary treatment systems inefficient to remove residues of pesticides and organ chlorinated pollutants or parasites, viruses, and other non-parasitic microorganisms. As a result, these residues of chemical and biological pollutants may persistently remain in drinking water and we still cannot obtain safe clean water, moreover the increase of levels of the usually added chlorine to filtered or raw water leads to increased concentration of organic-chlorinated compounds that are known as carcinogenic and mutagenic [11].

Drinking water in Alexandria is supplied from surface water sources of Mahmoudia and Noubaria canals. The water is transferred to the eight water treatment plants for treatment operations. The primary objective of conventional drinking water treatment and distribution is to deliver to the consumer water that satisfy the requirements and does not constitute human health risk. To achieve this, water utilities employ a range of physical and chemical barriers in order to reduce risks. The treatment process is comprised from different steps coagulation, flocculation, sedimentation, filtration and disinfection in the treatment plants, source waters are treated to comply with Egyptian water quality standards in reference to World Health Organization (WHO) standards [10,12].

The present study offered comprehensive water quality information of Mahmoudia canal. Pollution cannot be measured on an absolute scale. Pollution can be thought of as a unifying constraint or label that characterizes responses to groups of variables. Strongly correlated variables may lead to the conclusion that a 'pollution factor' is present. Therefore the magnitude of pollution can be inferred from the measurement of its directly observable variables.

Pollution can be considered as the exceeding of tolerable

	Temp. (0C)	pH	EC (µmosh/cm)	UV 254	DO (mg/l)	NH3 (mg/l)	Chloride (mg/l)	H. total (mg/l)	TOC(mg/l)	COD (mg/l)	BOD (mg/l)	TDS (mg/l)	Sulphate (mg/l)	Nitrite (mg/l)	Nitrate (mg/l)	PO4 (µg/l)	Al (mg/l)	Fe (mg/l)	Ba (mg/l)	B (mg/l)	a-BHC (µg/l)	b-BHC (µg/l)	Heptachlor	TGC (u/ml)	F. clog (u/ml)	T& O (u/ml)	P. W (u/ml)	TPC (CFU/ml)	TC (CFU/ml)	FC (CFU/ml)
Temp. (0C)	1																													
pH	0.4	1.0																												
EC (µmosh/cm)	-0.3	-0.4	1.0																											
UV 254	-0.4	-0.3	1.0	1.0																										
DO (mg/l)	0.4	0.3	-0.9	-0.9	1.0																									
NH3 (mg/l)	-0.1	-0.2	0.8	0.8	-0.9	1.0																								
Chloride (mg/l)	-0.3	-0.3	1.0	1.0	-0.9	0.8	1.0																							
H. total (mg/l)	-0.2	-0.3	1.0	0.9	-1.0	0.9	1.0	1.0																						
TOC(mg/l)	-0.4	-0.3	1.0	1.0	-0.9	0.8	1.0	0.9	1.0																					
COD (mg/l)	-0.3	-0.4	0.9	0.9	-1.0	0.9	1.0	1.0	0.9	1.0																				
BOD (mg/l)	-0.4	-0.4	0.8	0.8	-0.9	0.9	0.9	0.9	0.8	0.9	1.0																			
TDS (mg/l)	-0.3	-0.4	1.0	1.0	-0.9	0.8	1.0	1.0	1.0	0.9	0.8	1.0																		
Sulphate (mg/l)	-0.3	-0.4	1.0	1.0	-0.9	0.8	1.0	1.0	1.0	0.9	0.8	1.0	1.0																	
Nitrite (mg/l)	-0.1	-0.5	0.5	0.5	-0.3	0.0	0.4	0.3	0.5	0.4	0.3	0.5	0.5	1.0																
Nitrate (mg/l)	0.2	-0.2	0.3	0.1	-0.1	-0.2	0.1	0.1	0.1	0.1	0.1	0.3	0.3	0.8	1.0															
PO4 (µg/l)	-0.1	-0.5	0.6	0.4	-0.4	0.1	0.4	0.4	0.4	0.4	0.4	0.5	0.6	0.8	0.9	1.0														
Al (mg/l)	-0.5	-0.3	0.9	1.0	-0.9	0.8	1.0	0.9	1.0	0.9	0.8	0.9	0.9	0.4	0.1	0.4	1.0													
Fe (mg/l)	-0.4	-0.3	0.9	1.0	-0.8	0.8	1.0	0.9	1.0	0.9	0.8	0.9	0.9	0.5	0.1	0.4	1.0	1.0												
Ba (mg/l)	-0.4	-0.3	0.8	0.9	-0.8	0.8	0.9	0.8	0.9	0.8	0.8	0.8	0.8	0.2	-0.2	0.1	0.9	0.9	1.0											
B (mg/l)	-0.1	-0.1	0.7	0.7	-0.8	1.0	0.8	0.8	0.7	0.8	0.8	0.7	0.7	0.0	-0.3	0.1	0.8	0.7	0.9	1.0										
a-BHC (µg/l)	-0.4	-0.4	1.0	1.0	-0.9	0.8	1.0	0.9	1.0	0.9	0.8	1.0	1.0	0.5	0.2	0.4	1.0	1.0	0.9	0.7	1.0									
b-BHC (µg/l)	-0.5	-0.3	0.9	1.0	-0.8	0.7	0.9	0.9	1.0	0.9	0.8	0.9	0.9	0.5	0.2	0.4	1.0	1.0	0.9	0.7	1.0	1.0								
Heptachlor	-0.4	-0.4	1.0	1.0	-0.8	0.7	0.9	0.9	1.0	0.9	0.8	1.0	1.0	0.5	0.3	0.5	1.0	1.0	0.8	0.7	1.0	1.0	1.0							
TGC (u/ml)	-0.6	-0.3	0.9	1.0	-0.8	0.7	0.9	0.8	1.0	0.9	0.8	0.9	0.9	0.4	0.1	0.4	1.0	1.0	0.9	0.7	1.0	1.0	1.0	1.0						
F. clog (u/ml)	-0.6	-0.2	0.7	0.9	-0.7	0.6	0.8	0.7	0.9	0.8	0.7	0.7	0.7	0.3	-0.1	0.2	0.9	0.9	0.9	0.7	0.9	0.9	0.9	1.0	1.0					
T& O (u/ml)	-0.6	-0.3	0.8	0.9	-0.8	0.6	0.8	0.7	0.9	0.8	0.8	0.8	0.8	0.5	0.0	0.3	0.9	0.9	0.9	0.6	0.9	0.9	0.9	1.0	1.0	1.0				
P. W (u/ml)	-0.6	-0.4	0.6	0.8	-0.6	0.5	0.7	0.6	0.8	0.7	0.7	0.6	0.6	0.3	0.0	0.3	0.8	0.8	0.8	0.6	0.8	0.8	0.8	0.9	0.9	0.9	1.0			
TPC (CFU/ml)	-0.4	-0.2	0.7	0.8	-0.7	0.8	0.7	0.7	0.8	0.8	0.8	0.7	0.7	0.2	-0.2	0.2	0.8	0.8	0.9	0.8	0.8	0.8	0.8	0.9	0.9	0.9	0.9	1.0		
TC (CFU/ml)	-0.6	0.0	0.5	0.7	-0.6	0.5	0.6	0.4	0.7	0.6	0.7	0.5	0.5	0.2	-0.1	0.2	0.7	0.7	0.8	0.6	0.7	0.7	0.6	0.8	0.9	0.8	0.9	0.9	1.0	
FC (CFU/ml)	-0.6	-0.3	0.7	0.8	-0.8	0.7	0.7	0.7	0.8	0.8	0.8	0.7	0.7	0.3	0.0	0.3	0.8	0.7	0.8	0.6	0.8	0.8	0.7	0.8	0.9	0.9	0.7	0.8	0.8	1.0

Table 3: Correlation Matrix of Mahmoudia canal water parameters.

	Initial	*Extraction*
pH	1.000	0.302
EC	1.000	0.959
UV_{254}	1.000	0.961
DO	1.000	0.896
Amm	1.000	0.852
Cl	1.000	0.964
H.T	1.000	0.896
TOC	1.000	0.964
COD	1.000	0.963
BOD	1.000	0.829
TDS	1.000	0.959
SO_4	1.000	0.959
NO_2	1.000	0.960
NO_3	1.000	0.940
PO_4	1.000	0.918
Al	1.000	0.968
Fe	1.000	0.939
Ba	1.000	0.920
B	1.000	0.845
Heptachlor	1.000	0.962

Extraction Method: Principal Component Analysis.

Table 4: Factor analysis on the samples collected from Mahmoudia canal.

	Initial Eigenvalues		
Component	*Total*	*% of Variance*	*Cumulative %*
1	14.667	73.335	73.335
2	2.955	14.773	88.108

Extraction Method: Principal Component Analysis.

Table 5: Total variance of Mahmoudia canal parameters.

concentration limits of specific substances in a system. Identification of such underlying dimensions or factors greatly simplifies the description and understanding of complex phenomena [13]. Pollution can be estimated by observing the number of measurable parameters, viz. concentration of various elements in water, land and air, rise in temperature, variation in rain fall, changes in biological death rate of the consumers, etc. Since pollution is a dimensionless entity, it can be quantified by the measurement of dependent variables. Factor analysis is such a tool which can be applied to the measurable variables to identify certain factors to draw inferences. Based on these assumptions, factor analysis on the chemical data obtained on the water samples of Mahmoudia canal is carried out [13].

Initially, correlation matrices of the major and minor elements of all the water samples are calculated. Factor analysis of the above data is shown in Tables 4-6 and Figures 2 and 3 which include communalities, Eigen values, percent variance and cumulative percent variance.

A critical look at Tables 4-6, for factor 1 which explains 73.33% variance shows that variables such as Al, UV_{254}, NH_3, Cl, TH, TOC, BOD, COD, SO_4, Fe, Ba, B, a-BHC, total algal count (TGC), total plate count (TPC) and BOD have higher positive loadings (+0.70 and higher), and strong negative loading with DO; suggesting their 'deposition' or 'accumulation'. Therefore, by factor analysis, it has become clear that the previous parameters are entering the environment (Mahmoudia canal water) due to the discharge of pollutants by the Zarcon Drain and non- point sources. Factor 2 which explains 14.77% variance has negative loading for the variable trace metal B, ammonia and TPC. It means that all these elements are moving together. Therefore factor 1 for this set of samples would be 'Migration of pollutants' or 'Pollution'. This factor could be agricultural activity which includes the use of excess fertilizers by farmers.

Table 1 summarizes briefly the mean value and standard deviation of the 31 measured variables in Mahmoudia canal water samples from the selected ten sites. It must be noticed the high dispersion of most variables (high relative standard deviations), which indicates variability in chemical composition between samples, thus pointing to the presence of temporal variations caused likely by polluting sources and/or climatic factors. Recommended guide levels of these variables and maximum levels allowed by the law 48/1982 and its amendments and ministerial decree 458/2007 concerning the quality of water intended for human consumption [10].

	Component	
	1	2
Al	0.983	-3.495×10^{-2}
TOC	0.982	1.699×10^{-2}
Cl	0.982	7.535×10^{-3}
COD	0.981	-3.755×10^{-2}
Heptachlor	0.980	3.558×10^{-2}
UV_{254}	0.980	1.461×10^{-2}
Fe	0.969	2.131×10^{-2}
TDS	0.960	0.195
SO_4	0.959	0.196
EC	0.959	0.196
H.T	0.946	-8.910×10^{-3}
DO	-0.945	5.030×10^{-2}
TGC	0.930	-6.125×10^{-2}
BOD	0.909	-4.494×10^{-2}
Ba	0.893	-0.350
Amm	0.845	-0.371
TPC	0.823	-0.329
B	0.814	-0.427
FC	0.812	-0.175
pH	-0.469	-0.287

Extraction Method: Principal Component Analysis.

Table 6: Score coefficient of Mahmoudia canal parameters.

It must be emphasized that, the average concentrations of some variables such as chloride, COD, TDS, BOD, ammonia, nitrite, phosphate and sulphate are higher than those recommended by the Egyptian legislation, therefore this water resource is not adequate for human consumption or industrial purposes and needs to be purified. High levels of phosphate may originate from municipal wastewater discharges since it is an important component of detergents [2,14]. The presence of nitrate in the canals section sampled is suspected to originate from discharge from agricultural drain (Zarcon drain in Mahmoudia canal, Rahawy Drain which discharges in Rosetta branch) and the use of inorganic fertilizers (usually as ammonium nitrate) is rather frequent [15]. This practice could also explain the high levels of ammonia, but this pollutant may also originate from decomposition of nitrogen containing organic compounds such as proteins and urea occurring in municipal wastewater discharges. In the presence of high levels of organic matter, nitrate can be reduced in some extent to nitrite, what could explain the high concentration of this pollutant in some samples [16,17].

Because the ten sampling stations were combined to calculate the correlation matrix, the correlation coefficients should be interpreted with caution as they are affected simultaneously by spatial and temporal variations. Nevertheless, some clear hydro-chemical relationships can be readily inferred: High and positive correlation can be observed between ammonia, UV_{254} absorption values, BOD, COD, TOC, sulphate, chloride, total dissolved solids, conductivity, total plate count, total coliform, algal count and hardness (R^2=0.55 to 0.99) (Table 2). BOD, TOC and COD are strongly correlated (R^2=0.82 to 0.93) in Mahmoudia canal and also with ammonia, phosphate (closely related to contamination for organic matter), dissolved oxygen is weakly correlated with temperature because the solubility of oxygen in water decreases with increasing temperature; BOD, COD and nitrogen and phosphorous compounds are also anti-correlated with dissolved oxygen as organic matter is partially oxidized by oxygen, whilst nutrients are responsible for eutrophication of freshwater, thus causing a further increase in organic matter concentration and, hence, in oxygen demand (Table 3).

Iron, nitrate and pH showed no significant correlation with any other variables [18,19]. Rosetta branch, starting from Delta Barrage receives relatively high concentrations of organic compounds, nutrients and oil AND grease. Major sources of pollution are Rahawy

Site (km) / Season		Start of canal (0.0)	Zarcon town (9.0)	Khandak canal (15.2)	After mix (17.0)	Abou-Hommous (28.0)	Kafr Dwar (45.0)	Khorshied (59)	Siouf WTP (81.2)	Manshia WTP (96.C)
WQI	Winter	43.65	38.98	40.84	51.42	41.58	41.83	41.65	41.54	41.44
	Spring	44.49	39.49	41.44	49.08	41.62	41.75	41.76	41.77	41.43
	Summer	43.92	39.54	41.47	52.64	42.96	43.89	43.02	42.84	43.72
	Autumn	44.98	39.93	42.04	55.39	42.90	43.09	44.19	43.55	43.84
	Average	44.2	39.4	41.4	52.1	42.2	42.6	42.6	42.4	42.6

Table 7: WQI of Mahmoudia canal at 9 sampling points.

drain (which receives part of Greater Cairo wastewater), Sabal drain, El- Tahrrer drain, Zawiet El-Bahr drain and Tala drain. At Kafr El-Zayat, Rosetta branch receives wastewater from Maleya and Salt and Soda companies [1,2]. Mahmoudia canal receive water from Rosetta branch at Mahmoudia city (receive about 15 Mm^3/day) [2].

Dissolved oxygen concentrations, as indicated by the results of Mahmoudia canal during the period of study, ranged from 4.9 mg O_2/l at the beginning of canal after mixing fresh water with discharge of Zarcon drain (k.m 8.0) to 5.9 mg O_2/l at the Manshia WTP intake (k.m 96.0) while the DO ranged in Khandak canal, which inflow water in Mahmoudia canal at (k.m 15.2); 6 to 6.2 mg O_2/l. Nutrient concentrations are not comply with the permissible limits (Table 1).

Ammonia, COD and BOD values were exceeded the Egyptian standard (law 48/1982). Ammonia, 50 samples out of 90 samples were not complying with Egypt standard. All COD and BOD data obtained were exceeding of Egyptian standard. Drains are mainly used for discharge of predominantly untreated or poorly treated wastewater (domestic & industrial), and for drainage of agricultural areas. Therefore, they contain high concentrations of various pollutants such as organic matter (BOD, COD), nutrients, fecal bacteria, heavy metals and pesticides. Mahmoudia canal is suffering from pollution resulted by the agriculture drains (Zarcon Drain) which is feeding the canal at Km 6. The drainage water inflows to Mahmoudia canal are 0.5 Mm^3 / day at Zarcon town (Km 6.0) and about 0.25 Mm^3/ day from non-point sources, which acts 5.0 % of total flow of Mahmoudia canal on average (Table 1).

The affiliation among the stations were obtained through cluster analysis using Ward's method (linkage between groups), with Euclidian distance as a similarity measure and were amalgamated into Dendrogram plots as indicated in Figure 5. The physico-chemical Characteristics were used as variables to show the spatial heterogeneity among the stations as a result of sequence in their relationship and the degree of contamination.

Mahmoudia canal cluster analysis; there were four major groups obtained from the four seasons, the first group compromise from Kafr Dwar (km 45.0), Khorished (km 54), Siouf WTP intake (km 81.2) and Manshia WTP intake (km 96.0) which high polluted, the second group include Zarcon town (km 9.0), Khandak canal (km 15.2) and after mix point (km 17.0) which moderate polluted, the third group include the start of canal (km 0.0) and Abou-Hommous (km 28.0) which lower polluted and the fourth group which very high polluted include the Zarcon Drain (km 8.0) as indicated in Figure 5.

Water, a natural resource which has been used for different purposes, namely for drinking, domestic, irrigation and industrial, mainly depends on its intrinsic quality hence it is prime importance to have prior information on quality and quality of water resources available in the region. Water quality index (WQI) is regarded as one of the most effective way to communicate water quality. The WQI, which was developed in the nearly 1970s, can be used to monitor water quality changes in a particular water supply over time, or it can be used to compare a water supply's quality with other water supplies in the region or from around the world. The result can also be used to determine if a particular stretch of water is considered as "healthy" [8].

WQI as judged according to The Canadian Council of Ministries of the Environment (CCME) [20] Water quality Index as follows: Values of WQI of Mahmoudia canal ranged from 38.98 (poor water quality) to 55.39 (marginal water quality). The lowest value was observed at Zarcon town (Km 9.0) during winter, and the highest value was observed at after mix (Km 17.0) during autumn. WQI values of Khandak canal ranged from 40.84 (poor water quality) during winter to 42.04 (poor water quality) during autumn.

The water at almost all the sites showed that; the water quality was poor as a result of outflow of irrigation drainage.

Conclusions

Assessment of the available data obtained from both Mahmoudia canal indicates the following:

- The main water quality is exhibit high pollution levels that create health risks at present, indicates unsafe levels of pollution for direct use in drinking water, irrigation and fisheries.
- The major sources of pollution in Mahmoudia canal are discharge of Zarcon drain
- Major sources of pollution of Rosetta branch is El- Rahawy drain in the southern part and industry at Kafr El-Zayat. (Reverse impact on Mahmoudia canal water quality).
- Delta drains receive high concentrations of organic and inorganic pollutants from industrial, domestic as well as diffuse agricultural wastewater. High priority should be given to those drains receiving high loads of pollution such as: El-Rahawy.
- Increase the water flow in Khandak canal reduces the salinity and COD in Mahmoudia canal water quality.

Acknowledgement

The authors thank our AWCO central lab team for their scientific and technical assistance.

References

1. Egyptian Environmental Affairs Agency (EEAA) (2002) Water Quality Data of the Nile River.
2. Drainage Research Institute (2000) Monitoring and Analysis of Drainage Water Quality Project, Drainage Water Status in the Nile Delta Yearbook 97/98. Technical Report No. 52.
3. Galal NM (1983) Determination of pesticides in Alexandria Drinking Water. Master thesis, High Institute of Public Health, Alexandria University.
4. APHA-AWWA-WPCF (2005) Standard Methods for the Examination of Water and Wastewater (SMWW), (21st edn) American Public Health Association, American Water Works Association, Water Pollution Control Federation, U.S.A.
5. USEPA (1995) Methods for the Determination of Organic Compounds in Drinking Water. Supplement III, August 1995, EPA/600/R- 95/131.
6. European Commission (2007) Method Validation and Quality Control Procedures for Pesticide Residues Analysis in Food and Feed. Document No. SANCO/2007/3131. p. 35.
7. Magnusson B, Näykki T, Hovind H, Krysell M (2003) Handbook for Calculation of Measurement Uncertainty in Environmental Laboratories. Espoo 2003. Nordtest, NT Techn Report 537. 41 p. NT Project No. 1589-02.
8. Cude CG (2001) Oregon Water Quality Index: A tool for evaluating water quality management effectiveness. Journal of the American Water Resources Association 37: 125-137.
9. Egypt law No. 48/1982 (1982) Water Quality of the Nile River and Main Branches and Canals.
10. Ministerial Decree No. 485 (2007) Egypt Health Ministry, Ameryia Press.
11. Slavinskaya GV (1991) Chlorination effect on quality of drinking water. Khimiya I Tekhnologiya Vody, 13: 1013-1022.

12. WHO (2006) Guidelines for drinking water quality. (3rd edn), Geneva.

13. Aruga R, Negro G, Ostacoli G (1993) Multivariate data analysis applied to the investigation of river pollution. Fresenius J. Anal. Chem. 346: 968-975.

14. AWT (2001) Surface Water Quality Assessment of the Hawkesbury-Nepean Catchment 1995-1999, Australian Water Technologies, Department of Land and Water Conservation, Sydney, NSW.

15. Ghallab MHM (2000) Some physical and chemical changes on River Nile downstream of Delta Barrage at El-Rahawy drain. M. Sc. Thesis Fac. of Sci. Ain Shams Univ. Cairo, Egypt.

16. Elewa AA, Shehata MB, Issa YM, Abd El-Satar AM (1995) Study on major anions and cations of the river Nile water at El-Kanater El-Khyria region. Fac., Sci. Zagazig University, Zoology 144-165.

17. Harrison MR, De Mora JD (1996) Introductory chemistry for environmental science, Cambridge University, Press P. 301.

18. Essa M (2002) The impact of river Nile water quality on the south Cairo water supply. Ms. Thesis, chemical Dep. Fac., Cairo, p259-270.

19. Gohar MA (1998) Factors affecting the precipitation and dissolution of some chemical elements in River Nile at Damietta branch. M.Sc. Thesis fact. of Sci. Menofia Univ. Egypt.

20. Canadian Council of Ministers of the Environment (2001) Canadian water Quality Guidelines for the protection of aquatic life, CCME water quality index 1.0, Users' manual in Canadian Environment quality Guidelines.

4

Towards a Convenient Procedure to Characterize Polycyclic Aromatic Hydrocarbons in Sediments Receiving Industrial Effluents

Nadia Morin-Crini*, Coline Druart, Caroline Amiot, Frédéric Gimbert, Etienne Chanez and Grégorio Crini

Department of Chrono-Environment, UMR UFC/CNRS 6249 USC INRA, University of Franche-Comté, 16 Route de Gray, 25030 Besançon cedex, France

Abstract

Using pressurized liquid extraction and GC-MS/MS, Polycyclic Aromatic Hydrocarbon (PAH) levels were determined in pond and river sediments receiving effluents from a Chemicals Installation (CI) and a Surface Treatment Installation (STI), respectively. Maximum values were obtained for the STI site with total PAH concentrations of 3000-4000 ng g^{-1} compared to 200-2500 ng g^{-1} for the CI site. Furthermore, in the river (STI), for two PAHs (phenanthrene and acenaphthylene), one sample presented concentrations that exceeded the probable effect concentrations leading to this sediment being classified as toxic. However, PAH levels were higher upstream of the STI discharge water than downstream, indicating sediment PAH accumulation was not exclusively due to this industrial activity. At the CI site, the concentrations found at different points showed that in the pond, PAHs were derived from industrial activities but were rapidly diluted in the water mass. PAH fingerprinting following various guidelines, revealed the prevalence of a pyrolytic origin.

Keywords: Freshwater sediments; Hydrocarbons; Industrial sites; GC-MS/MS; Pyrolytic origin; Pressurized liquid extraction

Introduction

Polycyclic Aromatic Hydrocarbons (PAHs) are organic pollutants with two or more fused aromatic rings. They are introduced into the environment via natural and anthropogenic processes and contaminate all environmental compartments. Indeed, they have already been found in food or food supplements [1], marine organisms [2-4], water [5-7], sewage sludge [8], dust particles, soils and sediments [9-12]. Due to their high toxicity to humans [13,14] and aquatic organisms [15], they have become a focus for scientific research. Consequently, it was necessary to develop analytical methods able to quantify this pollutant family. Numerous methods have already been developed for PAH extraction in various environmental matrices, such as Pressurized Liquid Extraction (PLE), Supercritical Fluid Extraction (SFE), subcritical water extraction, Dispersive Liquid-Liquid Microextraction (DLLME), Solid-Phase Extraction (SPE) or Microextraction (SPME), Stir Bar Sorptive Extraction (SBSE) or soxhlet extraction [16-20].

Due to their chemical properties such as low water solubility, PAHs released into the aquatic environment tended to sorb onto suspended particles and sediments and could therefore impact both benthic organisms, directly in contact with sediments, and pelagic microphagous organisms feeding on suspended particles [21,22]. It could be interesting to determine the origin of PAHs which could be pyrogenic (incomplete combustion of organic material from anthropogenic activity or natural source such as forest fire), petrogenic (crude oil or coal) or biogenic (diagenesis) sources in freshwater sediments, given that their presence has already been shown in industrial discharge waters [3,23,24] usually flushed into the freshwater environment. Thus, it is necessary to develop selective and sensitive methods of extraction and analysis of freshwater sediments from industrial sites, because of the simultaneous presence of a large quantity of various pollutant types (salts, metal ions, organics) as described by Morin-Crini et al. [23].

Firstly, the present work aimed to develop, in freshwater sediments, a specific, rapid and sensitive analytical method for the quantification of the 16 PAHs which have been selected by the US Environmental Protection Agency (US EPA) as priority pollutants (Figure 1). Secondly, we investigated the presence of PAHs in sediments in contact with discharge waters from industrial sites carrying out chemistry and surface treatment activities. Finally, we propose to determine the PAH compositional origin through three indicators based on the number of rings, isomeric ratios and total index (based on weighted-isomeric ratios).

Material and Methods

Chemicals

Sixteen PAHs and three deuterated internal standards (ISTD) were purchased from Supelco Sigma Aldrich (Saint Quentin Fallavier, France) and Interchim (Montluçon, France): naphthalene (NAP), acenaphthene (ACE and ACE-d10), acenaphthylene (ACY), fluorene (FLU), phenanthrene (PHE), anthracene (ANT), fluoranthene (FLT), pyrene (PYR and PYR-d10), benzo[a]anthracene (BaANT), chrysene (CHY), benzo[b]fluoranthene (BbFLT), benzo[k]fluoranthene (BkFLT), benzo[a]pyrene (BaPYR and BaPYR-d12,), dibenzo[a,h] anthracene (dBahANT), indeno[1,2,3-cd]pyrene (IcdPYR), benzo[g,h,i]perylene (BghiPL). n-hexane and methanol Distol, a range of products for organic trace analysis, were obtained from Fisher Scientific (Illkirch, France). Acetone (analytical grade, Fisher Scientific, Illkirch, France), Decon-90 (Fisher Scientific, Illkirch, France) and ultra-pure water (Millipore Milli-Q Integral 3 system, Molsheim, France) were used for glassware clean-up.

PAH extraction

Different experiments were performed to improve PAH

***Corresponding author:** Nadia Morin-Crini, Department of Chrono-Environment, UMR UFC/CNRS 6249 USC INRA, University of Franche-Comté, 16 Route de Gray, 25030 Besançon cedex, France
E-mail: nadia.crini@univ-fcomte.fr

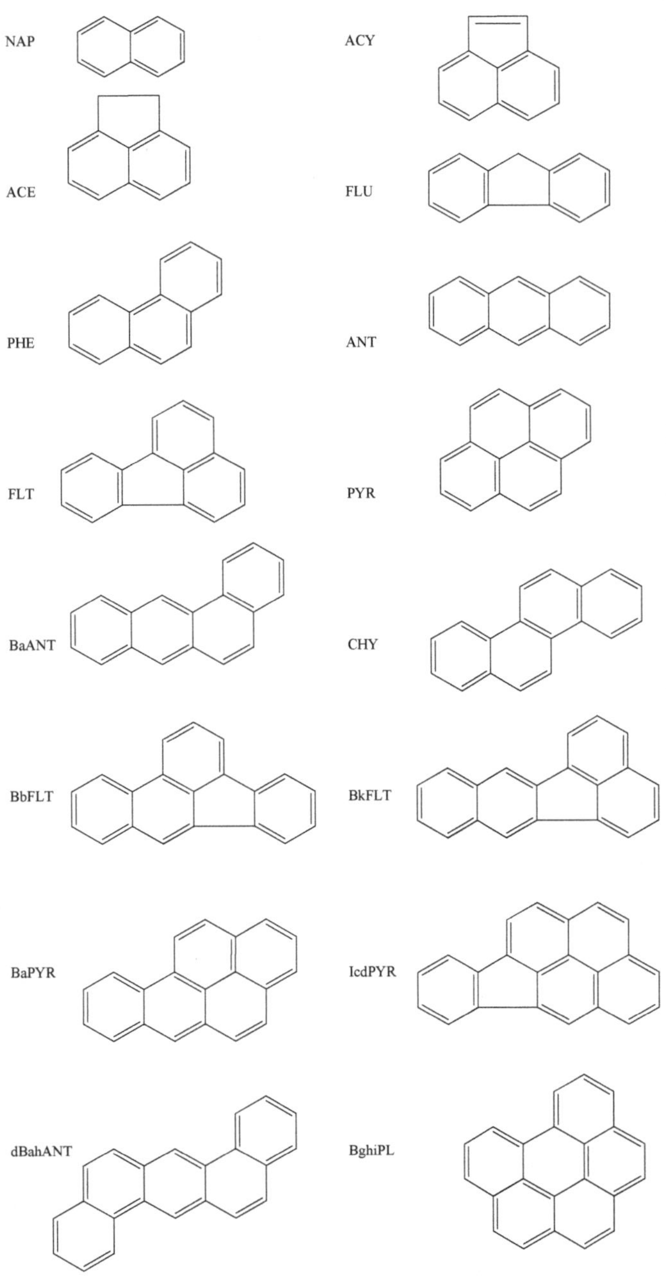

Figure 1: Chemical structure of the 16 PAHs defined as priority pollutants by the US EPA.

extraction efficiencies using sediment from a river located in a rural (forest) and weakly anthropized watershed, referred to as blank. The blank sediment was spiked at 50 ng g^{-1} dry mass. However, to allow equilibrium to be reached between PAHs and matrix, a contact time of one night in the dark and at 4°C (to prevent photodegradation) was applied. Comparison between PLE and ultrasonic extraction (USE) was investigated. For USE, 5 g of sediment was sonicated with 50 mL of solvent (hexane or methanol) for 15 min, conditions adapted from Martinez et al. [25]. The extract was centrifuged (5 min at 4500 rpm) then the sediment was retrieved and sonicated a second time in the same conditions. Both fractions were concentrated and injected separately into the GC-MS/MS apparatus and then summed. PLE was performed using a Büchi (Rungis, France) SpeedExtractor E-914 system equipped with stainless steel extraction cells (volume 40 mL). The extraction cell was filled from the bottom to the top with the following materials: first, a glass fiber filter, then a few grams of pure quartz sand followed by the dried sediment sample (5.000 ± 0.001 g), covered by glass beads and finally, a cellulose filter. The sediment samples were extracted with hexane or methanol under the following conditions: temperature: 100°C; pressure: 10 MPa; cycles: 1 or 2; heat-up: 1 min; hold: 10 min; discharge: 2 min; flush with solvent: 1 min; flush with nitrogen gas: 2 min. In both cases, USE or PLE, the extracts were concentrated using a Syncore Analyst (Büchi, Rungis, France). In the case of PLE, it was the same vessel which was used both for collected extracts and their concentration, avoiding the loss of organic compound during transfer. Moreover, a locally cooled appendix on the sample vessel prevented evaporation to dryness (residual volume of roughly 0.3 mL). The concentrated extracts of PLE and USE were transferred to a 1 mL volumetric flask and made up with hexane.

PAH analysis

Sample extracts were analyzed in the GC-MS/MS apparatus (Agilent, Massy, France) which included a 7890A GC system, a 7000 GC triple quadrupole mass spectrometer and an 80 GC Combipal autosampler. 1 µL of the extract was injected at a temperature of 300°C in splitless mode under a constant He flow (purity 99.9999 %) at 1.5 mL min^{-1} followed by a purge flow to split vent after 0.5 min. We used a (5 % phenyl)-methylpolysiloxane HP5MS column (30 m x 0.25 mm i.d, 0.25 µm film thickness, Agilent 19091J-433). The temperature gradient was from 45°C (hold 0.5 min) to 100°C at 30°C min^{-1} then from 100°C to 325°C at 10°C min^{-1} (hold 5 min). Total run time was 30 min. The temperatures of the transfer line, ion source and quadrupoles were 300, 300 and 150°C respectively. An electron-impact ion source at 70 eV was required. Nitrogen with a purity of 99.9 % was produced by a NiGen LC-MS 40-1 Claind generator (Gengaz, Wasquehal, France) and used as collision gas at 1.5 mL min^{-1}. For data analysis, Agilent MassHunter B.05.02.1032 software was used. For apparatus performance controls, an autotune was performed weekly at m/z 69, 264 and 502 from the electron-impact of perfluorotributylamine.

Validation of extraction method

A calibration graph was established for each of the 16 compounds using blank sediment spiked at five increasing concentrations (10, 20, 40, 60 and 100 ng g^{-1} dry mass). The spiked calibration samples were treated following the validated extraction procedure: PLE in hexane with 2 cycles. Calibration curves were constructed by plotting the analyte/ISTD peak area ratios versus the analyte concentration. Five experimental replicates for each concentration were carried out to validate the method through the following parameters: linearity, Limit of Detection (LOD), precision and trueness.

Field samples

The method was applied to field samples of sediments receiving industrial effluents. Sediments were manually collected at two sites using a modified pickaxe which sampled 0.05 m^2 to a depth of 5 cm. At the Chemical Installation (CI) site, sediments were sampled at six locations in the pond in which the wastewater was released (Figure 2). Points 2 to 6 were situated along a transect starting from the nearest to the discharge to the furthest whereas point 1 was isolated upstream of the discharge water by a dike (Figure 2). At the Surface Treatment Installation (STI) site, sediments were sampled at four locations in the river in which the discharge water was released. Points Cup and Dup were situated upstream of the discharge water, close (110 m) and distant (645 m) respectively, whereas Cdown and Ddown were situated

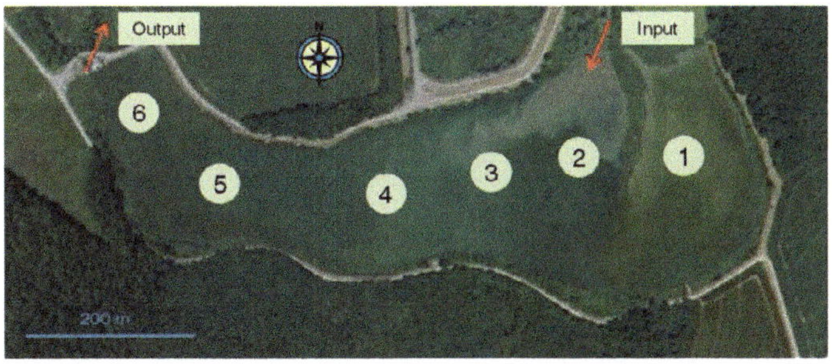

Figure 2: Location of the six sampling points along the length of the chemical industry pond.

downstream of the discharge water, close (130 m) and distant (545 m), respectively. An aliquot of each sample was immediately lyophilized until constant mass and then sieved through a 0.5 mm mesh. The other aliquot was stored at -20°C in case of need for further analyses. Indeed, additional analysis was performed on sediment from point 2 (site CI) which had been centrifuged for 15 minutes at 6100 rpm to remove interstitial water, before being lyophilized.

Regulatory standards to sediment toxicity classification

The concentrations determined in field samples were compared to different types of reference values which defined threshold concentrations regarding biological effects (Table 1). The effects range low (ERL) and effects range median (ERM) provided chemical concentration ranges that are rarely, occasionally, or frequently associated with adverse biological effects [26,27]. The threshold and probable effect concentrations (TEC and PEC, respectively) proposed by Zeng et al. [28] and derived from several guidelines [29-31] allowed us to judge whether sediment was toxic or nontoxic.

Methods for determining the composition and hence the origin of the PAHs

Several approaches were used to determine the source of PAH pollution (petrogenic, biogenic and pyrogenic). The first, which was described by several authors [5,32,33] stated that pyrogenic PAHs were dominated by heavier molecules with 4 to 6 rings. The second approach used analyses of selected PAH isomer ratios [22,34,35]:

$$\text{ratio178}=\frac{[\text{ANT}]}{[\text{ANT}]+[\text{PHE}]}$$

$$\text{ratio 202}=\frac{[\text{FLT}]}{[\text{FLT}]+[\text{PYR}]}$$

$$\text{ratio 228}=\frac{[\text{BaANT}]}{[\text{BaANT}]+[\text{CHY}]}$$

$$\text{ratio 276}=\frac{[\text{IcdPYR}]}{[\text{IcdPYR}]+[\text{BghiPL}]}$$

where [ANT], [PHE], [FLT], [PYR], [BaANT], [CHY], [IcdPYR], [BghiPL] were PAH concentrations.

Ratios above 0.10 at mass 178, 0.50 at mass 202, 0.35 at mass 228 and 0.20 at mass 276 represented dominance of pyrogenic origin.

The third approach used the total index defined by Orecchio [36] which stated that a value above 4 indicated a pyrogenic process:

$$\text{Total index}=\frac{\text{ratio 178}}{0.2}+\frac{\text{ratio 202}}{0.4}+\frac{\text{ratio 228}}{0.1}+\frac{\text{ratio 276}}{0.5}$$

where ratios 178 to 276 were defined in equations (1) to (4).

Results and Discussion

Investigations of extraction parameters

Whatever the solvent used (hexane or methanol) and the PLE cycle number, the extraction yields were still higher with PLE than with USE. For instance, in hexane and with two extraction cycles, an average efficiency increase of 36% was obtained with the use of PLE compared to both fractions summed in USE. Thus, all results presented hereunder were realized with the PLE method. Investigations on the choice of the extraction solvent, hexane or methanol, showed that, except for BaPYR, IcdPYR, dBahANT and BghiPL, the signal to noise ratio (S/N) was increased with the use of hexane up to 500-fold

	ERL	ERM	TEC	PEC
NAP	160	2100	176	561
ACY	44	640	5.87	128
ACE	16	500	6.71	88.9
FLU	19	540	77.4	536
PHE	240	1500	204	1170
ANT	85.3	1100	57.2	845
FLT	600	5100	423	2230
PYR	665	2600	195	1520
BaANT	261	1600	108	1050
CHY	384	2800	166	1290
BbFLT	nd	nd	nd	nd
BkFLT	nd	nd	240	13400
BaPYR	430	1600	150	1450
IcdPYR	nd	nd	200	3200
dBahANT	63.4	260	33	135
BghiPL	nd	nd	170	3200
Σ PAHs	4022	44792	1610	22800

nd non defined

Table 1: Reference values in sediments (ng g^{-1}) [26-31] and associated biological effects (rarely adverse effect < ERL < occasionally < ERM < frequently; nontoxic < TEC < neither toxic nor nontoxic < PEC < toxic).

	Correlation coefficient	LOD (ng g^{-1})	RSD (%)	Recovery (%)
NAP	0.634	18.6	27	141
ACY	0.915	1.1	19	119
ACE	0.957	2.1	14	114
FLU	0.961	2.0	14	116
PHE	0.941	10.6	15	123
ANT	0.935	1.7	17	122
FLT	0.935	5.5	16	121
PYR	0.951	3.5	15	111
BaANT	0.944	0.5	16	113
CHY	0.924	1.9	16	112
BbFLT	0.967	1.4	18	102
BkFLT	0.941	1.7	15	106
BaPYR	0.943	0.5	16	122
IcdPYR	0.886	0.6	27	106
dBahANT	0.938	0.2	26	119
BghiPL	0.941	0.7	24	98

Table 2: Correlation coefficient (R^2), Detection Limits (LOD), Relative Standard Deviations (RSD) and recoveries for the determination of the 16 PAHs.

at a given concentration. These results are in accordance with those of Martinez et al. [25] and Foan and Simon [37] who indicated that the use of polar solvent, such as methanol, caused chromatographic interference due to the co-extraction of polar compounds. Finally, it was demonstrated that performing the extraction in two cycles instead of only one increased the average efficiency by 13%.

Method validation

Correlation coefficients of each compound were always higher than 0.9 except for NAP and IcdPYR (Table 2). These results are in agreement with those of Burkhardt et al. [38] who obtained the lowest R^2 value for NAP. Precision, which was assessed through the Residual Standard Deviations (RSD), gave suitable values for all PAHs (RSD values ≤ 30%; Table 2). The trueness was acceptable, since recoveries were determined between 98 % (BghiPL) and 141 % (NAP) (Table 2). High sensitivity was obtained for most PAHs (LOD between 0.2 and 5.5 ng g^{-1} for 1 μL injection volume, 5 g sediment, Table 2). However, NAP and PHE gave higher LOD values (18.6 and 10.6 ng g^{-1} respectively). The LODs obtained in this study were better than those observed by the GC-MS analysis of 1 μL of extract obtained by PLE and SPE of 25 g of sediment [38]. Similar values were obtained for an injection volume of 10 μL and a mass of extracted sample of 0.5 g, by USE and SPE coupled to reversed phase liquid chromatography with UV detection [39] and for Soxhlet extraction of 3 g sediment and an injection volume of 1 μL in GC-MS [36]. Nevertheless, very high sensitivity was obtained by a longer, multi-step method composed of PLE followed by SBSE coupled with thermal desorption and GC-MS/MS analysis [40].

Analysis of field samples

The validated method was applied to PAH determination in sediments from two stations linked to industrial activities. At the CI site, note that among the 16 PAHs studied, 14 were systematically found in all samples (Table 3). The highest values were obtained for point 2, located closest to the discharge water release point from the chemical industry with a total PAH content of 2467 ng g^{-1}. Because two compounds (BaANT and CHY) presented concentrations higher than ERL (Tables 1 and 3), the sediment from point 2 could cause adverse biological effects. In return, considering TEC and PEC (Table 1), this sample was judged to be neither toxic nor nontoxic firstly because concentrations of 6 individual PAHs (BaANT, CHY, BkFLT, BaPYR, dBahANT and BghiPL) and secondly the sum of PAHs were between the TEC and PEC values. To determine if the high concentrations at point 2 came from the sediment itself or from the interstitial water, the sample was analyzed a second time with the addition of a centrifugation step to separate the liquid and solid phases of the sediment. The results indicated that concentrations decreased by a factor of 15 to 30 for BaANT, CHY, BbFLT, BkFLT, BaPYR, IcdPYR, dBahANT and BghiPL. Thus, PAHs present at point 2, were mainly concentrated in the interstitial water in colloidal or dissolved forms. The low concentrations found at point 1 (Table 3), always below ERL and TEC (Table 1), were expected since this sample was isolated from the discharge water by the dike. Thus, point 1 could be defined as a relative field control sample. Taking these considerations into account, the sample concentrations at points 3 to 6 seemed to regain this base level, reflecting a sharp decrease in the sum of PAH concentrations from the input point. Various abiotic and biotic phenomena may explain these results, from simple PAH dilution in the pond to more complex biogeochemical degradation processes [41,42].

In the STI site, all samples were highly contaminated, notably in comparison with the CI site (Tables 3 and 4). Indeed, they all exceed the ERL values for at least one compound and considering the sum of PAH concentrations, they were defined as neither toxic nor nontoxic (Tables 1 and 4). However, with regard to PHE (1337 ng g^{-1}) and ACY (231 ng g^{-1}) concentrations, Dup sediment can be classified as toxic since it overshot PEC values for these compounds (Table 1). Note that this point was the most contaminated among the four locations along the river. However, this sample originated from a location 545 meters

	Point 6	Point 5	Point 4	Point 3	Point 2	Point 1
NAP	31.12	< LOD	< LOD	< LOD	62.49	23.12
ACY	1.91	1.73	2.01	2.23	2.46	2.09
ACE	3.26	< LOD	< LOD	< LOD	5.10	< LOD
FLU	19.73	8.85	5.88	8.36	15.17	16.81
PHE	48.40	30.45	25.76	32.02	54.33	44.18
ANT	5.82	4.29	4.37	5.48	7.24	5.46
FLT	35.47	35.05	35.80	42.56	56.24	40.16
PYR	33.10	28.12	29.60	34.75	48.93	30.92
BaANT	11.38	12.81	13.53	15.69	286.88	12.77
CHY	21.35	23.39	24.10	28.46	528.68	24.01
BbFLT	15.55	15.56	16.59	18.87	344.30	15.10
BkFLT	16.86	17.31	19.23	21.19	359.22	18.51
BaPYR	13.68	12.91	13.7	15.96	275.61	13.74
IcdPYR	7.59	8.25	7.93	8.32	138.07	6.22
dBahANT	2.01	2.96	5.22	2.32	44.63	2.00
BghiPL	12.88	11.87	12.03	13.95	237.75	11.27
$\sum PAH_s$	280.1	213.6	215.8	250.2	2467	266.3
Ratio 178	0.11	0.12	0.15	0.15	0.12	0.11
Ratio 202	0.52	0.55	0.55	0.55	0.53	0.57
Ratio 228	0.35	0.35	0.36	0.36	0.35	0.35
Ratio 276	0.37	0.41	0.40	0.37	0.37	0.36
Total index	6.05	6.36	6.48	6.41	6.18	6.14

Table 3: Concentrations (ng g^{-1}), isomeric ratios and total index of PAHs from pond sediments from chemical installation site.

	Dup	Cup	Cdown	Ddown
NAP	294.00	<LOD	<LOD	<LOD
ACY	231.00	46.00	39.80	37.30
ACE	46.50	3.70	4.40	6.90
FLU	225.00	18.10	16.70	21.70
PHE	1336.5	255.20	259.40	225.80
ANT	467.50	79.70	69.80	60.80
FLT	1728.30	718.70	879.40	672.90
PYR	1496.50	468.70	578.80	441.50
BaANT	668.34	368.43	390.37	252.03
CHY	456.75	301.60	348.98	216.40
BbFLT	1122.70	391.20	498.80	399.50
BkFLT	929.20	388.67	432.47	329.87
BaPYR	1037.30	367.30	390.90	390.20
IcdPYR	296.37	107.40	115.50	105.70
dBahANT	227.37	81.64	88.14	72.80
BghiPL	365.40	145.90	172.30	143.10
$\sum PAH_s$	10929	3742	4286	3377
Ratio 178	0.26	0.24	0.21	0.21
Ratio 202	0.54	0.61	0.60	0.60
Ratio 228	0.59	0.55	0.53	0.54
Ratio 276	0.45	0.42	0.40	0.42
Total index	9.47	9.05	8.65	8.80

Table 4: Concentrations (ng g^{-1}), isomeric ratios and total index of PAHs from river sediments from surface treatment installation site.

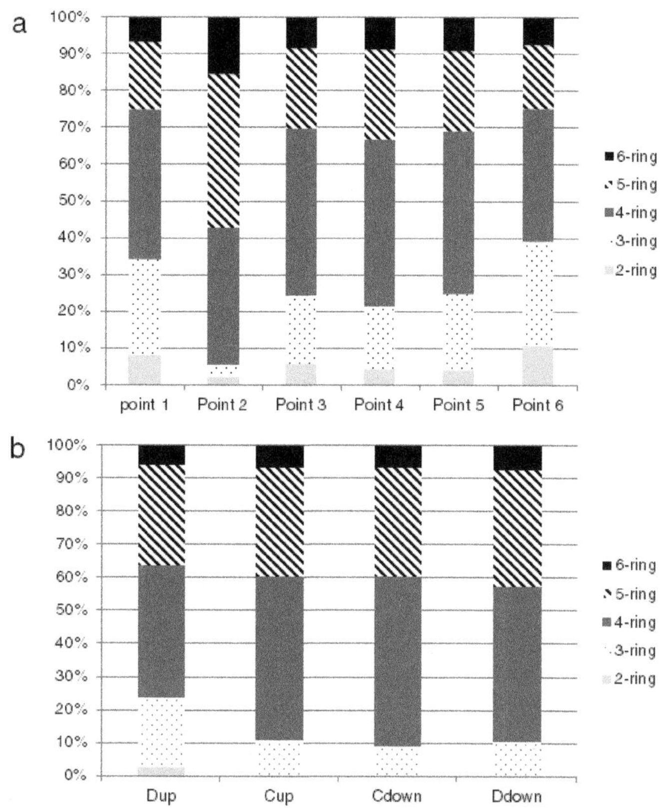

Figure 3: PAH (2-6 rings) distribution patterns of samples from (a) chemical installation site and (b) treatment surface installation site.

upstream of the STI site, so its high levels of contamination cannot be attributed to the STI activities. Therefore, at least a portion of the PAH levels measured downstream of the discharge waters must result from this upstream sediment contamination. However, considering the differences in PAH concentrations in sediments just before and just after the discharge water (Cup and Cdown, respectively, Table 4), the STI activities also contributed to the PAHs released in the river (+23% for the sum of PAHs).

As the two sites studied were close to industrial areas, we can assume that most PAHs were of anthropogenic pyrogenic origin (furnaces, industrial production, etc.) rather than from a petroleum source. It can be observed that PAHs with 4-6 rings made up the major part of total PAHs in both sites (Figure 3), which was 61-94% and 76-91% for CI and STI sites, respectively, confirming that the PAH arose from combustion processes. The average ratios 178 were 0.12 and 0.23 for sediments from CI and STI sites, respectively (Tables 3 and 4) stressing the dominance of combustion origin. For ratio 202, the values were situated in the range 0.52-0.61 for both sites (Tables 3 and 4) indicating rather biomass or coal combustion. An average value of 0.35 was obtained for ratio 228 at site CI (Table 3), corresponding to the limit between petrogenic and pyrolytic sources. The average value of 0.55 obtained at site STI (Table 4) pointed to a combustion origin. For CI and STI sediments, ratio 276 gave average values of 0.38 and 0.42 respectively, suggesting combustion processes. However, whatever the site and the location, all total indexes were above 4 (Tables 3 and 4), indicating high temperature combustion processes and thus confirming a mainly pyrolytic origin of the PAHs found in the sediments impacted by both industrial sites.

Conclusions

In this work we developed and validated a method based on PLE-GC-MS/MS for the analysis of PAHs, allowing their identification and quantification at ultra-trace levels in freshwater sediments. The results showed large concentration differences between the chemical installation and surface treatment installation sites, STI samples having total PAH values situated in the range 210-2500 ng g^{-1} against 3400-11000 ng g^{-1} for CI samples. After checking against the regulatory values which classified sediments as toxic or not, the present study showed that these media sometimes reached hazardous concentrations for the aquatic environment. Following fingerprinting to determine the PAH origin, the total index indicated that whatever the location, the PAHs were of pyrolytic origin.

Acknowledgments

The authors thank the Agence de l'Eau Rhône Méditerranée Corse, the Conseil Régional de Franche-Comté and the FEDER (Fonds Européens de Développement Régional) for financial support. The authors wish to thank Dr Peter Winterton (University Toulouse III, Toulouse, France) for his English grammar and syntax review and his critical reading.

Conflict of Interests

The authors declare that they have no conflict of interest.

References

1. Danyi S, Brose F, Brasseur C, Schneider YJ, Larondelle Y, et al. (2009) Analysis of EU priority polycyclic aromatic hydrocarbons in food supplements using high performance liquid chromatography coupled to an ultraviolet, diode array or fluorescence detector. Anal Chim Acta 633: 293-299.
2. Jaouen-Madoulet A, Abarnou A, Le Guellec AM, Loizeau V, Leboulenger F (2000) Validation of an analytical procedure for polychlorinated biphenyls, coplanar polychlorinated biphenyls and polycyclic aromatic hydrocarbons in environmental samples. J Chromatogr A 886: 153-173.
3. Naes K, Axelman J, Naf C, Broman D (1998) Role of soot carbon and other carbon matrices and the distribution of PAHs among particles, DOC, and the dissolved phase in the effluent and recipient waters of an aluminium reduction plant. Environ Sci Technol 32: 1786-1792.
4. Serpe FP, Esposito M, Gallo P, Serpe L (2010) Optimisation and validation of an HPLC method for determination of polycyclic aromatic hydrocarbons (PAHs) in mussels. Food Chem 122: 920-925.
5. Bihari N, Fafandel M, Piskur V (2007) Polycyclic aromatic hydrocarbons and ecotoxicological characterization of seawater, sediment, and mussel *Mytilus galloprovincialis* from the Gulf of Rijeka, the Adriatic Sea, Croatia. Arch Environ Contam Toxicol 52: 379-387.
6. Law RJ, Dawes VJ, Woodhead RJ, Matthiessen P (1997) Polycyclic aromatic hydrocarbons (PAH) in seawater around England and Wales. Mar Pollut Bull 34: 306-322.
7. Oros DR, Ross JR, Spies RB, Mumley T (2007) Polycyclic aromatic hydrocarbon (PAH) contamination in San Francisco Bay: a 10-year retrospective of monitoring in an urbanized estuary. Environ Res 105: 101-118.
8. Santos JL, Aparicio I, Alonso E (2007) A new method for the routine analysis of LAS and PAH in sewage sludge by simultaneous sonication-assisted extraction prior to liquid chromatographic determination. Anal Chim Acta 605: 102-109.
9. Christensen JH, Tomasi G, de Lemos Scofield A, de Fatima Guadalupe Meniconi M (2010) A novel approach for characterization of polycyclic aromatic hydrocarbon (PAH) pollution patterns in sediments from Guanabara Bay, Rio de Janeiro, Brazil. Environ Pollut 158: 3290-3297.
10. Nemirovskaya IA (2010) The concentration and composition of hydrocarbons in water, particulate matter, and bottom sediments of the Kara sea. Oceanology 50: 716-728.
11. Schulze T, Seiler TB, Streck G, Braunbeck T, Hollert H (2012) Comparison of different exhaustive and biomimetic extraction techniques for chemical and biological analysis of polycyclic aromatic compounds in river sediments. J Soil Sediment 12: 1419-1434.
12. Smith JN, Lee K, Gobeil C, Macdonald RW (2009) Natural rates of sediment

containment of PAH, PCB and metal inventories in Sydney Harbour, Nova Scotia. Sci Total Environ 407: 4858-4869.

13. Roos PH, Tschirbs S, Pfeifer F, Welge P, Hack A, et al. (2004) Risk potentials for humans of original and remediated PAH-contaminated soils: application of biomarkers of effect. Toxicology 205: 181-194.

14. Sánchez-Guerra M, Pelallo-Martínez N, Díaz-Barriga F, Rothenberg SJ, Hernández-Cadena L, et al. (2012) Environmental polycyclic aromatic hydrocarbon (PAH) exposure and DNA damage in Mexican children. Mutat Res 742: 66-71.

15. Cachot J, Geffard O, Augagneur S, Lacroix S, Le Menach K, et al. (2006) Evidence of genotoxicity related to high PAH content of sediments in the upper part of the Seine estuary (Normandy, France). Aquat Toxicol 79: 257-267.

16. Fatoki OS, Ximba BJ, Opeolu BO (2011) Polycyclic aromatic hydrocarbons (PAHs) in food and environmental samples: an overview. Fresenius Environ Bull 20: 2012-2020.

17. Hawthorne SB, Grabanski CB, Martin E, Miller DJ (2000) Comparisons of Soxhlet extraction, pressurized liquid extraction, supercritical fluid extraction and subcritical water extraction for environmental solids: recovery, selectivity and effects on sample matrix. J Chromatogr A 892: 421-433.

18. Itoh N, Numata M, Aoyagi Y, Yarita T (2008) Comparison of low-level polycyclic aromatic hydrocarbons in sediment revealed by Soxhlet extraction, microwave-assisted extraction, and pressurized liquid extraction. Anal Chim Acta 612: 44-52.

19. Kim JH, Moon JK, Li QX, Cho JY (2003) One-step pressurized liquid extraction method for the analysis of polycyclic aromatic hydrocarbons. Anal Chim Acta 498: 55-60.

20. Wang XW, Lin L, Luan TG, Yang LH, Tam NFY (2012) Determination of hydroxylated metabolites of polycyclic aromatic hydrocarbons in sediment samples by combining subcritical water extraction and dispersive liquid-liquid microextraction with derivatization. Anal Chim Acta 753: 57-63.

21. Cailleaud K, Forget-Leray J, Souissi S, Hilde D, Lemenach K, et al. (2007) Seasonal variations of hydrophobic organic contaminant concentrations in the water-column of the Seine Estuary and their transfer to a planktonic species *Eurytemora affinis* (Calanoïda, copepoda). Part 1: PCBs and PAHs. Chemosphere 70: 270-280.

22. Readman JW, Fillmann G, Tolosa I, Bartocci J, Villeneuve JP, et al. (2002) Petroleum and PAH contamination of the Black Sea. Mar Pollut Bull 44: 48-62.

23. Morin-Crini N, Druart C, Gavoille S, Lagarrigue C, Crini G (2013) Analytical monitoring of the chemicals present in the discharge water generated by the surface treatment industry. J Environ Protection 4: 53-60.

24. Umbuzeiro Gde A, Kummrow F, Roubicek DA, Tominaga MY (2006) Evaluation of the water genotoxicity from Santos Estuary (Brazil) in relation to the sediment contamination and effluent discharges. Environ Int 32: 359-364.

25. Martinez E, Gros M, Lacorte S, Barceló D (2004) Simplified procedures for the analysis of polycyclic aromatic hydrocarbons in water, sediments and mussels. J Chromatogr A 1047: 181-188.

26. Mai BX, Fu JM, Sheng GY, Kang YH, Lin Z, et al. (2002) Chlorinated and polycyclic aromatic hydrocarbons in riverine and estuarine sediments from Pearl River Delta, China. Environ Pollut 117: 457-474.

27. McCauley DJ, DeGraeve GM, Linton TK (2000) Sediment quality guidelines and assessment: overview and research needs. Environ Sci Policy 3, Supplement 1: 133-144.

28. Zeng S, Zeng L, Dong X, Chen J (2013) Polycyclic aromatic hydrocarbons in river sediments from the western and southern catchments of the Bohai Sea, China: toxicity assessment and source identification. Environ Monit Assess 185: 4291-4303.

29. CCME (1999) Canadian sediment quality guidelines for the protection of aquatic life: polycyclic aromatic hydrocarbons (PAHs), in, Canadian Council of Ministers of the Environment.

30. Ingersoll CG, MacDonald DD, Wang N, Crane JL, Field LJ, et al. (2001) Predictions of sediment toxicity using consensus-based freshwater sediment quality guidelines. Arch Environ Contam Toxicol 41: 8-21.

31. MacDonald DD, Ingersoll CG, Berger TA (2000) Development and evaluation of consensus-based sediment quality guidelines for freshwater ecosystems. Arch Environ Contam Toxicol 39: 20-31.

32. Boitsov S, Jensen HK, Klungsøyr J (2009) Natural background and anthropogenic inputs of polycyclic aromatic hydrocarbons (PAH) in sediments of South-Western Barents Sea. Mar Environ Res 68: 236-245.

33. Culotta L, De Stefano C, Gianguzza A, Mannino MR, Orecchio S (2006) The PAH composition of surface sediments from Stagnone coastal lagoon, Marsala (Italy). Mar Chem 99: 117-127.

34. Baumard P, Budzinski H, Garrigues P (1998) PAHs in Arcachon Bay, France: Origin and biomonitoring with caged organisms. Mar Pollut Bull 36: 577-586.

35. Yunker MB, Macdonald RW, Vingarzan R, Mitchell RH, Goyette D, et al. (2002) PAHs in the Fraser River basin: a critical appraisal of PAH ratios as indicators of PAH source and composition. Org Geochem 33: 489-515.

36. Orecchio S (2010) Contamination from polycyclic aromatic hydrocarbons (PAHs) in the soil of a botanic garden localized next to a former manufacturing gas plant in Palermo (Italy). J Hazard Mater 180: 590-601.

37. Foan L, Simon V (2012) Optimization of pressurized liquid extraction using a multivariate chemometric approach and comparison of solid-phase extraction cleanup steps for the determination of polycyclic aromatic hydrocarbons in mosses. J Chromatogr A 1256: 22-31.

38. Burkhardt MR, Zaugg SD, Burbank TL, Olson MC, Iverson JL (2005) Pressurized liquid extraction using water/isopropanol coupled with solid-phase extraction cleanup for semivolatile organic compounds, polycyclic aromatic hydrocarbons (PAH), and alkylated PAH homolog groups in sediment. Anal Chim Acta 549: 104-116.

39. Sun FS, Littlejohn D, Gibson MD (1998) Ultrasonication extraction and solid phase extraction clean-up for determination of US EPA 16 priority pollutant polycyclic aromatic hydrocarbons in soils by reversed-phase liquid chromatography with ultraviolet absorption detection. Anal Chim Acta 364: 1-11.

40. Camino-Sánchez FJ, Zafra-Gómez A, Pérez-Trujillo JP, Conde-González JE, Marques JC, et al. (2011) Validation of a GC-MS/MS method for simultaneous determination of 86 persistent organic pollutants in marine sediments by pressurized liquid extraction followed by stir bar sorptive extraction. Chemosphere 84: 869-881.

41. Bertilsson S, Widenfalk A (2002) Photochemical degradation of PAHs in freshwaters and their impact on bacterial growth - influence of water chemistry. Hydrobiologia 469: 23-32.

42. Haritash AK, Kaushik CP (2009) Biodegradation aspects of polycyclic aromatic hydrocarbons (PAHs): a review. J Hazard Mater 169: 1-15.

Effects of Crude Oil Contamination under the Controlled Conditions on the Physicochemical Properties of Water in Khurmala and Guwayar, Kurdistan Region, Iraq

Jadoon S[1]*, Amin AA[2], Malik A[3] and Khalel H[1]

[1]*Department of Natural Resources Engineering and Management, University of Kurdistan, Hewlêr, Iraq*
[2]*Erbil Technology Institute, Erbil Polytechnic University, Kurdistan, Iraq*
[3]*Institute of Molecular Biology and Biotechnology, University of Lahore, Lahore, Pakistan*

Abstract

During the last couple of decades, the economic sector related to hydrocarbon resources has been one of the most attractive for investors. As a result of different exploration activities for crude oil, ecosystems in general and the aquatic ecosystem in particular have suffered a huge disadvantage. These industrial activities have increased the risk of water contamination and have caused serious problems for water environmental systems and their species. The intent of this paper is to introduce promising ways for solving contamination problems of aquatic environment. Although our approach is simple, but it can help to achieve a considerable change in the contamination ranks. Our study proposes an easy method to decrease the crude oil pollution by using different instrumental techniques to investigate the percentages of pollution under a control condition contamination process by checking the concentration of different cations like (calcium Ca^{+2}, sodium Na^{+}, potassium K^{+}), anions like (chloride Cl^{-}, nitrate NO_3^{-}, sulphate SO_4^{-2}), heavy metals like (Pb^{+2}) , and different quality measurements such as, ionic composition, hydrogen ion concentration (P^{H}), total dissolved solid (TDS) mg/L, and electron conductivity (EC) μS/cm of water samples collected from Darbandikhan reservoir, and crude oil samples collected from Khurmala oil field located between Erbil and Kirkuk Province. After careful investigations, it is concluded that except Pb^{+2} and TDS, rest of the data are at the permissible limits. The record of ranges of these data for both river (samples) and drinking water (samples) were respectively Ca^{+2} (1-4.5) mg/L, Na^{+} (2.-13) ppm, K^{+}(0.47-0.89) mg/L, Pb^{+2} (0.434-1.166) mg/L, Cl^{-} (0.0009-1.5) mg/l, SO_4^{-2} (40.8-87.9) mg/l, NO_3^{-} (2.29-4.65) mg/L, EC (0-1697) μS/cm, Hardness(0.000096-0.02) mg/L, TDS(0-780.6) mg/L. Ca^{+2} (4.2-4.5) mg/L, Na^{+}(6.9-7) mg/L, K^{+}(0.07-0.15) mg/L, Pb^{+2} (1.16-1.24) ppm, Cl^{-} (0.0011-0.125) ppm, SO_4^{-2} (79.15-87.9) mg/L, NO_3^{-} (1.50-4.58) mg/L, EC (0-199.7) μS/cm, Hardness (0.00023-0.002) mg/L, P^{H} (6.6-7) mg/L. $t_{stat} < t_{critical}$, we are 95% confident and we rejected the null hypothesis, for chloride and sulphate ions, sulphate and nitrate ions. $t_{stat} < t_{critical}$, we are 95% confident that we accept the null hypothesis for nitrate and chloride ions.

Keywords: Crude oil; Contamination; Darbandikhan reservoir; Ifraz River water; Drinking water

Introduction

Natural resources especially hydrocarbons have become one of the most important resources of energy for mankind. However, the process of extraction of these natural resources is very complicated and most of the time the pollutant's products that accompany the process of oil exploration play a vital role to declare an inconsistency in the ecosystem [1]. This means an increase in the probability of environmental risk and as a result a widespread of hazardous material in the aquatic environment. An aquatic environment can be considered as one of the environments that are subjected more to the contamination by crude oil. This seems to be a big issue due to the spillage of some crude oil into the water. This spillage is due to the leakage which accompanies the process of oil exploration and transportation [2]. As a result, the aquatic environment becomes contaminated region. This situation is a big threat for the evolution of macrophyte, and for the life support balance of living organisms in that environment. Therefore, with world's increasing dependency on the production of crude oil products, crude oil water contamination problem becomes the major factor that can alter the natural consistency of the outstanding life source in a particular environment. Therefore, in order to decrease or remedied these effects, oil companies must adopt proper measure that are helpful to minimize the contamination rate to the normal wedge. This paper investigates the effects of crude oil on aquatic environment. Different instrumental techniques have been used in the process of measuring the physiochemical and ionic characteristics of a controlled condition contaminated water samples as a part of solution that helps to achieve chemically balanced aquatic environment.

Materials and Methods

Samples collection

Three types of samples were collected: crude oil, river water, and drinking water samples from the area under consideration. The crude oil samples were collected from Khurmala oil refinery which is located between Erbil and Kirkuk Province, while the river water samples were collected from three different places within Darbandikhan Reservoir (Figure 1). These places have been clearly identified by the coordinates 35.2009°N and 45.81295°E. Furthermore, about the drinking water samples, theses samples can be collected from any source of pure drinking water. The odour and the color of the samples were examined.

***Corresponding author:** Jadoon S, Department of Natural Resources Engineering and Management, University of Kurdistan, Hewlêr, Iraq
E-mail: saima.jadoon@ukh.edu.krd

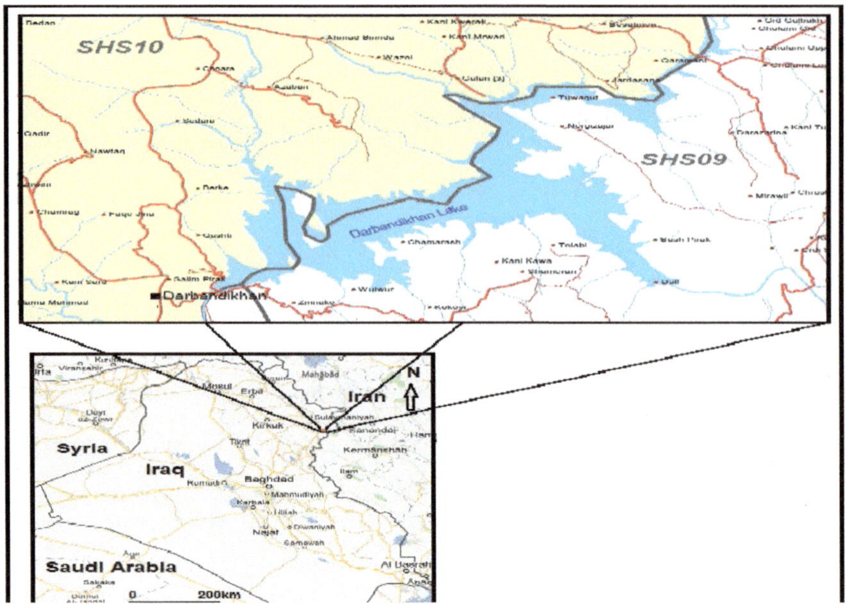

Figure 1: shows the location of Darbandikhan reservoir relative to its location on the Iraq map.

Different instrumental procedures were used to analyze different samples.

Apparatus and procedure of analysis

Field analysis: For field analysis the odor and the color of the samples were examined. The color of the derived crude oil sample was black, while the sample odor was strong similar to the normal oil odor.

Laboratory analyses: In order to reach the selected target goal, a number of instrumentation measurements were followed.

For the chloride ion concentration, Mohrs titration method was used [3]. Furthermore, for the determination of physiochemical properties of the contaminated water samples 600 modle Plaintest micro was used to determine the P^{H}, TDS, and electric conductivity respectively (600 modle Plaintest micro, UK). While for the total water hardness, EDTA titration method was employed to establish the hardness data of the contaminated samples [4]. In addition, for the determination of the anions concentration Sulphate and Nitrate in the contaminated water samples, UV/Visible Spectrophotometer was employed (modle 6705 UV/Vis Jenway, UK) [5]. While, for the Cations concentration which are (Ca^{+2}, Na^{+}, K^{+}, Pb^{+2}) instrumental techniques were employed to determine the concentrations. These instruments are respectively flame photometer for calcium, Sodium, and Potassium determination (model PFP7 Jenway, UK), and Atomic absorption spectrophotometer for Lead Concentration (model210/211/ Buck Scientific) [6].

Main Results

In Table 1 the recorded values of the common cations in the contaminated river water samples showed a range of (1-4.5) mg/L for Ca^{+2}, (2.-13) mg/L for Na^{+}, (0.47-0.89) mg/L for K^{+}, and (0.434-1.166) mg/L for Pb^{+2}. According to the World Health Organization (WHO) the concentrations were found at the permissible limits except for Pb^{+2} concentrations whose maximum value was recognized to fall in a range which exceeds the WHO [7] permissible limits.

In Table 2 the recorded values of the common cations in the contaminated drinking water samples showed a range of (4.2-4.5) mg/L for Ca^{+2}, (6.9-7) mg/L for Na^{+}, (0.07-0.15) mg/L for K^{+}, and (1.16-1.24) mg/L for Pb^{+2}. According to the World Health Organization (WHO) standards, the concentrations are at the permissible limits except for Pb^{+2} concentrations which are recognized to fall in a range which exceed the WHO permissible limits.

In Table 3 the recorded values of major anions in river water contaminated by crude oil showed a range of (0.001-1.5) mg/l for Cl^{-}, (40.8-87.9) mg/l for SO_4^{-2}, and (2.29-4.65) mg/l for NO_3^{-}. According to the World Health Organization (WHO) standards, the concentrations are at the permissible limits [7].

In Table 4 the recorded values of major anions in drinking water contaminated by crude oil showed a range of, (0.0011-0.125)mg/l for Cl^{-}, (79.15-87.9) mg/l for SO_4^{-2}, and (1.50-4.58)mg/l for NO_3^{-}. According to the World Health Organization (WHO) standards, the concentrations are at the permissible limits [7].

In Table 5 the recorded values of the physiochemical characteristics of river water samples contaminated with crude oil showed a range of (0-1697) µS/cm for EC, and TDS(0-780.6) mg/L. According to the World Health Organization (WHO) standards, the concentrations are at the permissible limits except the TDS of the sample that was recognized to fall in a range which exceed the WHO permissible limits. The concentration of this sample was prepared by mixing 99.5 ml of river water with 0.5 ml of crude oil.

In Table 6 the recorded values of the physiochemical characteristics of drinking water samples contaminated with crude oil showed a range of (0-199.7) µS/cm for EC, (0.00023-0.001646) mg/l for Hardness, and (6.6-7) for P^{H}. According to the World Health Organization (WHO) standards, the concentrations were at the permissible limits except the concentration of the first reading of the first and second sample of TDS that was recognized to fall in a range which exceed the WHO permissible limits. The concentration of these samples was prepared by mixing 97 ml of drinking water with 3 ml of crude oil, and 95 ml of drinking water with 5 ml of crude oil.

In Table 7 the recorded values of the total Water hardness of the

No.of sample	Location	Calcium (Ca^{+2}) mg/L	Sodium(Na^{+}) mg/L	Potassium(K^{+}) mg/L	Lead (Pb^{+2}) mg/L	Volume of river water + crude oil mL
1	Darbandikhan Reservoir	1	11.9	0.47	1.165716972	99.7+0.3
2	Darbandikhn Reservoir	4.5	13.1	0.89	0.4340246	99.98+0.2
3	Darbandikhan Reservoir	4.2	2	0.78	0.596626685	99.99+ 0.1

Table 1: Shows the concentration of common cations in River water samples contaminated with crude oil (Darbandikhan Reservoir) Sulaimaniya.

No of sample	Location	Calcium (Ca^{+2}) mg/L	Sodium (Na^{+}) mg/L	Potassium (K^{+}) mg/L	Lead (Pb^{+2}) mg/L	Volume of Drinking water + Crude Oil mL
1	Darbandikhan Reservoir	4.19	9.38	0.149	1.2455	99.7+0.3
2	Darbandikhan Reservoir	4.49	6.99	0.079	1.165	99.98+0.2
3	Darbandikhan Reservoir	4.195	6.89	0.069	1.245	99.99+ 0.1

Table 2: Shows the concentration of common cations in Drinking water samples contaminated with crude oil (Darbandikhan Reservoir) Sulaimaniya.

No of Sample	Location	Volume of river water + crude oil (ml)	Chloride (Cl-) mg/L	Sulphate ($SO4^{-2}$) mg/L	Nitrate (NO^{-3}) mg/L
1	Darbandikhan Reservoir	99.7+0.3	1.5	40.80308219	2.296256693
2	Darbandikhan Reservoir	99.98+0.2	1.35	79.15924658	4.561059612
3	Darbandikhan Reservoir	99.99+ 0.1	0.0009	87.92636986	3.122841903

Table 3: Shows concentration of major anions in river water samples contaminated by crude oil from (Darbandikhan Reservoir) Sulaimaniya.

No of samples	Volume of river water + crude oil (ml)	Chloride(Cl^{-}) mg/L	Sulphate(SO_4^{-2}) mg/L	Nitrate(NO_3^{-}) mg/L
1	99.7+0.3	0.00125	87.65239726	1.502117175
2	99.98+0.2	0.00116	79.15924658	4.587577726
3	99.99+ 0.1	0.00116	87.92636986	4.511143162

Table 4: Shows concentration of major anions in drinking water contaminated samples by crude oil (Darbandikhan Reservoir) Sulaimaniyah.

No. of samples	Location	Conductivity μS/cm	TDS mg/L	Volume of River water+ Crude oil mg/L
1	Darbandikhan Reservoir	0	0	99.7+3
2	Darbandikhan Reservoir	0	0	99.98+2
3	Darbandikhan Reservoir	161	161	99.99+1
4	Darbandikhan Reservoir	1697	1697	99.95+5
5	Darbandikhan Reservoir	169	169	99.97+3
6	Darbandikhan Reservoir	170	170	99.98+2
7	Darbandikhan Reservoir	171	171	99.9+1

Table 5: Shows Physiochemical Characteristics of River water samples contaminated with crude oil (Darbandikhan Reservoir) Sulaimaniya.

No. of samples		Conductivity μS/cm	TDS mg/L	TDS after 24h mg/L	pH	pH after 24h	Volume of Drinkingwater + crude oil (ml)
1	Darbandikhan Reservoir	199.7	717.18	126.65	6.9	7.17	97+3
2	Darbandikhan Reservoir	19.1	725.12	12.21	7	7.36	95+5
3	Darbandikhan Reservoir	127	516.4	81.187	6.8	6.88	90+10
4	Darbandikhan Reservoir	96	0	61.32	6.6	6.37	80+20
5	Darbandikhan Reservoir	0	0	0	6.7	6.87	70+30
6	Darbandikhan Reservoir	0	0	0	6.9	6.67	60+40
7	Darbandikhan Reservoir	0	0	0	6.7	6.38	50+50
8	Darbandikhan Reservoir	0	0	0	6.6	5.75	40+60

Table 6: Shows Physiochemical Characteristics of Drinking water samples contaminated with crude oil (Darbandikhan Reservoir) Sulaimaniyah.

No. of samples	Location	Volume of river water + crude oil ml	Total Hardness mg/L
1	Darbandikhan Reservoir	99.7+0.3	0.002
2	Darbandikhan Reservoir	99.98+0.2	0.002
3	Darbandikhan Reservoir	99.99+0.1	0.000096

Table 7: Shows the total water hardness of the River water samples contaminated with crude oil (Darbandikhan Reservoir) Sulaimaniyah.

River water samples contaminated with crude oil showed a range of (0.000096-0.02) mg/L. According to the World Health Organization (WHO) standards, the resulted concentration was under permissible standards of the WHO permissible limits [7].

In Table 8 the recorded values of the total water hardness of the Drinking water samples contaminated with crude oil showed a range of (0.00023-0.002) mg/L. According to the World Health Organization (WHO) standards, the resulted concentration was under permissive standards of the WHO permissible limits [7].

Discussion

In Table 9 the $t_{stat} < t_{critical}$, we are 95% confident that there is strong evidence so we are rejecting the null hypothesis. It means there is no relationship between chloride and sulphate ions concentration.

No. of river water sample	Location	Volume of river water + crude oil ml	Total Hardness mg/L
1	Darbandikhan Reservoir	99.7+0.3	0.00175
3	Darbandikhan Reservoir	99.98+0.2	0.001686
3	Darbandikhan Reservoir	99.99+0.1	0.000096

Table 8: Shows the total water hardness of the drinking water samples contaminated with crude oil (Darbandikhan Reservoir) Sulaimaniyah.

	Chloride (Cl^-) mg/L	Sulphate (SO_4^{-2}) mg/L
Mean	0.9503	69.29623288
Variance	0.68164527	628.1103397
Observations	3	3
Hypothesized Mean Difference	0	
df	2	
t Stat	-4.720845685	
P(T<=t) one-tail	0.021029925	
t Critical one-tail	2.91998558	
P(T<=t) two-tail	0.042059851	
t Critical two-tail	4.30265273	

Table 9: Shows t-Test: Two-sample assuming unequal variances.

In Table 10 $t_{stat} < t_{critical}$, we are 95% confident that there is strong evidence, and therefore we accept the null hypothesis. It means that there is strong relationship between nitrate and chloride ion concentration of the samples.

In Table 11 $t_{stat} < t_{critical}$, we are 95% confident that there exists strong evidence so we are rejecting the null hypothesis. It means that there is no relationship between sulphate and nitrate ion.

Conclusion

All in all, by checking the concentration of different water samples derived from Darbandikhan reservoir, and crude oil samples derived from Khurmala oil field, it is concluded that all the resulted data excepting Pb^{+2} and TDS are at the permissible limits. The record of ranges of these data for both river and drinking water are respectively Ca^{+2} (1-4.5) mg/L, Na^+ (2.-13) mg/L , K^+(0.47-0.89) mg/L, Pb^{+2} (0.434-1.166) mg/L, Cl^- (0.0009-1.5) mg/L, SO_4^{-2} (40.8-87.9) mg/L, NO_3^- (2.29-4.65) mg/L, EC(0-1697) µS/cm, Hardness (0.000096-0.02) mg/L, TDS(0-780.6) mg/L. Ca^{+2} (4.2-4.5) mg/L, Na^+(6.9-7) mg/L, K^+(0.07-0.15) mg/L, Pb^{+2} (1.16-1.24) mg/L, Cl^- (0.0011-0.125) mg/L, SO_4^{-2} (79.15-87.9) mg/L, NO_3^- (1.50-4.58) mg/L, EC (0-199.7) µS/cm, Hardness (0.00023-0.002) mg/L, P^H (6.6-7). We are 95% confident and we rejected the null hypothesis, for chloride and sulphate ions , sulphate and nitrate ions. We are 95% confident that we accept the null hypothesis for nitrate and chloride ions. Furthermore, according to the standard permissible limits of irrigation water system, it was concluded that all the investigated water parameters are at the standard normal range of the irrigation permissible limits.

Recommendation

From the above statement and by referring to the concluded results, the data showed that the concentrations of both Pb^{+2} and TDS were higher than the normal permissible values of WHO standard organization. For that reason, a number of remediation procedures needs to be conducted. These procedures are demonstrated as:

Pb^{+2}

To remediate the excessive amount of Pb^{+2} in water different procedures can be used such as using reverse osmosis method, using phosphate dosing, aware the residential areas of the major steps of how to identify lead in water contamination, and identify lead contamination source [8,9].

	Nitrate (NO_3^-) mg/L	Chloride(Cl^-) mg/L
Mean	3.326719403	0.9503
Variance	1.313507592	0.68164527
Observations	3	3
Hypothesized Mean Difference	0	
df	4	
t Stat	2.914040808	
P(T<=t) one-tail	0.021750948	
t Critical one-tail	2.131846786	
P(T<=t) two-tail	0.043501896	
t Critical two-tail	2.776445105	

Table 10: Shows t-Test: Two-sample assuming unequal variances.

	Sulphate(SO_4^{-2}) mg/L	Nitrate(NO_3^-) mg/L
Mean	84.91267123	3.533612688
Variance	24.84518668	3.096691074
Observations	3	3
Hypothesized Mean Difference	0	
Df	2	
t Stat	26.66523986	
P(T<=t) one-tail	0.00070172	
t Critical one-tail	2.91998558	
P(T<=t) two-tail	0.00140344	
t Critical two-tail	4.30265273	

Table 11: Shows t-Test: Two-sample assuming unequal variances.

TDS

The concentration of TDS in water has a direct relationship with the percentage of different chemical constituent such as Mg, Ca, and other compounds. To remediate the higher concentration of TDS a number of remediation methods can be adopted such as salt based softeners, reverse osmosis, and ion exchange [10,11].

Acknowledgment

Authors would like to thank referees for their valuable comments and suggestions to improve the quality of this research paper. Authors are also thankful to Mr. Sathyendar Sreepada for his assistance in applying statistical tests on data obtained through experimental work.

References

1. Omoregie E, Ufodike EBC, Onwuliri COE (1997) Effects of water soluble fractions of crude oil on carbohydrate reserves of Oreochromis niloficus (L). Journal of Aquatic Science 12: 1-7.
2. Mariano JB, La Rovere EL (1999) Environmental impacts of the oil industry. Petroleum engineering- Downstream.
3. University of Canterbury (2009) Determination of Chloride by titration method. College of Science, University of Canterbury.
4. Poonam, Rajesh K, Ashok Y (2013) Ground Water Quality in Mohendergarh Town, Haryana (India). International Journal of pharmaceutical and chemical sciences 2: 226-228.
5. Badiadka N, Kenchaiah S (2009) Spectrophotometric Method for the Determination of Nitrite and Nitrate. Eurasian J Anal Chem 4: 204-214.
6. Elarina ND, Paul SD, JashaMomo HA (2014) Trace Elements Analysis in Drinking Water of Meghalaya by Using Graphite Furnace-Atomic Absorption Spectroscopy and in Relation to Environmental and Health Issues. Journal of Chemistry 8.

7. WHO (2011) Drinking water guidelines: potable water standards. World health organization.

8. Provin TL, Pitt JL (2002) Description of Water Analysis Parameters. Extension Soil Chemist and Extension Associate, Soil and Crop Sciences Department.

9. Bruce ID, Sharon OS, Wayne W, Sreve D (2013) Drinking water: Lead. Extension Environment Engineering Specialist, Extension Water quality Educator, Extension Water and Environment Specialist, LEAD & Copper rule Manager, Nebraska Department of Health and Human Services. University of Nebraska- Lincoln Extension and the Department of Health and Human Services place a high priority on water quality and jointly sponsor thirds series of educational publication.

10. NSF International (n.d.) About Total Dissolved solid in water. Consumer fact sheet.

11. Agricultural and Consumer production (n.d.). Water Quality for agriculture. Fao Corporate Document Repostior.

6

Effect of Omumm Water Drainage on Noubaria Canal Water Quality and Alexandria Drinking Water Treatments Plants

Ali M Abdullah*

Reference Laboratory for Drinking Water, Holding Company for Water and Wastewater, Alexandria University, IGSR, Egypt

Abstract

Drainage water reuse in Egypt represents a major potential water resource for expanding the cultivated area. El-Noubaria canal irrigates 2.8 million feddans. Alexandria Drinking Water Treatments intakes; mainly El-Noubaria WTP and K.40 WTP take their water from El-Noubaria canal and El-Alamein WTP obtains raw water from El-Nasr canal branched at K. 55 from El-Noubaria canal, which suffers from pollution by agriculture drainage water. A main quantity of such drainage water is coming from Omumm Drain. The present study focuses on the impact of the drainage water of El- Omumm Drain by mixing of its water directly into El-Noubaria canal, which serves agriculture uses and drinking water. Electrical conductivity (EC), was increased in Noubaria canal after receiving point at Nasr canal intake (site III) by 83.9, 109, 118, 81.8%, at Noubaria WTP intake (site IV) by 43.8, 5, 69.5, 43.4% and at K40 WTP (site V) by 66.7, 49.7, 78.6, 67.8% during winter, spring, summer and autumn, respectively due to the attribute of Omumm Drain effluent.. a-BHC, b-BHC, heptachlor, diendrin and endrin, were increased after mixed fresh water with Omumm Drain effluent at Nasr canal intake (km 55), and the average observed values were 2.65, 1.1, 2.4 and 1.3 µg/l for a-BHC, b-BHC, heptachlor and diendrin while endrin not detected during summer and autumn. The ANOVA test showed effects of sites on EC were significant The ANOVA test showed the effect of sites on COD were significant and the seasonal effects on BOD were significant.

Keywords: Noubaria canal; Omumm drain; Water quality assessment

Introduction

In the 1970s, a policy was formulated for reusing drainage water for irrigation. Because irrigation water was scarce, reuse of drainage water became an essential element in water management starting in the late 1980s after a series of studies by the Drainage Research Institute and others [1]. This reuse was made possible by the extensive network of open drains and subsurface drains that was being developed. Of all the countries in arid regions, Egypt has gone farthest in reusing drainage water. In Upper Egypt, drainage water is reused by gravity flow. In Lower Egypt; reuse is by lifting water to the irrigation canals. In implementing the reuse policy, 25 centralized mixing stations were built on the main drains in the Nile Delta, managed by the MWRI Mechanical and Electrical Department (MED) [2]. Upper limits were set on the salinity of the water that can be used for mixing. The official reuse of drainage water was 2.9 BcM/yr in 1984–85, it increased to 4.4 BcM/yr in 1996–97 [1]. The Water Policy Reform Program of 1999 projected further increases in reuse to 8.4 BcM/yr by 2017. This water will be required to meet demand for new horizontal expansion projects, including those now coming on stream, as well the increased water demand from the domestic and industrial sectors. The national policy seeks a 20 percent increase the volume of water available by 1997–2017. Of this, a fifth would have to come from increased reuse. In the context of water quality management, agriculture must be seen as a widespread non-point source of pollution. Pollutants include leached salts, nutrients like nitrogen and pesticides [2].

These non-point sources will be collected in agricultural drains to form point sources of pollution for the River Nile, lakes and irrigation canals in case of mixing water for reuse. Although there are different mechanisms for retaining the pollutants by passing the polluted water through soil, the non-point sources of pollution may influence the groundwater quality.

In the downstream direction, the water quality gradually deteriorates due to the poorly treated wastewater discharges from both domestic and industrial activities and uncontrolled mixing with water from polluted drains. Therefore, they contain high levels of various pollutants, such as fecal bacteria, heavy metals and pesticides. Some drains should be considered as open sewage system that badly due to the production of hydrosulfide [3].

With reuse, all kinds of pollutants are diffused throughout the entire water network. Reuse of drainage water is no longer limited to irrigation and drainage only but involves wastewater disposal, drinking water supply, and aquatic ecology as well. A number of mixing stations were closed, mainly because the drinking water supply from the irrigation canals was jeopardized by the inflow of low-quality drainage water. Of the 25 mixing stations in the delta, five stopped operations and two were interrupted for extended periods. The closed mixing stations represent a capacity of 1.56 BcM/yr, about a third of total capacity. The largest mixing station no longer in operation is on the Omumm Drain in the West Delta, where water supplies are polluted not so much by agricultural production water as by untreated municipal and industrial wastewater from the Abu Hommos and Shereshtra area.

Omumm drain water is mixed with Noubaria canal at k. 46.5 to meet the increasing demand of irrigation water for new reclaimed lands. This mixing of such polluted water causes some problems to the drinking water treatment plants along the Noubaria canal. The pollution of mixed that irrigation water from Omumm drain result in adverse effect on physical, chemical and biological characteristics

***Corresponding author:** Ali Mahmoud Ahmed Abdullah, Chemist (M. Sc) Reference Laboratory for Drinking Water, Holding Company for Water and Wastewater, Alexandria University, IGSR, Egypt
E-mail: dr2252000@dr.com

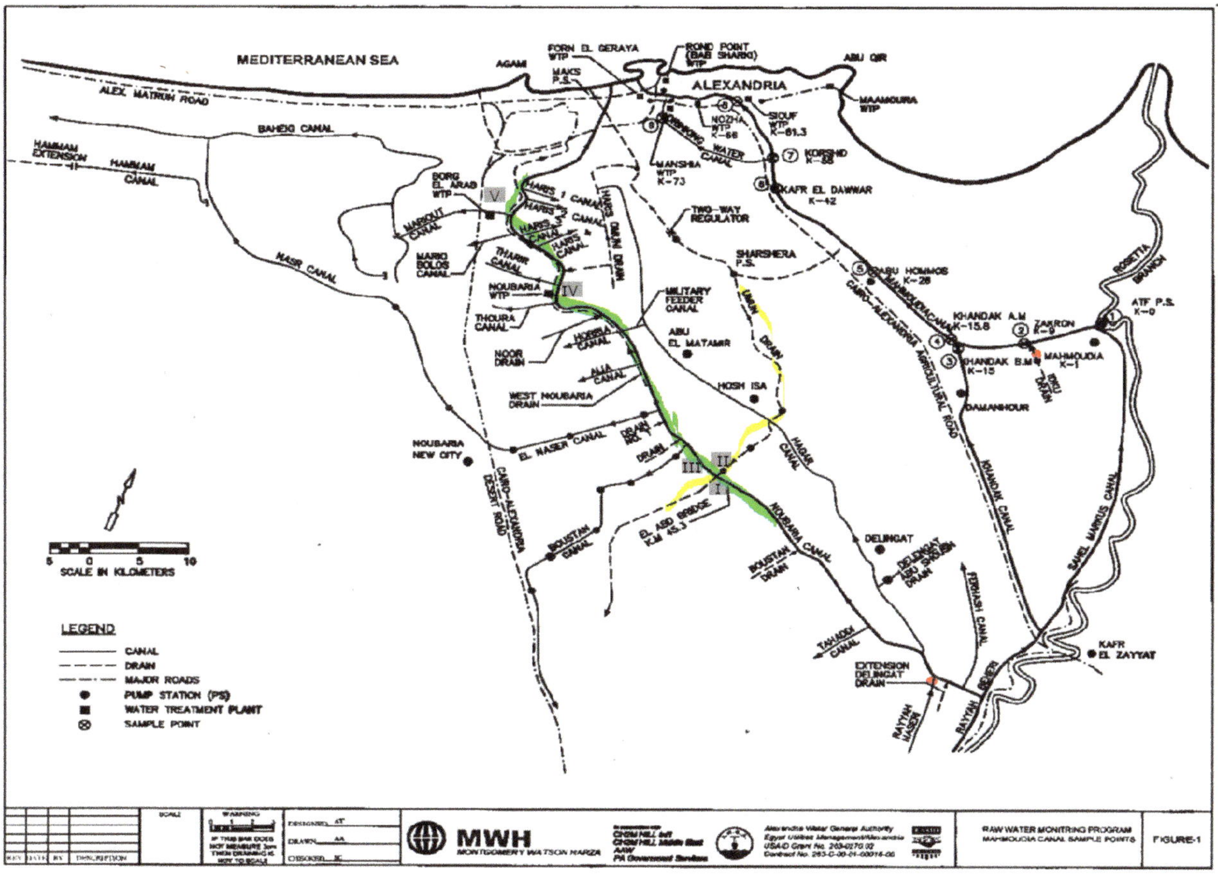

Figure 1a: Sampling sites of Noubaria canal and Omumm drain.

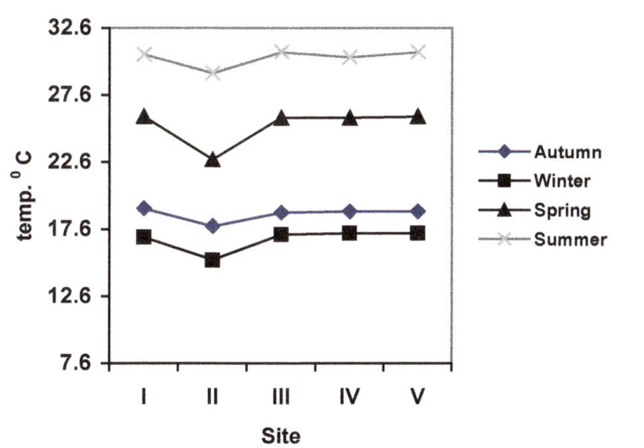

Figure 1b: Seasonal variation of Temp.

of Noubaria canal. Omumm drainage system receives untreated and semi-treated wastewater from Abou-Hommos city, Hosh Isa city and Abou-El Matameer city.

The enrichment and distribution of metals in water column depends on combination of environmental, physico-chemical and biological factors [4].

This study has been performed to follow the effect of El-Omumm Drain effluents on seasonal variation of pH: dissolved oxygen; heavy metals, pesticides, biological characteristics and major inorganic compounds in surface water of El-Omumm Drain; at the point of discharge, and also before and after the discharge point. Comparison with five selected locations along the Noubaria canal was carried out in order to determine the effect of waste disposal discharge into the Noubaria canal water and obtain information, which may help in giving information about variability in Noubaria canal quality.

Description of the area of investigation

The area of investigation is 46.5 Km to the North of Kanater Bolin and. five different sampling sites are chosen along the Omumm drain. Figure 1a shows the selected sites. It must be noted that, site (II) which represents, Omumm Drain, is the site which receives domestic and agriculture wastes from south Behara governorate then pour its content into the Noubaria canal at site III. Figure 1b shows the position of each site. Accordingly the sites under investigation are divided into three classes the first include station II which is at Omumm drain itself. The second include sites I, which are located before, and III, IV, V after Omumm Drain respectively.

Materials and Methods

Sixty water samples were collected from the five sites. Fifteen samples were taken per season, starting from June 2012 to May 2013. From subsurface of each sites two samples were taken using water collectors. The experimental procedures of the studied parameters were carried out according to standard methods (SMWW, 2012) as follows:

Physical parameters

The different parameters were determined according to the standard methods [5].

pH: The pH was measured using an analytical pH meter (CG- 840 SCOTT, Germany); the instrument was calibrated using standard pH buffers (4, 7, and 10) [5].

Dissolved Oxygen (DO): DO was determined by using the modified Winkler method according to standard methods (SMWW, 2012)[5] and expressed as (mg/l).

Electrical conductivity: The electrical conductance of water was measured by using EC-meter model (Janway 3000) after calibration by 0.1 KCl (1413 μmos/cm) (SMWW, 2012) [5].

Temperature: The temperature was measured using an analytical thermometer embedded in (Electrical Conductivity - Jenway 4320) [5].

Turbidity: Turbidity was measured by a nephelometric method (NTU) using turbidimeter (Hach model 2100N), Hach Co., USA. Formazin turbidity standard were prepared weekly and used as a reference according to Standard Methods for Water and Wastewater Treatment [5].

Inorganic parameters

Total hardness: Calcium and magnesium (total hardness) were determined using EDTA titimetric method according to APHA 1999. the ethylene diamine tetra acetic acid (EDTA) form a chelated soluble complex when added to a solution of certain metal cations. The calcium and magnesium were determined by using Eriochrome black T as indicator at pH 10. The calcium ions are determined alone by using Mureoxide as an indicator at high pH (12-13) by adding NaOH, the magnesium is estimated as the difference between the total hardness and calcium as $CaCO_3$. The concentrations of calcium and magnesium are expressed as mg/l [5].

Sulphate: It is determined by turbidimetric method according to standard methods. This method depends on the precipitation of sulphate ion in acidic medium using of barium chloride forming barium sulphate crystals. Barium sulphate suspension was carried out using HACH turbidimeter model 2100A (USA). Sulphate concentration was determined by comparison with standard values [5].

Chloride: Chloride ions was determined according to standard methods with titration of 50 ml sample against standard silver nitrate solution (Mohr's method) and expressed as mg/l [5].

Ammonia: Ammonia was determined by "Nesslerization method" according to standard methods. This method depends on the reaction of water samples containing ammonia as ammonia salt with Nessler`s reagent in the presence of Rochelle salt. The resultant yellow colour due to complexation between ammonia and Nessler`s reagent is proportional to the concentration of the present ammonia. This colour was measured spectrophotometrically at wave length 420 nm by using (UV/VIS) spectrophotometer Lambda 6 Perkin Elmer instrument [5] (Table 1).

Source of Variation	SS	df	MS	F	P-value	F crit
season	0.406561	3	0.13552	1.054605	0.4043	3.490295
sites	6.087091	4	1.521773	11.84228	0.000395	3.259167
Error	1.54204	12	0.128503			
Total	8.035692	19				

Table 1: ANOVA test for ammonia.

Nitrate: Nitrate was determined according to standard methods; 50 ml of sample were filtered and then acidified to less than pH 2, where measurements were done at 220 nm using a spectrophotometer Cecil 2040, (Perkin E. Co, Germany).The instrument was calibrated weekly by using nitrate standard solution at 0.5, 1.0, 2.0, 4.0, 6.0, 8.0, 10.0, 12.0, 14.0 mg/l to produce calibration curve [5].

Nitrite: Nitrite was determined according to standard methods; through formation of a reddish purple azo dye produced at pH 2.0 to 2.5 by coupling diazotized sulfanilamide with N-(1-naphthyl)-ethylenediamine dihydrochloride (NED dihydrochloride). 50 ml of sample were filtered and then acidified to less than pH 2, where measurements were done at 543 nm using a spectrophotometer Cecil 2040, (Perkin E. Co, Germany).The instrument was calibrated weekly by using nitrite standard solution at 0.5, 1.0, 2.0, 4.0 mg/l to produce calibration curve [5].

Chemical Oxygen Demand (COD): This method is called Open Reflux Method. Chemical oxygen demand is used as a measure of the oxygen equivalent to the organic matter content of the sample. A sample is refluxed in strongly acid solution with a known excess of potassium dichromate. After digestion, the remaining unreduced $K_2Cr_2O_7$ is titrated with ferrous ammonium sulphate to determine the amount of $K_2Cr_2O_7$ consumed and the oxidizable organic matter is calculated in terms of oxygen equivalent (SMWW, 2012) [5].

Biological Oxygen Demand (BOD): Determination of BOD was carried out using a 5-day BOD test according to standard methods. Basically the BOD was measured dissolved oxygen consumed by aerobic microorganisms as they metabolise the complex unstable molecules of pollutions e.g. proteins, carbohydrates, lipids….etc. into CO_2, H_2O, sulphate….etc. as simple stable inorganic compound [5].

Heavy metals: Four metals included Al, Fe, Ba and B were measured in water samples using (coupled plasma 400 emission spectrometer Perkin Elemer Emission Spectrometer) according. The Preservation and digestion were carried out as follows:

- **Preservation**: Samples were preserved immediately after collection by acidifying with concentrated HNO_3 to pH<2 by adding 5 ml nitric acid to 1 liter water samples. After acidification store in refrigerator.
- **Digestion**: Heavy metals were extracted in water samples using digestion by nitric acid method as follows below.

Mix well 500 ml of sample in a beaker at hot plate, add 10 ml nitric acid bring slow boil, evaporate on a hot plate till volume become 200–100 ml before precipitation occurs, continue heating add another 10 ml of nitric acid until the volume reached to 100–80 ml until digestion is complete. The samples do not dry during digestion, mix well and complete to 100 ml by distilled water, and take portion of their solution for required determination. Four metals included Al, Fe, Ba and B were determined in the digested solution using inductively coupled plasma emission spectrometry [5].

Organic parameters

TOC: TOC was determined according to standard methods; 50 ml of sample were filtered through 0.45 μm membrane filter and then acidified to less than pH 2, where measurements were done at 254 nm using a spectrophotometer Cecil 2040, (Perkin E. Co, Germany).The instrument was calibrated weekly by using dipotassium phthalate at 0.5, 1.0, 2.0, 4.0, 6.0, 8.0, 10.0, 12.0, 14.0, 16.0, 18.0, 20.0 mg/l to produce calibration curve [5].

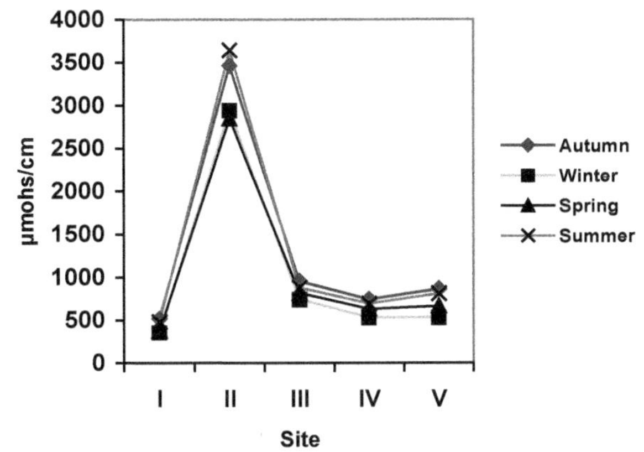

Figure 2: Seasonal variation of EC.

Source of Variation	SS	df	MS	F	P-value	F crit
season	342816.6	3	114272.2	5.740621	0.011313	3.490295
sites	21327686	4	5331921	267.8564	1.28E-11	3.259167
Error	238870.7	12	19905.89			
Total	21909373	19				

Table 2: ANOVA test for electrical conductivity.

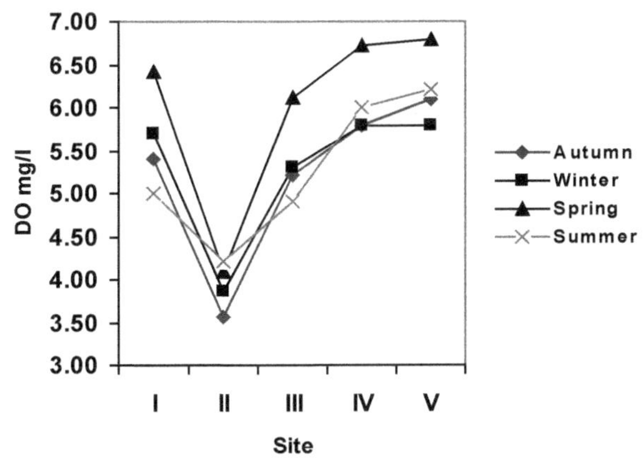

Figure 3: Seasonal variation of DO.

Chlorinated pesticides: The liquid-liquid extraction (LLE) gas chromatographic (GC) method is used to monitor both the PCBs and the organochlorine pesticides simultaneously. This method has excellent sensitivity. One liter of sample is extracted with methylene chloride. The extract is dried and exchanged to hexane during concentration. If other determinations having essentially the same extraction and concentration steps are to be performed, a single sample extraction is sufficient. The extract is separated by gas chromatography and the compounds are measured with an electron capture detector according to EPA 508.1 method [6].

Microbiological examination

Determination of bacterial count: The enumeration of total bacterial count (TBC) in water sample was done using the spreading plate technique over Difco agar medium (plate count Agar ref. 247940, USA). The medium was prepared by adding 23.5 g to one litter distilled water. The medium was dissolved by boiling for 1 min, and then 20 ml of dissolved medium were transferred into 20 ml clean test tube. The tubes were then autoclaved for 20 minutes at 121°C, and then each tube was poured while at 45°C into a sterile Petri-dish. Water sample was serially diluted to a suitable concentration to get 30 – 300 single separated colonies. One hundred microlitter of the suitable dilution were transferred into the middle of the medium surface under aseptic conditions, and then the sample was distributed on the surface using sterile spreader. Plates were left for 10 min, then inverted and incubated at 30°C for 24 hours. The resulting colonies were counted using CC 30, Ch, and Hern scientific colony counter [5].

Biological Examination

Determination of algal count: Algae were counted in 10 ml sample of raw water. The sample was centrifuged for 10 min at 4000 rpm. The supernatant was discarded, and then the pellet was dissolved in 1 ml of saline water, which was transferred to Sedgwick - rafter counting chamber that was covered by a slide cover. Algae were counted using Carl Zeiss Axiolab Optical Microscope fitted with 3CCD colour video camera attached to TV monitor. The count was done from the TV monitor [5].

Results and Discussion

Sixty samples were collected from five sites from June 2008 to July 2009. The first site at before mixing point by 1000 m (location I) at Abd-Bridge, the second point at Noubaria canal receive the discharge water from Omumm drain (location II) at k.m 46.5, the third point after receiving drainage water of Omumm drain to Noubaria canal by 8 k.m (location III) at Nasr canal intake, forth point at Noubaria WTP intake at K.m 81.3 along the Noubaria canal (location IV) and the end point at K.40 WTP intake at K.m 96.0 along the Noubaria canal (location V).

Water temperature showed noticeable seasonal trends with a lowest value (15.3°C) recorded in autumn and a highest (30.8°C) in summer (Figure 1). The slight variations between different sites were mainly due to different sampling times. Temperature is a key parameter which influences physical, chemical and biological transformation in the aquatic environment [7]. It is evident that in Figure 2, electrical conductivity (EC), was increased in Noubaria canal after receiving point at Nasr canal intake (site III) by 83.9, 109, 118, 81.8%, at Noubaria intake (site IV) by 43.8, 5, 69.5, 43.4% and at K40 WTP (site V) by 66.7, 49.7, 78.6, 67.8% during winter, spring, summer and autumn, respectively due to the attribute of Omumm Drain effluent.. The ANOVA test showed effects of sites on EC were significant (Table 2).

It is evident that in Figure 3, dissolved oxygen (DO), was decreased in Noubaria canal after receiving point at Nasr canal intake (site III) by 3.7, 7.0, 4.7, 2.0% during winter, spring, summer and autumn, respectively due to the attribute of Omumm Drain effluent and due to its consumption by oxidation of nitrogenous compounds, then the DO values return to increase at sites IV and V due to navigation activity and the dilution effect of the polluted water from Omumm Drain. DO was correlated negatively with nitrate, nitrite, sulphate and ammonium during the most seasons (Table 3). The dissolved oxygen content can be an indicator of organic loading nutrient input and biological activity. Cole and Saad [8] reported that, the decrease in DO during the hot seasons in surface water is due to the elevation of water temperature which decreases the solubility rate of atmospheric oxygen. The ANOVA test showed effects of sites on DO parameters were significant (Table 4).

pH values were in the alkaline side (7.66-8.34). The higher pH values during winter might be due to the blooming of algae, which were accompanied by photosynthetic activity and consumption of CO_2 with

	Temp	pH	EC	DO	TOC	COD	BOD	TH	SO_4	Cl	NH_3	NO_3	NO_2	Al	Fe	Ba	B	a-BHC	b-BHC	Heptachlor	TPC	FC	TAC	BGA
Temp	1.0																							
pH	-0.4	1.0																						
EC	-0.2	-0.6	1.0																					
DO	0.4	0.4	-0.8	1.0																				
TOC	0.0	-0.3	0.2	-0.1	1.0																			
COD	-0.5	-0.3	0.9	-0.8	0.2	1.0																		
BOD	-0.5	-0.2	0.2	-0.3	0.3	0.5	1.0																	
TH	-0.1	-0.6	0.1	-0.8	0.2	0.9	0.2	1.0																
SO_4	-0.2	-0.6	1.0	-0.8	0.2	0.9	0.2	1.0	1.0															
Cl	-0.2	-0.5	0.9	-0.8	0.1	0.9	0.1	0.9	0.9	1.0														
NH_3	0	-0.6	0.8	-0.8	0.2	0.7	0.3	0.9	0.8	0.7	1.0													
NO_3	0.4	-0.7	0.6	-0.4	0.0	0.2	-0.3	0.6	0.6	0.6	0.6	1.0												
NO_2	-0.4	-0.3	0.7	-0.5	0.2	0.6	0.0	0.6	0.7	0.6	0.3	0.4	1.0											
Al	-0.4	-0.4	0.9	-0.9	0.1	0.9	0.3	0.9	0.9	0.9	0.7	0.4	0.6	1.0										
Fe	-0.3	-0.3	0.8	-0.6	0.2	0.5	0.0	0.7	0.8	0.6	0.6	0.5	0.8	0.7	1.0									
Ba	-0.1	-0.5	0.8	-0.8	0.2	0.7	0.2	0.9	0.8	0.8	0.8	0.5	0.4	0.9	0.7	1.0								
B	0.0	0.4	0.8	-0.6	0.0	0.6	-0.2	0.8	0.8	0.9	0.6	0.7	0.5	0.7	0.6	0.6	1.0							
a-BHC	-0.5	-0.1	0.6	-0.6	0.2	0.7	0.4	0.5	0.6	0.5	0.6	0.3	0.6	0.6	0.5	0.5	0.4	1.0						
b-BHC	-0.6	0.0	0.7	-0.6	0.3	0.6	0.1	0.7	0.5	0.7	0.4	0.3	0.8	0.6	0.7	0.4	0.5	0.7	1.0					
Heptachlor	-0.5	0.3	0.3	-0.4	-0.1	0.5	-0.1	0.2	0.3	0.4	0.1	0.0	0.4	0.4	0.2	0.1	0.4	0.6	0.7	1.0				
TPC	-0.1	-0.6	0.9	-0.8	0.1	0.8	0.2	1.0	0.9	0.9	0.8	0.6	0.5	0.9	0.6	0.9	0.8	0.5	0.5	0.2	1.0			
FC	0.1	-0.6	0.8	-0.8	0.1	0.7	0.2	0.9	0.8	0.8	0.9	0.6	0.2	0.8	0.5	0.9	0.7	0.4	0.2	0.1	0.9	1.0		
TAC	0	0.0	-0.1	0.1	-0.4	-0.1	-0.1	-0.1	-0.1	-0.1	-0.2	0.0	-0.1	0.0	-0.1	-0.1	-0.1	-0.3	-0.3	-0.2	-0.1	-0.1	1.0	
BGA	0	-0.4	0.7	-0.8	0.2	0.5	0.0	0.7	0.7	0.7	0.6	0.5	0.3	0.7	0.4	0.7	0.6	0.5	0.5	0.4	0.7	0.7	-0.4	1.0

Table 3: Correlation coefficient matrix between physical and chemical parameters during the present study.

Source of Variation	SS	df	MS	F	P-value	F crit
season	2.28214	3	0.760713	9.535138	0.001688	3.490295
sites	13.38088	4	3.34522	41.93056	5.85E- 07	3.259167
Error	0.97536	12	0.07978			
Total	16.62038	19				

Table 4: ANOVA test for DO.

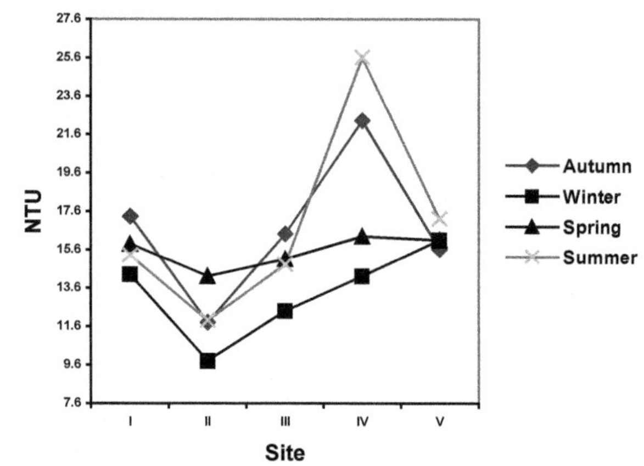

Figure 4: Seasonal variation of turbidity.

expect pH elevation [9] pH showed a negative correlation with most studied parameters, Table 3. It is evident that in Figure 4, turbidity in Omumm Drain effluents were lower than the Noubaria water turbidity during the period of the present study.

It is evident that in Figures 5-7, TOC, COD and BOD were increased in Noubaria canal after receiving point at Nasr canal intake (site III) by average increasing percentage (47.5, 20.7 and 32.6%), the maximum increasing percentage were 58.8, 26.6 and 44.4% and the minimum increasing percentage were 30.6, 10 and 13.8%, respectively. the maximum and minimum values of TOC and BOD were observed during autumn and summer, respectively. While the maximum and minimum values of COD were observed during winter and autumn, respectively due to the attribute of Omumm Drain effluent. The ANOVA test showed effects of sites on COD were significant (Table 5) and the seasonal effects on BOD were significant (Table 6). Most waters contain organic matter that can be measured as total organic carbon (TOC). The high value of TOC observed in cold period; this due to the decrease in the current water stream which facilitates the growth of algae and other microorganisms which the main source of TOC. During the study period, the most BOD and COD values in the study area of El-Noubaria canal after the point of receiving the effluent of Omumm Drain were exceeds the limit of 6 mg/l set by the law 48/1982 as maximum permissible limit for drinking water. As regard to spatial variation. Ali, 1998, reported that, the concentration of most physical and chemical parameters were increased northward more than southern region of river Nile, mainly due to agricultural and industrial, sewage wastes pour in the northern region [10-12].

It is evident that in Figures 8-10 the hardness total, sulphate and chloride were increased after receiving point (Nasr canal intake km 55) due to Omumm Drain effluent, the maximum increasing percentage of hardness total during winter while sulphate and chloride during autumn. Calcium and magnesium are the principle ions responsible for water hardness. The hardness in water is derived largely from contact with the soil and rock formations [10,11]. Rain water as it falls upon the earth is incapable of dissolving the tremendous amounts of solids found in many natural waters. The ability to dissolve is gained in the soil where carbon dioxide is released by bacterial action. The soil water becomes highly charged with carbon dioxide which exists in

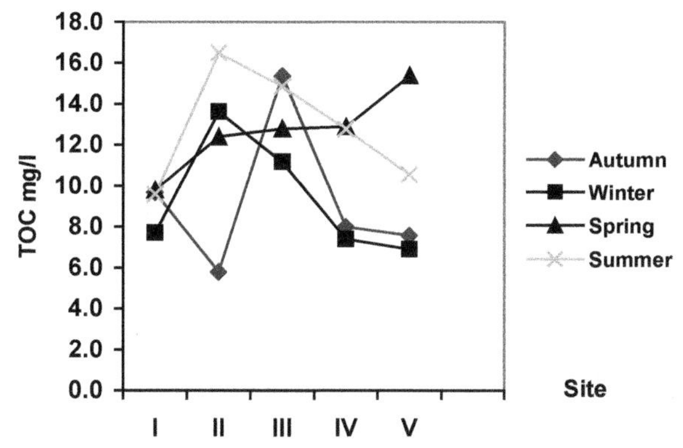

Figure 5 Seasonal variation of TOC.

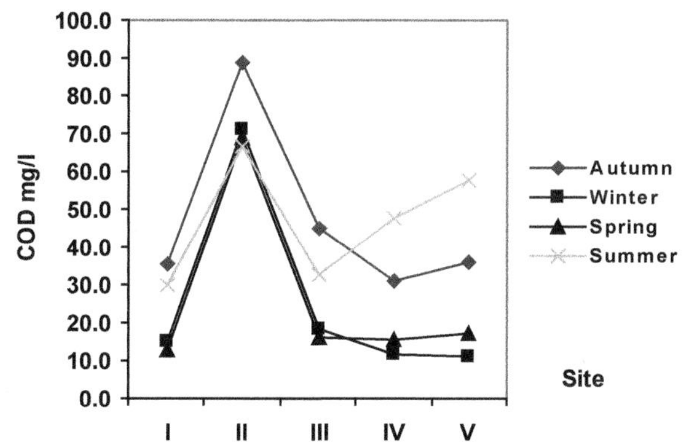

Figure 6 Seasonal variation of COD.

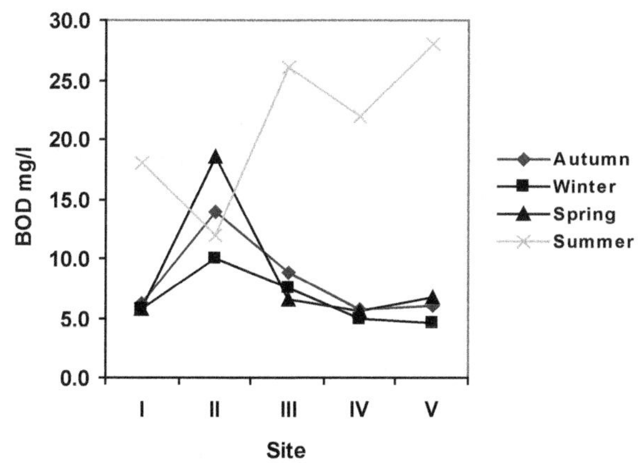

Figure 7 Seasonal variation of BOD.

Source of Variation	SS	df	MS	F	P-value	F crit
season	2299.97	3	766.6567	8.485252	0.0027	3.490295
sites	7123.732	4	1780.933	19.71113	3.3E-05	3.259167
Error	1084.22	12	90.35167			
Total	10507.92	19				

Table 5: ANOVA test for COD.

Source of Variation	SS	df	MS	F	P-value	F crit
season	683.948	3	227.9827	9.161081	0.001987	3.490295
sites	59.028	4	14.757	0.592984	0.674374	3.259167
Error	298.632	12	24.886			
Total	1041.608	19				

Table 6: ANOVA test for BOD.

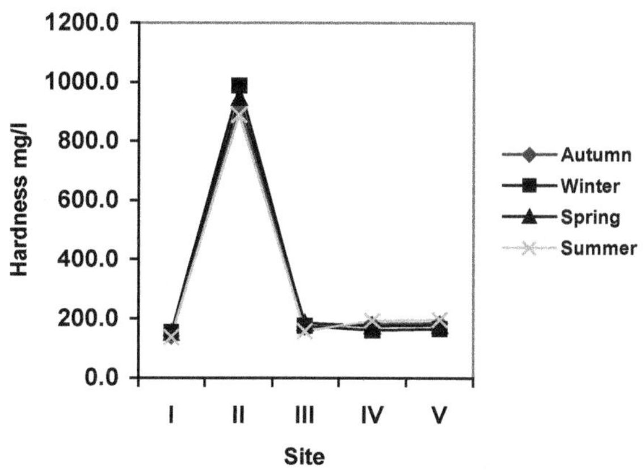

Figure 8: Seasonal variation of hardness total.

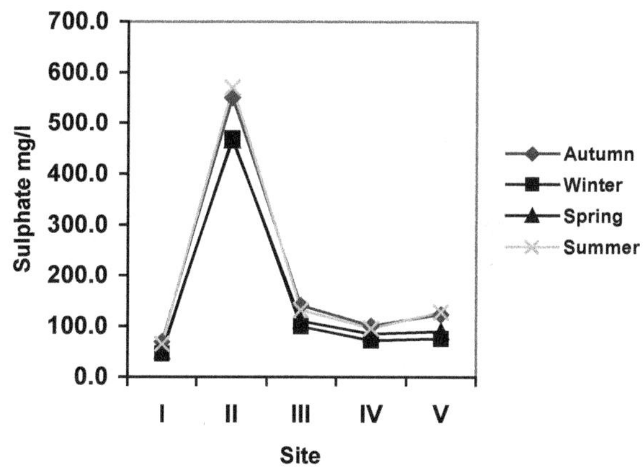

Figure 9: Seasonal variation of sulphate.

equilibrium with carbonic acid. Under low pH conditions that develop, basic materials, particularly limestone formations are dissolved [13]. In raw waters, calcium is the main source of hardness and found in large quantities in ground water versus surface water. Malik [14] reported that, the chloride ion is one of the major inorganic anion in wastewater because of the presence of sodium chloride as a common article of diet, as well as, chloride ion may be increase by industrial processes. The high values of hardness total, sulphate and chloride were observed in cold period; this due to the decrease in the current water stream and nearly the amount of Omumm Drain discharge on Noubaria canal. Hardness total, sulphate and chloride were showed a positive correlation with most studied parameters, Table 3.

It is evident that in Figures 11-13, ammonia, nitrate and nitrite were increased in Noubaria canal after receiving point at Nasr canal intake (site III) by average increasing percentage (29.3, 37.5 and 172.4%), the maximum increasing values were observed during summer, winter

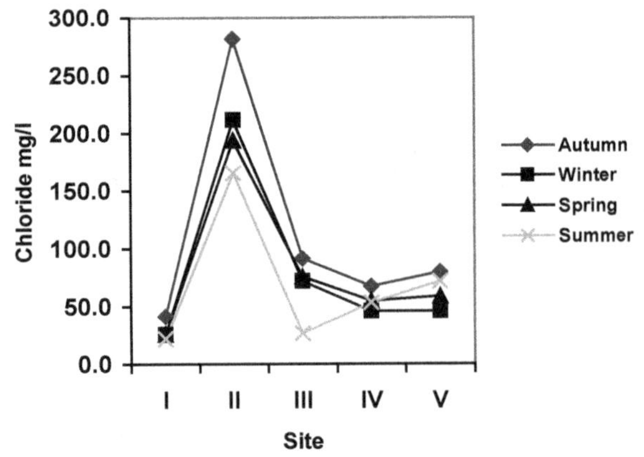

Figure 10: Seasonal variation of chloride.

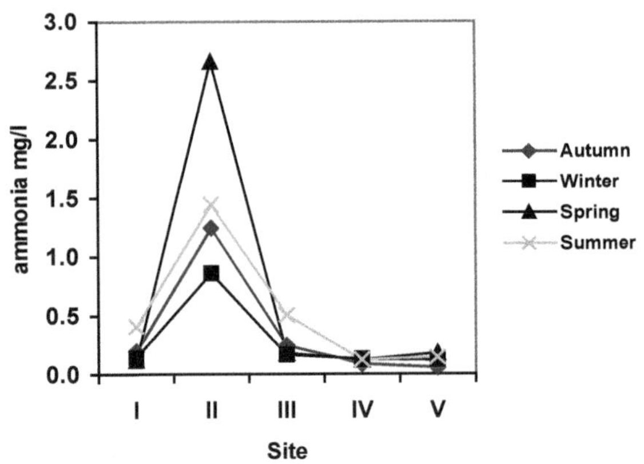

Figure 11: Seasonal variation of ammonia.

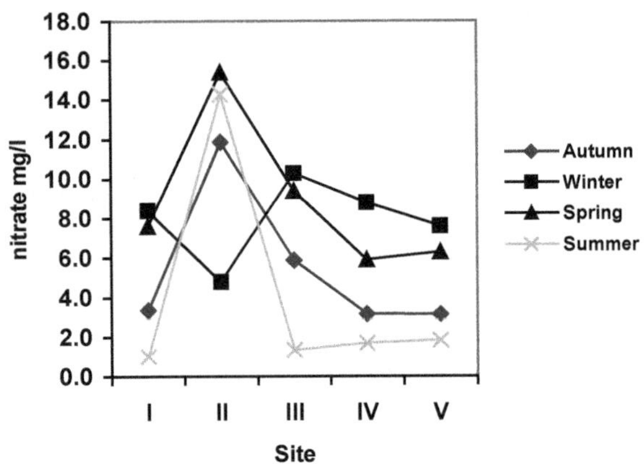

Figure 12: Seasonal variation of nitrate.

and spring respectively. Ammonia is rarely present in natural waters and originates mainly from fertilizers runoff. Bacteria usually convert ammonia in raw water to nitrite then to nitrate [15,16]. Train, 1979 [17] reported that, the toxicity of ammonia is increased with high alkalinity and pH values and in some cases ammonia may reach high toxic levels. High concentration of dissolved ammonia in cold session may be attributed to flushing agricultural land with irrigation water in order to plant winter crops. Considerable amounts of nitrogen fertilizers are leached from agricultural land into Noubaria canal. The high value of ammonia observed in cold period; this due to the decrease in the current water stream. Ammonia, nitrate and nitrite were showed a positive correlation with most studied parameters, Table 3.

Al, Fe, Ba and B are common pollutants which are widely distributed in aquatic environment. According to United Nation Environmental Program, the threshold toxicity values of these metals are 0.3, 5.00, 0.005, 0.05 mg/l, respectively [18]. Iron is a component of iron-porphyrins haem and ferroxins and it is essential for growth and well being of living organisms including man. However it likely to show toxic effects when organisms are exposed to levels higher than normally required. Heavy metals may enter the river water from the geology of catchments soil. Probable sources included: sewage plants industrial, agriculture, domestic effluents and urban run-off [4,19]. De-Groot and Allersma [20] reported that, metals can enter rivers from different sources: rocks and soils, dead and decomposed plants and animal matter, from man's activity including the discharge of various treated and untreated liquid wastes to the water. Data in represent graphically in Figures 12-15 indicated that, the metals were increased after fresh Noubaria water mixed with the Omumm Drain effluent during the most season, and the increasing percentage were ranged 28-100%, 0-78%, 10-210% and 32-455% for Al, Fe, Ba and B, respectively. ANOVA test showed spatial effect was significant on Al as shown in Table 7.

It is evident that in Figures 16-20 the a-BHC, b-BHC, heptachlor, diendrin and endrin, were increased after mixed fresh water with Omumm Drain effluent at Nasr canal intake (km 55), and the average observed values were 2.65, 1.1, 2.4 and 1.3 μg/l for a-BHC, b-BHC, heptachlor and diendrin while endrin not detected during summer and autumn. Pesticides are used for agricultural as well as public health purposes. Pesticide residues may seep from the soil into drains, irrigation water and finally into the main canal such as Noubaria canal, the leaching mainly depends on the type of pesticides, soil characteristics, hydrogeological conditions, climatic factors, agro-technical factors and human factors [21].

It is evident that in Figures 21-23, the average increasing percentage after receiving polluted water from Omumm Drain were 110, 45.6 and 55% for total plate count (TPC), total coliform TC) and fecal coliform

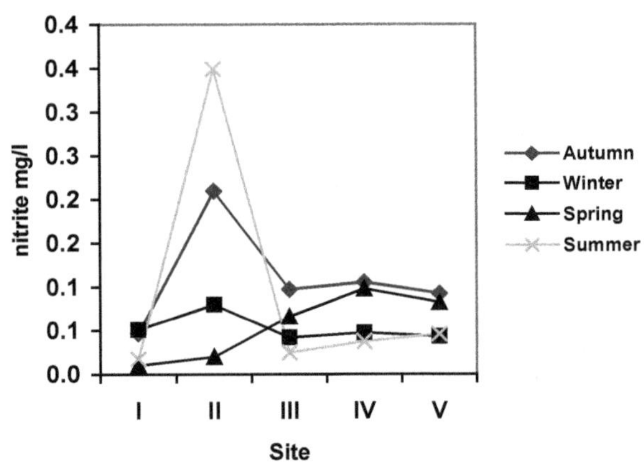

Figure 13: Seasonal variation of nitrite.

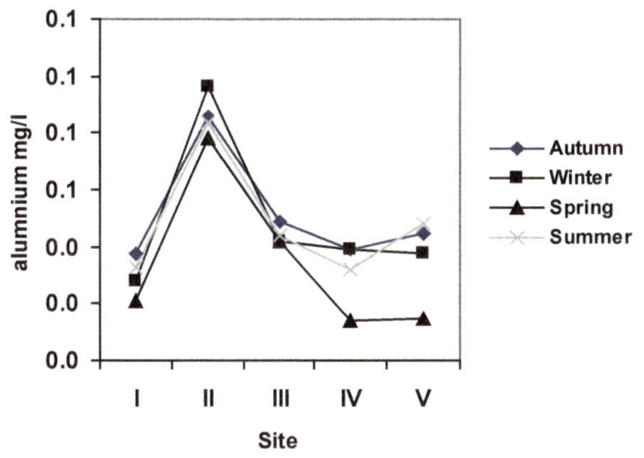

Figure 14: Seasonal variation of aluminum

Source of Variation	SS	df	MS	F	P-value	F crit
Season	0.00092	3	0.000307	6.994297	0.005639	3.490295
Sites	0.008693	4	0.002173	49.58099	2.3E-07	3.259167
Error	0.000526	12	4.38E-05			
Total	0.010139	19				

Table 7: ANOVA test for Al.

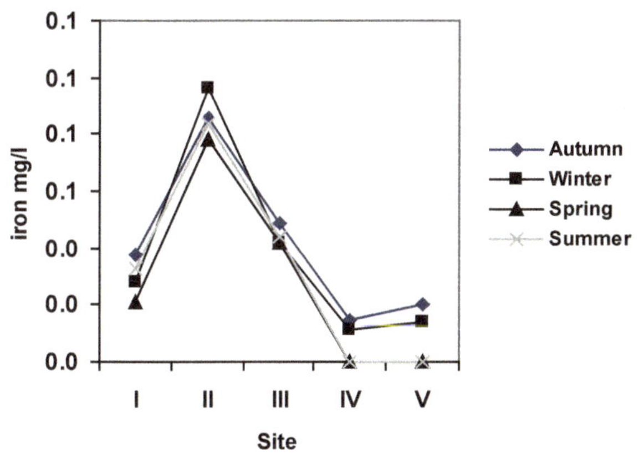

Figure 15: Seasonal variation of iron.

(FC), respectively. The high values were observed during winter due to the decrease in the current fresh water stream and nearly the amount of Omumm Drain discharge on Noubaria canal. As a result of poor wastewater treatment, high concentrations of coliform bacteria are found in the Noubaria canal, most of observed FC was exceed the Egypt standard [2].

It is evident that in Figures 24 and 25, the average increasing percentage after receiving polluted water from Omumm Drain were 15.4 and 14.4 for total algal count (TAC) and blue green algae (BGA), respectively. The high values were observed during autumn due to the decrease in the current fresh water stream and nearly the amount of Omumm Drain discharge on Noubaria canal.

Various water quality indexes have been developed in the past 40 years and the utility of a water quality index (WQI) were summarized as a follow: i) volumes of water quality data are summarized in a single index value in an objective, rapid, and reproducible manner; ii) the numerical scale of an index facilitates evaluation of "within class" variations, thereby allowing identification of changes in water quality at a site that would not precipitate a change within the classification system; iii) the index values may be related to a "potential water use" classification scheme to help determine the ecological potential of the water body; iv) the index and associated water body classification scheme may be used in operational management to identify surface waters requiring priority

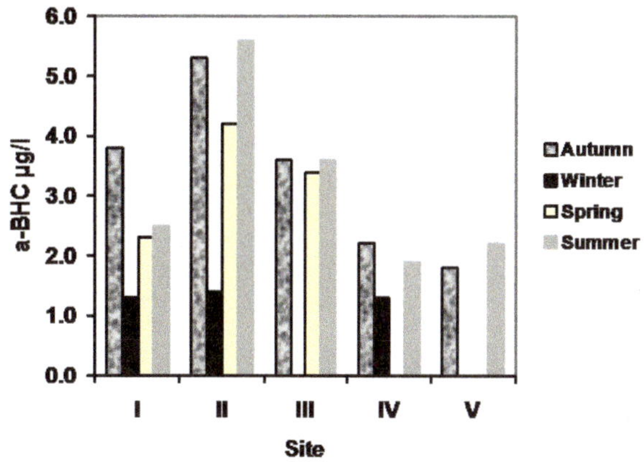

Figure 16: Seasonal variation of a-BHC.

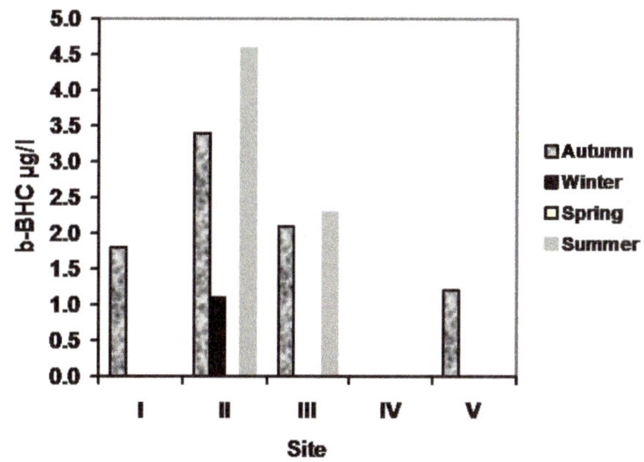

Figure 17: Seasonal variation of b-BHC.

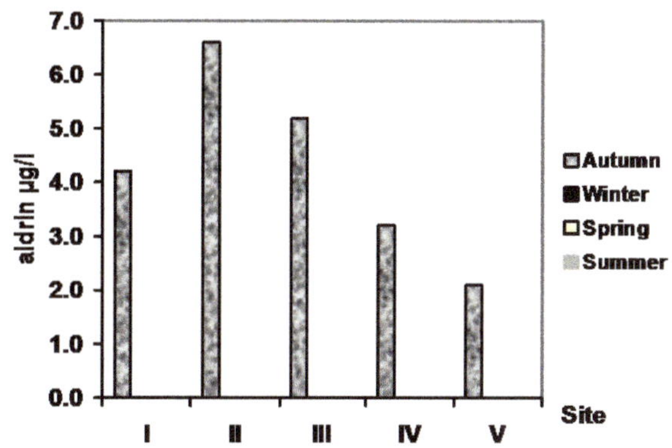

Figure 18: Seasonal variation of dieldrin.

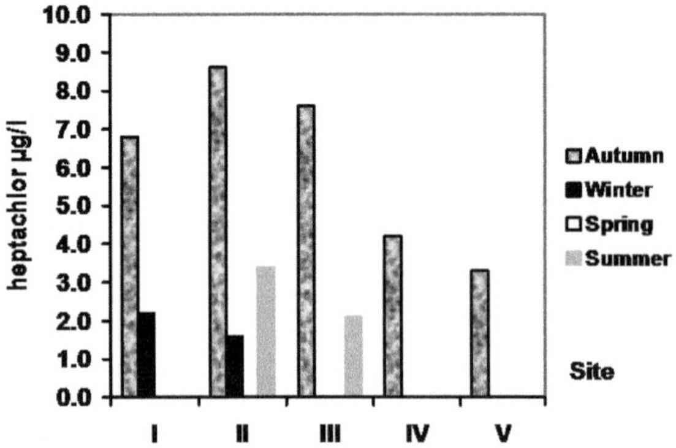

Figure 19: Seasonal variation of heptachlor.

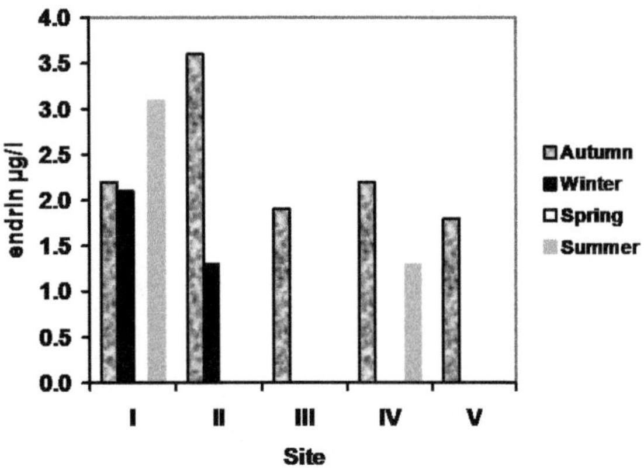

Figure 20: Seasonal variation of endrin.

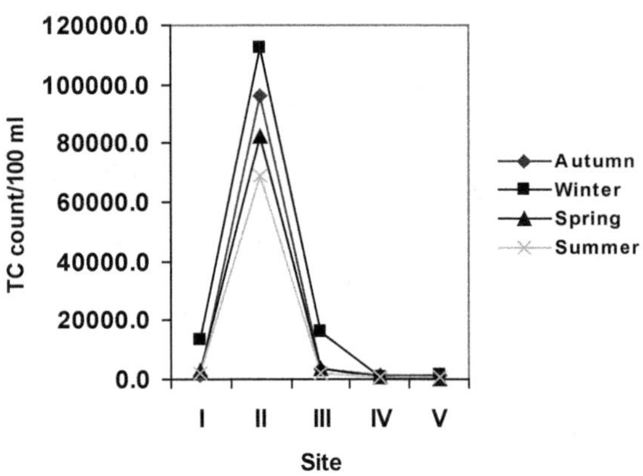

Figure 21: Seasonal variation of TC.

action; and v) the index facilitates communication with the layperson, while maintaining the initial precision of measurement [22].

It is evident that in Figure 26, the average value after receiving polluted water from Omumm Drain was 442 (poorest) for water quality index (WQI) and the increasing percentage ranged from 7.0 to 47.0% as a result of mixed polluted water from Omumm Drain to Noubaria canal at Nasr canal intake (km 55.0) [23]. Water quality index is the result of the sum of the products of water quality ratings (q_i) and weighting each individual parameter (w_i) divided by each Egypt standard value (S_i), according to the following equation:

$$WQI = \frac{1}{100} \times \left(\sum_i^n \frac{q_i w_i}{S_i}\right)$$

WQI, > 140% represents the poorest and <50% the highest water

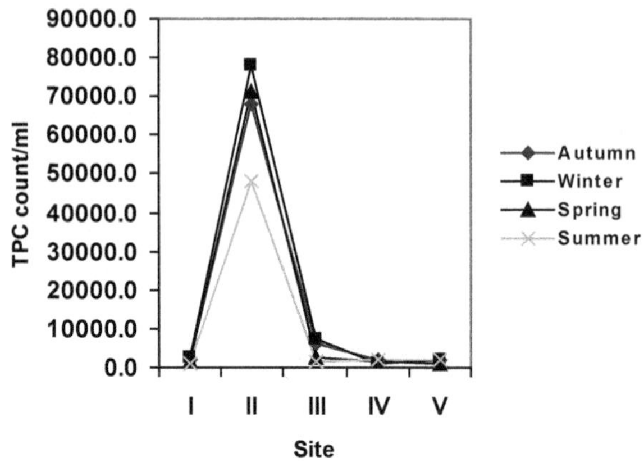

Figure 22: Seasonal variation of TPC.

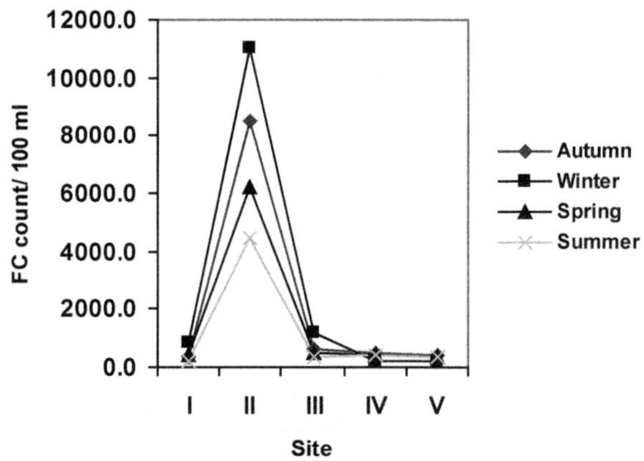

Figure 23: Seasonal variation of FC.

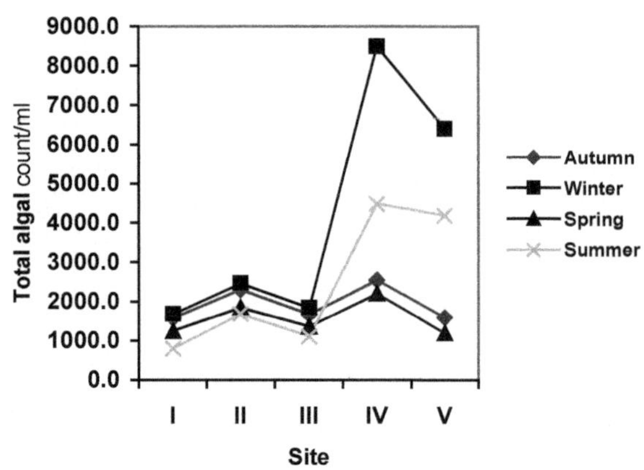

Figure 24: Seasonal variation of total algal count.

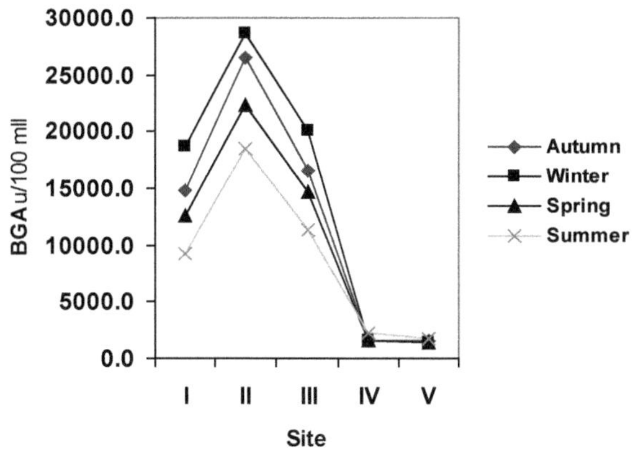

Figure 25: Seasonal variation of BGA.

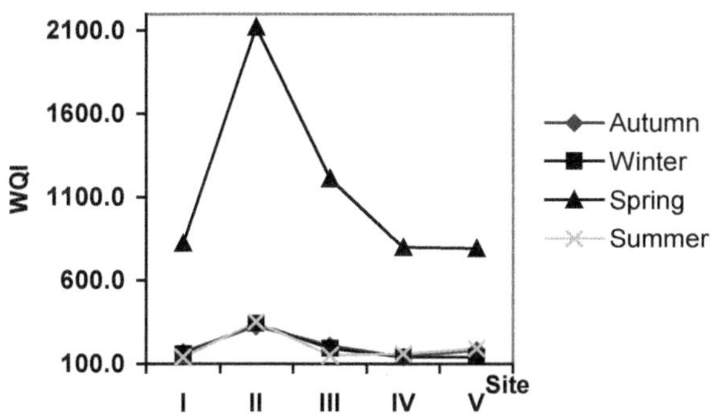

Figure 26: Seasonal variation of WQI.

quality. In this study, class rating was adopted: 51–80%, Good; 81–110%, Fair; 111–140%, Poor.

Recommendation

1. In order to conserve water quality in Noubaria canal; Ministry of Water Resource and Irrigation should be limit the mix of irrigation drainage water from Omumm drain with Noubaria canal and should be look for alternatives for using this water for agriculture only.
2. Boustan canal locate north of Omumm effluent point and feed from Noubaria canal from the other side and can be take the effluent of Omumm drain which useful in agriculture uses.
3. Water treatment plants intakes should be protected according to Ministerial law No. 301/1995.
4. Limit the discharge of wastewater on irrigation drain and fresh water canal by extend the sanitary services to all rural area.

References

1. Bazaraa AS (2002) Reuse of Agricultural Drainage Water: Revision of the Reuse Mixing Policy. Cairo: Advisory Panel Project on Water Management.
2. Soltan ME (1988) Study of River Nile water pollution. M.Sc. Thesis, Fac. Sci. Aswan, South Valley University, Egypt.
3. Barcament J (2003) Modeling of dissolved oxygen and biochemical oxygen demand in River Water using detailed.
4. Elewa AA (1995) Distribution of Mn, Cu, Zn, and Cd in water and sediment of River Nile at Aswan, Egypt. J. Appl. Sci, 8: 711–713.
5. SMWW (2012) Standard methods of the examination of water and waste water 22nd edn. APHA, New York.
6. USEPA (1995) Methods for the Determination of Organic Compounds in Drinking Water-Supplement III.
7. Radwan M, Willems P, Berlamont J (2003) Modeling of dissolved oxygen and biochemical oxygen demand in River Water using detailed and simplified models. Int. J. River Basin Management 1: 97-103.
8. Cole GA (1979) Text book of limnology (2nd edn), Mosby company, Toronto.
9. Booth DB, Jackson CR (1997) Urbanization of aquatic systems: degradation thresholds, stormwater detection, and the limits of mitigation. Journal of the American Water Resources Association 33: 1077-1090.
10. Hem JD (1992) Study and interpretation of the chemical characteristics of natural water. Water-Supply Paper 2254, U.S. Geological Survey, Washington, D.C.
11. Clarke KD, Scruton DA (1997) The benthic community of stream riffles in Newfoundland, Canada and its relationship to selected physical and chemical parameters. Journal of Freshwater Ecology 12: 113-121.
12. Ali MHH (1998) Chemical and physical studies on the river Nile at Damittea branch region. M. Sc. Fac. Of Sci. Menofiya University.
13. VADEQ, Virginia Department of Environmental Quality (1999) Virginia Ambient Water Quality Monitoring Report, Monitoring Year 1997. DEQ Technical Bulletin WQA/1999-004. Water Quality Assessments. Richmond, VA.
14. Malik CP, Seivastava AK (1982) Plant physiology. Kalyant Publisheres, New Delhi, India.
15. Allan JD (1995) Stream Ecology: Structure and Function of Running Waters. Chapman & Hall, London.
16. Ghallab MHM (2000) Some physical and chemical changes on River Nile downstream of Delta Barrage at El-Rahawy drain. M. Sc. Thesis Fac. of Sci. Ain Shams Univ. Cairo, Egypt.
17. Train ET (1979) Quality criteria for water. Castle House Publication.
18. USEPA, U.S. Environmental Protection Agency (1990) The Quality of our Nation's Water: A Summary of the 1988 National Water Quality Inventory. EPA Report 440/4-90-005. Washington, DC.
19. Kleene JW (1995) Watershed Nonpoint Source Management System: A Geographic Information System Approach. Virginia Polytechnic Institute and State University, USA.
20. De-Groot AJ, Allersma E (1975) Field observation on the transport of heavy metals in the aquatic environment. Pergamon Press, NY.
21. Thomas GS, William MS (1996) Chemistry of the environment. New Jersey.
22. CCME (2001) CCME Water Quality Index 1.0, User's Manual. In Canadian Environmental Quality Guidelines, 1999, Canadian Council of Ministers of the Environment, Winnipeg.
23. Bhargava DS (1983) Use of a Water Quality Index for river classification and zoning of Ganga River. Environmental Pollution, 6: 51-67.

7

Physico-chemical Parameters of Water in Bibi Lake, Ahmedabad, Gujarat, India

Qureshimatva Umerfaruq M* and Solanki HA

Department of Botany, University School of Sciences, Gujarat University, Ahmedabad-380 009, Gujarat, India

Abstract

Water as a universal solvent has the capability to dissolve many substances including organic and inorganic compounds. The quality of water generally refers to the component of water present at the optimum level for suitable growth of plants and animals. Temperature, turbidity, nutrients, hardness, alkalinity, dissolved oxygen, etc. are some of the important factors that determines the growth of living organisms in the water body. The Bibi lake, a natural lake of Vatva taluka, Gujarat, India is completely surrounded by slum area whom discharge their domestic waste directly into the lake water creating environmental hazard. To understand the situation, water samples were collected monthly to analyze different physical and chemical parameters such as temperature, pH, electric conductivity, turbidity, alkalinity, dissolved oxygen, etc. The comparison of statistics with the standard values of W.H.O guidelines (1998) revealed the high level of turbidity, total dissolved solids, pH, hardness, alkalinity and phosphate in the water content. The accumulation of these pollutants poses a dangerous threat to both aquatic and human lives.

Keywords: Water pollution; Bibi lake; Gujarat; India; Physico-chemical parameters; Freshwater

Introduction

Water as a universal solvent has the capability to dissolve many substances including organic and inorganic compounds. This outstanding property of water can be ascertained to the inconceivability to take in water in its pure form [1]. The quality of water generally refers to the component of water present at the optimum level for suitable growth of plants and animals. Aquatic organisms need a healthy environment to live and adequate nutrients for their growth; the productivity depends on the physicochemical characteristics of the water body [2,3]. The maximum productivity can be obtained only when the physical and chemical parameters are present at optimum level. Water for human consumption must be free from organisms and chemical substances and such large concentrations may affect health [4].

The pollution of water is increased due to human population, industrialization, the use of fertilizers in agriculture and man-made activity. Parameters such as temperature, turbidity, nutrients, hardness, alkalinity, dissolved oxygen, etc. are some of the important factors that determines the growth of living organisms in the water body [5]. Hence, water quality assessment involves the analysis of physico-chemical, biological and microbiological parameters that reflect the biotic and abiotic status of the ecosystem [6].

Materials and Methods

Study area

The Bibi lake, a natural lake of Vatva taluka, Ahmedabad, Gujarat, India, entirely surrounded by slum area and masses from the surrounding area channelize their domestic waste straight into the lake water.

Location: Vatva-Narol road, Vatva, Ahmedabad

Latitude: 22057'38.65" N

Longitude: 72036'37.09" E

Area covered: 11,425 m^2

Maintained by: Ahmedabad Municipal Corporation, Govt. of Gujarat.

Selection of sample sites

The water samples were collected from five different points of the lake (Figure 1). The points can be classified into three types,

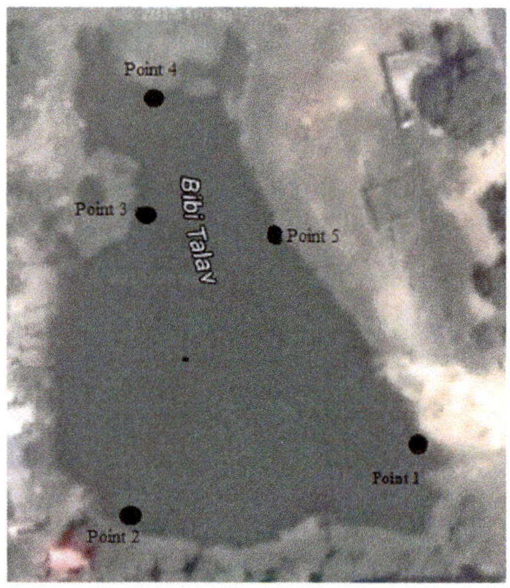

Figure 1: Satellite image of Bibi lake, Ahmedabad showing sampling sites denoted by dot.

***Corresponding author:** Qureshimatva Umerfaruq M, Department of Botany, University School of Sciences, Gujarat University, Ahmedabad-380 009, Gujarat, India, E-mail: ufmqureshi@yahoo.in

1. Inlet point - The feeder region of the lake for domestic animals

2. Centre point - The point which represents the general water quality of the lake

3. Outlet point - The place where the overflows occur

In addition, two points were randomly selected to enhance the sampling procedure.

Collection of samples

The water samples were collected in the polyethylene bottles. Initially, the prewashed bottles were rinsed with sample water. The closed bottle was dipped in the lake at the depth of 0.5 m, and then a bottle was opened inside and was closed again to bring it out at the surface. The samples collected in three replicates from five different points were mixed together to prepare an integrated sample.

Physico-chemical parameters

To study the physico-chemical properties of the lake water content, water samples were collected from the lake surface in a clean polythin container for the period of one year, September 2013 to August 2014. Samples were collected during morning hours in between 8.00 to 10.00 a.m. using one litre container. The physical and chemical parameters were analyzed in the seasons of monsoon, winter and summer, respectively. Parameters including temperature, pH, electrical conductivity, turbidity, total dissolved solid, total alkalinity, total hardness, calcium, magnesium, dissolved oxygen, biochemical oxygen demand, chloride, sodium, nitrate and phosphate, were analyzed. We adopted standard guidelines of water sampling and physico-chemical parameters evaluation [7]. Parameters such as temperature, pH, electrical conductivity, dissolved oxygen and turbidity, were directly evaluated in the study area whereas other parameters were analyzed in laboratory.

Results and Discussions

The physico-chemical parameters such as temperature, pH, electric conductivity, turbidity, alkalinity, dissolved oxygen, total dissolved solid, calcium, magnesium, sodium, chloride, phosphate, biological oxygen demand, nitrate and total hardness of water were analysed in the water samples taken from Bibi lake, Ahmedabad, Gujarat, India. These parameters were taken seasonally at the five points of the lake. All parameters were reported in mean value of the data with standard errors (Table 1) and its variations across the seasons are graphically presented in Figure 2.

Temperature

The temperature plays an important role for controlling the physico-chemical and biological parameters of water and considered as one among the most important factors in the aquatic environment particularly for freshwater [8]. The highest temperature in summer was recorded 27°C and can be due to high solar radiation, low water level, clear atmosphere and high atmosphere temperature [9]. The lowest temperature was reported during winter season was 18°C (Figure 2.1) due to cold low ambient temperature and shorter photoperiod [10].

Electrical conductivity

Water capability to transmit electric current is known as electrical conductivity and served as a tool to assess the purity of water [11]. The highest electrical condutivity was reported during summer was 3.54 Ω/cm and lowest in winter observed was 3.08 Ω/cm (Figure 2.2). During summer, a high level of conductivity indicates the pollution status as well as trophic levels of the aquatic body [12].

Turbidity

The turbidity of water can be related to the expression of optical property and reflects the intensity of light scattered by the particles present in the water. The high turbidity value reported in summer was 24 NTU due to the growth of aquatic vegetation and lowered the volume of water [13]. The lowest turbidity was found during monsoon was 17 NTU (Figure 2.3) due to dilution of water resulting from monsoon showers and comparatively low run off from the lake.

Total dissolved solids (TDS)

The highest total dissolved solids (TDS) in summer was observed as 1246 mg/L due to the addition of dead organic substances contributed by the decomposition of aquatic plants and animals which may be related to the water dilution caused by evaporation at high temperature during summer. The rain water may have decreased the TDS concentration to 958 mg/L (lowest TDS in the data) during monsoon (Figure 2.4). The maximum limit for TDS as suggested by W.H.O is 500 mg/L [14] which indicated that the recorded TDS signifies the polluted

Average with standard error values of physico-chemical parameters of Bibi Lake (2013-14)					
S. No.	Parameters	Monsoon	Winter	Summer	W.H.O. standards for Drinking water (annual max.)[a]
1	Temperature (°C)	23 ± 1.27	18 ± 1.2	27 ± 0.42	30-32
2	Electrical conductivity (Ω/cm)	3.15 ± 0.41	3.08 ± 0.15	3.54 ± 0.84	500
3	Turbidity (NTU)	17 ± 0.63	19 ± .25	24 ± 0.75	5
4	Total Dissolved Solid (mg/L)	958 ± 32.47	1073 ± 24.63	1246 ± 22.69	259-500
5	pH	8.9 ± 0.13	8.2 ± 0.84	9.3 ± 0.34	6.5-8.5
6	Alkalinity (mg/L)	242 ± 6.47	194 ± 4.15	274 ± 2.32	100
7	Total hardness (mg/L)	278 ± 6.10	255 ± 4.16	231 ± 3.15	200
8	Calcium (mg/L)	46 ± 0.36	37 ± 1.54	30 ± 0.86	75
9	Magnesium (mg/L)	33 ± 0.54	34 ± 2.58	31 ± 5.41	150
10	Dissolved oxygen (mg/L)	4.71 ± 0.47	5.41 ± 0.88	3.40 ± 1.63	7.5
11	Chloride (mg/L)	98 ± 4.1	99 ± 6.24	103 ± 4.63	200
12	Sodium (mg/L)	45 ± 2.47	41 ± 2.63	55 ± 2.12	200
13	Nitrate (mg/L)	8.7 ± 0.3	8.2 ± 0.24	7.5 ± 0.56	11
14	Phosphate (mg/L)	1.04 ± 0.73	0.96 ± 0.03	1.7 ± 0.08	0.5
15	Biochemical oxygen demand (mg/L)	1.18 ± 0.47	2.24 ± 0.66	2.41 ± 0.09	6.9

Table 1: Average with standard error values of physico-chemical parameters of Bibi Lake.

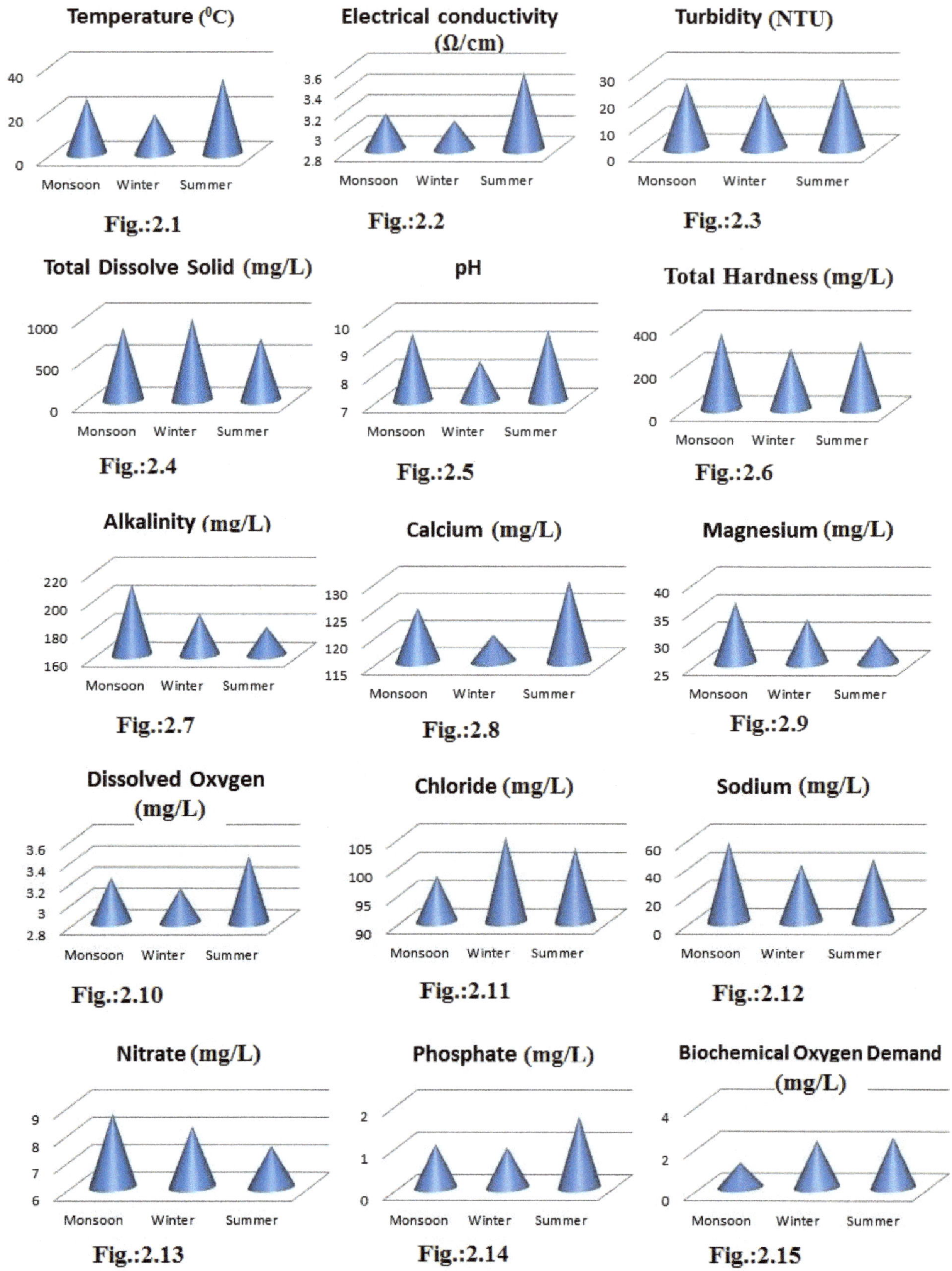

Figure 2: Graphs depicting the variations in physic-chemical parameters observed seasonally at Bibi lake, Ahmedabad, Gujarat, India during September 2013 to August 2014.

lake water. The contamination of domestic waste water, garbage and other related wastes in the surface water body can be one among the reasons for increase in TDS measure [8,9]. Our on-field observations were also similar in this case.

pH

We recorded a high pH of lake water (9.3) associated with the high decomposition activities of biotic and abiotic factors [15]. The low pH of 8.2 was noticed during winter (Figure 2.5) due to the production of CO_2 from biological oxidation process and may have ultimately contributed to the reduction of pH [16].

Alkalinity

The highest value of alkalinity was reported during summer was 274 mg/L due to the accumulation of organic matters produced by decay and decomposition of vegetation and in turn, added carbonate and bicarbonate concentrations in the lake water content [17]. The lowest alkalinity of 194 mg/L concentration was noticed during winter season (Figure 2.6) and can be related to the inflow of freshwater and dissolution of calcium carbonate ions in the water column [18].

Total hardness

The highest amount of total hardness in the water was recorded during monsoon as 278 mg/L due to the presence of high content of calcium and magnesium ions in addition to sulphate and nitrate in the sewage waste added during monsoon [19]. The lowest amount of total hardness was recorded during summer season as 231 mg/L (Figure 2.7) because of low volume of water and high rate of vegetation in the lake [20].

Calcium

The highest amount of calcium content in water was recorded during monsoon season as 46 mg/L by the addition of sewage waste which may be responsible for the increase in amount of calcium whereas the lowest amount of calcium was 30 mg/L (Figure 2.8) recorded during summer season due to calcium being absorbed by a large number of organisms. Jhingran suggested that calcium is one among the most abundant ions in freshwater and plays a pivotal role in shell construction, bone building and plant precipitation [21]. The magnesium act similar to calcium in ion exchange reaction and influence the absorption of sodium equally [22].

Magnesium

The highest amount of magnesium was 33 mg/L recorded during the winter season as the amount of magnesium is essential for chlorophyll bearing plant for photosynthesis and act as a limiting factor for the growth of phytoplankton [23]. The lowest value was 31 mg/L (Figure 2.9) estimated during the summer season.

Dissolved oxygen

The highest amount of dissolved oxygen recorded during the winter season was 5.41 mg/L because of the increased solubility of oxygen at lower temperature whereas the lowest dissolved oxygen was recorded as 3.40 (Figure 2.10) during the summer season which can be related to the high temperature and the addition of sewage and other wastes and drastically reduced the dissolved oxygen content.

Chloride

The highest chloride concentration of 103 mg/L was reported in summer due to frequent run-off loaded with contaminated water from the surrounding slum area and evaporation of water. The high chloride concentrations indicates the presence of organic matter, presumably of animal origin [24]. The lowest value of chloride was 98 mg/L (Figure 2.11) during monsoon season and can be connected to the dilution of lake by rain water. According to Solanki and Pandit, the concentration of chlorides can be related to purity or impurity of water [20].

Sodium

The highest amount of sodium was recorded during the summer season (55 mg/L) by the addition of waste water containing soap solution and detergent from the surrounding slum area [25]. The highest volume of sodium content during summer is due to shrinkage of water volume [26]. The lowest amount of 41 mg/L was recorded during winter season (Figure 2.12) because of bioaccumulation by living organisms.

Nitrate

The highest amount of nitrate was recorded during monsoon season (8.7 mg/L) due to the possible influx of nitrogen rich flood water into the lake water from the large amount of contaminated sewage water [27]. The lowest amount of nitrate in water was recorded during summer was 7.5 mg/L (Figure 2.13) and can be due to the utilization by plankton and aquatic plants for metabolic activities.

Phosphate

The highest amount of phosphate (1.7 mg/L) was recorded during summer season and associated by the entry of domestic sewage in the lake water. Similar result was observed by Benjamin et al., in the study of fish mortality in the lake waters of Bangalore, India [1]. Hastler observed that the constant addition of even low levels of nitrogen and phosphorous to an aquatic environment could greatly stimulate algal growth [28]. The lowest amount phosphate was recorded as 0.96 mg/L during winter season (Figure 2.14) due to increased uptake of phosphate for luxuriant growth of macrophytes.

Biochemical oxygen demand (BOD)

The highest biochemical oxygen demand was recorded during summer season as 2.41 mg/L which can be attributed to the high bacterial activity and heavy input of organic matter in the lake water [29]. The lowest demand of 1.18 mg/L was estimated during monsoon season (Figure 2.15) due to less vegetation and low decay of organic matter at low temperature.

The comparison of various physico-chemical parameters deduced from Bibi lake, Gujarat allowed us to study the pollution status of this water body. Quantities such as turbidity, total dissolved solids, alkalinity, pH, hardness and phosphate contents are significantly high. The comparison of estimated quantities from neighbour lake, Chandola lake, Ahmedabad [3] with the presented data revealed the content of magnesium and chloride ions and biological oxygen demand is high during the monsoon season. Verma et al., also noted that the Chandola lake is surrounded by industrial and slum areas and wastages from both of these places are deposited here [3]. During winter, electrical conductivity, turbidity, total dissolved solids and chloride, sodium, nitrate ions and biological oxygen demand are high according to Chandola lake statistics. The concentrations of chloride, nitrate and sodium ions and biological oxgen demand seems to decrease during summer monsoon. The pollution rate of Bibi lake is comparatively less than Chandola lake. This may be one among the reasons to implant cleaning water plant in the Bibi lake to procure drinking water to the local communities. The estimation of physic-chemical parameters at par with the W.H.O guidelines [14] may guide the civic authorities to modify the sustainable techniques to enhance the water quality.

Conclusion

The statistics obtained from the physic-chemical analysis of the water quality in the Bibi lake, Ahmedabad clearly indicates that most of the important quantities such as turbidity, total dissolved solids, pH, hardness, alkalinity and phosphate contents in the lake water are above the upper threshold of the W.H.O guidelines. This present situation may drastically affect the aquatic and terrestrial organism growth in the water repository and significant pollutants emerge from domestic sections pose an additional threat to the water quality in the near future. To sustain the ecology and aquatic life in the lake, certain measures and planning must be taken by the civic body to combat the pollution rate in the lake.

References

1. Benjamin R, Chakrapani BK, Devashish K, Nagarathna AV, Ramachandra TV, et al. (1996) Fish mortality in Bangalore lakes, India. Electronic Green Journal, 6. Retrieved from http://egj.lib.uidaho.edu/egj06/ramachandra.html.

2. Agbaire PO and OBI CG (2009) Seasonal Variation of some physico-chemical properties of River Ethiope water in Abraka, Nigeria. Journal of Applied Sciences and Environmental Management 13: 55-57.

3. Verma P, Chandawat D, Gupta U, Solanki HA (2012) Water quality analysis of an organically polluted lake by investigating different physical and chemical parameters. International Journal of Research in Chemistry and Environment 2: 105-112.

4. Uduma AU (2014) Physicochemical analysis of the quality of sachet water consumed in kano metropolis. American Journal of Environment, Energy and Power Research 2: 1-10.

5. Smitha (2013) Physico-chemical analysis of the freshwater at River Kapila, Nanjangudu Industrial Area, Mysore, India. International Research Journal of Environment Sciences 2: 59-65.

6. Verma PU, Purohit AR, Patel NJ (2012) Pollution status of Chandlodia Lake located in Ahmedabad Gujarat. International Journal of Engineering Research and Applications 2: 1600-1610.

7. Reasoner DJ (2004) Heterotrophic plate count methodology in the United States. Int J Food Microbiol 92: 307-315.

8. Singh RP, Mathur P (2005) Investigation of variations in physicochemical characteristics of a fresh water reservoir of Ajmer city, Rajesthan. Indian Journal Environmental Sciences 9: 57-61.

9. Swarnalatha N, Narasingrao A (1998) Ecological studies of Banjaralake with reference to water pollution. Journal of Environmental Biology, 19: 179-186.

10. Bohra OP, Bhargava SC (1977) Abiotic factor, chlorophyll pigment and primary production in two lakes of Jodhpur. Geobios 4: 215-216.

11. Murugesan A, Ramu A, Kannan N (2006) Water quality assessment from Uthamapalayam municipality in Theni District, Tamil Nadu, India. Pollution Research 25: 163-166.

12. Ahluwalia AA (1999) Limnological Study of wetlands under SardarSarovar command area. Ph.D. Thesis, Gujarat University, Ahmedabad.

13. Kumar SM, Ravindranath S (1998) Water Studies-methods for monitoring water quality. Center for Environment Education (CEE), Banglore, Karnataka, India, 191.

14. World Health Organization (W.H.O.) (1998) Guideline for drinking water quality. Health criteria and other supporting information (2ndedn) Geneva, 2: 231 -270.

15. Chaudhary S, Anuradha, Sastry KV (2004) Pollution of River Yammuna at Faridabad Hariyana. Pollution Research 23: 749-756.

16. Solanki HA (2007) Ecological studies of phytoplankton of Mini Mahi River, Gujarat, India. Vidya 2: 47-57.

17. Jain CK, Bhatica KKS, Vijay T (1997) Ground water quality in coastal region of Andraapradesh. Indian Journal of Environmental Health 39: 182-190.

18. Padma S, Periakali P (1999) Physicochemical and geochemical studies in Pulicatlake, east coast of India. Indian Journal of Geo-Marine Sciences 28: 434-437.

19. Angadi SB, Shiddamallayya N, Patil PC (2005) Limnological studies of Papnash pond, Bidar (Karnataka). J Environ Biol 26: 213-216.

20. Solanki, Pandit BR (2006) Trophic status of lentic waters of ponds water of Vadodara, Gujarat state, India. Inernational Journal of Bioscience Reporter 4: 191-198.

21. Jhingran VG (1975) Fish and Fisheries of India. Hindustan Publication Corporation, New Delhi, 954.

22. Paliwal PV, Yadav BR (1976) Irrigation water quality and crop management in the union territory of Delhi report. Water Technology Centre, IART, New Delhi.

23. Dagaonkar A, Saksena DN (1992) Physicochemical and biological characterization of a temple tank, Kaila Sagar, Gwalior, Madhya Pradesh. Hydrobiologia, 8: 11-19.

24. Tresh JC, Suckling EV, Beale JF (1944) Chemical and zoopklankton studies of lentic habitats in North Eastern New South Wales. Australian Journal Marine Freshwater Research, 21: 11-33.

25. Mishra SR, Saxena DN (1991) Pollution ecology with reference to physciochemical characteristics of Morar (Kalpi) river Gwalior (M.P). In: Shastree NK (ed.), Current Trends in Limnology 159-184.

26. Solanki HA (2001) Study on pollution of soils and water reservoirs near industrial areas of Baroda. Ph.D Thesis submitted to Bhavnagar University, Bhavnagar.

27. Zimba PV, Khoo L, Gaunt PS, Brittain S, Carmichael WW (2001) Confirmation of cat fish, IctaluruspunciatusRalfinesque, mortality from microcystis toxins. Journal of Fish Diseases 24: 41-47.

28. Hastler AD (1947) Eutrophication of lakes by domestic drainage. Journal of Ecology, 28: 383-395.

29. Campbell IC (1978) A biological investigation of an organically polluted urban stream in Victoria. Australian Journal of Marine & Freshwater Research 29: 275-291.

8

Community-based Water Quality Improvement in a Rehabilitated Slum Building

Archita Bharti[1]* and Devansh Jalota[2]

[1]*Green Health Foundation, NGO, Mumbai, India*
[2]*Dhirubhai Ambani International School, Mumbai, India*

Abstract

The Indian mega-city of Mumbai is undergoing a major makeover through large-scale shifting of slum dwellers from hutments to proper buildings with fixed size of dwelling units. This rehabilitated community of the urban poor is facing a high disease burden, apparently due to its poor sanitation and hygienic practices. This study conducted a baseline health survey, especially with regard to water borne diseases, along with water quality assessment, in a slum rehabilitated building of Mumbai. The health survey and water assessment indicated that while the rehabilitated slum dwellers had poor hygienic practices, the contamination of municipality-supplied water was causing significant health hazards for the residents. Some residents did use basic water disinfection processes like boiling, straining and chlorination, but with little positive impact on health quality. This study highlights the importance of sanitation and hygiene along with the significance of provision of better water quality. Quality of water supplied by the Brihanmumbai Municipal Corporation (BMC) being fairly good, the contamination of drinking water with sewer lines is indicated. For such problems, community-based treatment should augment household-level treatment of water.

Keywords: Household Survey; Drinking water; Contamination; Sanitation; Hygiene

Introduction

Water being an indispensable necessity of life, it is imperative that every human being has access to clean drinking water. However, for a large population of the globe, this remains either difficult to attain or actually impossible. According to a 2009 United Nations World Water Development Report, only 54% of the world's population had a piped connection to their dwelling, plot or yard, and 33% used other improved drinking water sources. The remaining 13% (884 million people) relied on unimproved sources. The rising populations are making matters worse. More than 60% of the world's population growth between 2008 and 2100 will be in sub-Saharan Africa (32%) and South Asia (30%) where the water stress is already severe. According to this report, an estimated 90% of the 3 billion people who are expected to be added to the population by 2050 will be in developing countries, many in regions that are already experiencing water stress and the current population does not have sustainable access to safe drinking water and adequate sanitation. The same report also mentions the fact that more than 5 billion people, i.e., 67% of the world population – may still not be connected to public sewerage systems in 2030 [1].

According to a WHO/UNICEF joint monitoring report of 2012, 11% of the global populations, i.e., 783 million people, are still without access to safe drinking water. The case of India is no different. India is the second most populated country in the world with over 1.2 billion people (Census of India, 2011). As the sources of freshwater are very limited, to provide water for all is an enormous challenge.

India has 97 million people without access to improved sources of drinking water, second only to China [2]. The 2001 Census reported that only 68.2% households in India have access to safe drinking water. Along with this, non-uniformity in the levels of awareness, socio-economic development, education, poverty, rituals, lack of accountability, and lack of a good civic culture also add to the problem of clean drinking water supply.

This speaks volumes about the need of a planned administration in providing clean drinking water at the government level. Article 47 of the Constitution of India speaks about "The duty of providing clean drinking water and improving public health standards to the State". The government has initiated various different types of programs since independence to provide safe drinking water to the masses. Till the tenth five year plan, 2002-2007, INR 1,105 billion has been spent on providing safe drinking water [3]. The global importance of water, sanitation and hygiene for development, poverty reduction and health has also been specifically mentioned in the eight Millennium Development Goals of the United Nations Millennium Declaration, in the reports of the United Nations Commission on Sustainable Development and at many international platforms.

The water distribution system in India is ridden with challenges. In the urban areas, especially in metro cities, the challenge is colossal due to the vast geographical area, high population density, and inaccessibility to many parts, like slums and *chawls*, resulting in poor water supply. Urban water delivery is also characterized by irregular delivery and pressure, lack of skilled manpower and solid infrastructure. The municipal boards are already over-subsidized and further expansion is beyond their capacity. The situation is especially complicated in a rapidly mushrooming city like Mumbai. Mumbai has often been called Slumbai or Slumbay as nearly 41.3% of its population resides in slums, according to the 2011 census [4].

Details of the present water supply system in Mumbai

According to the Chitale committee report (Table 1) [5], only 6222 MLD is available to Mumbai, while the demand is of 8316 MLD.

***Corresponding author:** Archita Bharti, Green Health Foundation, NGO, Mumbai, India, E-mail: architabharti85@gmail.com

This shows that the Mumbai Metropolitan Region (MMR) region has a deficit of 2094 MLD of water. Importantly, if we look at the water consumption pattern in slum and non-slum areas, as depicted in Table 2, it is evident that slums consume less water as compared to the non-slum area, despite a larger population density. The "Study of Mumbai MMR Sustainability: Housing and Transport" report also figures that though the slum population is much higher, their contribution towards wastewater generation is relatively less. The other major issue which emerges is that though the estimated water supply per capita is much higher, the actual per capita water available at the household level is lower. It indicates that this gap is due to leakage of water, pilferage or other reasons [6].

Water can be contaminated by organic materials such as animal carcasses, animal and human feces and sewage, food waste, plant matter, oil, petrol and grease or inorganic materials such as scrap metal, junk, sand and chemicals. The water supply also gets contaminated due to the entry of disease-causing germs or due to leakage in the sewer lines. USEPA report on 'Potential Contamination Due to Cross-Connections and Backflow and the Associated Health Risks' has mentioned the problems faced by the distribution system in typical cities. They are: microbial growth and biofilms, cross-connections, backflow, intrusion, corrosion, aging infrastructure, decay of water quality over distribution system residence time, contamination during infrastructure repair and replacement, nitrification, covered storage, permeation and leaching. Through cross-connections, backflow of non-potable water can mix with the potable one. Backflow can occur either due to reduced pressure in the distribution system (termed backsiphonage) or the presence of increased pressure from a non-potable source (termed backpressure). Backsiphonage may be caused by a variety of circumstances, such as main breaks, flushing, pump failure, or emergency firefighting water drawdown. Backpressure may occur due to heating/cooling or improper waste disposal. It may also occur when industrial manufacturing systems are connected to potable supplies and the pressure in the external system exceeds the pressure in the distribution system. During incidents of backflow, these chemical and biological contaminants have caused illness and deaths, with contamination affecting a number of service connections [7] (Table 3).

Regions	Name of the Dams	Available Water (mld)
Greater Mumbai	Modak sagar, Tansa, Upper vaitarna, Bhatsa,	3392
	Vihar and Tulsi	
Rest of Corporations		
Western Mumbai	Usgaon, Tansa river, Pelhar, Ulhas river	119
North eastern region	Ulhas river, Barvi, Chikhloli	1550
Navi Mumbai	Patalganga, Ransai, Hetawane, Barvi	479
Municipal Council		
Panvel- Uran	Ulhas river, Morbe, Dehrang	474
Pen- Alibag	Hetawane, Amba (Nagothane)	208
Total Water Supply (MMR)		6222

Source: Compiled from Dr. Chitale Committee Report 2003 and Environment Status of Brihanmumbai 2008-2009 [5]

Table 1: Water Supply System in Mumbai

	Non-slum	Slum
Greater Mumbai	127	50
Rest of Corporations Region	139	44
Municipal Councils	170	39

Source: NEERI Survey 2009 [5].

Table 2: Water Consumed Per Capita / Day (lpcd) in Mumbai (from base line survey)

Problems related to water

Apart from water supply, sanitation and hygiene are other decisive parameters for good health, survival, overall progress and reduction in the burden of disease from sanitation and hygiene. Hygiene refers to acts that can lead to good health and cleanliness, such as frequent hand-washing and bathing with soap and clean water. Safe drinking water together with proper hygienic practices is vital for reducing the burden of disease from sanitation and hygiene-related diseases. According to a WHO Report [1], almost one-tenth of the global disease burden can be prevented by improving the water supply, sanitation, hygiene and management of water resources. Such improvements reduce child mortality and improve health and nutritional status in a sustainable way.

According to the Centers for Disease Control and Prevention (CDC), water also plays a critical role in the spread of insect-borne diseases because many insects such as mosquitoes breed around water. Worldwide, over one million people die each year due to mosquito-borne diseases, most of them being young children in sub-Saharan Africa [8].

Around 10% of the total burden of disease can be prevented by improvements related to drinking water, sanitation, hygiene and water resource management. These diseases are diarrhea, malnutrition, intestinal nematode infections, lymphatic filariasis, trachoma, schistosomiasis, malaria, and drowning. Other quantifiable diseases linked to water resource development and management are dengue, Japanese encephalitis and onchocerciasis [9]. Diarrhea is the most important public health problem directly related to water and sanitation. About 4 billion cases of diarrhea per year cause 1.8 million deaths, over 90 per cent of them (1.6 million) among children under five. Repeated episodes of diarrheal disease make children more vulnerable to other diseases and malnutrition [10]. In 2000, diarrhea accounted for 17% of the 10.6 million deaths in children younger than five, and malaria for 8%. Ordinary diarrhea remains the major killer among water, sanitation and hygiene-related diseases, contributing to 43% of deaths. Sub-Saharan Africa and South Asia are the most affected regions [10]. It is estimated that around 37.7 million Indians are affected by water borne diseases annually, 1.5 million children are estimated to die of diarrhea alone and 73 million working days are lost due to waterborne disease each year as discussed by Khurana and Sen [11]. The World Health Organization (WHO) estimates that about 1.1 billion people globally drink unsafe water [12] and the vast majority of diarrheal disease in the world (88%) is attributable to unsafe water, sanitation and hygiene [13]. Poor hygiene is third on the list of the 20 leading risks factors for health burden in developing regions after unsafe water and sanitation [14].

Globally, improving water, sanitation and hygiene has the potential to prevent at least 9.1% of the disease burden (in disability-adjusted life years or DALYs, a weighted measure of deaths and disability), or 6.3% of all deaths. Children, particularly those in developing countries, suffer a disproportionate share of this burden, as the fraction of total deaths or DALYs attributable to unsafe water, inadequate sanitation or insufficient hygiene is more than 20% in children up to 14 years of age [10].

Malaria is a serious disease caused by the *Plasmodium* parasite

	Non-slum	Slum
Greater Mumbai	516	304
Rest of Corporations Region	563	37
Municipal Councils	200	5

Source: NEERI Survey, 2009 [5].

Table 3: Waste water generated by total population (MLD) in regions of MMR

carried by the *Anopheles* mosquitoes. Humans are infected when bitten by the mosquitoes. Each year, there are 300 million to 500 million cases of malaria throughout the world and about 1 million child deaths. Reducing the mosquito population in households and communities by eliminating standing water (caused by poor drainage and uncovered water tanks) can be an important factor in reducing malaria cases [10].

Investment in water is a long-term benefit, not only in terms of better health and good life span, but also with respect to the amount of money returned. For every $1 (USD) invested in safe drinking water and sanitation, the World Health Organization estimates returns of $3-34 USD [15]. A recent World Water Development Report ranked India 133rd among 180 countries in terms of water availability and 120th among 122 countries in terms of water quality [16].

This paper highlights the health and water quality assessment in the Network Park Compound (NPC) of the Shivaji Nagar Area, MHADA Colony. Shivaji Nagar is a slum rehabilitated building (SRB) in the eastern suburbs of Mumbai, India. The residents here have not received due attention from the town planners considering that though they have been provided with decent living apartments, they face the challenging problem of safe drinking water.

The Government of Maharashtra has tackled the problem of slum rehabilitation only partially. The population living here has frequently complained about water sanitation problems. According to them, drinking water is reportedly mixed with discharge from the drainage pipeline, and clean water is not being supplied to them.

To get an insight into the matter and for better understanding of the situation, these authors visited the site and conducted initial health assessment along with measurement of physico-chemical parameters of the drinking water samples from selected buildings of the location. To the best of the authors' knowledge, this is an exercise that has not been attempted before in the Mumbai rehabilitated buildings.

Objectives of the study

With this background, the study focuses on two primary objectives. First is to conduct a baseline survey and collect primary data about factors contributing towards the water problem. Currently, there is no updated information available on what people know, do and would like to do in relation to water, sanitation and hygiene. Secondly, the information collected through this baseline survey will aid in adopting the right approaches towards hygiene, education and environment, in particular for the access to adequate sanitation in the target area.

Methodology

The entire study has been divided into two broad parts - the health assessment study and water quality assessment. The main aim is to also assess the possible health burden reduction through community-based water treatment system.

Health assessment survey

4.1.1 Study area details: The Network Park compound has 60 buildings. Each of the 60 buildings has seven floors, and each floor has 12 apartments, making a total of 84 apartments per building. Out of these, 12 buildings where residents had complained of gastrointestinal ailments or waterborne diseases were randomly selected, along with two controls. Each household of the 12 selected buildings was assessed for health as per a standard questionnaire-based survey in accordance with WHO norms (Table 4).

Out of the total 12 buildings surveyed, 10 were taken as samples while two buildings represented the controls. The 10 buildings selected for target were named from B1 to B10. The criteria of selection of these buildings were their unhygienic conditions, maximum number of people affected and willingness of the residents. Here, the term people affected refers to the number of people that were affected by water borne diseases like diarrhea, typhoid, and jaundice or by unhygienic conditions like malaria and dengue. In buildings C1 and C2, it was observed that the walking space between the buildings was clean and hygienic with no fecal matter visible in the area. Hence, these two buildings were chosen as control. C1 and C2 buildings will henceforth be referred as control buildings while the others will be mentioned as target buildings in the rest of the paper (Figure 1).

Water quality assessment

To check the physicochemical properties of the water, water samples were collected from five buildings, based on availability of water in tanks, approachability of the tanks with safe staircases, willingness to participate of the residents, and complaints from the residents about quality of water. The samples were collected from B1, B3, B7, B9 and B10. From each building, the samples were collected from three points: two tanks at the ground level (L.H.S. and R.H.S.) and one overhead tank placed upon the roof of the building. These samples were tested for essential parameters as per Drinking Water Standards of BIS (IS: 10500: 1991). Sample collection, preservation and analysis were as per APHA guidelines (APHA 2005).

To ascertain the microbial contamination of the drinking water, samples were chosen randomly using a stratified random sampling process (stratified as per streets) and tested for the presence of Total Coliforms (TC) as well as Fecal Coliforms (FC) using the membrane filtration (MF) technique in 100 ml of sample. Water samples were collected in the pre - monsoon as well as in monsoon period. Water samples were collected aseptically in sterile PVC sampling bottles. The samples were transported within 1 hour of collection in an ice box to the laboratory for analysis. All samples were collected in triplicates.

Importantly, a few building tanks could not be approached because of poor infrastructure, i.e., safe staircases were not available for approaching the tanks or due to unavailability of samples. Water from such tanks could not be sampled, and these have been indicated by 'No Sample Collection' or NSC in the Tables.

Sr No.	Questions asked to Residents.
1	Number of family members in a family?
2	What is the main source of drinking water for members of your household?
3	What is the main source of water used by your household for other purposes such as cooking and hand washing?
4	Do you treat your water in any way to make it safer to drink?
5	What kind of toilet facility do members of your household usually use?
6	Has any of your family member been sick from waterborne diseases in the last 6months/ two weeks?
7	What and how many hand-washing facilities are available?
8	What water source is used for hand washing?
9	Is soap/ash/mud available?
10	What is condition of the house or nearby house?
11	Are feces visible in the area?
12	What is done with collected garbage?
13	What is the condition inside of water containers?
14	Are drinking water containers properly covered?

Table 4: Questions asked to the residents for baseline survey

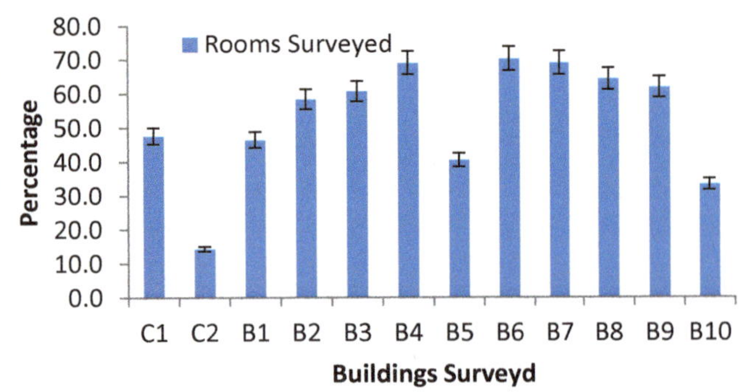

Figure 1: Percentage of total rooms surveyed in target buildings

Results and Discussions

As discussed, the Network Park compound area houses a total of 60 buildings. Of these, only 12 buildings along with 2 controls were selected as the study area.

The number of rooms surveyed in the target buildings ranged from 33% to 70%. All rooms could not be surveyed due to unavailability of the residents. In control buildings, it ranged from 14% to 47%. In these buildings, most of the rooms were vacant. These two buildings followed the strict regulation of not letting the rooms to ineligible candidates, i.e., only government-listed residents were allowed in. This is, in fact, a very good rule to follow as this in turn makes residents more responsible towards their society and makes them contribute more towards its welfare and growth.

However, in the target buildings, it was observed that there were many tenants. One common observation made during the visit was that the people living in the upper floors showed little concern about the ground floor residents. They were in the appalling habit of throwing garbage, kitchen waste and even fecal matter below, making the streets unfit for walking. Piled garbage could be easily noticed in this area.

For water supply, two tanks of 20,000 liter capacity each are placed at the two corners of the building at the ground level. For convenience, we have named them as Left hand side (L.H.S) tanks, and Right hand side (R.H.S) tanks. Water is pumped in the overhead tank with a 7.5HP pump. The water in each ground-level tank gets divided into four overhead tanks. From these four overhead tanks, water is supplied to the seven floors.

Health survey analysis

- **Source of drinking water:** The source of drinking water in all the buildings was piped water. The water is supplied from the Brihanmumbai Municipal Corporation (BMC) in the morning time for 3-4 hours. After that, the water is pumped into the four overhead tanks from where it gets distributed to all the households.
- **Main source of water used by the household for other purposes such as cooking and hand washing:** In this survey question, it was found that almost all the households use piped water. Only few homes bought packed drinking water, but its percentage was negligible. People generally stored the water in vessels and then used it for their daily activities.
- **Water collection time:** As the water was supplied directly through pipe, no one was involved in water collection from outside sources. Only when the water tasted foul or a clear contamination could be observed, few households preferred buying bottled water from the nearby shops. Else they opted for normal household disinfection treatment methods.
- **Kind of toilet facility and its sharing:** Each household had its own toilet with no sharing. The type of the toilet was pit latrine with slab.
- **Soap/ash/mud availability:** All the households used soap solution for hand wash as it is easily available. Also, being city residents, they were more exposed to media like T.V and radio and showed high awareness levels regarding the importance of hand washing and sanitation.
- **Condition of nearby Houses:** In the survey, we found that all the floors inside the apartments were very clean. However, the streets were piled up with garbage.
- **Collection of Garbage:** Daily door-to-door garbage collection is managed by the BMC, which also has the responsibility of cleaning the streets. Apart from the municipal people, most of the buildings have employed private staff for door-to-door garbage collection, considering the relative irregularity of the municipality staff. The collected garbage is then transferred to the nearby solid waste dumping ground. However, as people throw the garbage on the streets, it discourages many workers in their cleaning activity.
- **Conditions inside water containers:** The survey found that residents cleaned their water-collection vessels on a daily basis and also covered them properly.

Different disinfection methods employed at the household level

Treating water at the household level has been shown to be one of the most effective and economic measures for preventing waterborne diseases. Household water treatment and safe storage (HWTS) helps the populations to take charge of their own water in a better way. Because HWTS prevents recontamination of water at home, treating water at the household level is more effective than conventional improvement in water supply system for ensuring the microbiological quality of drinking water at the point of consumption, as discussed by Sobsey [17]. A more recent and comprehensive Cochrane review covering more than 38 randomized, controlled trials and 53,000 people in 19 countries found that household-based interventions were about twice as effective in preventing diarrheal disease (47%) than improved wells, boreholes and communal stand pipes (27%) [18,19].

In India, the common disinfection method to treat water ranges from boiling, adding bleach/chlorine, straining it through a cloth, use of water filter (ceramic, sand, composite, etc.) or allowing water to stand and settle.

Boiling of water: It was observed that many residents preferred the boiling of water. In target buildings, the percentage of people using this disinfection method ranged from 15.7% to 40.7%. In control buildings, this figure was from 25% to 42% approximately (Figure 2a). Overall, 28.6% test households boiled their water, which is comparable with that of control buildings (28.8%) (Figure 2b). As frequent boiling of water consumes electricity and gas, and many households are not financially sound enough, they prefer other modes of treating the water. Also, boiled water tastes slightly different, which is how some households explained their choice of other methods.

Straining of water through cloth: It is one of the commonest methods of treating water since it is cost free and convenient. In target buildings, 15% to 66% residents used this method while in control C1, more than half of the households opted for it (Figure 3a). In C2, no one reported using this method. As seen in Figure 3b, overall percentage in target and control buildings was similar –around 44.2% and 42.3%, respectively. In a few cases, straining was not through cloth but via a plastic filter fitting into the tap.

Use of water filter: Water filters are mostly popular in urban areas, but as they require frequent changing of the filter candles and are relatively more expensive, they are not common among the low income group residing in slum-rehabilitated buildings. Our observations supported this, with only 9%-32.9% of the target population opting for it. In the two controls, it ranged from 33%-62%, which is fairly good (Figure 4a). The overall percentage of households using the water filter was a mere 19.5% while in the two controls, it was approximately 56%, which is a very significant difference (Figure 4b). This shows that the residents of these buildings were more concerned and committed towards water issues apart from inculcating hygienic behavior.

Let the water stand and settle: Few people living in the rehabilitated buildings are not much inclined towards spending money for water-related issues. They use it directly, i.e., let it settle for 20-30 minutes and then use it for drinking and other purposes. This may be due to poor financial conditions or apathy. 19% to 47% (Figure 5a) people directly used the water. Most of the surveyed households cited economic constraints as an explanation for this attitude. In the control population, a mere 3.8% (Figure 5b) went for this option.

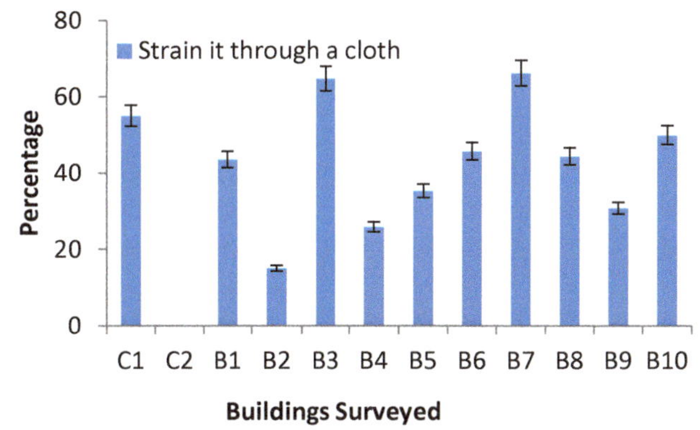

Figure 3a: Percentage of target and control cluster opting for water treatment by straining through cloth

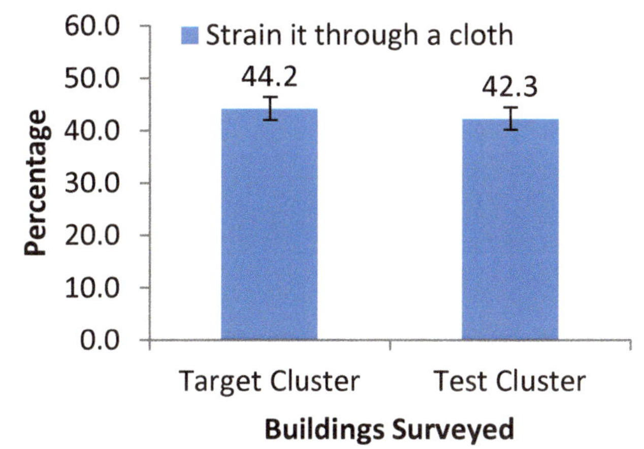

Figure 3b: Comparison between target and control households using cloth strained water as disinfection method

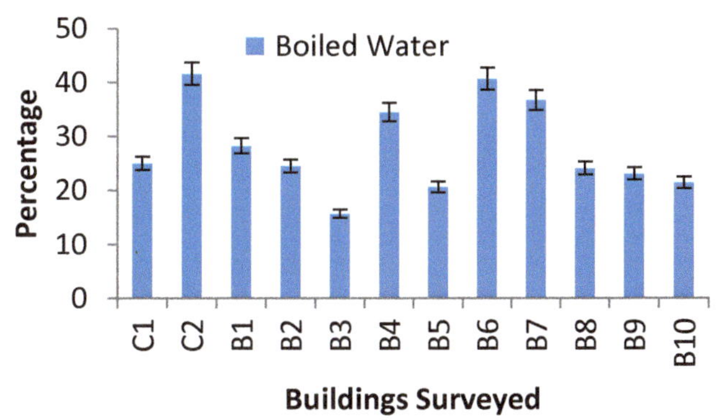

Figure 2a: Percentage of surveyed households using boiled water as disinfection method

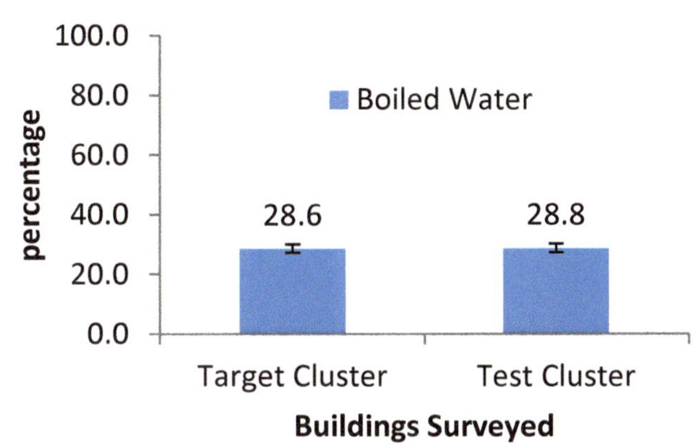

Figure 2b: Comparison between target and control households using boiled water as disinfection method

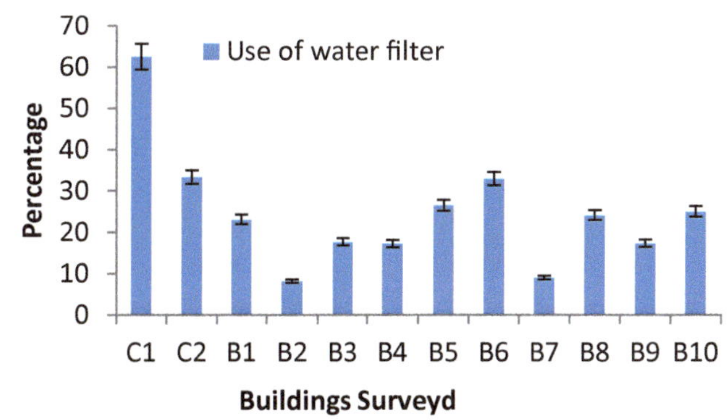

Figure 4a: Percentage of target and control cluster using water filter

Addition of bleach/chlorine: For more than 100 years, chlorine has been used around the world to disinfect drinking water supplies [20]. Chlorine is simple, effective and relatively inexpensive. It also provides protection against contamination by micro-biological organisms. This makes it a very popular disinfectant worldwide. A recent WHO-sponsored analysis also concluded that household-based chlorination

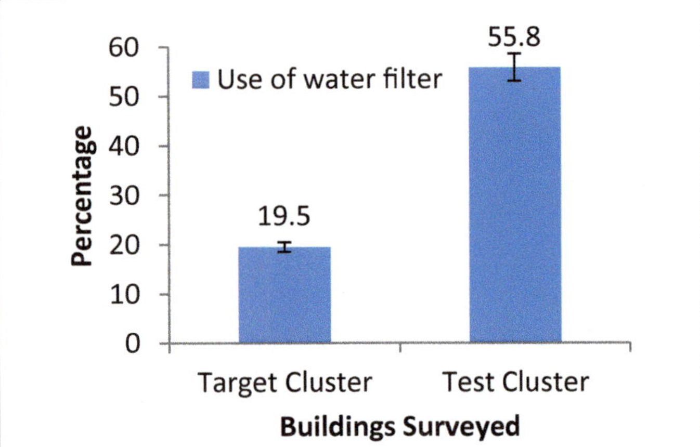

Figure 4b: Comparison between target and control households using water filter

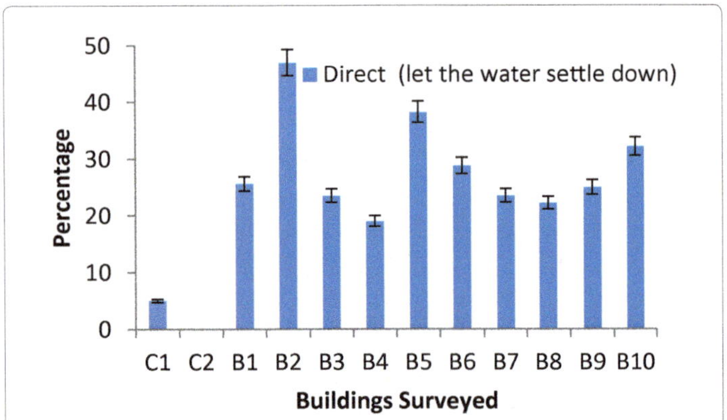

Figure 5a: Percentage of target and control cluster going for direct use of water

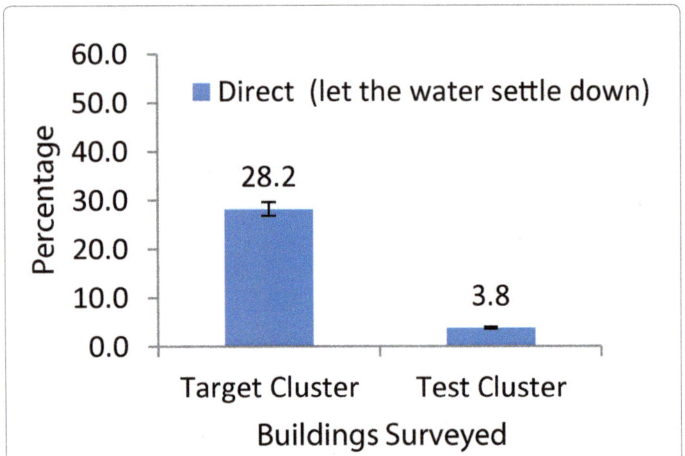

Figure 5b: Comparison between target and control households going for direct use of water

was among the most cost-beneficial of the various options for pursuing the Millennium Development Goals (MDG) water and sanitation targets, yielding high returns on every dollar invested, mainly from lower health care costs and also increased productivity and value of school attendance [21].

However, it was observed that only 1%-15% (Figure 6a) of the target population went for this option. In the control buildings, a mere 7.5% of the population was using this disinfection method. This may be due to a lack of awareness, or unwillingness to pay (Figure 6b).

Water quality analysis

All the samples were colorless, odorless and tasteless. The pH, conductivity, turbidity and total hardness of the water were within the permissible limits of drinking water. In the summer season, residual chlorine could not be traced in any of the water samples, leading to the presence of microbial contamination, as indicated by the presence of Total Coliform (TC) and Fecal Coliform (FC). However, in the monsoon water sample collection, few buildings showed the presence of residual free chlorine, which can be correlated with the absence of TC and FC in those buildings, as shown in the above results (Table 5).

In the study area, there is only a single line of separation between the collection tank (L.H.S and R.H.S) and the nearby sewerage line. Due to reduced or increased pressure in the sewerage line, backflow or backsiphonage occurs, as discussed above in the introduction section. This may result in cross-contamination of the collection tank with the sewerage line, leading to the presence of total and fecal coliforms in the sample. There is also a probability of household contamination from dirty vessels, and unhygienic water handling practices. At the household level, contamination of stored water is even more common. In one of the pilot countries, only 43.6% of samples from stored water were in compliance with the WHO guideline value and national standards, and more than half of the household samples showed post-

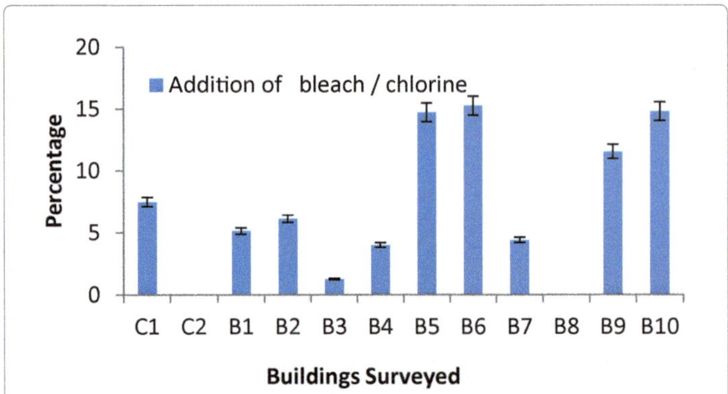

Figure 6a: Percentage of target and control cluster using Chlorine/ Bleaching powder

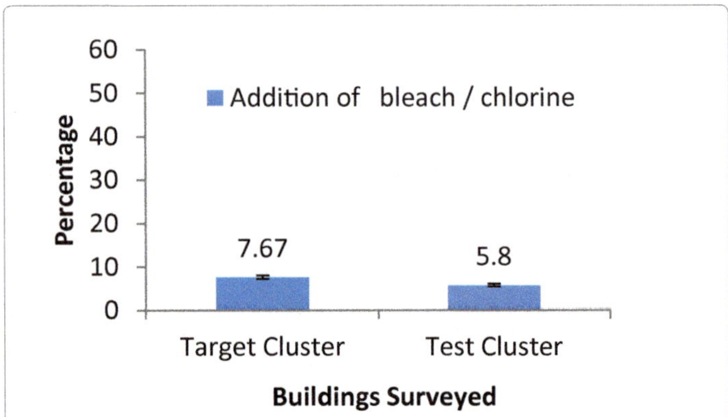

Figure 6b: Comparison between target and control households using Chlorine/bleaching powder

Building No	Place of Collection	pH	Conductivity	Turbidity (mg/l)	Chloride (mg/l)	TC/ 100ml	FC/ 100ml
Pre-Monsoon Season							
B1	L.H.S	6.5 ± 0.06	143 ± 1.00	1.32 ± 0.01	8.9 ± 0.15	12 ± 1.00	4 ± 1.00
	R.H.S	6.5 ± 0.10	281 ± 1.00	1.14 ± 0.02	7.9 ± 0.10	10 ± 0.58	5 ± 0.5
	Overhead Tank	6.5 ± 0.21	149 ± 1.53	1.22 ± 0.01	7.9 ± 0.21	16 ± 1.53	6 ± 1.15
B3	L.H.S	6.6 ± 0.10	96 ± 0.5	2 ± 0.15	8.9 ± 0.10	30 ± 1.5	6 ± 1.53
	R.H.S	6.5 ± 0.00	101 ± 1.73	2.03 ± 0.02	8.9 ± 0.25	32 ± 2.52	4 ± 1.00
	Overhead Tank	6.7 ± 0.17	104 ± 1.00	9.67 ± 0.05	8.9 ± 0.17	40 ± 2.08	8 ± 1.53
B7	L.H.S	6.7 ± 0.10	83 ± 2.00	0.94 ± 0.02	8.9 ± 0.10	42 ± 2.5	16 ± 2.52
	R.H.S	6.8 ± 0.26	93 ± 3.21	1.44 ± 0.02	7.9 ± 0.10	Nil	Nil
	Overhead Tank	NSC	NSC	NSC	NSC	NSC	NSC
B9	L.H.S	NSC	NSC	NSC	NSC	NSC	NSC
	R.H.S	NSC	NSC	NSC	NSC	NSC	NSC
	Overhead Tank	6.5 ± 0.3	93 ± 2.00	1.22 ± 0.00	7.9 ± 0.20	4 ± 0.10	1 ± 0.00
B10	L.H.S	NSC	NSC	NSC	NSC	NSC	NSC
	R.H.S	6.5 ± 0.10	91 ± 1.50	0.72 ± 0.00	10.9 ± 0.20	50 ± 2.10	16 ± 1.50
	Overhead Tank	6.5 ± 0.15	201 ± 1.00	2.3 ± 0.21	8.9 ± 0.21	38 ± 1.53	12 ± 0.58

Monsoon Season									
Bldg.No	Place of Collection	pH	Cond.	Turbidity (mg/l)	Chloride (mg/l)	ResidualChlorine (mg/l)	Total-Hardness (mg/l)	TC/ 100 ml	FC/100 ml
B1	L.H.S	6.5 ± 0.06	281 ± 1.00	2.51 ± 0.10	9.9 ± 0.15	Absent	0.8 ± 0.01	16 ± 1.08	4 ± .0.10
	R.H.S	6.7 ± 0.00	142 ± 2.00	1.32 ± 0.02	8.9 ± 0.30	Absent	0.7 ± 0.10	20 ± 2.00	5 ± 0.60
	Overhead Tank	6.6 ± 0.10	147 ± 3.1	1.38 ± 0.05	10.9 ± 0.1	Absent	0.6 ± 0.2	2 ± 0.60	Nil
	Room	6.5 ± 0.10	146 ± 3.00	2.86 ± 0.04	8.9 ± 0.40	Absent	0.7 ± 0.1	6 ± 1.00	1 ± 0.60
B3	L.H.S	6.7 ± 0.10	102 ± 1.50	1.49 ± 0.02	9.9 ± 0.2	Absent	1.1 ± 0.1	35 ± 1.5	8 ± 1.0
	R.H.S	6.7 ± 0.2	98 ± 2.00	0.85 ± 0.02	8.9 ± 0.2	Absent	1.3 ± 0.2	30 ± 1.0	8 ± 1.2
	Overhead Tank	7.0 ± 0.1	102 ± 3.50	1.21 ± 0.02	9.9 ± 0.1	Absent	1.0 ± 0.0	25 ± 3.5	5 ± 1.5
	Room	7.0 ± 0.2	99 ± 1.5	0.91 ± 0.02	8.9 ± 0.2	Absent	1 ± 0.2	13 ± 2.1	3 ± 1.00
B7	L.H.S	7.0 ± 0.00	93 ± 2.1	1.24 ± 0.02	8.9 ± 0.3	0.25 ± 0.02	0.9 ± 0.2	Nil	Nil
	R.H.S	6.9 ± 0.1	84 ± 2.00	0.71 ± 0.02	10.9 ± 0.2	0.25 ± 0.05	0.8 ± 0.1	Nil	Nil
	Overhead Tank	7.0 ± 0.1	82 ± 0.6	0.86 ± 0.03	11.9 ± 0.2	0.25 ± 0.01	1.0 ± 0.0	Nil	Nil
	Room	7.0 ± 0.0	93 ± 0.60	1.1 ± 0.02	11.5 ± 0.30	0.25 ± 0.01	0.5 ± 0.10	Nil	Nil
B9	L.H.S	6.4 ± 0.30	88 ± 2.6	0.4 ± 0.04	10.9 ± 0.20	0.25 ± 0.0	0.5 ± 0.1	Nil	Nil
	R.H.S	6.3 ± 0.20	81 ± 1.0	3.13 ± 0.03	8.9 ± 0.20	0.25 ± 0.02	0.4 ± 0.1	Nil	Nil
	Overhead Tank	6.0 ± 0.0	93 ± 1.5	2.39 ± 0.03	9.9 ± 0.10	0.25 ± 0.0	1.2 ± 0.1	Nil	Nil
	Room	6.0 ± 0.10	98 ± 2.60	3.36 ± 0.03	10.9 ± 0.20	0.25 ± 0.01	1.2 ± 0.2	Nil	Nil
B10	L.H.S	6.5 ± 0.1	91 ± 2.60	1.09 ± 0.1	9.9 ± 1.20	0.25 ± 0.0	2 ± 0.1	Nil	Nil
	R.H.S	6.6 ± 0.1	144 ± 1.5	0.67 ± 0.02	9.9 ± 0.1	0.25 ± 0.02	1.9 ± 0.0	Nil	Nil
	Overhead Tank	7.0 ± 0.1	201 ± 1.00	4.08 ± 0.03	9.9 ± 0.1	0.25 ± 0.03	1.1 ± 0.1	Nil	Nil
	Room	6.2 ± 0.1	104 ± 1.5	0.48 ± 0.04	8.9 ± 0.1	0.25 ± 0.01	0.9 ± 0.1	Nil	Nil

All samples collected were colorless, odorless and tasteless.
Symbols used: NSC means No Sample collection
L.H.S: left hand side tank
R.H.S: Right Hand side tank
Overhead Tank: tank at the overhead, collected form one overhead tank only, out of four

Table 5: Physico-Chemical Analysis of water samples carried at Network Park Compound for the pre-monsoon and monsoon Season

source contamination. This is consistent with a large body of research worldwide that has shown that even drinking water which is safe at the source is subject to frequent and extensive fecal contamination during collection, storage and use at home [22].

People affected due to water borne diseases and unhygienic conditions

Interestingly, the percentage of affected people varied from 19% to 47% in the target buildings while in the two controls it was in the vicinity of 40% (Figure 7a). In other words, both test and control populations were almost equal as far as overall effect was concerned (Figure 7b). Though the control population followed very good hygiene practices and maintained their society fairly well, a significant percentage of people were affected by gastrointestinal infections. This might be due to the contamination of water by the nearby sewer lines.

Intervention

Studies have shown that household water handling, clean utensils, type of utensils, together with good sanitation practices play a significant role in the risk associated with outbreak of water borne illnesses. An intervention study in Zimbabwe found that homes where traditional drinking water containers are replaced with covered, narrow mouthed urns with a tap outlet have significantly less contamination than the control group [23]. A combination of special storage vessels with point of use treatment has given very promising results [24]. The study also found that fecal contamination in households using a specially designed safe water storage container alone, but not in households using both the container and a 5% calcium hypochlorite solution [24]. In Calcutta, India, the introduction of a narrow-mouthed and covered container from which water was poured significantly reduced cholera

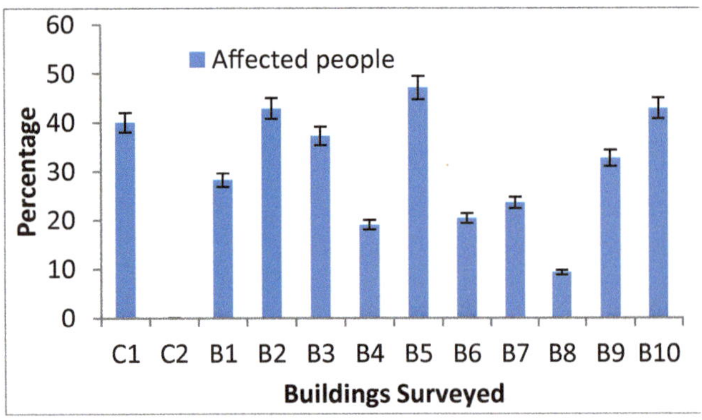

Figure 7a: Percentage of affected people in target and control cluster Bleaching

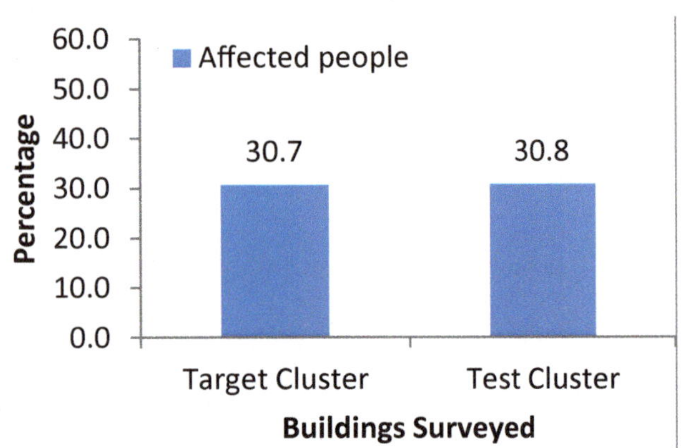

Figure 7b: Comparison between Percentage of affected people in target and control cluster

contamination [25]. In a study of randomized control trials in Pakistan, it was observed that hand washing initiatives and the introduction of point-of-use disinfection can reduce diarrheal incidence [26]. A Cochrane review of the efficacy of hand washing interventions concluded that diarrheal incidences may be reduced by about 30% [27]. Other factors such as number of residents in a household and presence of sewage in streets have been associated with feco-orally transmitted parasitic diseases [28].

Nevertheless, in this case, it is a clear case of high level microbial contamination. Hence, to provide clean drinking water on a long-term basis, community level water disinfection treatment would be very economical and effective, to augment household level water treatment process. To provide clean drinking water to the residents, on a pilot scale, water purifying devices like Ozonator, NaDCC (Sodium Dichloroisocyanaurate)-based water filters; sand and activated carbon-based filter, and UV irradiation and chlorinators were installed. Based on the economic feasibility and efficiency in water treatment process, these filters will be compared and the best one will be implemented in the entire study area (Table 6) [29].

Conclusion and Policy Recommendations

It is evident from the health survey assessment that it is the habit of inculcating cleanliness and hygienic condition that is responsible for a clean neighborhood and better health quality. . Control buildings were observed to be much cleaner as compared to the target buildings due to better hygienic practices. This may only be attributed to the practice of clean habits and a will power to keep the society clean. During monsoon season, drainage pipeline should be cleaned on a regular basis and any scope of mixing with the drinking water supply should be completely avoided.

Sr. no	Water treatment Method	Cost/person/ year
	Solar Disinfection	US$0.63
	Chlorination	US$0.66
	Ceramic Filters	US$3.03
	Combined	US$4.95
	Flocculation/Disinfection	US$4.95
	Installing & Maintaining Wells, borehole and Communal Tap Stands in Africa	US$1.88

* The combination of lower cost and higher effectiveness renders household-based chlorination the most cost effective of water quality interventions to prevent diarrhoea, with a cost effectiveness ratio in Africa of US$53 per disability-adjusted life year (DALY) averted, compared to US$123 for conventional source-based interventions [29].

Table 6: The cost of implementing water quality interventions varies as per table*

Besides, basic treatment of the water at the community or household level by chemical disinfection using chlorine, filtration using simple household filters, and boiling should also be promoted. These interventions may have a great impact on the health of the residents.

The findings from this study act as an eye-opener regarding the quality of water in the rehabilitated buildings of Mumbai. However, more sampling of different water sources in other such buildings is highly recommended. Proper measures should be drafted for periodic monitoring and stricter implementation of sanitation and hygienic activities. Moreover, at definite time intervals, the old and rusted pipelines should be replaced by the new ones. At the administrative level, there should be proper planning during the construction of SRA buildings. Sewerage lines should not be placed near the drinking water collection tank. Apart from this, concrete steps should be taken for significantly improving the situation. The need of the hour is to change the mindset and overall inculcation of civic behavior. All this will certainly help in improving the current situation. Qualitative and quantitative measures along with better reformed, time-suited administrative reforms, and the introduction of advanced technology is required to provide better quality of drinking water to the inhabitants.

Acknowledgement

The authors deeply acknowledge the hard work and efforts of the volunteers who contributed in the completion of this study. Our heartfelt thanks to Dr. Deepti Sharma of the Poison Pen Science Editors, without whose help and support the manuscript could not have been prepared. The authors also pay special thanks to Saheer Sawar and Sagar Patadiya for their kind co-operation in completing this report. The help of Ms. Anita Kumari and Ms. Lakshmi Kumari is also gratefully acknowledged.

References

1. http://www.unesco.org/new/fileadmin/MULTIMEDIA/HQ/SC/pdf/WWDR3_Facts_and_Figures.pdf.
2. http://www.who.int/water sanitation_health/monitoring/jmp2012/fast_facts/en/
3. www.wateraid.org/.../drinking-water-q.
4. Deshmukh MS (2013) Conditions of Slum Population of Major Sub-Urban Wards of Mumbai in Maharashtra, Voice of Research 2: 34-40.
5. Source: Compiled from Dr. Chitale Committee Report (2003) and Environment Status of Brihanmumbai 2008-2009.
6. National Environmental Engineering Research Institute & All India Institute of Local Self Government Mumbai Transformation Support Unit (2011) Study of Mumbai MMR Sustainability: Housing and Transport. Final report.

7. EPA Report: Potential Contamination Due to Cross-Connections and Backflow and the Associated Health Risks, September 27 (2001).

8. CDC. The Impact of Malaria, a Leading Cause of Death Worldwide.

9. Prüss-Üstün A, Bos R, Gore F, Bartram J (2008) Safer Water, Better Health: Costs, Benefits and Sustainability of Interventions to Protect and Promote Health. World Health Organization, Geneva.

10. World Health Organization. Safer Water, Better Health: Costs, benefits, and sustainability of interventions to protect and promote health.

11. Khurana I, Sen R (2008) Drinking water quality in rural India: Issues and approaches. Water Aid, India.

12. Kindhauser MK (2003) Global defence against the infectious disease threat. Communicable Diseases 2002. World Health Organization, Geneva.

13. http://www.who.int/water_sanitation_health/publications/factsfigures04/en/.

14. WHO (2003a) Quantifying selected major risks to health. The World Health Report 2002. World Health Organization, Geneva (Chapter 4).

15. http://www.unesco.org/new/en/natural-sciences/environment/water/wwap/facts-and-figures/all-facts-wwdr3/fact-38-investments-economic-health/).

16. http://www.vpmthane.org/publication-wms/WMScenario202025.pdf.

17. Sobsey MD (2002) Managing water in the home: accelerated health gains from improved water supply. Geneva: The World Health Organization (WHO/SDE/WSH/02.07).

18. Clasen T, Boisson S (2006a) Household-based ceramic water filters for the treatment of drinking water in disaster response: an assessment of a pilot programme in the Dominican Republic. Water Practice & Tech. 1: 2.

19. Clasen T, Roberts I, Rabie T, Schmidt W, Cairncross S (2006a) Interventions to improve water quality for preventing infectious diarrhoea (a Cochrane Review). In: The Cochrane Library, Issue 3, 2006. Oxford: Update Software.

20. http://www.public.health.wa.gov.au/cproot/4933/2/chlorinated%20drinking%20water%20finalv2.pdf.

21. Hutton G, Haller L, Bartram J (2007) Global cost-benefit analysis of water supply and sanitation interventions.

22. Wright J, Gundry S, Conroy (2003) Household drinking water in developing countries: a systematic review of microbiological contamination between source and point-of-use. Trop. Med. & Int'l Health 9: 106-117.

23. Mazengia E, Chidavaenzi MT, Bradley M, Jere M, Nhandara C, et al. (2002) Effective and culturally acceptable water storage in Zimbabwe: maintaining the quality of water abstracted from upgraded family wells. J. Environ. Health 64: 15-18.

24. Mintz ED, Reiff FM, Tauxe RV (1995) Safe water treatment and storage in the home. A practical new strategy to prevent waterborne disease. JAMA 273: 948-953.

25. Deb BC, Sircar BK, Sengupta PG, De SP, Mondal SK, et al. (1986) Studies on interventions to prevent eltor cholera transmission in urban slums. Bull. World Health Organisation. 64: 127-131.

26. Luby SP, Agboatwalla M, Painter J, Altaf A, Billhimer W, et al. (2006) Combining drinking water treatment and hand washing for diarrhoea prevention, a cluster randomised controlled trial. Trop. Med. Int. Health 11: 479-489.

27. Ejemot RI, Ehiri JE, Meremikwu MM, Critchley JA (2008) Hand washing for preventing diarrhoea. Cochrane Database of Syst. Rev.

28. Teixeira JC, Heller L (2006) Impact of water supply, domiciliary water reservoirs and sewage on faeco-orally transmitted parasitic diseases in children residing in poor areas in Juiz de Fora, Brazil. Epidemiol. Infect. 134: 694-698.

29. Clasen T, Haller L, Walker D, Bartram J, Cairncross S (2007) Cost-effectiveness analysis of water quality interventions for preventing diarrhoeal disease in developing countries. J. Water & Health 5: 599-608.

Nitrogen Sources and Cycling in the Ecosystem and its Role in Air, Water and Soil Pollution

Ghaly AE* and Ramakrishnan VV

Department of Process Engineering andApplied Science, Dalhousie University, Halifax, Nova Scotia, Canada

Abstract

The natural cycle of nitrogen involves several biological and non-biological process including: mineralization, nitrification, denitrification, nitrogen fixation, microbial and plant uptake of nitrogen, ammonia volatilization, leaching of nitrite and nitrate and ammonia fixation. Nitrogen exists naturally in the environment and is constantly being converted from organic to an inorganic form and vice versa. Production of commercial fertilizer adds up to the natural source of nitrogen. The main source of nitrogen include: atmospheric precipitation, geological sources, agricultural land, livestock and poultry operations and urban waste. Agricultural emissions show a strong increase due to the application of fertilizer to agricultural soils, grazing of animals and spreading of animal manure. Emissions from agricultural practices and animal manure wastes are the major source of nitrogen pollution in surface and underground water. Soil erosion and runoff from fertilized land as well as domestic and industrial wastes contribute to the enrichment of lakes and streams with nutrients. Nitrates concentration exceeding certain limits in drinking water is toxic to animals and humans, especially infants. Nuisance of algal bloom and fish kills in lakes and rivers occurs due to eutrophication. Obnoxious colours and smells are developed as a result of organic matter decay and are destroying the natural beauty of the environment. The water born contaminants affect human health from both recreational use of contaminated surface water and from ingestion of contaminated drinking water derived from surface or ground water sources. The methods for abatement of nitrogen pollution must follow multi pathways. First, the source and amount of pollution must be detected and defined. Second, the possible ways to treat animal and domestic wastes should be carefully investigated. Third, better agricultural practices should be developed that include: proper storage and application of slurry and solid manure, rapid incorporation of slurry and solid manure into the soil, use of band spreading machineries such as trailing house and trailing shoe and sub-surface applicators, use of specifically made round covers fitted to above ground tanks and slurry lagoons, applying fertilizers during periods of greatest crop demand at or near the plant roots in smaller amounts with frequent applications, using multiple cropping systems such as using crop rotations or intercropping to increase the efficiency of nitrogen uses and changing current livestock production techniques.

Keywords: Nitrogen cycle; Organic nitrogen; Nitrogen dioxide; Ammonium; Nitrification; Volatilization; Denitrification; Nitrogen fixation; Leaching; Fertilizer; Manure; Air; Water; Soil

Introduction

Water is an indispensable resource on which human existence depends. About 73% of the fat free body weight in healthy human adults is water which is equivalent to 60% of body weight for non-obese subjects [1]. Although three quarters of the earth is covered with water, only one percent of which is fresh water that is considered safe for use to meet our daily needs. Water that cannot be used because it is polluted: (a) reduces present and future supply, (b) inhibits local, regional and national economic growth, (c) poses known and unknown dangers to the public health, (d) curtails expansion of recreational activities and (e) further dislocates the already badly disrupted balance of nature [2].

Water pollution may be defined according to the usage of water and the presence of certain constituents and their effects on human health and the surrounding environment. One of the widely accepted definition of water pollution is "the introduction into water substances of such character and in such quantity that alter its natural quality as to impair its usefulness or render it offensive to the human senses of sight, taste or smell" [3,4]. A substance may not appear by itself as pollutant, but may indirectly (by its effect upon other materials or living organisms) lead to an offensive effect. The determination of whether a certain constituent creates a nuisance, impairs the usefulness of the water or interferes with nature's balance depends on the subsequent use of the water. Water can be polluted from many human activities and pollutants not only reduce the water quality and quantity but may act as harmful media to other living organisms. Polluted water can contain excess nutrients, pathogens, toxic materials and dangerous chemical which can affect human health, destroy marine species and disturb the ecological cycle [5,6].

Nitrogen in its elemental form is the major component of the air constituting about 78% of the gases in the earth atmosphere. There are also different nitrogen gaseous compounds that exist in the atmosphere including NH_3, NO and N_2O. The importance of nitrogen to life is that it constitutes (with carbon, hydrogen and oxygen) the major part of proteins of all living materials. All living organisms, except few microorganisms (nitrogen fixing bacteria, algae and fungi), cannot use N_2 as a source of nitrogen and need different forms of fixed nitrogen (NH_4 and NO_3 for plants and organic nitrogen for animals and humans) as a supply for their requirement of protein synthesis.

The elemental form of nitrogen (N_2) is very unreactive and hard to

***Corresponding author:** Abdel Ghaly, Department of Process Engineering and Applied Science, Dalhousie University, Halifax, Nova Scotia, Canada
E-mail: abdel.ghaly@dal.ca

extract from the environment although there are approximately five billion metrics tons of nitrogen around the earth [7]. Some forms of nitrogen are considered poisonous to plants, animals and humans if they exceed certain concentration in their environment. The presence of higher quantities of fixed nitrogen (NH_4 and NO_3) in the water or soil will encourage vegetative growth of plants beyond the favourable level. Nitrogen pollution means pollution due to the fixed forms of nitrogen (bonded with carbon, oxygen or hydrogen) also known as reactive nitrogen [8].

Nitrogen fixation is being done in nature in several ways and passed to the rest of living organism. Nitrogen will return back to the soil after the death of these organisms through the activities of soil microorganisms which provide the nitrogen in its ready form again to plants and then to animals and humans. Certain amounts of this fixed nitrogen are liable to be lost to the atmosphere in elemental form. However, the significant increase in human population increased the demand for food which leads to mass production of synthetic nitrogen fertilizers for agricultural activities. Human activities such as production and use of commercial fertilizer, production and use of fossil fuels in industrial processes, energy generation and transportation altered the nitrogen cycle and caused disturbance to the total environment (Air, soil and water) [9].

Natural Cycle of Nitrogen

The importance of nitrogen from the standpoint of fertility of the soil has long been recognized and our knowledge concerning the nature, distribution and transformations of nitrogen compounds in soil is extensive. A schematic diagram depicting the cycle of nitrogen in nature is shown in Figure 1. The nitrogen cycle includes several biological and non- biological processes. The biological processes are: ammonification/ mineralization, nitrification, denitrification, nitrogen fixation, nitrogen assimilatory reduction and microbial synthesis of ammonium and organic nitrogen into microbial cells, plant uptake and conversion of ammonium and nitrate nitrogen into plant proteins. The non-biological processes are ammonia volatilization, leaching of nitrite and nitrate nitrogen to ground water, ammonium fixation into soil clay minerals, precipitation of nitrate and ammonium nitrogen.

Mineralization

Mineralization (or ammonification) of soil nitrogen is the term used for the process by which nitrogen in organic compounds ($C_aH_bO_cN_d$) is converted by soil microorganisms into ammonium ion (NH_4^+) as follows [10,11].

$$\text{Complex organic nitrogen} \rightarrow \text{Ammonium} \qquad (1)$$

The soil microflora typically produces ammonium from organic compounds when they set free more nitrogen from the organic matter on which they are living than they can assimilate into their own protoplasm. This concept of the ammonium production (being the nitrogen waste product in the conversion of organic matter into microbial tissue and use of vital energy) is fundamental for understanding the effect of adding different types of organic matter on the mineralization of nitrogen in the soil [12]. Thus, when an animal protein (such as dried blood) is added to a soil, about 80% of the added nitrogen is liberated as ammonium and the remainder of the nitrogen is retained in microbial tissue. However, by increasing the quantities of a carbohydrate, such as cellulose mixed in with the protein, the amount of microbial tissue that can be built up is increased, with the consequence that the proportion of nitrogen liberated as ammonium

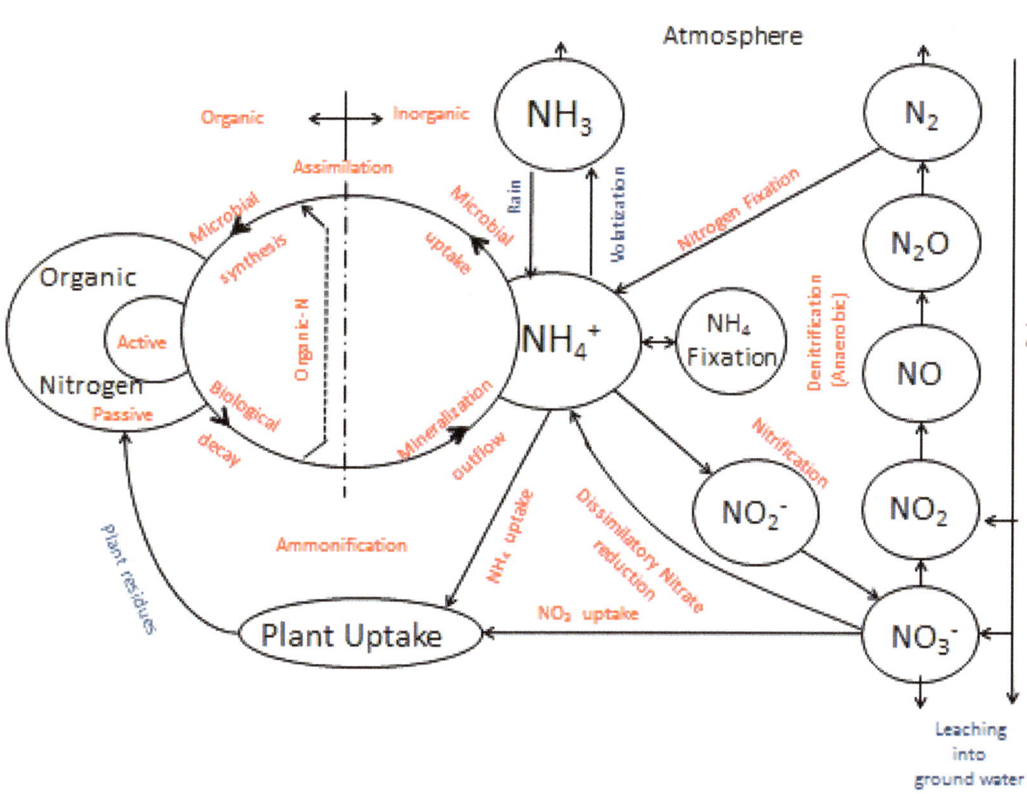

Figure 1: Nitrogen cycle in nature.

decreases until the ratio of carbohydrate to protein reaches a value of about 5:1 when all the nitrogen in the dried blood is needed by the microorganisms [13].

The accumulation of ammonium in the soil is affected by: (a) the rate of mineralization of organic nitrogen in the soil, (b) uptake of ammonium by microbes as a source of nitrogen for growth, (c) uptake of ammonium by plants as a source of nitrogen for growth, (d) volatilization of ammonia, (e) nitrification (biological conversion of ammonium to nitrate), (f) loss of nitrate by leaching (which increases the rate of nitrification) and (g) plant uptake of nitrate as a source of nitrogen for growth (which increases the rate of nitrification). Therefore, a low concentration of ammonium in the soil does not indicate low mineralization and may indicate high rates of nitrification, volatilization or microbial and plant uptake. Nonetheless, net mineralization will be directly affecting the organic nitrogen content in the soil and the availability of carbon for microbial growth. Thus, vegetation with high C: N ratio will result in a low rate of nitrogen mineralization [14].

Nitrification

Nitrification is the oxidation of ammonium nitrogen to nitrites and nitrates. It is the result of metabolism by chemoautotrophic (or chemolithotrophic) organisms. The two groups of organisms that are considered to be the primary nitrifying bacteria are *Nitrosomonas* Sp. and *Nitrobacter* Sp. *Nitrosmonas* carry out the oxidation of ammonium to nitrite to obtain energy (E) and *Nitrobacter* oxidizes nitrite to nitrate for the same purpose. The general oxidative processes involved can be represented by the following equations [15,16]:

$$NH_4^+ + 1\tfrac{1}{2}O_2 \xrightarrow{Nitrosomonas} NO_2^- + H_2O + 2H^+ + Energy \qquad (2)$$

$$NO_2^- + \tfrac{1}{2}O_2 \xrightarrow{Nitrobacter} NO_3^- + Energy \qquad (3)$$

Both genera use CO_2 as their sole carbon source for growth as follows, as they are obligate autotrophs and strict aerobes [17].

$$NH_4^+ + 5CO_2 + 2H_2O \rightarrow C_5H_7O_2N + H^+ + 5O_2 \qquad (4)$$

Nitrifying bacteria have a low activity level compared with heterotrophs. For example, it takes *Nitrosomonas* nearly an hour to produce twice its weight in NO_2 whereas some heterotrophic bacteria produce one thousand times their own weight in products in the same time. Under aerobic conditions, the endogenous respiration of the organisms results in the breaking down of cellular nitrogen and its release as ammonium [18,19].

Denitrification

Denitrification (or nitrate reduction) is a more complex and less understood process than nitrification. Nitrification and denitrification are redox processes involving nitrogen compounds to obtain energy. There are two processes of nitrate reduction: assimilatory and dissimilatory.

Assimilatory nitrate reduction: Assimilatory nitrate reduction (ANR) is one of the main processes in the nitrogen cycle in which nitrate (NO_3^-) is used as nitrogen source for the growth of new cells. At first, the nitrate (NO_3^-) is incorporated into the cells by a high affinity transporter which is then reduced to nitrite (NO_2^-) by the enzymes nitrate reductase which is further reduced to ammonium (NH_4^+) by the enzyme nitrite reductase. The ammonium which is produced in incorporated into the carbon skeltons by glutamine synthetase or glutamine synthase pathway as shown in Figure 2 [20].

Bonete et al [20] reported that the assimilatory nitrate reduction takes place in three steps: (a) nitrate (NO_3^-) uptake, (b) reduction of nitrate (NO_3^-) to nitrite (NO_2^-) and (c) reduction of nitrite (NO_2^-) to ammonium (NH_4^+). In the first step, the nitrate (NO_3^-) uptake into the cells takes place using ATP-dependent ABC transporter. In the second step, after the nitrate is being imported into the cells, it is reduced to nitrite (NO_2^-) by ferredoxin dependent assimilatory nitrate reductase (NR). In the third step, the nitrite is reduced to ammonium (NH_4^+) by ferredoxin dependent assimilatory nitrite reductase (NiR). The process is described as follows:

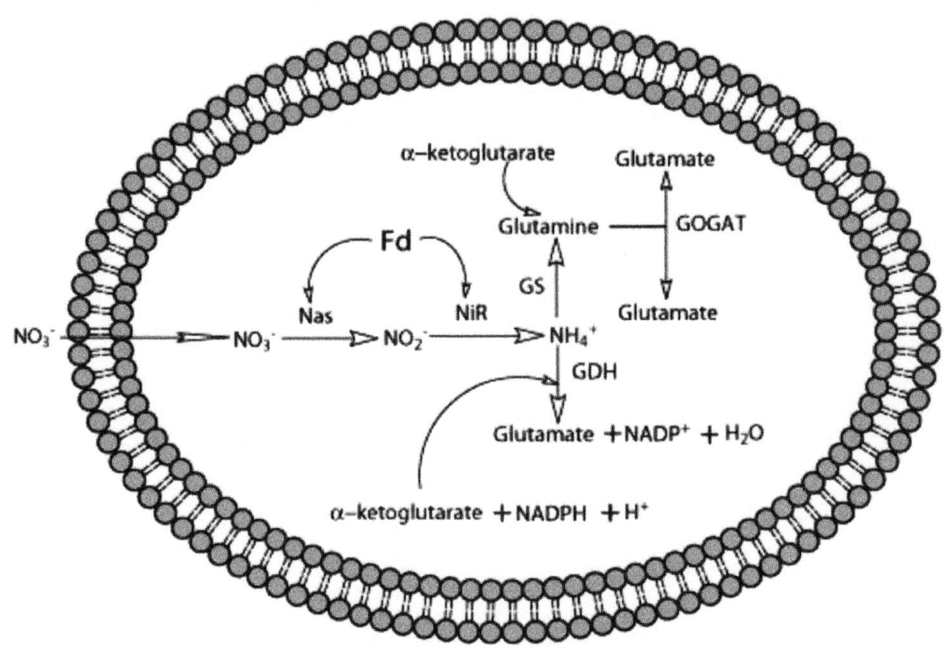

Figure 2: Assimilatory nitrate reduction [20].

$$NO_3^- uptake \xrightarrow{\text{ABC transporter}} NO_3^- \xrightarrow{\text{Nitrate reductase}} NO_2^- \xrightarrow{\text{Nitrite reductase}} NH_4^+ \quad (5)$$

Dissimilatory nitrate reduction: Dissimilatory (or respiratory nitrate) reduction is the process in which nitrate (specifically the oxygen in nitrate) serves as the terminal hydrogen acceptor in energy yielding reactions. It has been shown that nitrate reduction by this process closely follows the same steps as when molecular oxygen is used. It is tied in with the cytochrome system for electron transport. Thus, when the dissolved oxygen level drops to low levels for the aerobic metabolism of facultative organisms they can turn to nitrate reduction for oxygen fairly easily. The term denitrification is assigned to dissimilatory nitrate reduction. The general simplified equation for the denitrification process is postulated as follows [21-24]:

$$NO_3^- \xrightarrow{\text{Nitrate reductase enzyme}} NO_2^- \xrightarrow{\text{Nitrite reductase enzyme}} N_2O_2 \xrightarrow{\text{Nitrous oxide reductase enzyme}} N_2O \xrightarrow{\text{Nitric oxide reductase enzyme}} N_2 \quad (6)$$

The nature of the intermediates is not generally known although many workers have suggested N_2O_2 and N_2O which themselves may involve other complicated reactions [14,25]. There are a great many genera of bacteria that can accomplish denitrification among them are *Pseudomonas* Sp., *Thiobacillus denitrificans* (an autotroph) and *Micrococcus denitrificans*. The presence of dissolved oxygen has been found to inhibit the denitrification process to various degrees. This point out the need for low or zero dissolve oxygen levels before denitrification can occur [26,27].

Nitrogen fixation

Although atmospheric nitrogen (N_2) is abundance in the atmosphere, it is not in a ready from of nitrogen to be used by most organisms. Strong triple bond which links the nitrogen element in N_2 is hard to break. In Harber process, in which chemically synthesized nitrogen is formed, red hot magnesium or a catalyst is used at elevated pressures and temperatures to make N_2 reactive [28]. However, in the biological nitrogen fixation process, nitrogen is fixed by the microorganisms that are capable of breaking the bond at ambient temperatures and pressures and is called diazothrophic [28,29]. The dizaotrophs are available in soil both as free living and in symbiotic association with plants. Diazothrophic microorgainsms use the enzyme nitrogenase to carry out the fixation process. There are about 100 species of enzymes but they are very similar in their activities [28]. Nitrogenase reduces dinitrogen to ammonium (NH_4^+). The enzymatic reaction can be described as follows:

$$N_2 + 4H_2O \xrightarrow{\text{Nitrogenase}} 2NH_4^+ + 2O_2 \quad (7)$$

Soils can gain small amounts of nitrogen from the rain which falls on them. However, the most important natural process for increasing the nitrogen content of soils is nitrogen fixation by microorganisms living in the soil on and around the roots (Figure 3). A functional classification of the range of nitrogen fixing bacteria is soil is shown in Table 1 [30]. The estimates of nitrogen fixed by different microorganisms are shown in Table 2 [31]. Nitrogen fixation is inhibited in the presence of high level of available nitrogen (NH_4^+ or NO_3^-). The process is

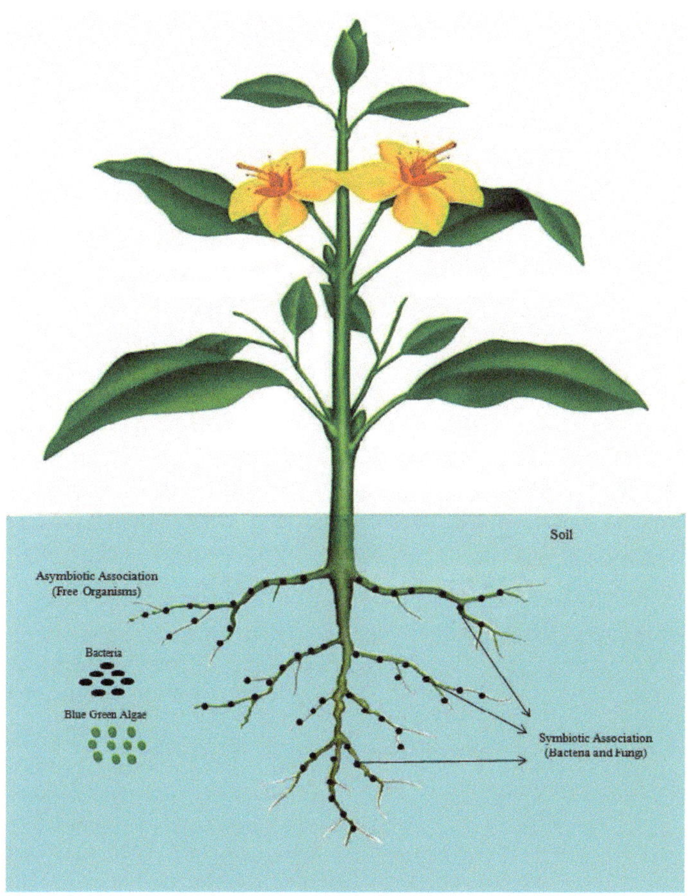

Figure 3: Nitrogen fixation in soil.

also controlled by N: P rations as phosphorous activate the gene for synthesis of [31]. Nitrogen can be fixed in the soil through symbiotic fixation and asymbiotic fixation. The rate of nitrogen fixation is related to the rate of photosynthesis (plant growth). A comparison between nitrogen fixation and photosynthesis is shown in Table 3.

Asymbiotic fixation of nitrogen: Soils contain a number of free-living, nitrogen-fixing organisms such as bacteria (*Azotobacter*, *Bejerinckia*, some *Clostridium*, and *Aerobacter*, *Achromobacter* and *Pseudomonas*), blue-green algae and yeast which possess the enzyme nitrogenase needed for nitrogen fixation [32]. The processes take place in soils high with organic matter that provide a ready source of energy. Nitrogen fixation goes on more actively under conditions of poor rather than good aeration, when the level of available nitrogen salts is low [33]. Asymbiotic nitrogen fixing bacteria are extremely diverse occurring in nine subdivision of the eubacteria, four subdivision of archeabacteria. Free living prokaryotes such as diazotrophs which have the ability to fix atmospheric free nitrogen are ubiquitous in the soil. The ability of free living diazotrophs to perform nitrogen fixation depends upon various conditions such carbon, nitrogen and oxygen partial pressures. The diazotroph populations depend upon the C: N ratio which is potentially used as bioindicators of nitrogen status of the soil. The contribution of assymbiotic nitrogen fixing bacteria in terrestrial ecosystem is much higher than symbiotic nitrogen fixing bacteria [34-36]. It is estimated that free living prokaryotes are capable of fixing nitrogen in the range of 0 to 60 kg/ha/y.

Type	Nitrogen fixing bacteria	Symbiont
	Heterotrophs	
Free living		
Anaerobic	*Clostridium*	
Microaerophilic	*Frankia, Azospirullum*	
Aerobic	*Bradyrhizobium, Azotobacter, Derxia*	
	Root-associated	
Microaerophilic	*Azospirullum, Herbaspirullum*	
Endophytic	*Acetobacter*	Sugar cane (*Saccharum sp.)*
Symbiotic	*Frankia*	*Casuarina sp. Alnus sp.*
	Rhizobium, Bradyrhizobium	Many legumes
	Azorhizobium	*Sesbania rostrata*
	Autotrophs	
Free living		
Anaerobic		
Microaerophilic	*Rhodospirillum, Bradyrhizobium*	
Aerobic	*Cyanobacteria*	
Symbiotic	*Cyanobacteria*	*Fungi* (lichens), Cycads
	Anabaena azollae	*Azolla sp.*
	Bradyrhizobium	*Aeschynomene sp.*

Table 1: Classification of nitrogen fixing bacteria which contribute to agriculture [30].

Nitrogen fixing system	Nitrogen (N_2) fixed (kg N ha^{-1})
Free living	
Rice-blue green algae	10-80
Rice-bacterial association	10-30
Sugarcane bacterial association	20-160
Symbiotic	
Rice-*Azolla*	20-100
Legume-Rhizobium	
Leucaena leucocephala	100-300
Glycine max	0-237
Trifolium repens	13-280
Sesbania rostrata	320-360
Non-legume-*Frankia Casuarina sp.*	40-60

Table 2: Estimation of dinitrogen fixed by different nitrogen fixing systems [31].

Parameters	Nitrogen fixation	Photosynthesis
	Differences	
Medium	In soil	Above soil
Substrate	N_2	CO_2
Delivery point	Root	Leaves
Machinery	Enzyme nitrogenase	Chlorophyll
End Product	NH_4	CH_2O
Energy	Chemical	Solar
Organism	Bacteria, Fungi, Blue green algae	Plant, Algae
Organic-C	Decomposed	Produced
	Similarities	
By product	O_2	O_2
Need for P	For nitrogenase	For ATP
Rate	Affected by nutrients	Affected by nutrients

Table 3: A comparison between nitrogen fixation and photosynthesis.

Symbiotic nitrogen fixation: Certain leguminous plants and a few other non-legumes possess nodules in their roots. These nodules contain bacteria (belong to the genus Rhizobium) and fungi living symbiotically with the plant. They have the ability to fix nitrogen and provide the plant with nitrogen compounds while they receive their source of energy (carbohydrate) from the plant.

Some of the symbiotic nitrogen fixing bacteria such as *Azolla* in symbiosis with *Anabaema azollae* can fix 2-4 kg N/ha/day. There are other benefits of *Azolla* including: (a) they can be used as weed suppressor, (b) potassium (K) scavenger from floodwater, (c) can be used in animal feed and fish feed, (d) phosphorous (P) scavenger in sewage treatment and (f) suppressor of ammonia volatilization [31]. Heterotrophs such as *Azotobacter chroococcum* and *Azotobacter vinelandii* are excellent symbiotic nitrogen fixing bacteria but they have strict requirement for neutral pH conditions and, therefore, their availability in tropical soils is very rare. However, other symbiotic bacteria such as *Beijerinckia indica, Bejierinckia fluminensis, Azospirullum sp.* and *Herbaspirullum sp.* are more tolerant to low pH and have wide range of ecological adaptation to fix nitrogen [30].

The nitrogen uptake by free living bacteria and blue green algae takes place in soils high in organic matter that provided ready source of energy. The reduction of N2 to NH_4 has large metabolic costs (respiration of organic carbon in the soil) as follows:

$$Organic-C+O_2 \rightarrow CO_2+H_2O+Energy \qquad (8)$$

The energy cost of nitrogen fixation links the biogeochemical process to the availability of organic carbon provided by photosynthesis. Asymbiotic are an important source of nitrogen for terrestrial ecosytems. The requirement of molybdenum (Mo) and iron (Fe) as structural components of nitrogenase links nitrogen fixation to availability of these elements in nautral ecosystems. Phosphorous is also required for the production of the enzyme nitrogenase.

The symbiotic association between fungi and higher plants are found in most ecosystems and is important for the nutrition of plants. Many trees harbor the ecotrophic mycorrhizal fungi which form a sheath around the active fine roots and extend their hypae into surrounding soil. Because of their large surface area, fungi obtain soil nutrient and transfer then to plants thereby contributing to weathering of soil minerals through the release of organic acids. Fungi in return, depend on plants for carbohydrate as source of carbon energy [37].

Nodulated legumes are used in cropping systems for centuries and they are used as primary source of food, fuel, fiber and fertilizer to other plants or to enrich the soil with nitrogen [31]. After removing the crop from the field, plant nodules will be a good source of nitrogen in the soil.

Microbial uptake of ammonium

The ammonium uptake across the biological membranes is generally facilitated by ammonium transporters (Amt). The ammonium transporters (Amt) are a class of membrane-integral transport proteins which are found in archaea and bacteria and even in eukaryotes [38,39]. The ammonium produced from nitrite reduction is assimilated by several microorganisms using glutamine synthetase-glutamate synthase (GS-GOGAT) pathway or glutamate dehydrogenase (GDH) pathway. The GS-GOGAT pathway requires ATP but has higher affinity towards ammonium. The GDH pathway does not require ATP and is less effective in growing cells in nitrogen limited conditions [20].

Most microorganisms use both pathways for ammonia assimilation. However, when the microorganisms which are able to assimilate nitrate are exposed to ammonium, the nitrate assimilation by the cells are drastically inhibited. Ammonium inhibition can be long or short term. Prolonged incubation of cells to ammonium can lead to repression of transporter genes and nitrate reductase (Nas) and nitrite reductase (Nir) enzymes. Shorter incubation of cells in ammonium inhibits nitrate and nitrite uptake without affecting the genes or enzymes [40].

Plant uptake of ammonium and nitrate

Most of the plant species absorb and assimilate nitrate (NO_3^-), ammonium (NH_4^+), urea and amino acids as nitrogen sources, but the specifity of these sources vary with different plants. Crawford and Glass [41] reported that the optimal ratio of nitrate : ammonium is 3:1 for tomato roots and the growth is inhibited if the ratio of ammoium is increased, but for white spruce prefer more ammonium in the soil and some artic sedges prefer more amino acids in the soil. Both nitrate (NO_3^-) and ammonium (NH_4^+) possess some common features including: (a) both ions (NO_3^- and NH_4^+) are absorbed by the roots at low external concentrations, (b) there are high affinity transport systems (HATS) in the roots for both nitrate (NO_3^-) and ammonium (NH_4^+) (c) the HATS for nitrate is double the HATS for ammonium and (d) the influx of both ions is responsive to plant N status and is subject to diurnal regulation [42].

Meyer et al. [43] and Stitt [44] reported that plants, unlike bacteria or fungi use ammonium as a nitrogen source and show better growth in the presence of nitrate. Nitrate can get accumulated in the leaves of the edible plants or in drinking water and may affect both the environment and human health. After nitrogen uptake, nitrate is either stored in the plant root system or translocated to aerial parts via the xylem. High concentrations are usually found in the vacuoles as a nutrient source and plays an important role in the maintanence of osmoticum. Nitrate reduction in plants takes place both in roots and shoots. The nitrate reduction takes place in cytoplasm and the nitrite reduction takes place in plastids/chloroplasts. The reduction of nitrate to nitrite is catalyzed by the enzyme nitrate reductase (NR) enzyme. The nitrite after nitrate reduction is translocated to the chloroplast where it is reduced to ammonium by nitrite reductase (NiR) enzyme.

Ammonium obtained after nitrite reduction and also from photorespiration or amino acid recycling is assimilated in the plastid/chloroplast by GS/GOGAT cycle. The glutamine synthetase (GS) fixes ammonium on a glutamate molecule to form glutamine. The glutamine formed reacts with 2-oxuglutarate to form two molecules of glutamate in the presence of glutamine 2-oxyglutrate amino transferase (GOGAT) enzyme [45,46]. There are two different forms of GOGAT enzymes present in plants: Fd-GOGAT and NADH-GOGAT. Fd-GOGAT uses ferredoxin and NADH-GOGAT uses NADH as electron donors, respectively. Fd-GOGAT is usually localized in the plant leaves, whereas NADH-GOGAT is localized in plastids of non-photosynthetic tissues such as roots, etiolated leaf tissues and companion cells. The ammonium formed in the plastids are converted to carbomylphosphate which is a precursor of citrulline and arginine in the presence of carbamoylphosphate synthase (CPase) as shown in Figure 4 [46,47].

Both NO_3^- and NH_4^+ are absorbed by plants in the form of amino group (- NH_2). The availability of NH_4^+ or NO_3^- depends on the environmental conditions that affect the production of NH_4^+ and the conversion of NH_4^+ to NO_3^-. However, many plants species show preferences to NO_3^- over NH_4^+, although thermodynamic analysis suggests that the metabolic energy cost of reducing NO_3 to NH_2 is significantly greater. There are several reasons for plants showing preference towards NO_3^- over NH_4^+: (a) NH_4^+ interacts with soil cation exchange where NO_3^- is slightly soluble and mobile, (b) the rate of NO_3^- delivery by diffusion is higher than that of NH_4^+, (c) plant that uses NH_4^+ have to compensate for the difference in diffusion by investing more energy in the root growth, (d) uptake of NO_3^- avoids the competition that occurs in root enzyme carriers between NH_4^+ and other positively charged ions and (e) relatively high concentration of NH_4^+ is toxic to plants.

The amino group (-NH_2) is attached to soluble organic compounds and the nitrogen absorbed by the roots is transported to the xylem in the steam (capillary system) as amides, amino acids or uried before its eventually incorporated into proteins in leaves. There are two types of amides: (a) metallic derivatives of NH_4^+ in which NH_2 group is retained and (b) organic derivative. The following are examples of the different types of amides, amino acids and uride.

(a) Metallic Amides:

$$K^+ + NH_4^+ \rightarrow K\,(NH_2)\; + \;2\,H^+ \quad (9)$$

$$Na^+ + NH_4^+ \rightarrow Na\,(NH_2)\; + \;2\,H^+ \quad (10)$$

(b) Organic Amides:

$$COOH\; + NH_4 \rightarrow COO(NH_2)\; + \;2\,H^+ \quad (11)$$

(c) Amino Acids:

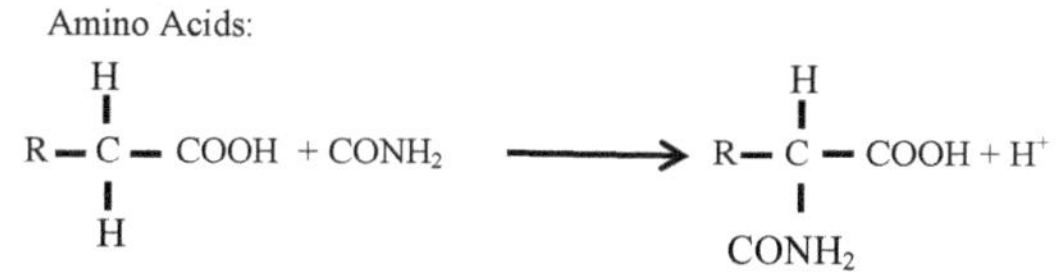

(12)

(d) Uride:

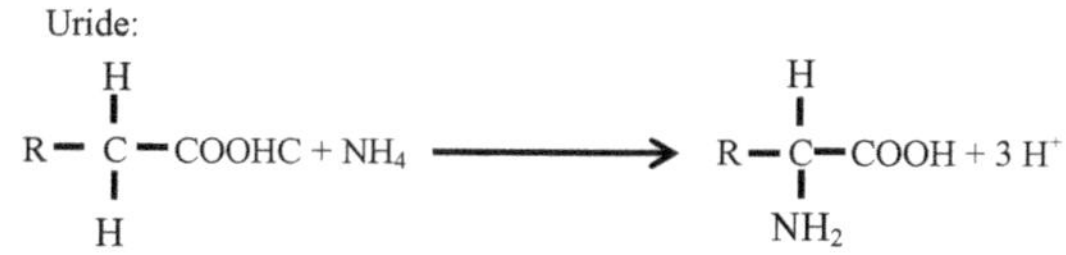

(13)

In case of NO_3 uptake by plant, it is first reduced to amino group which is then interacts with CO_2 to form amine. The amine interacts with organic acid to form aminoacid as follows.

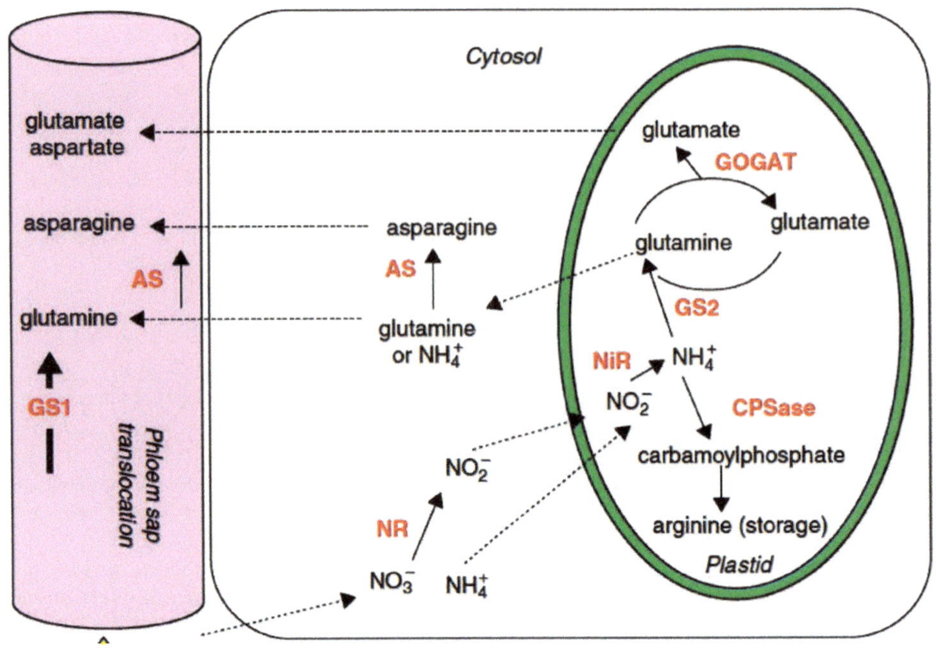

Figure 4: Uptake of nitrogen in plants [46].

(a) $NO_3 + H_2O \rightarrow NH_2 + 2O_2$ (14)

(b) $NH_2 + CO_2 \rightarrow COONH_2 + \frac{1}{2}\, O_2$ (15)

(c) $$R-\overset{\displaystyle H}{\underset{\displaystyle H}{C}}-COOH + CONH_2 \longrightarrow R-\overset{\displaystyle H}{\underset{\displaystyle CONH_2}{C}}-COOH + H^+ \quad (16)$$

Ammonia volatilizaiton

Ammonium exists in two forms: free or unionized form (NH_3) and ionized form (NH_4^+). Ionised ammonium is soluble in water while un-ionised ammonia is volatile and could be easily removed from water. The process of volatilization carries the free ammonium from the water into the atmosphere. un-ionized ammonia is volatile and could be easily removed. Ammonium volatilization occurs when ammonium ions are present in an alkaline medium and gets dissociated into gaseous ammonia which then gets released into the atmosphere [48].

$$NH_4^+ + OH^- \rightarrow NH_3 \uparrow + H_2O \quad (17)$$

The overall process of ammonia volatilization could be comprised into four major steps: (a) conversion of iodized ammonium into free ammonium by dissociation, (b) diffusion of free ammonia to the water-air interface, (c) volatilization or releasing the free ammonium to the atmosphere at the interface and (d) diffusion of free ammonium from the water-air interface into the atmosphere which is carried out by mass transfer [48-50].

Various agricultural activities including livestock production and fertilizer application are the main sources of atmospheric ammonia emisson. Ammonia emission occur from livestock buildings, open feedlots and maure storage facilities as well as manure handling and application to the land. In livestock facilities, the urea present in the urine is broken by the enzyme urease which is present in the feces to release ammonia into the atmosphere. There are several factors affecting ammonia volatilization including: (a) manure type and characteristics, (b) temperature, (c) wind speed, (d) pH and (e) method and timing of manure application [51,52].

Rochette et al. [53] reported that the impact of urea application rate on the nitrogen loss in the form of ammonia (NH_3) is variable as shown in Table 4 [54-64]. The large variability in the results reported from several studies has been due to the nonlinear response of the ammonia to variation in pH and the availability of ammonia in the soil. Ammonia volatilization occurs in a soil when the pH is high (>7.5) [48]. Sommer et al. [65] reported that pH affects the ammonia loss and is affected by climatic and soil conditons.

Manure type and characteristics such as total nitrogen (TN), ammonium nitrogen (NH_4-N) and percentage dry matter (%DM) play an important role in the ammonia volatilization during manure application. Liquid manure (slurry) used for agricultural purposes have high rate of ammonia loss when compared to soil manure. Slurries with lower solids have greater fluidity and can infiltirate soil more readily, where ammonium is protected from volatilization by adsorption onto soil colloids [52,66]. Sommer and Hutchings [66] reported that the composition of animal maure varies widely between animal species and manure types. The animal manure contains a mixture of faeces, urine, blood, split feed, split drinking water and water used in washing and it is collected as slurries. In some animal housing systems, the soild manure containing only faeces and straw and liquid manure containing urine, water and dissolvable faecal components are separated. The typical composition of different types of animal manure are shown in Table 5 [66].

Ammonia volatilization increases with increases in temperature because at higher temperature the solubiliy of ammonia (NH_3) gas is reduced in soil by increasing the proportion of total ammonia nitrogen (TAN) as NH_3 gas [67,68]. Sommer and Hutchings [66] reported that TAN is produced by the hydrolysis of urea by ureic acid. Therefore, the ammonia (NH_3) emissions from the polutry manure is influenced by

Conditions	Soil pH	N rate (kg ha⁻¹)	NH_3 loss (% appl. N)	N placement	References
Laboratory	9.0	15	56	Surface broadcast	[54]
		30	48		
		60	41		
		120	34		
		240	29		
Field	8.7	89	14	Surface broadcast	[55]
		178	16		
		255	20		
Field	8.6	75	44	Surface broadcast	[56]
		200	48		
Greenhouse	8.0	55	58	Surface broadcast	[57]
		110	58		
		275	61		
		550	67		
		1100	63		
		2200	61		
Greenhouse	8.0	55	55	Surface broadcast	[57]
		165	67		
Greenhouse	8.0	55	68	Surface banded	[57]
		165	66		
Greenhouse	7.6	25	22	Surface broadcast	[58]
		50	21		
		100	22		
Field	7.2	135	15	Surface broadcast	[59]
		225	19		
Field	6.3	40	11	Surface broadcast (moist)	[60]
		80	9		
		120	8		
Field	6.1	30	10	Surface broadcast	[61]
		100	17		
		300	33		
Laboratory	6.1	50	13	Surface broadcast	[52]
		100	14		
		150	16		
		200	17		
		300	16		
Field	5.9	70	24	Sub-surface Banded (2 cm)	[53]
		140	20		
		250	15		
Field	5.2	80	14	Surface broadcast	[54]
		160	18		

Table 4: Impact of urea application rate on ammonia losses [53].

water content and temperature. Sommer and Hutchings [65] reported that there is a linear relationship between ammonia volatilization rate and temperature in the range of 15-25°C. Agrifacts [51] reported that 50% of the total nitrogen is volatilized as ammonia at 30°C compared to 25% at 25°C. Therefore, manure applications should not be carried out at higher temperatures.

Thompson et al. [69] reported that wind speed had a positive effect on ammonia (NH_3) volatilization but less significant compared to other factors affecting the loss of nitrogen. Increasing the wind speed from 0.5 to 3.0 m/s increased the total 5 day loss by 29% and the loss of nitrogen was more pronounced in the first 24 h after application of manure. Meisinger and Jokela [52] reported a linear relantionship between ammonia volatilization and wind speeds to about 2.5 m/s.

Application of slury on the soil plays an important role in the ammonia volatilization. There are four types of application techniques (Figure 5) [67]: (a) surface broadcast, (b) surface band-spreading (via trailing hoses), (c) surface placement (via trailing shoes) and (d) shallow

Manure	Animal	Dry matter (g/Kg)	N. tot (g/Kg)	TAN (g/Kg)	Ureic acid-N (g/Kg)	P (g/Kg)	K (g/Kg)	pH
Slurry	Cattle	74.23	3.95	1.63		0.63	3.46	7.20
	Pig	34.50	9.35	3.66		0.74	3.62	6.72
	Poultry	218.00	12.00	5.93				7.23
Solid Manure	Cattle	181.50	4.85	1.33		1.45	3.85	7.80
	Pig	222.00	10.45	4.40		3.70	5.25	7.70
	Poultry	574.60	29.60	5.49	6.0	5.98	6.53	8.50
Deep Litter	Cattle	261.00	5.20	0.90		1.40	9.70	8.60
	Pig	412.00	11.20	2.80				8.90
	Poultry	570.00	27.10	6.48	7.54	9.25	15.50	9.1
Liquid Manure	Cattle	1.68	2.60	2.05		0.03	4.33	8.70

Table 5: Typical composition of animal manure [66].

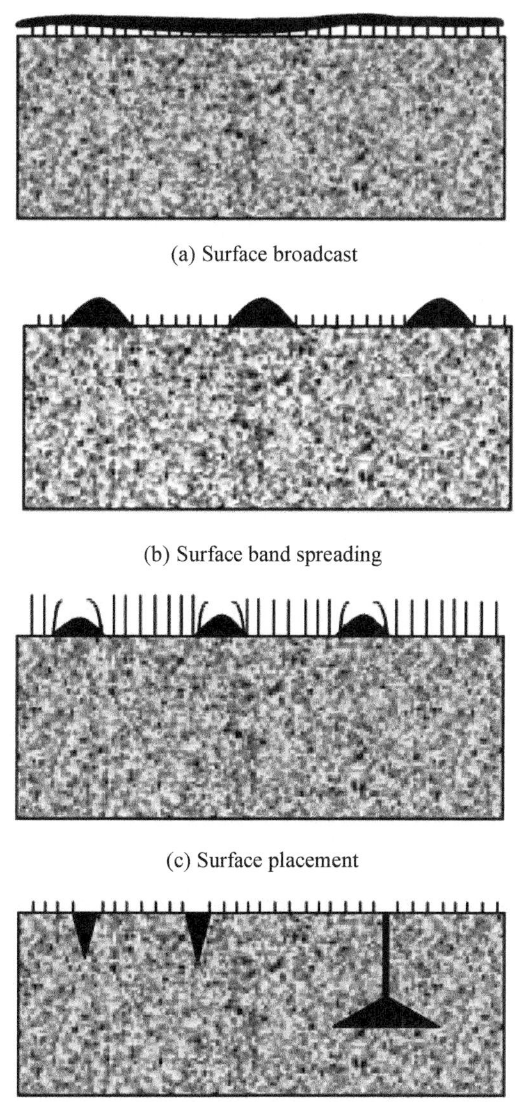

Figure 5: Schematic representaion of slurry application techniques [70].

slot injection (500 mm depth). During surface broadcaset application the slurry is applied by tanker with a single outlet and splash plate. The surface band spreading involves multiple hoses depositing slury bands 5-10 cm wide on the ground with approximately 30 cm between each bands. The surface placement technique involves placing slurry in bands approximately 3 cm wide and 20 cm apart, between and below the crop canopy. The shallow injection techniquee involves the injected beneath the soil surface either via open slot, shallow injection (to 50 mm) or deep tines (to>150 mm). Misselbrook et al. [70], Smith et al. [71], Huijsmans et al. [72] and Bussink and Oenema [73] observed direct loss of nitogen fertilizer in the form of ammonia during manure application on soil surface which could be prevented by carrying out shallow injection technique or band spreader technique. Bussink and Oenema [73] stated that there are other factors such as moisture content, rainfall, soil texture and cation exhange capacity of the soil that affect the volatilization of ammonia druing manure application. The range of measured losses of ammonia at different steps is shown in Table 6 [73].

Leaching of NO_2 and NO_3

Leaching of nitrite or nitrate refers to the removal of nitrite or nitrate from the plant root zone by the movement of water through the soil. Since nitrite (NO_2^-) and nitrate (NO_3^-) are negatively charged, they are found to move freely with the water unless soils have a significant anion exchange capacity. It was estimated that 55Tg of nitrate are leached from agricultural soils every year [74]. Leaching of nitrogen form soil reduces the bioavailability to plants and impacts the environmental quality.

Ammonium fixation

Soils are found to have the ability to bind ammonium (NH_4^+) in such a way that it would not be readily recovered by extraction with alkali or dilute acids. This form of ammonium is referred to as non-exchangeable or fixed and is held in between the lattice of clays. It is generally not affected by cations on the clay surfaces or by addition of KCl [75]. However, if ammonium gets exposed due to clay expansion, the exposed ammonium is made available for plant growth [76].

Sources of Nitrogen

Nitrogen exists naturally in our environment and is constantly being converted from organic to an inorganic form and vice versa. Production of fertilizer adds up to the natural source of nitrogen [77]. Naturally occurring and anthropogenic production of nitrogen make up the whole nitrogen cycle today. Human activities are the biggest contributor of nitrogen and have a major influence on the nitrogen cycle nowadays. The changes in the cycle can affect other natural cycle such as carbon, sulfur and oxygen cycles. Global nitrous oxide emissions are mainly caused by agricultural activities and large scale biomass burning. Agricultural emissions show a strong increase due to the application of fertilizer to agricultural soils and due to grazing of animals and spreading of animal manure. The inter annual variability of agricultural N_2O emission is caused by the annual savannah burning. Table 7 summarize the global N_2O emissions from number of resources around the world [78]. Table 8 shows the N_2O emission by source in Canada in 2010 [79]. Overall, approximately 140TgNy^{-1} was fixed by

Compartment	Ammonia loss	Factors affecting NH_3 loss
Artificial fertilizer	6-42% of urea N applied	T, SWC, CEC, pH, fertilizer type, N input, $CaCO_3$, vegetation
Application slurry	1-100% of NH_4-N applied on grassland 3-70% NH_4-N applied on arable land	T, rainfall, CEC, SWC, pH, %DM, amount applied, application technique, vegetation
Farm yard manure	45-100% of NH_4-N applied	T, CEC, SWC, pH, %DM, amount applied, application technique, vegetation
Grazing	0-18% of the N excreted per grazing or 3.1-8.5% of the N excreted per year	T, CEC, SWC, N input grassland management, urease activity, rainfall
Housing slurry	0-70% of N excreted	T, stable type, N content of urine, residence time, ventilation
Storage of slurry	0-20% of total N	Time, T, N content, storage type, aeration
Crops	<0-156 g N ha^{-1} d^{-1}	N status, crop type

*T: Temperature; CEC: Cation Exchange Capacity; SWC: Soil Water Content; N: Nitrogen

Table 6: Range of ammonia (NH_3) losses [73].

Source	1992	1995	1997	2002	2008
Manure management	317.81	320.40	304.14	313.00	336.00
Direct soil emissions	2231.80	2296.46	2370.49	2480.00	2690.00
Manure in pasture/range/paddock	1672.96	1727.17	1731.01	1810.00	1980.00
Indirect N_2O emission from agriculture	746.26	767.57	780.44	819.00	894.00
Savanna burning	897.03	629.10	582.99	503.00	907.00
Agricultural waste burning	31.81	32.45	33.86	35.30	40.90
Forest fires	424.66	287.48	305.94	285.00	185.00
Grassland fires	97.92	96.32	87.02	117.00	23.80
Peat fires and decay of drained peatland	85.28	17.26	310.67	70.70	2.22
Forest Fires-Post burn decay	275.18	290.19	285.81	292.00	276.00
Wastewater handling	268.56	285.83	297.43	320.00	342.00
Waste incineration	6.16	6.36	6.49	6.77	7.39
Other waste handling	8.49	11.74	13.83	19.10	25.70
Fossil fuel fires	0.75	0.75	0.75	0.75	0.75
Indirect N_2O from non-agricultural NOx	536.54	520.04	536.71	536.00	603.00
Indirect N_2O from non-agricultural NH_3	273.97	137.00	558.50	224.00	117.00
Total	7875.18	7426.12	8206.09	7831.62	8430.76

Table 7: Global N_2O emissions in Tg N_2O [78].

Source	Nitrous Oxides (Kilotonnes)
Transportation	680.0
Off-road vehicles	458.1
Oil and Gas industries	457.5
Fuel for electricity and heating	277.5
Other industries	174.6
Home firewood burning	10.1
Incineration and miscellaneous	2.4

Table 8: N_2O pollution in Canada in 2010 [79].

human activity in terrestrial ecosystem [9]. The sources and pathways of nitrogen that result in direct and indirect N_2O emissions from soils and water are shown in Figure 6 [80].

Atmospheric precipitation

Most of the nitrogen in the atmosphere is in the molecular form of N_2. However, there are small amounts of ammonia (NH_3) as well as various nitrogen oxides (NO_2), such as nitric oxide (NO) and nitrous oxide (N_2O) and their hydration products nitric acid (HNO_3), present in the atmosphere [81]. Most atmospheric nitrogenous compounds are attributed to industrial air pollution. Small amounts are released from the decomposition of organic matter in the soil and photochemical reactions in the atmosphere [80,82]. Most of dissolved inorganic N in rain water is ammoniac nitrogen. Cape et al. [83,84] reported that organic nitrogen makes up 24-40% of the total amount of dissolved nitrogen in rain and snow across UK (Table 9). Maximum concentration of ammonium and nitrate occur during spring while late summer rain contains more water-soluble organic nitrogen. Goldberg [85] stated contribution from rainwater to nitrogen added to soil or runoff water vary greatly depending upon time of the year and the location which have great implication on the amount of fertilizer used by farmers and the consequent air and water pollution.

Production of nitrous oxide (N_2O) is one of the greatest concerns today as it is one of the main greenhouse gases. Earlier estimation of the global budget of N_2O production, suggested that the emission of N_2O to the atmosphere by agricultural activities was relatively small [86-88]. However, recent investigation suggested that agricultural activities can actually contribute a lot more N_2O emission than was thought earlier [89]. If total agricultural emission is to be considered,

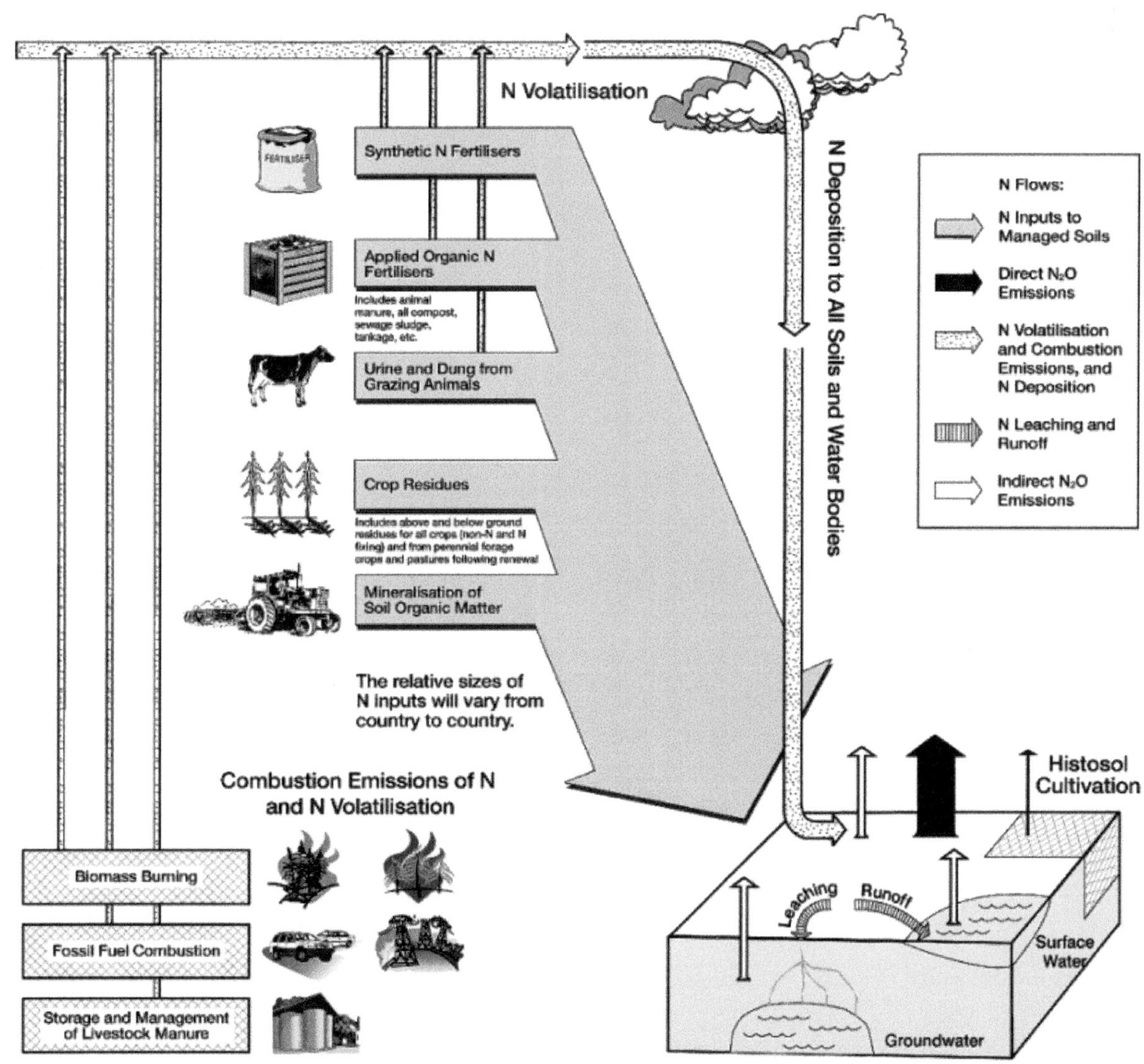

Figure 6: Sources of nitrogen that results N_2O emissions [80].

Site	Nitrate-N		Ammonium-N		Organic-N		Total
	(µM)	(%)	(µM)	(%)	(µM)	(%)	(µM)
			Cairngorm				
North-east Scotland	30	52	14	24	14	24	58
			Bush				
East Scotland	15	33	16	35	15	33	46
			Merlewood				
North-west England	17	19	27	31	35	40	87
			Moor house				
North-west England	20	24	32	38	31	37	84
			Climoor				
North Wales	22	30	33	45	29	39	73
			Norwich University				
South-East England	38	29	58	45	3	26	129
			Winfrith				
South-west England	23	31	26	35	23	31	75

Table 9: Concentration (µM) of inorganic and total nitrogen in precipitation across United Kingdom during the period of 2000-2002 [84].

N_2O emission can be produced not only from fields fertilized with synthetic nitrogen fertilizer but also includes animal production system [90]. Volatilization, runoff and leaching are some of the mechanisms through which nitrogen is being release back to the environment from agricultural activities. However, with human population exceeding seven billion people in 2012, the nitrogen emission from agricultural activities needs a further investigation [91]. As human population continues to increase rapidly, it is only a matter of time that nitrogen pollution from agricultural production systems will be our biggest concern of all.

The most abundant oxide of nitrogen is probably nitrogen tetroxide (N_2O_4), produced by internal combustion engines. More than 20 Tg of N is fixed or mobilized during fossil fuel combustion and other high temperature process which is emitted in a form of NO. Ion molecule reactions, which occur in the stratosphere and upper ionosphere, account for formation of nitrogen molecules other than N_2 [92].

$$N_2 + O_2 \xrightarrow{Heat} 2\,NO \quad (18)$$

$$NO + \tfrac{1}{2}O_2 \xrightarrow{Heat} NO_2 \quad (19)$$

The formation of ammonia in the atmosphere is mainly due to the effect of lighting. Some of this ammonia is oxidized to some oxides of nitrogen. Butcher et al. [93] reported that formation of two molecules of NO from collision of N_2 and O_2 in a direct heat process is relatively slow reaction compared to the formation of NO in the presence of lightening which results in a series of simple steps comprising of Zeldovich mechanism as shown in the following reaction:

$$N_2 + O \xrightarrow{Lightning} NO + N \quad (20)$$

$$N + O_2 \xrightarrow{Lightning} NO + O \quad (21)$$

Geological sources

A very good example of geological sources of nitrate is the sediments of nitrate of soda on the plateau of Tarapacá, a part of Atacama Desert in Chile. The nitrate deposits in Chile have probably been developed since the ancient times. Evans et al. [94], Noble [95] and Ehleringer et al. [96] stated that Atacama Desert had to go through decades of intervals of rainfall. The almost no-existence moisture in that area causes Atacama Desert to be known as the driest places on Earth. Since nitrate is soluble in water, the absence of water allow nitrate to deposits in layers forming *caliche*, type of sedimentary rock that consists layers of sodium nitrate. Holloway et al. [97] and Blatt et al. [98] reported that 75% of rocks exposed at the Earth's surface are sedimentary in nature containing 10^{21} g of fixed nitrogen globally which is a lot more than the amount of fixed nitrogen in the total biosphere of 10^{19} g. The formation of nitrogen in the bedrocks was derived from burial of organic matter in marine and freshwater sediments [99].

Other sources of nitrate are igneous rocks, metamorphic rocks and coal. These fixed forms of nitrogen in rocks may amount in some estimation to a total 50 times greater than the amount of fixed nitrogen in the atmosphere. Although rocks contain significant amount of fixed nitrogen, it was not thought as a ready source due to access problem [97,99]. Morford [99] suggested that nitrogen rich bedrocks may influence the forest ecosystem ability to store carbon and nitrogen. Bedrocks rich in nitrogen might have been feeding the forest, enhancing its growth. Nitrogen content of soils and forest foliage on nitrogen rich sedimentary rocks is 50% more than the soil and forest foliage on nitrogen poor bedrocks. Forest associated with nitrogen rich bedrocks also have 42% more carbon in above ground tree biomass and 60% more carbon in the upper 30 cm of the soil in comparison to similar sites on top of poor nitrogen bedrocks. These findings may explain why the nitrogen budget for forest is always imbalanced.

The role of nitrogen as a nonpoint sources to the contamination of surface water was reported by Holloway et al. [97] who observed increased nitrogen content in stream water in certain California watersheds caused by bedrock lithology. The Mokelumne River watershed in the central Sierra Nevada of California experienced consistently low nitrogen concentration in its stream water at its lower watershed that has diorite lithology. While for other streams that have biotite schist lithology, the concentration of nitrogen in the streamwater was highest during the rainy season and decreasing during the spring period before the end of the stream flow. Lower watershed that contained geological nitrogen was the source of 90% of nitrogen in the streamwater in comparison to upper watershed. Although upper watershed had 90% of watershed area, the nitrogen flux was only 0.12 kg N $ha^{-1}\,yr^{-1}$ while the lower watershed had nitrogen flux of 10-20 kg N $ha^{-1}\,yr^{-1}$ during the 1995-96 water year. Significance difference between the two parts of watershed suggested that geological nitrogen would be the only source.

Agricultural land

Agricultural sources of nitrogen result primarily from organic and inorganic materials added to soils as a source of crop nutrition. Since fertilizers are now one of the lowest-cost "inputs" in crop production, the use of synthetic nitrogen is growing rapidly in areas where natural

Nitrogen	2008		2009		2010		2011		2012	
	Fertilizer	Nitrogen	Fertilizer	Nitrogen	Fertilizer	Nitrogen	Fertilizer	Nitrogen	Fertilizer	Nitrogen
Total supply	206431	131106	212225	136252	219930	140732	230334	147748	240711	154199
Total demand	197004	127820	201482	130409	205947	133059	211230	136198	216019	139140
Surplus	9427	3286	10473	5843	13983	7673	19104	44550	24692	15059

Table 10: Supply and demand of fertilizer and nitrogen (thousand tonnes) in the world for the duration of 2008-2012 [103].

soil phosphorus, potassium and calcium are inadequate and limit growth [100]. Kjaergaard [101] reported that synthetic nitrogen was added to the soil in Europe when the area of fallow land was decreasing, limiting the available nitrogen for the crops.

Increasing world population causes more than ever a demand for food and boosts up the production of synthetic nitrogen for agricultural land. Galloway [102] stated that for food production per capita basis to increase, additional nitrogen must be added to agro-ecosystem to keep up with the demand. Supply and demand of fertilizer in the world has steadily increased since 2007 and the trend is predicted to be the same for the next fifty years [103]. The supply and demand balance for the nitrogen has also increasing (Table 10) [103]. The total world nitrogen fertilizer supply increased form 131.106 Tg/year in 2007 to 154.199 Tg/year in 2012.

Table 11 shows the top ten countries with highest fertilizers application per a hectare of land [104]. China, one of the largest agricultural countries in the world, increased the use of chemical fertilizer from 1.74×10^7 Mg N in 1990 to 2.2×10^7 Mg N in 1995 and used 54 million tonnes of fertilizers in 2011 [105]. The fertilizers used in China alone accounted for one fourth of the world total fertilizers [103,106]. The rapid increase in fertilizer use in China is polluting ground water and surface water and causing many environmental and health problems. Besides synthetic fertilizer, China still uses animal manure and human faeces as organic fertilizers.

In the 17th century, the animal manure collected from animal in grazing season and the biological nitrogen fixation were enough to provide about 20 Kg N ha^{-1} for grain production of 1999 kg ha^{-1} in Western Europe [107]. Currently, synthetic fertilizers have been used to provide much higher crop yield. Artificial fertilizers are widely used to provide three types of primary plant nutrients: nitrogen (N), phosphate (P_2O_5) and potash (K_2O). Table 12 shows the use of fertilizer in USA from 2007 to 2010 for agricultural activities [108].

Different forms of added nitrogen as fertilizers could be lost from the soil to the surface water and underground water causing pollution problems. The nitrogen salts (NO_2 and NO_3) carried by the runoff are in a direct relation to the amount of water applied and land use practice. Drainage water contains nitrogen concentrations of 1-60 mg/L, mostly in the form of nitrate [109]. Sediment suspended in the flowing water may carry relatively high amounts of ammonium nitrogen as well as particulate organic nitrogen [110-112].

The lateral and vertical movements of nitrogen in the soil are quite different in their effect as a source of pollution. Iqbal [113] indicated that the vertical movement of nitrogen is more significant than the lateral movement as a source of pollution. Nitrate dissolves easily in water and moves through soil pores, contaminating ground water as shown in Table 13 [113]. Nitrate horizontal potential movement in soils is negligible and can be negative in shallow subsoil. Platzer [114] stated that the vertical flow has a very high nitrogen capacity while the horizontal flow associates effectively with denitrification.

Country	Amount of fertilizers used (Kg/ha)
Ireland	594.5
Netherlands	450.2
Egypt	385.8
Costa Rica	385.0
Slovenia	369.4
Japan	301.0
United Kingdom	285.8
Vietnam	285.3
Israel	256.0
China	255.6

The fertilizer use was calculated by dividing fertilizer consumption in kilograms by hectares of arable and permanent cropland. The fertilizer consumption includes nitrogenous, phosphate and potash fertilizers.

Table 11: List of countries with the highest consumption of fertilizers in the world [104].

Year	1000 tonnes			
	Nitrogen	Phosphate (P_2O_5)	Potash (K_2O)	Total
2007	13194	4572	5133	22899
2008	12561	4247	4660	21468
2009	11461	3138	3090	17689
2010	12285	4099	4458	20843

Table 12: Fertilizer use in USA for the years of 2007-2010 [108].

Subsurface drainage of gravitational water from the soil through tiles is commonly used in the agriculture to improve crop production in poorly drained soils. The drainage water from the crops contains significant amounts of nitrate-N (NO_3-N) [115]. Bolton [116] conducted a tile drainage experiment at Woodslee, Ontario, Canada on a Brookston clay soil and estimated the nutrient losses through tile drains under three different cropping systems mainly in the nitrate form with and without addition of fertilizers. They found that the greatest losses of nitrogen occurred with corn grown continuously or in rotation. They reported that the losses were increased by fertilizer application and were associated with the total and average effluent flows as shown in Table 14 [116].

Randall et al. [117] studied the nitrate losses through subsurface tile drainage in four cropping systems including: continuous corn, a corn-soybean rotation, alfalfa and conservation reserve program. The conservation reserve program (CRP) was carried out in United States of America as part of converting highly erodible and environmentally sensitive agricultural land to permanent grassland cover to reduce soil erosion and produce row crops [118]. The results from the study indicated that average NO_3-N concentrations from continuous corn, corn-soy bean rotation, alfalfa and conservation reserve program

Fertilizer application kg ha	Nitrate vertical leaching (mg)								Nitrate lateral leaching	
	30 cm				60 cm				Amount (mg/L)	Movement (mg/m)
	I	F	S	T	I	F	S	T		
0	50.13	134.74	43.95	101.13	57.12	75.39	43.95	123.63	5.79	1.22×10^{-3}
90	102.62	58.87	-	128.24	72.60	119.54	47.84	143.24	8.97	3.24×10^{-3}
180	142.23	102.49	-	185.17	112.47	61.49	85.27	147.31	5.06	-1.50×10^{-3}
270	129.00	125.73	92.38	209.35	63.57	89.17	77.88	212.39	13.43	7.58×10^{-3}
360	230.28	83.11	88.43	196.71	146.37	43.88	88.43	214.89	4.92	-1.06×10^{-3}

*I: Irrigation; F: First rain; S: Second rain; T: Third rain

Table 13: Vertical and lateral nitrate leaching in a wheat farm [113].

Crop	N Losses (Kg/ha/y)		Effluent from drains (ppm)	
	No Fertilizer	Fertilizer	No Fertilizer	Fertilizer
Rotation				
Corn	5.6	15.1	8.5	14
Oats and alfalfa	4.3	5.7	6.4	8.5
Alfalfa (1st year).	4.8	3.9	6.3	5.8
Alfalfa (2nd year)	4.7	8.6	9.3	10.1
Continuous				
Corn	6.6	14	4.4	8.9
Bluegrass sod	0.3	0.7	3.5	1.1
Mean	4.4	8.1	6.4	8.1

Table 14: Nitrogen annual losses through tile drains and concentration in effluent of tile drains on Brookston clay soil [116].

(CRP) were 32, 24, 3 and 2 mg/L, respectively. Nitrate losses from continuous corn and corn-soybean were 37 times and 35 times more than alfalfa and CRP systems, respectively.

Kladivko et al. [119] summarises the results from the study conducted over 15 years on nitrate leaching to subsurface drains. The results indicated that the mean nitrate concentration decreased from 28 mg/L in the period of 1986-1988 to 8 mg/L in the period of 1997-1999. The results also showed that the annual nitrate leaching decreased from 38 kg/ha in the period of 1986-1988 to 15 kg/ha in the period of 1997-1999, respectively. The reduction in the concentration of nitrate loss was due to reduction in the fertilizer N rates and addition of a winter cover crop as a trap crop after corn-soybean rotation. The author concluded that nitrate concentrations and nitrate losses in the tile drains vary with soil organic matter, yearly weather conditions, fertilizer N rates and timing, drain spacing, cover crop growth, cash crop yield and water table control practices.

Randall and Goss [120] reported that some non-controllable factors such as precipitation and mineralization of soil have great impact on the drainage volume and nitrate loads which put limits on the concentrations and loads that can be achieved by row-crop agriculture.

Livestock and poultry operations

In 2009, Luxembourg recorded the highest meat consumption per capita (194.2 kg/year) followed by U.S (190.6 kg/year). Canada was in the 16th place with meat consumption per capita of 152 kg/year. The top 20 meat consumption countries are shown in Table 15 [121]. Currently, chicken is the highest available stocks in the world, followed by cattle and pigs as shown in Table 16 [122]. Today, China leads the total meat consumption in the worl by consuming 71 million tonnes of meat per year. Half of the pork in the world is in China. China's pork consumption rose to 52 million tonnes in 2012 [123]. Canada has 12.5 million cattle on their farms in 2012. The beef cows were 4.2 million, the beef replacement heifers were 554,300, the dairy cows were 1.4 million and the cattle and calf slaughter totalled 3.5 million [124]. The Canadian cattle inventories from 2006-2012 are shown in Table 17.

Animals are not breed solely for the meat but also for milk, eggs and fibers. The high demand for meat based foods changed the way food is produced today [125,126].

The system known as concentrated animal feeding operation (CAFO) allows for growing thousands of animals on strict diets. Phang [127] reported that animal feeding operation is an agricultural operation where animals are kept and raised in confined areas. Feed is bought to the animals rather than the animals grazing for their feed. A farm reaches the status of CAFO when it houses at least 300 cattle, 750 pigs or 25,000 chickens. In the CAFO system, animals are fed with high nutrients diet including nitrogen rich foods.

Olson et al. [128] reported that increasing meat consumption around the world justifies the needs of having CAFO as this system provide the meat needed in large amount with lowest cost. There are several advantages of CAFOs including: (a) low-cost source of meat, milk and eggs, (b) efficient feeding and housing of animals, (c) increased facility size and (d) animal specialization. There are also major environmental risks involved in this method of housing animals. The most health related issue is the production of large amounts of manure which contains plant nutrients such as nitrogen and phosphorous, pathogens such as *E. coli*, growth hormones, anitbiotics, chemical additives used to clean manure, animal blood, leachate from corn feed and copper sulfate used in the footbath for cows [13,125,129,130].

Rank	Country	Per Captia (kg/yr)
1	Luxembourg	194.2
2	United States of America	190.6
3	Austria	185.1
4	Australia	183.4
5	New Zealand	179.0
6	Oceania	173.4
7	Denmark	170.6
8	Spain	163.0
9	Argentina	161.8
10	Bahamas	160.3
11	Italy	158.4
12	Portugal	157.0
13	Bermuda	156.8
14	Germany	156.3
15	French Polynesia	155.4
16	Canada	152.0
17	Mongolia	151.1
18	Slovenia	149.6
19	Ireland	147.8
20	France	146.5

Table 15: Meat consumption per captia of top 20 countries in 2009 [121].

Animal	Number in × 10^4
Asses	4,323.0700
Beehives	7,841.1600
Buffaloes	19,539.7500
Camelids, other	839.1900
Camels	2,663.5400
Cattle	1,42,638.9000
Chickens	2,070.8000
Ducks	132.38500
Geese and guinea fowls	38.1200
Goats	92,414.5900
Horses	5,847.2100
Mules	1,045.7100
Pigeons, other birds	3.2500
Pigs	96,716.4600
Rabbits and hares	89.5000
Rodents, other	1.8400
Sheep	1,09,356.6700
Turkeys	46.7600

Table 16: World stocks of live animals in 2011 [122].

Cattle	2006	2007	2008	2009	2010	2011	2012
Bulls	264.0	244.6	246.8	243.9	232.1	222.2	221.2
Beef cows	5,247.2	5,020.1	4,981.9	4,649.5	4,391.0	4,273.0	4,228.4
Dairy cows	1,019.1	994.8	984.3	978.5	981.0	983.1	985.3
Heifers, dairy	495.1	480.1	471.1	450.6	450.7	443.1	444.1
Heifers, beef	628.3	587.1	595.0	537.0	516.4	531.6	554.3
Heifers, slaughter	986.8	963.5	982.9	834.5	899.8	844.0	839.5
Steers	1,146.8	1,145.2	1,101.6	1,067.6	1,141.7	1,081.7	1,098.1
Calves	4,867.7	4,719.6	4,506.4	4,433.4	4,292.3	4,078.3	4,144.1
Total	14,655.0	14,155.0	13,870.0	13,195.0	12,905.0	12,457.0	12,515.0

Table 17: Cattle inventories of Canada from 2006-2012 [124].

Burkholder et al. [131] stated that growing the animal in CAFO systems is unnatural and unhealthy. The animals are confined in a very limited space where they cannot even turn around. They are fed with enormous amount of corn and soy beans to increase their weight significantly in the shortest time. High density, limited space and continuous supply of food will produce large volume of manure. Sherman [132] stated that manure confined feeding operations is heavier because the manure is a mixture of faeces and urine with additional water used to pump it to the storage facility. The heavy manure is hard to be transported and disposing it has caused a major form of concern because of its impact to the environment).

Table 18 shows the number of farm animals slaughtered for human consumption in the world [133]. Table 19 shows the amount of manure produced by different animals [134-136]. USDA [137] estimated the manure production to be more than 335 million tonnes/year on the farms in United States, which was 100 times more than the manure produced by humans. It is estimated that the amount of manure produced by 2,500 cows are equal to the waste produced by 411,000 humans (1 cow=164.4 persons).

Manure from concentrated animal feeding operations is usually collected in a lagoon near the feedlots. The manure is disposed either by pumping the liquefied manure onto nearby "sprayfields" (crops, pasture or forage fields fertilized and irrigated by nutrient rich liquid) or being transported to another location with a high cost of transportation [138,139]. These practises increase the amount of nitrogen released freely into the environment which results in air, soil and water pollution. However, in recent years, the economic advantages of commercial fertilizers have made the use of animal wastes unprofitable to many farmers. The cost of handling and treating manure reduced the popularity of using it as organic fertilizer. The large automated feeding operations further magnified the problem by concentrating thousands of animals in localized areas [125].

Air pollution is mainly caused by various emissions from CAFOs such as ammonia, hydrogen sulfide, volatile organic acids, methane and particulate matter. The type of air pollutants surrounding CAFOs are shown in Table 20 [125]. Varel [140] reported on the nitrogen loss from feedlots via volatilization of nitrogen gases, primarily ammonia, into the atmosphere. Ammonia volatilization has contributed to soil and water pollution, where 90% of the nitrogen in urine exerted in cattle feed yards may escape into the air as ammonia. The authors found that a significant amount of ammonia volatilized from cattle feedlots is returned from the air to water bodies in the vicinity of the feedlot, and so the magnitude of nitrogen enrichment of lakes via this pathway can be significant.

Ndegwa et al. [141] reported that ammonia volatilization is one of the major pathways for nitrogen loss from agricultural feeding operations. This process of losing nitrogen not only affects agricultural production but can adversely affect the environment. Ammonia deposited in the atmosphere in lower amounts are beneficial to plants as a nutrient source for growth but when excess N is deposited, it gets oxidized and reduced and causes various environmental issues including: (a) exposure to high concentrations of fine particulate aerosols (PM2.5) causes respiratory diseases, (b) contamination of drinking water with nitrate, (c) eutrophication of surface water bodies causing algal blooms and decreased water quality; (d) higher concentrations of N causes changes in vegetation and ecosystem, (e) climatic changes associated with increases in nitrous oxide (N_2O), (f) saturation of forest soils with nitrogen and (g) acidification of soil through nitrification and leaching.

Water pollution results mainly due to the insufficient land used to contain the volume of manure. Runoff and leaching from feedlots have always been a threat to the nearby water bodies but a lagoon failure can send tens of millions of liters of untreated manure directly into water bodies. Minimal leaching and accumulation of nitrogen below the feedlot pens can possibly occur from denitrification of the poorly aerated layers [131,142]. In a study of soil and groundwater quality under a cattle feedlot in Southern Alberta, Olson et al. [128] reported that denitrification may have prevented an increase in NO_3-N concentration in the groundwater. However, concentration of NH_4-N increased in the shallow groundwater beneath the pens although it did not accumulate in the sub soil layer during the four years of study.

Westerman et al. [138] evaluated swine lagoon seepage in sandy soils. They investigated two swine manure anaerobic lagoons on coastal sandy soil. Both lagoons had significant seepage after 3.5-5 years. The authors reported that manure lagoons constructed in sandy soils with high water tables have resulted in some groundwater pollution and about 3-6 mg nitrate NO_3/L were found in surface runoffs from spray fields. They concluded that too much variability of NH_3-N and NO_3-N with time which made it difficult to determine the factors affecting the pollutant transport.

Stone et al. [139] conducted a 5 year research study on a watershed stream with intensive agricultural practices of crop, swine, poultry and cattle production. The results from the study indicated that most of the streams in the watershed had acceptable water quality. The nitrogen mass balance calculations and comparisons of the N loadings in the streams showed that 26% of the excess nitrogen applied is lost to stream water. The streams adjacent to swine effluent spray fields had 6-8 mg total inorganic N/L and 0.7-1.3 mg P/L.

Mahmoodabadi et al. [143] studied the effect of sheep manure leaching on soybean root characteristics and soil salinity. The results from the study indicated that the sheep manure significantly increased N, Fe, Cu, Zn and Mn concentrations. There was no significant increase in the pH of the soil (from 7.8 to 7.9). The leaching manure helped in

Animals	Heads
Buffalo	57,681,750.00
Cattle	609,729,971.76
Ass	2,461,800.00
Camel	2,031,008.00
Goat	879,124,344.00
Horse	4,504,308.00
Mule	542,600.00
Camelids	655,712.00
Pig	1,382,927,239.95
Sheep	1,046,814,221.65

Table 18: Total number of farmed animals slaughtered for human consumption in 2011 [133].

Animal	Average body weight (kg)	Urinary N (g/d)	Fecal N (g/d)	Total N (g/d)
Cattle	400	90.0	70.0	160.0
Sheep	40	7.0	3.0	10.0
Goats	15	5.0	3.0	8.0
Horses	400	49.0	27.0	76.0
Pigs	80	15.0	3.0	18.0
Chicken	4	7.2	3.4	10.6

Table 19: Nitrogen content in animal excretion [134-136].

increasing the number and dry weight of root nodules. Applying sheep manure increased the net nitrogen fixation in root nodules to 1.81 mg/Kg compared to 0.54 mg/Kg in the control pot. The sheep manure also reduced the salinity in the soil by 2%.

Dikinya et al. [144] studied the effect of chicken manure on three different soils including: luvic calcisol, ferralic arenosol and vertic luvisol that were classified as sandy clay loam, loamy sand and clay, respectively. The experiment was conducted for 63 days and the results indicated that there was substantial increase in pH. The highest pH recorded was 7.75 for ferralic arenosol soil and the lowest pH recorded was 7.01 for vertic luvisol soil. For most of the time the pH for the soil treated with chicken manure remained neutral or slightly alkaline. Lopez-Mosquera et al. [145] also reported that soil pH treated with chicken manure remained in the range of 7-7.9.

Whalen et al. [146] studied the effect of cattle manure on the pH of two acid soils obtained from Beaverlodge and Fort Vermillion in the Peace River Region of Alberta, Canada. The results obtained from the 8 week study indicated that the pH of Beaverlodge and Fort Vermillion soils increased from 4.8 to 6.0 and 5.5 to 6.3, respectively. The availability of minerals such as N, P, K, Ca and Mg increased immediately after manure application and were 3 to 4 times higher than untreated soils.

Plaza et al. [147] studied the effect of annual additions of pig slurry on soil pH at the rate of 30, 60, 90, 120 and 150 $m^{3h} a^{-1} y^{-1}$ over a 4 year period under semiarid conditions. The control soil had a pH of 6.0. After slurry application the pH of soil increased significantly and the highest pH of 7.6 was observed. The soil had greater P and K contents and slightly higher total N content in it. However a significant decrease in the organic C was observed in soil treated with high slurry rates. The effect of animal manure of the pH of the soil from several other studies is shown in Table 21 [148-157].

Urban waste

Domestic waste water effluents without specific treatment for nitrogen removal have nitrogen concentrations in the range of 20-85 mg N/L [158,159]. A typical composition of domestic wastewater is shown in Table 22 [160]. Ammonium is the predominant from of nitrogen in effluents from primary and high rate treatment plants. Storm water from residential areas can supply some nitrogen to rivers. Analysis of Wei River in China showed that it contains 7.31 mg NH_4/L [161]. Zhang et al. [162] noted down ammonium concentration of 10-28 mg/L when examined the Yellow River in China.

Several substances containing nitrogen are commonly found in industrial wastes. Ammonia is a waste material from gas and coke manufacturing and other chemical manufacturing processes. Cyanide (CN^-) is evolved during gas manufacture, plating, case hardening and metal cleaning. Nitrogen compounds also originate from explosive factories and other chemical works. There are some cases recorded of severe nitrogen pollution from industrial sources. Haber Bosch process does not only produce ammonia for fertilizer, it also provides ammonia as a raw material to create multiple products. Production of nylon, resins, plastics and melamine require ammonia [163]. Approximately 23 Tg N produced from Haber Bosch process was used for chemical production [164]. Unfortunately, little is known about the fate of this nitrogen in the environment.

Severe ammonia pollution in the Tolka River in Ireland is due to poor discharge qualities of Clonee, Co [165]. The pollution caused chemical burn to the vegetation, absence of aquatic flora and fauna, shortness of breath, skin and eye irritation and live fish and invertebrates were not observed in that area. Thousands of dead fishes were also observed. Nitrogen inputs to rivers, lakes and ocean originate either from point or non-point sources (Table 23) [166]. The point sources of nitrogen pollution include effluent pipes from municipal sewage treatment plant and factories. The non-point sources include the runoff from urban runoff from areas having a population of less than 100,000. These point sources are continuous and can be easily identified, monitored and controled. The non-point sources arise from suite of activities across large areas and are difficult to control [166,167].

Rouse et al. [168] reported that industrial effluents and wastewater treatment plant discharges from urban areas are a substantial source of nitrogen to aquatic ecosystems. Initially, these discharges contribute

CAFO emissions	Source	Traits	Health risks
Ammonia	Formed when microbes decompose undigested organic nitrogen compounds in manure	Colorless, sharp pungent odor	Respiratory irritant, chemical burns to the respiratory tract, skin and eyes, severe cough, chronic lung disease
Hydrogen Sulfide	Anaerobic bacterial decomposition of proteins and other sulfur containing organic matter	Odor of rotten eggs	Inflammation of the moist membranes of eye and respiratory tract, olfactory neuron loss, death
Methane	Microbial degradation of organic matter under anaerobic conditions	Colorless, odorless, highly flammable	No health risks. Is a greenhouse gas and contributes to climate change
Particulate matter	Feed, bedding materials, dry manure, unpaved soil surfaces, animal dander, poultry feathers	Comprised of fecal matter, feed materials, pollen, bacteria, fungi, skin cells, silicates	Chronic bronchitis, chronic respiratory symptoms, declines in lung function, organic dust toxic syndrome

Table 20: Air Pollutants from CAFO emissions [125].

Type of manure	Characteristics (gKg^{-1} DM)							Time	pH change	Reference
	C	N	P	K	Ca	Na	Mg			
Poultry(L)	250.0	39.0	205.0	26.9	110.7	5.4	11.0	6 weeks	4.0→6.5	[148]
Poultry(B)	343.0	25.0	188.0	19.0	34.8	3.5	17.0	6 weeks	4.0→5.0	[148]
Poultry	211.5	13.7	13.0	5.2	69.5	2.0	19.5	3 weeks	4.7→5.8	[149]
Poultry	239.0	63.0	21.0	20.0	77.0	2.6	5.7	6 weeks	4.2→5.4	[150]
Poultry	252.0	39.0	20.5	26.9	110.7	--	11	3 days	4.1→6.3	[151]
Poultry	131.0	32.8	43.0	29.3	27.0	--	5.7	30 days	4.3→4.6	[152]
Poultry	296.3	3.6	--	--	12.4	--	8.9	3 weeks	4.6→6.4	[153]
Poultry	252.0	38.5	20.5	26.9	110.7	5.4	11.0	7 weeks	4.1→6.5	[154]
Poultry	252.0	38.5	20.5	26.9	110.7	5.4	11.0	25 weeks	4.1→5.0	[154]
Cattle	114.0	10.0	1.8	0.60	0.3	--	0.1	3 days	4.1→4.6	[151]
Cattle	249.3	22.8	7.0	22.7	--	--	--	25 weeks	4.8→6.0	[155]
Cattle	313.0	4.9	--	--	11.2	--	19.9	3 weeks	4.6→5.9	[153]
Cattle	194.0	16.3	3.8	4.8	2.4	0.1	0.8	7 weeks	4.1→4.6	[154]
Cattle	194.0	16.3	3.8	4.8	2.4	0.1	0.8	25 weeks	stable	[154]
Pig	340.0	25.4	34.5	12.4	50.9	4.0	11.5	7 weeks	4.1→5.0	[154]
Pig	340.0	25.4	34.5	12.4	50.9	4.0	11.5	25 weeks	4.1→4.2	[154]
Pig (RL)	--	28.8	0.66	1.1	--	--	--	28 days	5.8→5.3	[156]
Pig (RL)	--	28.8	0.66	1.1	--	--	--	120 days	5.8→5.2	[156]
Pig (ADL)	--	30	0.78	1.1	--	--	--	28 days	5.8→5.3	[156]
Pig (ADL)	--	30	0.78	1.1	--	--	--	120 days	5.8→5.2	[156]
Pig	271.3	5.2	--	--	13.7	--	13.0	3 weeks	4.6→6.4	[153]
Pig	--	24.2	39.0	16.5	38.6	--	9.9	8 weeks	4.0→4.8	[157]

L: Layer; B: Broiler; RL: Raw Liquid; ADL: Anaerobically Digested Liquid

Table 21: Effect of animal manure application on soil pH.

Contaminants	Concentration (mg/L)		
	Low strength	Medium strength	High strength
Solids, total (TS)	390	720	1230
Dissolved, total (TDS)	270	500	860
Fixed	160	300	520
Volatile	110	200	340
Suspended solids, total (TSS)	120	210	400
Fixed	25	50	85
Volatile	95	160	315
Settleable solids	5	10	20
Biochemical oxygen demand			
5-d, 20°C (BOD_5, 20°C)	110	190	350
Total organic carbon (TOC)	80	140	260
Chemical oxygen demand (COD)	250	430	800
Nitrogen (total as N)	20	40	70
Organic	8	15	25
Free ammonia	12	25	45
Nitrites	0	0	0
Nitrates	0	0	0
Phosphorous (total as P)	4	7	12
Organic	1	2	4
Inorganic	3	5	10
Chlorides	30	50	90
Sulfate	20	30	50
Oil and grease	50	90	100
Volatile organic compounds (VOCs)	<100	100-400	>400
Total coliform (no./100 mL)	10^6-10^8	10^7-10^9	10^7-10^{10}
Fecal coliform (no./100 mL)	10^3-10^5	10^4-10^6	10^5-10^8
Cryptosporidum oocysts (no./100 mL)	10^{-1}-10^0	10^{-1}-10^1	10^{-1}-10^2
Giardia lambia cysts(no./100 mL)	10^{-1}-10^1	10^{-1}-10^2	10^{-1}-10^3

Table 22: Typical composition of untreated domestic wastewater [160].

Point Sources	Nonpoint Sources
Wastewater effluent, both municipal and industrial	Runoff from agriculture (including return flow from irrigated agriculture)
Runoff and leachate from waste disposal site	Runoff from pasture and range
Runoff and infiltration from animal feed lots	Urban runoff from unsewered areas and sewered aread with a population of less than 100,000
Runoff from mines, oil fields and unsewered industrial sites	Septic leachate and runoff from failed specifications
Strom sewer outfalls from cities with a population of greater than 100,000	Runoff from construction sites smaller than two hectares and abandoned mines
Runoff from construction sites larger than two hectares	Atmospheric deposition over a water surface
Overflows of combined storm and sanitary sewers	Activities on land that generate contaminants, such as logging, wetland conservation, construction and development of land or waterways

Table 23: Sources of point and nonpoint pollution [166].

only a small percentage of total nitrogen released to the environment but long term discharges into the water systems cause detrimental effect on stream ecosystems.

Another source of nitrogen contaminations in the urban areas is precipitation. Rain, snow and fog contain various amounts of nitrogen. Motor vehicles and industrial exhausts contribute nitrogen oxides (NO and NO_2) to the atmosphere which are deposited in the water systems through precipitation. Precipitation causes problem in the watersheds which do not have an extensive ground cover of natural vegetation such as in urban areas [169].

Effect of Nitrogen Pollution

Lake eutrophication

Eutrophication was considered a natural biological aging process of aquatic ecosystems where nutrients increase produce more plants leading to the formation of pond and finally into a marsh [170]. However, the term eutrophication is currently used to describe a rapid nutrient enrichment process that take place in water bodies as a result of human activities that add up the nutrients. Present knowledge indicates that the fertilizing elements most responsible for Lake Eutrophication are nitrogen and phosphorus. Iron, silica and certain trace element are also important [171].

Although the increase in yield of a crop after fertilization is desirable in terrestrial situations, the effects of eutrophication of waters are undesirable. Generally, the aesthetic value of a lake is lowered through excessive growth of aquatic weeds and algae and production of floating algal scums which are a nuisance to those who used the water for recreational purposes. Other effects include undesirable odors and tastes, depletion of dissolved oxygen, destruction of aquatic life and impairment of water treatment operations such as clogging of filters by algae [172,173].

The stage of lake eutrophication is not controlled solely by the quantities of nutrients present or entering the receiving body of water, but also the interrelationship of climatic, physical, biological and chemical factors which affect lake metabolism are very important and significant and must be considered. However, nitrogen is the most important nutrient that limits primary production of photosynthetic organisms in temperate and coastal marine waters [174,175]. At the estuarine area, nitrogen often becomes the first nutrient to limit primary production. Excess inorganic nitrogen that flows to the estuarine will support a larger bloom in that area [176-177].

Different forms of nitrogen compound (organic or inorganic) reach the streams and subsequently the lakes through: (a) surface runoff, (b) groundwater interflow and (c) domestic and industrial wastes disposed of into the river. In the last four decades, nitrogen flux entering Mississippi River in USA had increased four folds and more than ten folds into the North Sea [77,150,178].

Besides agricultural activities, aquaculture industry (growing resources) also contributes nutrients to streams. Fish farms in developing countries are often located in a shallow bay area, allowing nutrients to accumulate and stimulate algal growth [179]. Romdhane et al. [180] described aquaculture activities in lagoons as a hazardous business as lagoons may function as a trap for toxins or other exudates from algal.

Other source of nutrients includes nitrous oxides in wet and dry deposition. About 50-100% primary production of harmful algal blooms in Yellow Sea, China is generated by typical rain event containing nitrogen, phosphorus and silica [181].

The level of nitrogen which will produce algae blooming is quite varied and depends on many factors, the most important of which are the available phosphorus and organic carbon. Nutrient ratios play an important role in influencing the algal growth. Hodgkiss et al. [182] reported algal bloom in Tolo Harbour in Hong Kong increases whenever the ratio N: P decreases. The authors recorded that the ration of N: P between 1982 and 1989 in Tolo Harbour and found it to decreases from 20:1 to 11:1 as the number of dinoflagellate red tides increased. Hodgkiss [183] demonstrated that whenever N: P ratio fell below approximately 10:1 in Tolo Harbour, the dinoflagellate cell number increased. Algal bloom of dinoflagellate red tides result in prominent colouration of the water bodies. Microflagella dinoflagella blooms when nitrogen concentration is low while some other species such as marine diatoms exploits nitrate-rich conditions through physiological adaptations [184-187].

Smith et al. [188] reported that in Northern Ireland, soil P reserves have accumulated at a rate of 1000 kg P $km^{-2}yr^{-1}$ over the past 50 years and these increases have resulted in increases in the losses of inorganic P runoff at a rate of 2 mg $m^{-3}yr^{-1}$. The average of characteristics of water bodies with various levels of nitrogen and phosphorous loading is shown in Table 24 [188]. McGarrigle [189] reported that annual dissolved inorganic P concentration should be lessern tha 47 mg m^{-3} to prevent the growth of algae and to preserve water quality suitable for salmonoid fish, in Irish rivers. Miltner et al. [190] reported that the total inorganic nitrogen (TIN) and dissolved inorganic phosphorus

Water bodies	Trophic state	Total nitrogen (mg m^{-3})	Total phosphorus (mg m^{-3})	Chl a (mg m^{-3})
Lakes	Oligotrophic	<350	<10	<3.5
	Mesotrophic	35-650	10-30	3.5-9
	Eutrophic	650-1200	30-100	9-25
	Hypertropic	>1200	>100	>25
Streams	Oligotrophic	<700	<25	<10
	Mesotrophic	70-1500	25-75	10-30
	Eutrophic	>1500	>75	>30
Marine	Oligotrophic	<260	<10	<1
	Mesotrophic	260-350	10-30	1-3
	Eutrophic	350-400	30-40	3-5
	Hypertropic	>400	>40	>5

Table 24: Average characteristics of lakes, streams and marine waters [188].

(SRP) concentrations exceeded 610 and 60 mg m^{-3}, respectively causing deleterious effects to marine organisms in Ohio streams, USA.

Xiangcan et al. [172] studied the eutrophication in various freshwater lakes in China and reported that a total nitrogen concentration of 5.45 mg/L in Lake Nanhu which is higher than the permitted level of 0.6 mg/L. The author also reported that the total phosphorous in Lake Dianchi was 0.529 mg/L which is higher than the permitted level of 0.02 mg/L.

Anderson et al. [191] reported that nitrogen and phosphorus are two major limiting factors in water bodies. In freshwaters, phosphorous are least available in large quantities for the photosynthetic organisms and can limit or co-limit the growth of algae in estuarine and marine environments that are sustaining high nitrogen inputs. In some tropical and highly eutrophic temperate lakes, nitrogen may be a more important limiting factor than phosphorus. Duan et al. [192] reported that the nutrient sources for eutrophication are mainly sewerage, livestock drainage, soil nutrients and loss of fertilizers in drained agricultural lands, which could be associated with human population and economic development.

Mackenthum [193] reported that a concentration of 0.30 mg/L inorganic nitrogen is considered critical for stimulation of algal growth in the presence of adequate phosphorus in Lake Michigan, where total inorganic nitrogen averaged 1.56 mg/L and reached as high as 3.14 mg/L in Indiana Harbor. The source of pollution in this example was from the discharged domestic waste. However, excessive growth of plants and algae in polluted water can be avoided if the concentration of nitrate nitrogen is kept below about 0.3 mg/L and the concentration of total nitrogen is not allowed to rise much more above 0.6 mg/L. Certain species of algal are classified as harmful algal blooms because toxic compounds are released from this type of algal. Anderson [194] suggested that there is a relation between the toxicity of the harmful algal blooms with nutrients content in the water. Toxicity can increase or decrease significantly depending on the limiting nutrients of the species. For example, saxitoxin production by A. tamarense can be 5-10 folds higher in phosphorus limited environment in comparison to nitrogen limited environment [194,195].

The algal bloom may affect fish, by lowering the dissolved oxygen (DO) in water during the night. This algae oxygen demand can deplete the DO sufficiently and lead to fish kills. Fish may also find it hard to feed if the algae color the water and obscure their vision. Algal blooming at the surface of water will block the sunlight from reaching the bottom of the water bodies. Lack of sunlight can disturb the photosynthesis process of plankton and may lead to another ecological problems such as loss of biodiversity [6,196].

Carpenter et al. [166] reported that the adverse effects of eutrophication are: (a) increased phytoplankton biomass, (b) shifts in phytoplankton to bloom forming species which are toxic or inedible, (c) increases in blooms of gelatinous zooplankton in marine environment, (d) increases in biomass of benthic and ephiphytic alage, (e) changes in macrophyte species composition biomass, (f) death of coral reefs, (g) decreased transparency of water, (h) depleted oxygen levels, (i) increased incidence of fish kills and (j) reduced harvest of fish and shellfish.

Human and animals health

Webb et al. [197] reported that animal wastes are rich in organics and high in biochemical oxygen-demand (BOD). The treated human sewage contains 20–60 mg BOD/L, raw sewage contains 300–400 mg BOD/L, and swine waste slurry contains 20,000–30,000 mg BOD/L. Skinner et al. [198] reported that agricultural wastes such as cattle slurry contain 50 times greater BOD (10,000-20,000 mg/l) than domestic sewage. The ranges of BOD concentrations for various wastes in shown in Table 25 [111].

Burkholder et al. [131] and Mellon et al. [199] reported that animal wastes carry parasites, viruses, and bacteria as high as 1 billion/g. Swine wastes contain >100 microbial pathogens that can cause human illness and disease. About one-third of the antibiotics used in the United States each year is routinely added to animal feed to increase growth. This practice is promoting increased antibiotic resistance among the microbial populations present and, potentially, increased resistance of naturally occurring pathogens in surface waters that receive a portion of the wastes. The pathogens in the animal manure are capable of causing various diseases to humans. Some of the select pathogens found in animal manure are shown in Table 26 [125].

Burkholder et al. [131] reported that waterborne contaminants affect human health from both recreational use of affected surface water and from ingestion of drinking water derived from either contaminated surface water or groundwater. Generally, the very young, elderly, pregnant women and immune compromised individuals are at great risk of infection. Accidental ingestion of contaminated water that may result in diarrhoea or other gastrointestinal tract distress from waterborne pathogens and dermal contact during swimming that may cause skin, eye, or ear infections. Drinking water exposures to pathogens could occur in vulnerable private wells; under normal

Source	BOD (mg/L)
Silage effluents	30,000-80,000
Pig slurry	20,000-30,000
Cattle slurry	10,000-20,000
Liquid effluents draining from slurry stores	1000-12000
Dilute dairy parlour and yard washing (dirty water)	1000-5000
Milk	140,000
Untreated domestic sewage	300-00
Treated domestic sewage	20-60
Clean river water	<5

Table 25: Ranges of BOD concentrations for various wastes [111].

Pathogen	Disease	Symptoms
Bacillus anthracis	Anthrax	Skin sores, headache, fever, chills, nausea, vomiting
Escherichia coli	Colibacilosis, Coliform mastitis-metris	Diarrhea, abdominal gas
Leptospira pomona	Leptospirosis	Abdominal pain, muscle pain, vomiting, fever
Listeria monocytogenes	Listerosis	Fever, fatigue, nausea, vomiting, diarrhea
Salmonella species	Salmonellosis	Abdominal pain, diarrhea, nausea, chills, fever, headache
Clostirdum tetani	Tetanus	Violent muscle spasms, lockjaw, difficulty breathing
Histoplasma capsulatum	Histoplasmosis	Fever, chills, muscle ache, cough rash, joint pain and stiffness
Microsporum and *Trichophyton*	Ringworm	Itching, rash
Giardia lamblia	Giardiasis	Diarrhea, abdominal pain, abdominal gas, nausea, vomiting, fever
Cryptosporidium species	Cryptosporidosis	Diarrhea, dehydration, weakness, abdominal cramping

Table 26: Pathogens found in animal manure [125].

circumstances community water utilities disinfect water sufficiently before distribution to customers.

WHO [200] and Carmichael et al. [201] reported that cyanobacteria (blue green algae) released from animal manure into the surface can produce toxins such as microcystins that are neurotoxins and hepatotoxins. Acute and chronic health impacts are bound to occur on exposure to these toxins.

High concentrations of nitrites in water consumed by infant causes methemoglobinemia. Heldin [202] reports that methemoglobinemia was found in young bottle-fed infants in parts of Neapawa and Portage le Prairie in Manitoba, Canada, and Health Service traced it to be caused by high nitrate content of water in local wells. Also, at about the same time, nitrate poisoning of cattle was reported by local veterinarian. High nitrate in forage can be contributing factor in livestock production. The drinking water standards (1962) of the U.S. Public Health Service list 10 ppm nitrate nitrogen (45 ppm NO_3) as the level that should not be exceeded for infants.

Cornell et al. [203] reported that applying synthetic nitrogenous fertilizer onto land has pronounced effect on the local and global environment. Ammonia (NH_3), nitrous oxide (N_2O) and nitric oxide (NO) are the main nitrogen gas species emitted to the atmosphere from agricultural activities. Nitrous oxide (N_2O) and nitric oxide (NO) are very minor contributors, but ammonia (NH_3) contributes significantly to the atmospheric nitrogen deposition. Ammonia and nitrogen oxides present in the atmosphere are removed by rainfall and dry deposition [204].

Camargo et al. [205] reported that some algal species can cause toxicity to aquatic and terrestrial animals by synthesis of certain toxin such as anatoxins, microcystins, nodularins, brevetoxins, hemolysins and saxitoxins (Table 27) [205]. These toxins remain can inside the algal cells (intracellular toxins) or can be released into the water (extracellular toxins) during active algal growth or when cells lyse. These toxins are consumed by animals by direct exposure to water, drinking of water or injestion of algal cells from water.

Methods of Abatements

Animal waste treatment

In order to reduce the amount of nitrogen compound in animal waste, manure is to be collected and treated in such a way as to make aerobic conditions prevail. Aerobic treatment will encourage the conversion of ammonium nitrogen into nitrate nitrogen through the process of nitrification. When the aerobic process is followed by anaerobic conditions, denitrification will be encouraged resulting in stripping out the elemental nitrogen to the atmosphere leaving the treated manure with a lower content of nitrate [141].

Another method is the ammonia removal from units holding animal wastewater. The ammonia release to the atmosphere is a function of the agitation. Aeration towers can be used to degasify animal waste. Because holding tanks are prevalent at animal production facilities, aeration systems can be used for ammonia release, where the amount of ammonia stripped from a solution is a function of the air flow rate, temperature, pH of the liquid, time and characteristics of waste water [206].

The ultimate utilization of animal manure will be land application. If nitrogen reduction is a goal in waste disposal, a rather wet schedule should be followed [207]. The applied waste water should contain some organic load so that a carbon substrate would be available for denitrifying bacteria. Once application of wastewater has begun on an area, it should be continued for as long as possible to ensure conversion of ammonia nitrogen to nitrate so that maximum reduction of nitrogen can be achieved by denitrification.

Proper storage and application of slurry and solid manure can help reduce emissions [208]. Rapid incorporation of slurry and solid manure into the soil is the simple way to reduce ammonia emission. Band spreading machineries such as trailing house and trailing shoe and sub-surface applicators reduces ammonia emissions by 30-40% compared with broadcast spreading. Shallow injection and deep injection can reduce ammonia emissions by 30-40% and 90%, respectively. The trailing house and subsurface applicators are shown in Figure 7 [208,209]. Some other additional methods of ammonia abatement are shown in Table 28 [208].

Slurry storage plays an important role in ammonia emission. Specifically made round covers fitted to above ground tanks and slurry lagoons can reduce ammonia emissions by up to 80%. In addition potentially cheaper covers such as LECA (light expanded clay aggregate) or UV stabilized plastic sheets provide significant reduction in ammonia emissions stored from pig slurry. The different emission

Toxins	Chemical Structure	Site and mode of action	Characteristic species
Anatoxin-a	Secondary amine alkaloid	Attacks nervous system by mimicking the action of acetylcholine and over stimulating muscle cells	*Anabaena circinalis, Anabaena flos-aquae, Anabaena planctonica, Aphanizomenon flos-aquae*
Anatoxin-a (s)	Organophosphate	Attacks nervous system by inhibiting acetyl-cholinesterase, and over stimulating muscle cells	*Anabaena flos-aquae, Anabaena lemmermannii*
Brevetoxins	Polycyclic ethers	Attacks nervous system by binding to sodium channels and disrupting nerve conduction	*Karenia brevis*
Domoic acid	Tricarboxylic amino acid	Attacks nervous system by binding to kainate glutamate receptors, and causing depolarization of the neurons, with a subsequent increase in intracellular $Ca^{2+,}$ neuronal swelling, and cell death	*Pseudo-nitzschia australis, Pseudo-nitzschia delicatissima, Pseudo-nitzschia multiseries, Pseudo-nitzschia seriata*
Hemolysins	Fatty acids	Target cells by altering membrane functions and causing cell lysis	*Alexandrium monilatum, Gymnodinium aureolum, Karenia mikimotoi*
Microcystins	Cyclic heptapeptides	Affects liver, hepatopancreas by shrinking the cytoskeleton, distorting cells and causing hepatic haemorrhages, and by inhibiting protein phosphatases, and causing tumor promotion	*Anabaena flos-aquae, Microcystis aeruginosa, Microcystis viridis, Planktothrix agardhii, Planktothrix rubescens*
Nodularins	Cyclic pentapeptides	Affects liver, hepatopancreas by shrinking the cytoskeleton, distorting cells and causing hepatic haemorrhages, and by inhibiting protein phosphatases, and causing tumor promotion	*Nodularia spumigena*
Saxitoxins	Carbamate alkaloids	Attacks nervous system by blocking sodium channels and disrupting nerve conduction	*Anabaena circinalis, Aphanizomenon flos-aquae, Alexandrium catenella, Alexandrium tamarense, Gymnodinium catenatum*

Table 27: Impact of toxins from algae on aquatic and terrestrial animals [205].

(a) Trailing house

(b) Sub-surface

Figure 7: Two types of applicators to reduce ammonia emissions [208,209].

control options for cattle and pig slurry storage is shown in Table 29 [208-210].

Domestic waste treatment

The use of nitrification-denitrification as a method for eliminating nitrogen contamination by waste water was examined by Mulbarger [211]. The process alternatives for the nitrification-denitrification process included: (a) substituting trickling filter for the high rate and/or nitrifying activates sludge systems, (b) adding inorganic or organic coagulants to primary clarifier to maximise solids removal and eliminate the high rate activated sludge systems and (c) utilizing anaerobic column or beds for denitrification. The whole process was carried out in a single, dual and three sludge systems. The author reported that an influent with 25.2 mg/L total nitrogen after treatment using the system produced an effluent with 5.4 mg/L total nitrogen (1.4 mg/L nitrate).

Strous et al. [212] studied the ammonium removal from concentrated waste streams with anaerobic ammonium oxidation (anammox) process. In this process, the ammonium was converted to dinitrogen gas with nitrite as the electron acceptor. The process is autotrophic and there was no need for COD addition to support denitrification. When the anammox process was preceded by a nitrification step, only a part of ammonium was nitrified and the anammox process combines the ammonium and nitrite to form dinitrogen gas. This reduced the oxygen demand and the biomass yield was very low. This approach reduces the total operation costs [213,214]. Strous et al. [212] studied the effects of sludge digestion effluent on waste streams in a fluidised bed reactor using anammox process. The whole process was carried out for 150 days at a pH of 8 and a temperature of 36°C. The results indicated that during 150 days of analysis, 82% of ammonium and 99% nitrite were removed in fluidized bed reactor. Dong et al. [215] studied the anaerobic digestion of poultry manure using anammox and denitrification processes. The results obtained from the study indicated that the anammox microorganisms developed at a very low rate and could not compete with the denitrifying bacteria for nitrite production.

Liao et al. [216] studied the removal of nitrogen from swine manure wastewaters by air (ammonia) stripping and aeration methods. The air stripping process took place in a stripping tower in which the wastewater was pumped in from the top which passed through the packing material and the air was blown from the opposite direction to release volatile ammonia in the air. The direct aeration involved pumped air directly into the reactor containing swine manure

Method	Mode of action	Disadvantages
Reduce crude protein in livestock diet	Reduces amount of N excreted and hence potential for ammonia emission	More research needed
Use extra straw for bedding cattle	Locks up ammonia	Research ongoing
Modify design for livestock housing	Reduces time and area of slurry exposed to air	Mainly for pigs. Difficult to implement for existing houses
Equip livestock houses with air scrubbers or filters	Removes ammonia from air ventilated from houses	Only for mechanical ventilated houses and it is very expensive
Cover surface of store with straw plastic sheet, clay granules, oil etc.	Physical barrier to ammonia emission	Can be unreliable. Granules or straw may block pumps etc.
Additives (eg. Zeolites)	Physically locks up ammonia or form stable chemical compound	Large quantities needed and not reliable
Dilute slurry with water	Speeds up infiltration into the soil	Effectiveness varies with soil conditions. Increases volume of slurry to be managed
Add acid to slurry	Lowers pH and so ammonia stays in solution	Hazardous and can release nitrous oxide
Remove proportion of solids from slurry with a mechanical separating machine	More free flowing liquid infiltrates into the soil more rapidly	Specialist machinery required and there are ammonia emissions form the remaining solid material
Irrigate with water after spreading slurry	Washes slurry into the soil	Needs large amounts of water and may cause run off to surface waters
Choose optimum time	Cool, humid weather discourages ammonia releases rain washes slurry into the soil	Difficult to quantify effectiveness. May cause odour problems
Match fertilizer type and use of crop need, soil and environmental conditions	Urea is one of fertilizers with highest ammonia emissions	Requires more time on fertilizer planning

Table 28: Additional methods for ammonia abatement [208].

Methods	Emission reduction efficiency (%)	Applicability
Store with no cover or crust	0	
Tight lid, roof or tent structure	80	Concrete or steel tanks and silos. May not be suitable on existing storage tanks
Plastic sheeting (floating cover)	60	Small earth-banked lagoons
Allowing formation of natural crust by reducing mixing and manure input below the surface	40	Only for slurries with higher content of fibrous material. Not suitable on farms where it is necessary to mix and disturb the crust in order to spread slurry frequently. Crust may not form on pig manure in cool climates
Replacement of lagoon, etc. with covered tank or tall open tanks (depth >3 m)	30-60	Only new build and subject to any planning restrictions concerning taller structures
Storage bag	100	Available bag sites may limit use on larger livestock farms
Floating LECA balls (Hexa covers)	60	Not suitable for crusting manures
Plastic sheeting (floating cover)	60	Large earth blended lagoons and concrete or steel tanks. Management and other factors may limit use of this technique
Low technology floating covers (eg.: chopped straw, peat, bark)	40	Concrete or steel tanks and silos. Probably not practicable on large earth banked lagoons. Not suitable if materials likely cause slurry management problems

Table 29: Storage methods for the reduction of ammonia emission [210].

wastewater to release ammonia to the air. The results indicated that the air stripping method achieved 90% ammonia removal at a pH of 11.5 after 1 h of treatment and the aeration method achieved 90% removal at a pH of 11.5 after 150 h of treatment.

Crop production practice

The use of nitrogen fertilizers is recommended to be at the optimum level required by the crop and soil type in order to eliminate addition of any excess. Land use practice and soil conservation methods that reduce the amount of erosion will consequently reduce the amount of organic nitrogen carried to the surface waters.

Flynn [217] recommended management strategies to improve the nitrogen use efficiency of crops, thereby reducing fertilizer requirements and associated greenhouse gas emissions. The author suggested that both environmental factors such as soil conditions and climate and management factors such as tillage play an important role in the amount of fertilizer to be applied. Appropriate nitrogen application rates are required to limit the buildup of nitrates in soil. Organic nitrogen sources such as animal manure, crop residues and nitrogen fixing crops can be used instead of synthetic fertilizer.

Tilman et al. [218] reported that nutrient use efficiency must be improved by better matching temporal and spatial nutrient supply with plant demand. Practices such as applying fertilizers during periods of greatest crop demand at or near the plant roots and smaller and frequent application of nutrients have the potential to reduce the nitrogen emissions and improving the yield and quality of the crops. This type of agriculture is called precision agriculture and it is currently carried out in large scale intensive farming which could be implemented on small level farming.

Multiple cropping systems such as using crop rotations or intercropping (two or more crops grown simultaneously) can help to improve pest control and increase nutrient and water use efficiency. Crops such as clover and other legumes can be in rotations to reduce fertilizer requirements by adding biologically fixed nitrogen into soils. Addition of post-harvest plants remains into the soil can increase the levels of soil organic matter and help in storing soil carbon. Agroforestry, in which growing trees can improve nutrient availability and efficiency and can reduce erosion in the soil [217,218].

Changing from fall nitrogen application to a spring application will reduce nitrous oxide emissions [219]. In Alberta cropping systems, between 30-50% nitrogen emissions have been reduced by changing fertilizer application methods. Increasing the acreage of minimum tillage also reduced the carbon di-oxide emissions from the soil and increased the carbon sequestration in soil. Naylor [220] reported that technologies such as drip irrigation can improve water use efficiency and decrease salinization in the soil. The sustainable growing of crops will require increased efficiency of nitrogen and water use along with ecologically based best management practices and effective use of pesticides.

Conclusion

The natural cycle of nitrogen involves several biological and non-biological process including: mineralization, nitrification, denitrification, nitrogen fixation, microbial and plant uptake of nitrogen, ammonia volatilization, leaching of nitrite and nitrate and ammonia fixation. Nitrogen exists naturally in the environment and is constantly being converted from organic to an inorganic form and vice versa. Production of commercial fertilizer adds up to the natural source of nitrogen. The main source of nitrogen include: atmospheric precipitation, geological sources, agricultural land, livestock and poultry operations and urban waste. Agricultural emissions show a strong increase due to the application of fertilizer to agricultural soils, grazing of animals and spreading of animal manure. Emissions from agricultural practices and animal manure wastes are the major source of nitrogen pollution in surface and underground water. Soil erosion and runoff from fertilized land as well as domestic and industrial wastes contribute to the enrichment of lakes and streams with nutrients. Nitrates concentration exceeding certain limits in drinking water is toxic to animals and humans, especially infants. Nuisance of algal bloom and fish kills in lakes and rivers occurs due to eutrophication. Obnoxious colors and smells are developed as a result of organic matter decay and are destroying the natural beauty of the environment. The water born contaminants affect human health from both recreational use of contaminated surface water and from ingestion of contaminated drinking water derived from surface or ground water sources. The methods for abatement of nitrogen pollution must follow multi pathways. First, the source and amount of pollution must be detected and defined. Second, the possible ways to treat animal and domestic wastes should be carefully investigated. Third, better agricultural practices should be developed that include: proper storage and application of slurry and solid manure, rapid incorporation of slurry and solid manure into the soil, use of band spreading machineries such as trailing house and trailing shoe and sub-surface applicators, use of specifically made round covers fitted to above ground tanks and slurry lagoons, applying fertilizers during periods of greatest crop demand at or near the plant roots in smaller amounts with frequent applications, using multiple cropping systems such as using crop rotations or intercropping to increase the efficiency of nitrogen uses and changing current livestock production techniques.

Acknowledgements

The research was supported by the Natural Sciences and Engineering Research Council (NSERC) of Canada.

References

1. Kotler DP, Thea DM, Heo M, Allison DB, Engelson ES, et al. (1999) Relative influences of sex, race, environment, and HIV infection on body composition in adults, Am J Clin Nutr 69: 432-439.
2. UNEP (2008) Vital water graphics: An overview of the state of the world's fresh and Marine waters, 2nd edn. United Nations Environment Programme, Nairobi, Kenya.
3. Peavy HS, Rowe DR, Tchobanoglous G (1985) Environmental Engineering McGrawHill, New York, USA.
4. Hammer MJ (2008) Water and wastewater technology. Upper Saddle River, New Jersey, USA.
5. Wu RSS (1999) Eutrophication, water borne pathogens and xenobiotic compounds, Environmental risk and challenges. Marine Pollution Bulletin 39: 1-12.
6. Smith VH, Joye SB, Howarth RW(2006) Eutrophication of freshwater and marine ecosystems. Limnol Oceanogr 51: 351-355.
7. Mackenzie FT (1998) Our Changing Planet. An introduction to earth system science and global environmental change, Second Edition. Prentice Hall, Upper Saddle River, New Jersey, USA.
8. Driscoll C, Whitall D, Aber J, Boyer E, Castro M, et al. (2003) Nitrogen Pollution in the Northeasterns United States: Sources, Effects and Management Options. Bioscience 53: 357-374.
9. Vitousek P, Aber J, Howarth R, Likens G, Matson P, et al. (1997) Human Alteration of the Global Nitrogen Cycle: Sources and Consequences. Ecol Appl 7: 737-750.
10. Schimel JP, Bennett J (2004) Nitrogen mineralization: Challenges of a changing paradigm. Ecology 85: 591-602.
11. Guggenberger G (2005) Humification and mineralization in Soils. In: Buscot F, Varma A (eds.) Microorganisms in soils: Roles in genesis and functions. Springer Berlin Heidelberg, New York, USA.
12. O'Leary, M., G. Rehm, M. Schmitt. 2002. Understanding Nitrogen in Soils. University of Minnesota. Extension Service. P5.
13. Rotz CA (2004) Management to reduce nitrogen losses in animal production. J Anim Sci 82 E-Suppl: E119-137.
14. Parton W, Stewart J, Cole CV (1988) Dynamics of C, N, P and S in Grassland Soils: A Model. Biogeochemistry 5: 109-131.
15. Anthonisen AC, Loehr RC, Prakasam TB, Srinath EG (1976) Inhibition of nitrification by ammonia and nitrous acid. J Water Pollut Control Fed 48: 835-852.
16. Ward BB1 (2011) Measurement and distribution of nitrification rates in the oceans. Methods Enzymol 486: 307-323.
17. Liu G Wang J (2012) Probing the stoichiometry of the nitrification process using the respirometric approach. Water Res 46: 5954-5962.
18. Stark JM, Firestone MK (1995) Mechanisms for soil moisture effects on activity of nitrifying bacteria. Appl Environ Microbiol 61: 218-221.
19. van der Heijden MG Bardgett RD, van Straalen NM (2008) The unseen majority: soil microbes as drivers of plant diversity and productivity in terrestrial ecosystems. Ecol Lett 11: 296-310.
20. Bonete MJ Martínez-Espinosa RM, Pire C, Zafrilla B, Richardson DJ (2008) Nitrogen metabolism in haloarchaea. Saline Systems 4: 9.
21. Knowles R (1982) Denitrification. Microbiol Rev 46: 43-70.
22. Tallec G, Garnier J, Billen G, Gousailles M (2008) Nitrous oxide emissions from denitrifying activated sludge of urban wastewater treatment plants, under anoxia and low oxygenation. Bioresource Technol 99: 2200-2209.
23. Kim SW, Miyahara M, Fushinobu S, Wakagi T, Shoun H (2010) Nitrous oxide emission from nitrifying activated sludge dependent on denitrification by ammonia-oxidizing bacteria. Bioresour Technol 101: 3958-3963.
24. Wunderlin P Mohn J, Joss A, Emmenegger L, Siegrist H (2012) Mechanisms of N2O production in biological wastewater treatment under nitrifying and denitrifying conditions. Water Res 46: 1027-1037.
25. Alluisetti GE Almaraz AE, Amorebieta VT, Doctorovich F, Olabe JA (2004) Metal-catalyzed anaerobic disproportionation of hydroxylamine. Role of diazene and nitroxyl intermediates in the formation of N2, N2O, NO+, and NH3. J Am Chem Soc 126: 13432-13442.
26. Robertson LA Kuenen JG (1990) Combined heterotrophic nitrification and aerobic denitrification in Thiosphaera pantotropha and other bacteria. Antonie Van Leeuwenhoek 57: 139-152.
27. Kemp MJ, Dodds WK (2002) The influence of ammonium, nitrate and dissolved oxygen concentration on uptake nitrification and denitrification rates associated with prairie stream substrata. Limnology and Oceanography 47: 1380-1393.
28. Postgate J (1998) Nitrogen fixation (3rd edn), Cambridge University Press, Cambridge, UK.
29. Kennedy IR, Choudhury ATMA, Kecskes ML (2004) Non symbiotic bacterial diazotrophs in crop farming systems: can their potential plant growth promotion be better exploited? Soil Biology and Biochemistry 36: 1229-1244.
30. Kahindi JHP, Woomer P, George T, de Souza Moreira FM, Karanja NK, et al. (1997) Agricultural intensification, soil biodiversity and ecosystem function in the tropics: the role of nitrogen-fixing bacteria. Applied Soil Ecology 6: 55-76.
31. Bohlool BB, JK Ladha, DP Garrity, T George (1992) Biological nitrogen fixation for sustainable agriculture: a perspective. Plant and Soil 141: 1-11.
32. Orr CH, James A, Leifert C, Cooper JM, Cummings SP, et al. (2011) Diversity and activity of free-living nitrogen-fixing bacteria and total bacteria in organic and conventionally managed soils. Appl Environ Microbiol 77: 911-919.
33. Belnap J (2003) Factors influencing nitrogen fixation and nitrogen release in biological soil crusts. In: Belnap J, Lange OL (eds.), Biological soil crusts: structure, function and management. Ecological Studies. Springer-Verlag Berlin, Germany. 150: 241-261.
34. Bürgmann H Widmer F, Von Sigler W, Zeyer J (2004) New molecular screening tools for analysis of free-living diazotrophs in soil. Appl Environ Microbiol 70: 240-247.
35. Vitousek PM, Cassman K, Cleveland C, Crews T, Field CB, et al. (2002) Towards an ecological understanding of biological nitrogen fixation. Biogeochem 57/58: 1-45.
36. Cleaveland CC, Townsend AR, Schimel DS, Fisher H, Howarth RW, et al. (1999) Global patterns of terrestrial biological nitrogen (N2) fixation in natural ecosystems. Global Biochemical Cycles 13: 623-645.
37. Mylona P Pawlowski K, Bisseling T (1995) Symbiotic Nitrogen Fixation. Plant Cell 7: 869-885.
38. Andrade SLA, Dickmanns A, Ficner R, Einsle O (2005) Expression, purification and crystallization of the ammonium transporter Amt-1 from Archaeoglobus fulgidus. Acta Crystallogr Sect F Struct Biol Cryst Commun F61: 861-863.
39. Andrade SL Einsle O (2007) The Amt/Mep/Rh family of ammonium transport proteins. Mol Membr Biol 24: 357-365.
40. Martínez-Espinosa RM Lledó B, Marhuenda-Egea FC, Bonete MJ (2007) The effect of ammonium on assimilatory nitrate reduction in the haloarchaeon Haloferax mediterranei. Extremophiles 11: 759-767.
41. Crawford NM, Glass ADM (1998) Molecular and physiological aspects of nitrate uptake in plants. Trends in Plant Science Reviews 3: 389-395.
42. Glass AD Britto DT, Kaiser BN, Kinghorn JR, Kronzucker HJ, et al. (2002) The regulation of nitrate and ammonium transport systems in plants. J Exp Bot 53: 855-864.
43. Meyer C, Stitt M (2001) Nitrate reductase and signalling. In: Lea PJ, Morot-Gaudry JF (eds.) Plant Nitrogen, Springer, New York, USA, pp 37-59.
44. Stitt M (1999) Nitrate regulation of metabolism and growth. Curr Opin Plant Biol 2: 178-186.

45. Lea PJ, Forde BG (1994) The use of mutants and transgenic plants to study amino acid metabolism. Plant, Cell and Environment 17: 541-556.

46. Masclaux-Daubresse C Daniel-Vedele F, Dechorgnat J, Chardon F, Gaufichon L, et al. (2010) Nitrogen uptake, assimilation and remobilization in plants: challenges for sustainable and productive agriculture. Ann Bot 105: 1141-1157.

47. Suzuki A Knaff DB (2005) Glutamate synthase: structural, mechanistic and regulatory properties, and role in the amino acid metabolism. Photosynth Res 83: 191-217.

48. Abrol Y, Raghuram N, Sachdev M (2007) Agricultural Nitrogen use & its Environmental Implications. I.K. International Publishing House, S-25, Uphaar Cinema Market, New Delhi; ISBN: 978-81-89866-33-4.

49. Kadlec R, Wallace S (2009) Treatment Wetlands. (2nd edn) CRC Press, Boca Raton, Florida, USA.

50. Rengel Z (2003) Handbook of Soil Acidity. Marcel Dekker AG, Hutgasse, Basel, Switzerland.

51. Agrifacts (2008) Ammonia volatilization from manure application. Practical Information for Alberta's Agriculture Industry. Soil Fertility Fact Sheet, Ag. Index No. 538-3. Alberta Agriculture and Rural Development, Edmonton, Alberta.

52. Meisinger JJ, Jokela WE (2000) Ammonia volatilization from dairy and poultry manure. Reproduced with permission from "Managing Nutrients and Pathogens from Animal Manure". Natural Resource, Agriculture and Engineering Service, Ithaca, New York, USA.

53. Rochette P, Angers DA, Chantigny MH, Gasser MO, MacDonald JD, et al. (2013) NH3 volatilization, soil $NH4^+$ concentration and soil pH following subsurface banding of urea at increasing rates. Canadian Journal of Soil Science 93: 261-268.

54. du Preez CC, Burger RDT (1987) Effect of application methods on ammonia volatilization from soils in a controlled environment. South African Journal of Plant and Soil 4: 57-60.

55. Musa MM (1968) Nitrogen fertilizer transformations in the Sudan Geriza soil. Plant and Soil 28: 413-421.

56. Cai GD, Chen RE, White XH, Fan A, Pacholski ZL, et al. (2002) Gaseous nitrogen losses from urea applied to maize on a calcareous fluvo-aquic soil in the North China Plain. Australian Journal of Soil Research 40: 737-748.

57. Fenn LB, Matocha JE, Wu E (1980) A comparison of calcium carbonate precipitation and pH depression on calcium-reduced ammonia loss from surface-applied ammonia. Soil Science Society of America Journal 44: 1128-1131.

58. Stumpe, JM, Vlek PLG, Lindsay WL (1984) Ammonia volatilization from urea and urea phosphates in calcareous soils. Soil Science Society of America Journal 48: 921-927.

59. Fan XH, Song YS, Lin DX, Yang LZ, Zhou JM (2005) Ammonia volatilization losses from urea applied to wheat on a paddy soil in Taihu region, China. Pedosphere 15: 59-65.

60. Tian G Cai Z, Cao J, Li X (2001) Factors affecting ammonia volatilisation from a rice-wheat rotation system. Chemosphere 42: 123-129.

61. Black AS, RR Sherlock, NP Smith (1987) Effect of timing of simulated rainfall on ammonia volatilization from urea, applied to soil of varying moisture content. Journal of Soil Science 38: 679-687.

62. Watson CJ, Kilpatrick DJ (1991) The effect of urea pellet size and rate of application on ammonia volatilization and soil nitrogen dynamics. Fertilizer Res 28: 163-172.

63. Rimski-Korsakov H, Rubio G, Lavado RS (2012) Fate of nitrogen from fertilizers in field-grown maize. Nutrient Cycling in Agroecosystems 8: 253-263.

64. Rojas CAL, Bayer C, Fontoura SMV, Weber MA, Vieiro F (2012) Ammonia volatilization from urea as affected by tillage systems ad winter cover crops in the south central region of Parana. RevistaBrasileira de Ciência do Solo 36: 261-270.

65. Sommer SG, Olesen JE (1991) Effects of dry matter content and temperature on ammonia loss from surface-applied cattle slurry. J Environ Qual 20: 679-683.

66. Sommer SG, Hutchings NJ (2001) Ammonia emission from field applied manure and its reduction-invited paper. Eur J Agron 15: 1-15.

67. Sommer SG, Jensen C(1994) Ammonia volatilization from urea and ammoniacal fertilizers surface applied to winter wheat and grassland. Fert Res 37: 85-92.

68. Svensson L (1994) Ammonia volatilization following application of livestock manure to arable land. J Agr Eng Res 58: 241-260.

69. Thompson RB, Pain BF, Rees YJ (1990) Ammonia volatilization from cattle slurry following surface application to grassland. II. Influence of application rate, wind speed and applying slurry in narrow bands. Plant and Soil 125: 119-128.

70. Misselbrook, TH, Smith KA, Johnson RA, Pain BF (2002) Slurry application techniques to reduce ammonia emissions: results of some UK field-scale experiments. Biosystems Engineering 81: 313-321.

71. Smith KA, Jackson DR, Misselbrook TH, Pain BF, Johnson RA (2000) Reduction of ammonia emission by slurry application technique. Journal of Agriculture and Engineering Research 77(3): 277-287.

72. Huijsmans JFM, Hol JMG, Vermeulen GD (2003) Effect of application method, manure characteristics, weather and field conditions on ammonia volatilization from manure applied to arable land. Atmospheric Environment 37: 3669-3680.

73. Bussink DW, O Oenema (1998) Ammonia volatilization from dairy farming systems in temperature areas: a review. Nutrient Cycling in Agroecosystems 51: 19-33.

74. Nieder R, Benbi D (2008) Carbon and Nitrogen in the Terrestrial Environment. Springer. Dordrecht, South Holland, Netherlands.

75. Carter M, Gregorich E (2008) Soil Sampling and Methods of Analysis. CRC Press, Boca Raton, Florida, USA.

76. Prasad R, J Power (1997) Soil Fertility Management for Sustainable Agriculture, CRC Press LLC, Florida, USA.

77. Smil V (2001) Enriching the Earth: Fritz Haber, Carl Bosch, and the transformation of world food production. Massachusetts Institute of Technology, Cambridge, Massachusetts, USA.

78. EDGAR (2013) Nitrous Oxide. Emission Database for Global Atmospheric Research. Emission Database for Global Atmospheric Research. European Commission, London.

79. EC (2013) Nitrogen oxide emission by source in Canada. Environment Canada, Quebec, Canada.

80. IPCC (2006) N_2O Emissions from Managed Soils and CO2 Emissions from Lime and Urea Application. In: IPCC Guidelines for National Greenhouse Gas Inventories. Intergovernmental Panel for Climate Change (IPCC), Geneva, Switzerland. Agricultural, Forestry and Other Land Use 4: 11.1-11.54.

81. Aneja V, Roelle P, Murray G, Southerland J, Erisman J, et al. (2001) Atmospheric Nitrogen Compounds II: Emissions, Transport, Transformation, Deposition and Assessment. Atmospheric Environment 35: 1903-1911.

82. Chadwick M, Huryn A (2005) Response of Stream Mactoinvertebrate Production to Atmospheric Nitrogen Deposition and Channel Drying. American Society of Limnology and Ocenography 50: 228-236.

83. Cape JN, Kirika A, Rowland AP, Wilson DR, Jickells TD, et al. (2001) Organic nitrogen in precipitation: Real problem or sampling artefact? Optimizing Nitrogen Management in Food and Energy Production and Environmental Protection: Proceedings of the 2nd International Nitrogen Conference on Science and Policy. The Scientific World 1: 230-237.

84. Cape JN, Anderson M, Rowland AP, Wilson D (2004) Organic nitrogen in precipitation across the United Kingdom. Water Air and Soil Pollution Focus 4: 25-35.

85. Goldberg MC (1970) Source of nitrogen in water supplies. In: Agricultural Practice and Water Quality. Iowa State University Press, Iowa, USA.

86. IPCC (1992) Climate Change. In: Houghton JT, Callander BA, Varney SK (eds.) The Supplementary Report to the IPCC Scientific Assessment. Published for the Intergovernmental Panel for Climate Change (IPCC), Geneva, Switzerland. Cambridge University Press, Cambridge, UK.

87. Mosier AR (1994) Nitrous oxide emissions from agricultural soils. Fertilizer Research 37: 191-200.

88. Mosier AR., Duxbury JM, Freney JR, Heinemeyer O, Minami K, et al. (1996) Nitrous oxide emissions from agricultural fields: Assessment, measurement and mitigation. Plant and Soil 181: 95-108.

89. Grosso S (2008) Estimating Agricultural Nitrous Oxide Emissions. EOS Transactions American Geophysical Union 89: 529-540.

90. Moiser A, Kroeze C, Nevison C, Oenema O, Seitzinger, et al. (1998) Closing the global N2O budget: nitrous oxide emissions through the agricultural nitrogen cycle. Nutrient Cycling in Agroecosystems, 52: 225-248.

91. World Bank (2013) Population Statistics, World Bank, Washington DC, USA.

92. EPA (1999) Nitrogen oxides (NOx), why and how they are controlled. Clean Air Technology Center (MD-12) Information Transfer and Program Integration Division, Office of Air Quality Planning and Standards, U.S. Environmental Protection Agency, Research Triangle Park, North Carolina.

93. Butcher SS, RJ Charlson, GH Orians, GV Wolfe (1992) Global biogeochemical cycles. Academic Press, New York, USA.

94. Evans RD, Ehleringer JR (1994) Water and nitrogen dynamics in arid woodland. Oecologia 99: 233-242.

95. Noble IR (1995) Deserts in a changing climate: Impacts. In: Watson RT, Zinyowera MC, Moss RH (eds). Climate Change, Scientific-technical analyses of impacts, adaptations, and mitigation of climate change, The Intergovernmental Panel on Climate Change (IPCC) second assessment report, Geneva, Switzerland.

96. Ehleringer JR, Mooney HA, Rundel PW, Evans RD, Palms B, et al. (1992) Lack of nitrogen cycling in the Atacama Desert. Nature 359: 316-318.

97. Holloway JM, Dahlgren RA, Hansen B, Casey WH (1998) Contribution of bedrock nitrogen to high nitrate concentrations in stream water. Nature 395: 785-788.

98. Blatt H, Jones RL (1975) Proportions of exposed igneous, metamorphic, and sedimentary rocks. Geological Society of American Bulletin 86: 1085-1088.

99. Morford SL Houlton BZ, Dahlgren RA (2011) Increased forest ecosystem carbon and nitrogen storage from nitrogen rich bedrock. Nature 477: 78-81.

100. Mulvaney RL, Khan SA, Ellsworth TR (2009) Synthetic nitrogen fertilizers deplete soil nitrogen: a global dilemma for sustainable cereal production. Journal of Environmental Quality 38: 2295-2314.

101. Kjaergaard T (1994) The Danish Revolution: An ecohistorical interpretation. Translated by D. Hohnen. Cambridge University Press, Cambridge, UK 1500-1800.

102. Galloway JN (1998) The global nitrogen cycle: changes and consequences. Environmental Pollution 1: 15-24.

103. FAO (2012) World stocks 2010. Food and Agricultural Organization, Rome, Italy.

104. World Resources Institute (2012) Fertilizer use (most recent) by country. World Resources Institute, Washington DC, USA.

105. Epoch (2011) Fertilizer Overuse Damage Agriculture and Environment in China.

106. Xing GX, Zhu JL (2000) An assessment of N loss from agricultural field to the environment in China. Nutrient Cycling in Agroecosystem 57: 67: 73.

107. Oomen, GJM, Lantinga EA, Goewie EA, Van der Hoek KW (1998) Mixed farming systems as a way towards a more efficient use of nitrogen in European Union agriculture, Environmental Pollution 51: 697-704.

108. USDA (2013) Fertilizer Use. United States Department of Agriculture, Washington DC, USA.

109. Baird J (1997) Nitrogen Management and Water Quality. Soil Facts. The North Carolina Agricultural Extension Service. North Carolina State University, Raleigh North Carolina. AG-439-2.

110. Byrnes BH (1990) Environmental Effects of N Fertilizer use- An Overview. Fertilizer Research 26: 209-215.

111. Hooda PS Edwards AC, Anderson HA, Miller A (2000) A review of water quality concerns in livestock farming areas. Sci Total Environ 250: 143-167.

112. Bouman AF, G Van Drecht, KW Van der Hoek (2005) Global and regional surface nitrogen balances in intensive agricultural production systems for the period 1970-2030. Pedosphere 15: 137-155.

113. Iqbal MT (2006) Study on vertical and lateral leaching of nitrate from a wheat field in China. Turkish Journal of Agriculture and Forestry 30: 59-65.

114. Platzer C (1999) Design recommendations for subsurface flow constructed wetlands for nitrification and denitrification, Water Science and Technology 40: 257-263.

115. Randall GW Mulla DJ (2001) Nitrate nitrogen in surface waters as influenced by climatic conditions and agricultural practices. J Environ Qual 30: 337-344.

116. Bolton EF, JW Aylesworth, FR Hore (1970) Nutrient losses through the drains under tree cropping systems and two fertility levels on a Brookston clay soil. Canadian Journal of Soil Science 50: 275-279.

117. Randall GW, Huggins DR, Russelle MP, Fuchs DJ, Nelson WW, Anderson JL (1997) Nitrate losses through subsurface tile drainage in conservation reserve program, alfalfa and row crop systems. Journal of Environmental Quality 26: 1240-1247.

118. USDA-FSA (2013) Conservation Reserve Program. United States Department of Agriculture-Farm Service Agency, Kansas City, Missouri, USA.

119. Kladivko EJ Frankenberger JR, Jaynes DB, Meek DW, Jenkinson BJ, et al. (2004) Nitrate leaching to subsurface drains as affected by drain spacing and changes in crop production system. J Environ Qual 33: 1803-1813.

120. Randall GW, Goss MJ (2008) Nitrate losses to surface water through subsurface tile drainage. In: Follett RF, Hatfield JL (eds.) Nitrogen in the environment: Sources, problems, and management. 2nd edn), Elsevier Science B.V., Amsterdam, Netherlands.

121. FAO (2013) FAOSTAT Database-Livestock Primary. Meat Consumption. Food and Agriculture Organization of the United Nation, Rome, Italy.

122. FAO (2013a) FAOSTAT Database-Live Animals. Food and Agriculture Organization of the United Nation, Rome, Italy.

123. Larsen J (2012) Meat consumption in China now double that in the United States. Earth Policy Institute, Washington DC, USA.

124. Statistics C (2012) Cattle Statistics.Statistics Canada, Ottawa, Ontario, Canada.

125. Hribar C (2010) Understanding concentrated animal feed operations and their impact on communities. National Association of Local Boards of Health, Bowling Green, Ohio, USA.

126. Sundrum A (2001) Organic live stock farming: A review. Livest Prod Sci 67: 207-215.

127. Phang J (2013) Concentrated animal feeding operations. Seton Hall University School of Law, Newark, New Jersey, USA.

128. Olson BM, Miller JJ, Rodvang SJ, Yanke LJ (2005) Soil and groundwater quality under a cattle feedlot in southern Alberta, Water Quality Research Journal of Canada, 40: 131-144.

129. Mackie RI Stroot PG, Varel VH (1998) Biochemical identification and biological origin of key odor components in livestock waste. J Anim Sci 76: 1331-1342.

130. Socolow RH1 (1999) Nitrogen management and the future of food: lessons from the management of energy and carbon. Proc Natl Acad Sci U S A 96: 6001-6008.

131. Burkholder J, Libra B, Weyer P, Heathcote S, Koplin D, et al. (2007) Impacts of waste from concentrated animal feeding operations on water quality. Environmental Health Perspectives 115: 308-312.

132. Sherman DG (2008) The untold costs of confined animal feeding operations. Union of Concerned Scientists (USC) Food Environment Program, Cambridge, Massachusetts, USA.

133. FAO (2013b) FAOSTAT Database-Livestock primary. Producing Animals/ Slaughtered (Head). Food and Agriculture Organization of the United Nation, Rome, Italy.

134. Kohn RA Dinneen MM, Russek-Cohen E (2005) Using blood urea nitrogen to predict nitrogen excretion and efficiency of nitrogen utilization in cattle, sheep, goats, horses, pigs, and rats. J Anim Sci 83: 879-889.

135. Lopez G, Leeson S (1995) Nitrogen content of manure from older broiler breeders fed varying quantities of crude protein. Journal of Applied Poultry Research 4: 390-394.

136. Shannon JA (1938) The excretion of uric acid by the chicken. Journal of Cellular and Comparative Physiology 11: 135-148.

137. USDA (2013b) National Program 206: Manure and Byproduct utilization annual report. United States Department of Agriculture, Washington DC, USA.

138. Westerman PW, Huffman RL, Feng JS (1995) Swine-lagoon seepage in sandy soil. Trans Am Soc Agricult Engineers 38: 1749-1760.

139. Stone KC, Hunt PG, Coffey SW, Matheny TA (1995) Water quality status of A USDA water quality demonstration project in the Eastern Coastal Plain. J Soil Water Conserv 50: 567-571.

140. Varel V (1997) Use of Urease Inhibitors to Control Nitrogen Loss from Livestock Waste. Bioresour Technol 62: 11-17.

141. Ndegwa PM, Hristov AN, Arogo J, Sheffield RE (2008) A review of ammonia emission mitigation techniques for concentrated animal feeding operations, Biosystems Engineering 100: 453-469.

142. Saint-Fort R, Raina RM, Prescott B (1995) Subsurface environmental quality under a cattle feedlot and adjacent cropfield. Journal of Environment Science and Health, Part A 30: 637-650.

143. Mahmoodabadi MR, Amini S, Khazaeepoul K (2010) Using animal manure for improving soil chemical properties under different leaching conditions. Middle East Journal of Scientific Research 5: 214-217.

144. Dikinya O, Mufwanzala N (2010) Chicken manure enhanced soil fertility and productivity: effects of application rates. Journal of Soil Science and Environmental Management 1: 46-54.

145. López-Mosquera ME Cabaleiro F, Sainz MJ, López-Fabal A, Carral E (2008) Fertilizing value of broiler litter: effects of drying and pelletizing. Bioresour Technol 99: 5626-5633.

146. Whalen JK, Chang C, Clayton GW, Carefoot JP (2000) Cattle manure amendments can increase the pH of acid soils. Soil Sci Soc Am J 64: 962-966.

147. Plaza C, Hernandez D, Garcia Gill JC, Polo A (2004) Microbial activity in pig slurry-amended soils under semiarid conditions. Soil Biology and Biochemistry 36: 1577-1585.

148. Naramabuye FX, Haynes RJ (2006a) Effect of organic amendments on soil pH and Al solubility and use of laboratory indices to predict their liming effect. Soil Science 177: 754-763.

149. Duruigbo CI, Obiefuna JC, Onweremadu EU (2007) Effects of poultry manure rates on soil activity in an Utisol. International Journal of Soil Science 2: 154-158.

150. Mokolobate MS, Haynes RJ (2002) Increases in pH and soluble salts influence the effect that additions of organic residues have on concentrations of exchangeable and soil solution aluminium. European Journal of Soil Science 53: 481-489.

151. Naramabuye FX, Haynes RJ (2006a) Effect of organic amendments on soil pH and Al solubility and use of laboratory indices to predict their liming effect. Soil Science 177: 754-763.

152. Tang Y, Zhang H, Schroder JL, Payton ME, Zhou D (2007) Animal manure reduces aluminum toxicity in an acid soil. Soil Science Society of America Journal 71: 1699-1707.

153. Ano AO, Ubochi CI (2007) Neutralization of soil acidity by animal manures: mechanism of reaction. African Journal of Biotechnology 6: 364-368.

154. Naramabuye FX, Haynes RJ (2007) The liming effect of five organic manures when incubated with an acid soil, Journal of Plant Nutrition and Soil Science 170: 615-622.

155. Whalen JK, Chang C, Clayton GW (2002) Cattle manure and lime amendments to improve crop production of acidic soils in northern Alberta. Canadian J Soil Sci 82: 227-238.

156. Loria ER, Sawyer JE (2005) Extractable soil phosphorus and inorganic nitrogen following application of raw and anaerobically digested swine manure. Agronomy Journal 97: 879-885.

157. Shen QR Shen ZG (2001) Effects of pig manure and wheat straw on growth of mung bean seedlings grown in aluminium toxicity soil. Bioresour Technol 76: 235-240.

158. Pescod M (1992) Wastewater Treatment and use in Agriculture – FAO Irrigation and Drainage Paper No. 47. Wastewater Characteristics and Effluent Quality Parameters. FAO Corporate Document Repository, Food and Agriculture Organization of the United Nations, Rome, Italy.

159. Guyer PE (2010) An Introduction to Domestic Wastewater Treatment. Guyer Partners, El Macero, California, USA.

160. Metacalf, Eddy (2004) Wastewater Engineering: Treatment and Reuse (4thedn) McGraw Hill publication, New York, USA.

161. Wen D, Huai-en (2011) Analysis on Nitrogen Pollution Characteristics of Weihe River. Xian University, Xian, China.

162. Zhang X, Xia X, Yang Z (2007) Reasons of High Concentration Ammonium in Yello River, China. Environmental Simulation and Pollution Control, School of Environment, Beijing University, 28: 1437-41.

163. Galloway JN Townsend AR, Erisman JW, Bekunda M, Cai Z, et al. (2008) Transformation of the nitrogen cycle: recent trends, questions, and potential solutions. Science 320: 889-892.

164. Galloway JN, Dentener FJ, Capone DG, Boyer EW, Howarth RW, et al. (2004) Nitrogen Cycles: past, present and future. Biogeochemistry 70: 153-226.

165. ERFB (2005) River Pollution Case Study, Summary Report on Pollution of Tolka River Near Clonee, Co. Eastern Regional Fisheries Board, Dublin, Ireland.

166. Carpenter S, Caraco NF, Correll DL, Howarth RW, Sharpley AN, et al. (1998) Nonpoint pollution of surface waters with phosphorous and nitrogen. Issues in Ecology 3: 1-12.

167. Puckett L J (1994) Nonpoint and point sources of nitrogen in major watersheds of the United States, Water Resources Investigations Report, U. S. Geological Survey, Reston, Virginia, USA 94-4001: 1-12.

168. Rouse JD Bishop CA, Struger J (1999) Nitrogen pollution: an assessment of its threat to amphibian survival. Environ Health Perspect 107: 799-803.

169. Kean AJ, Harley RA, Littlejohn D, Kendall GA (2000) On road measurement of ammonia and other motor vehicle exhaust emissions. Environmental Science and Technology 34: 3535-3539.

170. Wetzel RG (1983) Limnology. Saunders College Publishing, Philadelphia, Pennsylvania, USA.

171. Burkholder JM (2000) Eutrophication and oligotrophication. In: Levin S (ed.) Encyclopedia of Biodiversity, Academic Press, New York, USA, 2: 649-670.

172. Xiangcan J, Qiujin X, Changzhu H (2005) Current status and future tendency of lake eutrophication in China. Science in China Series C: Life Sciences 48: 948-954.

173. Vadeboncoeur Y, Jeppesen E, Zanden MJV, Schierup HH, et al. (2003) From Greenland to green lakes: cultural eutrophication and the loss of benthic pathway in lakes. Limnol Oceanogr 48: 1408-1418.

174. Dugdale RC, Goering JJ (1967) Uptake of new and regenerated forms of nitrogen in primary productivity. Limnology and Oceanography 12: 196-206.

175. Glibert PM, Magnien R, Lomas MW, Alexander J, Fan C, et al. (2001) Harmful algal blooms in the Chesapeake and coastal bays of Maryland, USA: Comparison of 1997, 1998, and 1999 events. Estuaries 24: 875-883.

176. Fisher TR, Carlson PR Barber RT (1982) Carbon and nitrogen primary productivity in three North Carolina estuaries. Estuarine, Coastal, and Shelf Science 15: 621-644.

177. Mallin MA, Paerl HW, Rudek J, Bates PW (1993) Regulation of estuarine primary production by watershed rainfall and river flow. Marine Ecology Progress Series 93: 199-203.

178. National Research Council (2000) Glean Coastal Waters—Understanding and Reducing the Effects of Nutrient Pollution, National Academy Press Washington, DC, USA.

179. Wu RSS, Lam PKS, Mackay DW, Lau TC, Yam V, et al. (1994) Impact of marine fish farming on water quality and bottom sediment: A case study in the subtropical environment. Marine Environmental Research 38: 115-145.

180. Romdhane MS, Eilertsen HC, Yahia OKD, Yahia MND (1998) Toxic dinoflagellate blooms in Tunisian lagoons: Causes and consequences for aquaculture, In Reguera B, Balance J, Ferndez ML, Wyatt T (eds.), Harmful Algae. Xunta de Galicia and Intergovernmental Oceanographic Commission of United Nations Educational Scientific and Cultural Organization, Paris, France.

181. Zhang J (1994) Atmospheric wet depositions of nutrient elements: Correlations with harmful biological blooms in the Northwest Pacific coastal zones. Ambio 23: 464-468.

182. Hodgkiss IJ, Ho KC (1997) Are changes in N:P ratios in coastal waters the key to increased red tide blooms? Hydrobiologia 852: 141-147.

183. Hodgkiss IJ (2001) The N: P ratio revisited. In: Ho KC, ZD Wang (eds.)

Prevention and Management of Harmful Algal Blooms in the South China Sea., School of Science and Technology, The Open University of Hong Kong, Kowloon, Hongkong.

184. Takahashi M, Koike I, Iseki K, Bienfang PK, Hattori A (1982) Phytoplankton species responses to nutrient changes in experimental enclosures and coastal waters. In: Marine Mesocosms: Biological and Chemical Research in Experimental Ecosystems. Springer-Verlag, New York, USA.

185. Goldman JC (1993) Potential role of large oceanic diatoms in new primary production. Deep Sea Research Part I: Oceanographic Research Papers 40: 159-168.

186. Glibert PM Terlizzi DE (1999) Cooccurrence of elevated urea levels and dinoflagellate blooms in temperate estuarine aquaculture ponds. Appl Environ Microbiol 65: 5594-5596.

187. Glibert PM, O'neil JM (1999b) Dissolved organic nitrogen release and amino acid oxidase activity by Trichodesium spp., In: Charpy L, Larkum AWD (eds.) Marine Cyanobacteria. Bulletin de l'Instimte Oceanographique, Monaco Musee Oceanographique, Monaco. pp. 265-271.

188. Smith VH Tilman GD, Nekola JC (1999) Eutrophication: impacts of excess nutrient inputs on freshwater, marine, and terrestrial ecosystems. Environ Pollut 100: 179-196.

189. McGarrigle ML (1993) Aspects of river eutrophication in Ireland. Annals of Limnology 29: 355-364.

190. Miltner R.J, Rankin ET (1998) Primary nutrients and the biotic integrity of rivers and streams. Freshwater Biology 40: 145-158.

191. Anderson DM, Glibert PM, Burkholder JM (2002) Harmful algal blooms and eutrophication: nutrient sources, composition and consequences. Estuaries 25: 704-726.

192. Duan H Ma R, Xu X, Kong F, Zhang S, et al. (2009) Two-decade reconstruction of algal blooms in China's Lake Taihu. Environ Sci Technol 43: 3522-3528.

193. Mackenthum KM (1969) The practice of water pollution biology. Federal Water Pollution Control Administration, Washington DC, USA.

194. Anderson DM Kulis DM, Sullivan JJ, Hall S (1990) Toxin composition variations in one isolate of the dinoflagellate Alexandrium fundyense. Toxicon 28: 885-893.

195. Boyer GL, JJ Sullivan, RJ Andersen, PJ Harrison, FJR Taylor (1987) Effects of nutrient limitation on toxin production and composition in the marine dinoflagellate Protogonnyaulax tamarensis. Marine Biology 96: 128-128.

196. Biggs BJF (2000) Eutrophication of streams and rivers: dissolved nutrient-chlorophyll relationships for benthic algae. Journal of the North American Benthological Society 19: 17-31.

197. Webb J, Archer JR (1994) Pollution of soils and watercourses by wastes from livestock production systems. In: Dewi IA, Axford RFE, Marai IFM, Omed HM (eds) Pollution in Livestock Production Systems. CABI Publishing, Oxfordshire, UK.

198. Skinner JA, Lewis KA, Bardon KS, Tucker P, Catt JA, et al. (1997) An overview of the environmental impact of agriculture in the UK. Journal of Environmental Management 50: 111-128.

199. Mellon MC, Benbrook C, Benbrook KL (2001) Estimates of antimicrobial abuse in livestock. Union of Concerned Scientists, Cambridge, Massachusetts, USA.

200. WHO (1999) Human health aspects. In: Toxic cyanobacteria in water: a guide to their public health consequences, monitoring and management, World Health Organization, Geneva, Switzerland.

201. Carmichael WW Azevedo SM, An JS, Molica RJ, Jochimsen EM, et al. (2001) Human fatalities from cyanobacteria: chemical and biological evidence for cyanotoxins. Environ Health Perspect 109: 663-668.

202. Heldin RA (1971) Nitrate contamination of ground water in the Neepawa-Langruth area of Manitoba. Canadian Journal of Soil Science 51: 75-84.

203. Cornell SE, Jickells TD, Cape JN, Rowland AP, Duce RA (2003) Organic nitrogen deposition on land and coastal environments: a review methods and data. Atmospheric Environment 37: 2173-2191.

204. Raes F, Van Dingenen R, Vignati E, Wilson J, Putaud JP, et al. (2000) Formation and cycling of aerosols in the global troposphere. Atmospheric Environment 3 4215-4240.

205. Camargo JA Alonso A (2006) Ecological and toxicological effects of inorganic nitrogen pollution in aquatic ecosystems: A global assessment. Environ Int 32: 831-849.

206. Koneswaran G Nierenberg D (2008) Global farm animal production and global warming: impacting and mitigating climate change. Environ Health Perspect 116: 578-582.

207. Koelliker JK, Miner JR (1970) Use of soil to treat anaerobic lagoon effluent renovations as a function of depth and application rate. Transactions of American Society of Agricultural Engineers 13: 496-499.

208. DEFRA (2002) Ammonia in the UK. Department for Environment Food & Rural Affairs. DEFRA Publications, London, UK.

209. Gay SW, Knowlton KF (2009) Ammonia emissions and animal agriculture. Virginia Cooperative Extension. Publication No. 442-110. Virginia Polytechnic Institute and State University, Blacksburg, Virginia, USA.

210. Oenema O, Velthof G, Klimont Z, Winiwarter W (2012) Emissions from agriculture and their control potentials. Thematic Strategy on Air Pollution Review (TSAP). International Institute for Applied Systems Analysis, Laxenburg, Austria.

211. Mulbarger MC (1971) Nitrification and denitrification in activated sludge systems. Water Pollution Control Federation 43: 2059-2070.

212. Strous M, Gerven EV, Zheng P, Kuenen JG, Jetten MS (1997) Ammonium removal from concentrated waste streams with the anaerobic ammonium oxidation (anammox) process in different reactor configurations. Water Res 31: 1955-1962.

213. Mulder A, van de Graaf AAA, Robertson LA, Kuenen JG (1995) Anaerobic ammonium oxidation discovered in a denitrifying fluidized bed reactor. FEMS Microbiology Ecology 16: 177-184.

214. Jetten MS Wagner M, Fuerst J, van Loosdrecht M, Kuenen G, et al. (2001) Microbiology and application of the anaerobic ammonium oxidation ('anammox') process. Curr Opin Biotechnol 12: 283-288.

215. Dong X Tollner EW (2003) Evaluation of Anammox and denitrification during anaerobic digestion of poultry manure. Bioresour Technol 86: 139-145.

216. Liao PH, Chen A, Lo KV (1995) Removal of nitrogen from swine manure wastewaters by ammonia stripping. Bioresource Technology 54: 17-20.

217. Flynn HC (2009) Agriculture and climate change: an agenda for negotiation in Copenhagen. The role of nutrient management in mitigation. International Food Policy Research Institute, Washington DC, USA.

218. Tilman D Cassman KG, Matson PA, Naylor R, Polasky S (2002) Agricultural sustainability and intensive production practices. Nature 418: 671-677.

219. AESAC (2000) Greenhouse gas emissions Alberta's cropping industry. Alberta Environmentally Sustainable Agriculture Council, Edmonton, Alberta.

220. Naylor RL (1996) Energy and resource constraints on intensive agricultural production, Annual Review of Energy and Environment 21: 99-123.

Impact of Dyeing Industry Effluent on Groundwater Quality by Water Quality Index and Correlation Analysis

David Noel S* and Rajan MR

Department of Biology, Gandhigram Rural Institute-Deemed University, Gandhigram-624302, Tamilnadu, India

Abstract

Ground water from Chinnalapatti, TamilNadu, India, was collected from the habitations at the vicinity of dyeing industries to study the impact of dyeing industry effluent on ground water quality. Physico-chemical parameters, namely EC, pH, Total hardness, Ca, Mg, Na, Cl, TDS, Potassium and Sulphate were analyzed and compared with drinking water standard. pH, Ca and Mg were within the desirable and permissible limit of drinking, whereas sulphate, Sodium and Potassium were within the desirable limit only. Water Quality Index was calculated to assess the level of pollution and it is 61, indicating the pollution rate of ground water between slight and moderate. Correlation of various physico-chemical parameters with EC, TDS, and Total hardness was also calculated to establish the nature of relationship between them. EC is positively correlated with sodium, similarly TDS and Total hardness are positively correlated with calcium.

Keywords: Physico-chemical parameters; WQI; Correlation and regression; Water pollution; Standard deviation

Abbreviations: EC: Electrical conductivity; Na: Sodium; Ca: Calcium; Cl: Chloride; Mg: Magnesium; TDS: Total soluble solids; WQI: Water quality index; SPSS: Statistical Package for the Social Sciences; BIS: Bureau of Indian standard

Introduction

Environmental pollution is a global phenomenon because of its adverse effects on human health, plants, animals and exposed materials [1]. Industrialization and urbanization are greatly attributed to the pollution issues. Because, a large number of organic and industrial effluents has been introduced into the environment that has increased water and land pollution problems manifold. Water pollution due to dyeing industry is the matter of great concern since large quantity of effluent is discharged into the water bodies. Central pollution control board has listed the dyeing industry as one of the heavily polluting industries [2]. The dye effluent is highly toxic in nature as it contains high suspended solid, COD, dye and chemicals along with high concentration of heavy metals like Cu, Cd, Zn Ni and Pb. The dye effluent contaminates the surface and ground water, thereby, making it unfit for irrigation and drinking [3]. The dye effluent contains certain chemicals that could be toxic, carcinogenic or mutagenic to living organisms [4]. The study area Chinnalapatti is located in Dindigul district of Tamil Nadu, there are around 80 to100 dyeing units running in the township. Most of the dyeing units discharge the untreated effluent into the land which ultimately pollutes the ground water and makes it unfit for drinking. The main objective of the present investigation was to assess the water quality index of ground water and to interpret the water quality of ground water by a statistical calculation called correlation co-efficient.

Materials and Methods

Study area

Chinnalapatti is a town at Dindigul district of Tamilnadu,India, geographically lies between 10.3000°N latitude and 77.9300°E longitude. The basin is characterized by an undulating topography with hills in the southern parts. The highest elevation in this area is Sirumalai that is 1350m (amsl).Ground water mostly occurs in the weathered as well as in the fractured zones, which are unconfined, semiconfined and confined conditions [5]. These aquifer conditions may change rapidly and vary over a wide range from place to place. The thickness of the weathered/ fractured zone varies even over a small region. The shallow aquifers are in phreatic condition, which may not be a stable source for large demands of ground water, but the deeper aquifers are partly confined, i.e., they are being recharged from the shallow unconfined aquifers through dug – cum bore wells/bore wells [6].

Ground water sampling and analysis

Ground water samples were collected in clean plastic containers 2L capacity from March to December 2012. The samples were immediately transferred to the laboratory and analyzed for various physico-chemical parameters, namely EC, pH, Total Hardness, Ca, Mg, Na, Cl, TDS, Potassium and Sulphate, using standard methods [7]. Mean, standard deviation and correlation study were calculated using SPSS software.

Water Quality Index of the collected effluent was calculated to arrive at the level of pollution. The standards for drinking water recommended [8] by BIS for the 10 parameters chosen for the analysis along with the assigned weights [9]. Water Quality Index calculation was carried out as per Harton, as modified by Tiwari [10]. Weights (wi) were assigned to various water parameters which ranged from 1 to 4. According to the role of various parameters on the overall quality of water, the rating scales were fixed. For example, sodium, chloride and sulphate were important parameters in dye industry effluent and hence 4 and 3 were assigned. The other parameters were assigned according to their importance and incidence in irrigation water. Even if they were present, they might not be the ruling factor. Hence they were assigned low weights. The weight (wi) for the ith parameter (i=1,2 ....10 in our case) was calculated. The

***Corresponding author:** David Noel S, Research Scholar, Department of Biology, Gandhigram Rural Institute-Deemed University, Gandhigram-624302, Tamilnadu, India, E-mail: davidnoel_22@yahoo.co.in

Parameters	Minimum	Maximum	Mean	Std. Deviation
pH	6.2	7.9	7.1	0.522
EC	908	1800	1102.30	254.94
TDS	871	1628	1281.60	297.91
Total Hardness	281	392	320.930	36.27
Calcium	57	92	71.875	11.77
Magnesium	20	44	32.75	6.942
Sodium	4.24	6.34	5.263	0.67
Potassium	0.46	1.70	0.996	0.49
Chloride	210	360	282.18	55.82
Sulphate	82	171	111.375	27.26

Table 1: Physico-chemical quality of Ground water (descriptive statistics).

Substance Characteristic	IS 10500:1991		WHO (1993) Recommendation	Undesirable effect outside the desirable limit	% of Sample exceeding recommended value
	Desirable Limit	Permissible Limit			
pH	.5 to 8.5	No relaxation	8.0	Mucous membrane will be affected	100% within the range of 6.5-.5
Total hardness mg/l	300	600	-	Encrustation in water supply system	62.5% above 300
Chloride mg/l	250	1000	250	Taste, corrosion and palatability are affected	62.5% below and 37.5%
TDS mg/l	500	2000	1000	Palatability decreases gastro intestine affected	31.25% between 500-1000,68.75% above 1000 mg/l
Calcium mg/l	75	200	250	Encrustation in water supply System	62.5% below 75,37.5% between 75-200.
Magnesium mg/l	30	100	50	Encrustation in water supply System	25% >30, 75% >50
Sodium mg/l	-	-	200	-	100% within the range
Potassium mg/l		-	12	-	100% within the range
Sulphate mg/l	200	400	250	Gastro intestine affected	100% within the range

Table 2: Comparisons of water samples with Drinking water.

Salinity Zone	EC(mhos/cm)	No of samples	% of samples
Low	250	0	-
Medium	250-750	0	-
High	750-2250	10	62.5
Very high	2250-5000	6	37.5

Table 3: Classification of Salinity Zone based on Richards [13].

TDS(mg/l)	Class	No of samples	% of samples
<500	Desirable for drinking	0	0
500-1000	Permissible for drinking	5	31.2
1000-3000	Useful for irrigation	11	68.75
>3000	Unfit for drinking and irrigation	0	0

Table 4: Classification of Water based on TDS by Davis and Dewiest [15].

rating scales for the 10-water quality parameters considered here. Each parameter has been divided into 5 intervals according to the ranges. The quality index (qi) is corresponding to each range (varying from 0 to 100) and the extent of pollution corresponding to various value ranges, in descriptive terms.

qi-100-Ideal limit (BIS)

0-Severe value (BIS)

Other ratings, namely qi – 25, 50 and 75 are intermediate scales between ideal and severe values of BIS for irrigation water. The Water Quality Index (WQI) is the aggregate of the multiplication of qi and wi of the ith parameters. Based on WQI value the quality status is assigned, i.e. If WQI is 75-100 the parameters are in "ideal" limit.

Results and Discussion

The minimum, maximum, mean and standard deviation values generated from the analysis of the 16 samples are presented in Table 1. The data was compared with drinking water guidelines of India ISI, WHO [11,12] and tabulated in Table 2. It was observed that pH of ground water was between 6.2 and 7.7 and met drinking water criteria. EC and TDS are important parameters to measure the salinity hazard of water. EC in the present study varies between 908-1800 μhos/cm. As per the classification of water samples by Richards [13] presented in Table 3, 62.5% of samples fell under high salinity range. David Noel et al [14] reported that EC of ground water in Chinnalapatti was 1200 μmhos. TDS values varied between 871-1628 mg/l. Groundwater classification based on TDS given by Davis [15] is presented in Table 4. This table clearly shows that only 31.25% of total samples are permissible for drinking and no samples are desirable for drinking. TDS concentration was high due to the presence of bicarbonates, carbonates, sulphates and chlorides of calcium [16] and TDS value of 500 mg/l is the desirable limit and water containing more than 500 mg/l causes gastrointestinal and irritation [17]. High value of TDS influences the taste, hardness and corrosive property of the water [18]. Chloride in the present study ranged between 210-391 mg/l. 37.5% of samples exceeded the desirable limit of drinking. Calcium and magnesium salts are essential plant nutrients and their concentration play a vital role in reducing sodium hazard [19]. Calcium in the present study varied between 57-92 mg/l Magnesium varied between 20 and 44 mg/l. Mg occurs in lesser concentration than calcium due to the fact that the dissolution of Mg rich minerals is slow process and that Ca is more abundant in the earth crust [20]. All the samples of Ca and Mg were within the desirable and permissible level of drinking as per ISI 1991 and WHO1993 classification of drinking water. Sulphate in the ground water was found in the range 82-171 mg/l. All the samples are within the desirable limit for drinking. Among the Cations Sodium is most dominant in water. In the present study Sodium varied between 4.24 and 6.34 mg/l within the desirable range of drinking. Potassium in the samples ranged between 0.23 and 1.70 mg/l that is within the desirable range of drinking. Total hardness of the samples ranged between 274 and 392 mg/l According to classification of water based on Total hardness proposed by [21]

TH(mg/l)	Water type	No of samples	% of samples
0-75	Soft	-	-
75-150	Moderately hard	-	-
150-300	Hard	6	37.5
>300	Very hard	10	62.5

Table 5: Classification of Water based on hardness [21].

S. No	Parameters	Values (BIS)	Rating (pi)	Unit weight	Product (piwi)
1	pH	7.39	100	0.04	4
2	Electrical Conductivity (μmhos/cm)	2,250	0	0.09	0
3	Total Dissolved Solids (mg/l)	2,100	0	0.09	0
4	Chloride (mg/l)	350	0	0.18	0
5	Total Hardness (mg/l)	300	100	0.04	4
6	Sodium (meq/l)	40	100	0.09	9
7	Potassium (meq/l)	2	100	0.18	18
8	Calcium (meq/l)	20	100	0.13	13
9	Magnesium(meq/l)	5	0	0.04	0
10	Sulphate(mg/l)	200	100	0.18	13
Total Water Quality Index					61

Table 6: Water Quality Index (WQI) of the Groundwater Samples.

	pH	EC	Ca	Mg	Na	K	Cl	SO	TH	TDS
pH	1									
EC	4.72	1								
Ca	-17.8	0.39	1							
Mg	-15.7	0.43	0.25	1						
Na	-2.04	0.69	-0.33	-251	1					
K	-80.97	-0.42	0.08	0.09	0.04	1				
Cl	-8.73	0.0002	-2.21	-0.46	-0.23	0.41	1			
SO	-8.90	0.37	0.21	0.32	-0.16	-0.62	0.23	1		
TH	-2.61	-0.28	0.41	0.82	-0.14	0.40	-0.40	0.39	1	
TDS	-12.48	-0.021	0.58	-0.17	-27.88	-0.64	-0.08	-0.13	-0.12	1

Table 7: Correlation Matrix.

presented in Table 5, 62.5% of samples are very hard water type. The Water Quality Index (WQI) of the ground water was 61 that indicates the pollution level between slight and moderate and presented in Table 6.

Correlation analysis

Correlation coefficient analysis is useful for interpreting groundwater quality data and relating them to specific hydro geological processes. The correlation of EC, TDS and Total hardness with different constituents of 16 water samples has been calculated and correlation matrix is represented in Table 7. There existed significant positive correlation of EC with pH r=4.72 and sodium r=0.69, TDS with calcium r=0.58 and Total hardness with calcium r=0.41 and potassium r=0.40 (Figures 1-5). However, significant negative correlation of EC with chloride r=0.0002, TDS with EC r=0.021 and Total hardness with EC r=-0.28. These results indicate that pH sodium and magnesium shows major contribution than other physico-chemical constituents towards causing salinity in ground water. Cyanobacterial species were isolated from textile industry effluent and identified as *Anabaena variabilis, Oscillatoria salina, Nostoc muscorum* and *Lyngbya majusculai*.

Conclusion

EC of water samples 2 and 14 have high salinity. Other samples also have high EC values unfit for drinking. Though the variation is less it is worth mentioning that the reason behind this may be continuous discharge of the chemicals and salts along with dyes from industries. Very high level of Total hardness gives an alarm to the future utility of ground water for drinking and irrigation purpose. No water samples are desirable for drinking as per the classification of water based on TDS by Davis and Dewiest. The ground water contamination of the investigated sites was due to dyeing industrial discharge of the effluent onto the land from Chinnalapatti town. Other geological factors such as soil porosity,

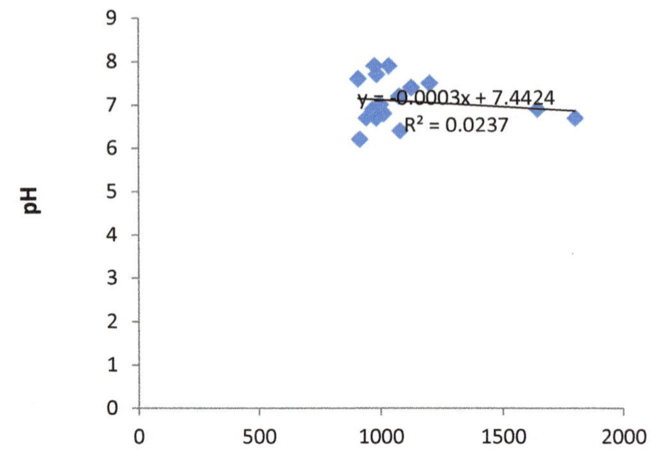

Figure 1: EC with pH r = 4.72.

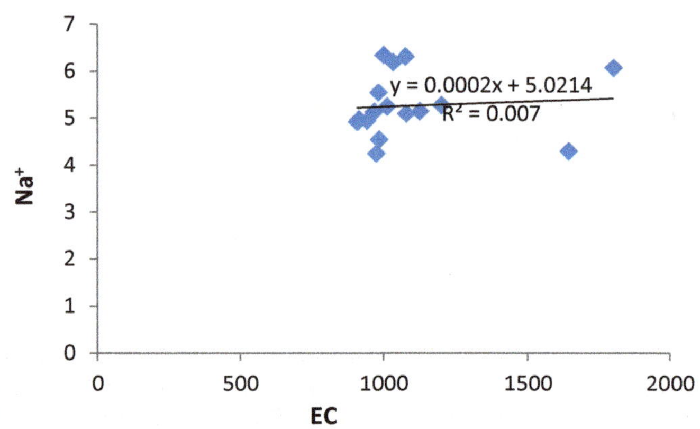

Figure 2: EC with sodium r = 0.69.

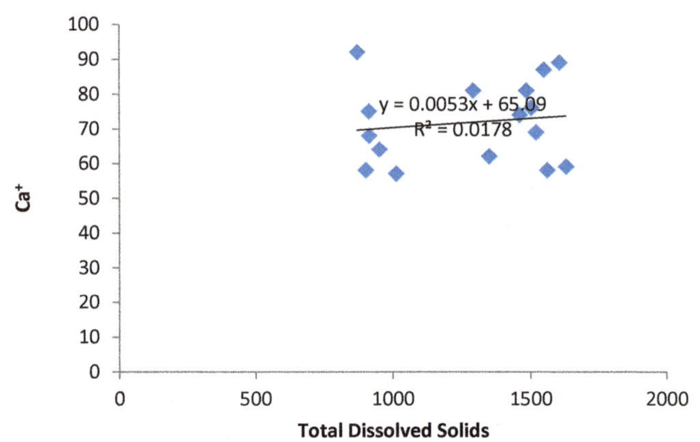

Figure 3: TDS with calcium r = 0.58.

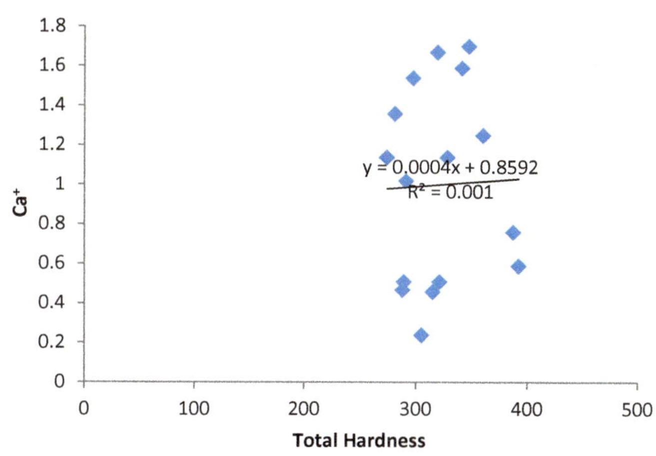

Figure 4: Total hardness with calcium r = 0.41.

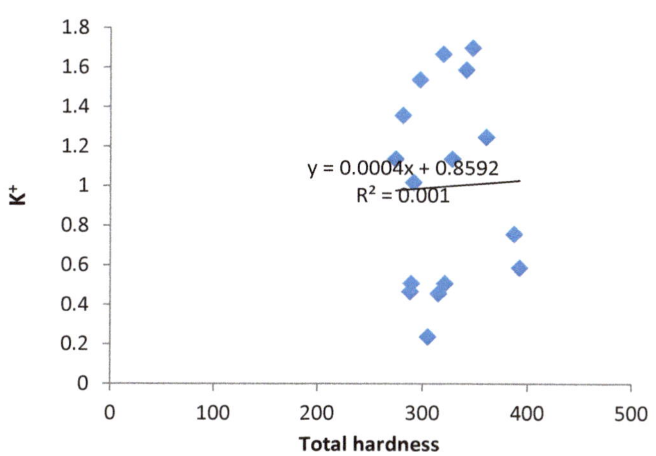

Figure 5: Total hardness with potassium r = 0.40.

permeability and the nature of rocks are also attributed to the ground water pollution. EC is positively correlated with sodium. Similarly TDS and Total hardness are positively correlated with calcium. Therefore it is inferred that enriched caustic soda and sodium chloride salts are added excessively during the dyeing process.

References

1. Irshad A, Ali S, Jan MR (1997) Physico-chemical studies of industrial pollutants. Environ. Poll. Islamabad, Pakistan.
2. CPCB (Central Pollution Control Board) (1990) Minimal National standards: Dye and Dye
3. Intermediate industry, Comprehensive Industry Document. Series: COINS/34/1990 (2007).
4. Mathur N, Bhatnagar P (2007) Mutagenicity assessment of textile dyes from Sanganer
5. (Rajasthan) J Environ Biol 28: 123-126.
6. Suzuki T, Tinolei S, Kurunczi L, Dietze U, Schuuirmann G (2001) Correlation of aerobic biodegradability of sulfonated azo dye with the chemical structure. Chemosphere 45: 1-9.
7. Mondal NC, Singh VS (2004) A new approach to delineate the groundwater recharge zone in hard rock terrain. Cur Sci 87: 658–662.
8. Mondal NC, Saxena VK, Singh VS (2005) Assessment of groundwater pollution due to tannery industries in and around Dindigul, Tamilnadu, India. Environ Geol 48: 149–157.
9. American Public Health Association (1998) Standard methods for the examination of Water and waste water. Washington DC
10. Trivedy RK (1988) Organic source effects and treatment. IndianJ.Env.Prot.8: 362- 371.
11. Punmia BC (1977) Water supply engineering, standard book house. New Delhi, India.
12. Tiwari TN, Misra N (1985) Preliminary assignment of water quality index to major indian rivers. Indian J. Environ. Protection.5: 276-279.
13. ISI (Indian Standards Institution) (1991). Indian standard Specification for drinking water, IS 10500.
14. WHO (World Health Organization) (1993). Guidelines for drinking water quality, 2nd (edn).
15. Richards LA (1954) Diagnosis and improvement of saline and alkaline soils. In USDA Handbook. U.S Salinity laboratory Staff.
16. David Noel S, Rajan MR (2012) Assessment of dye industry effluent for irrigation and its impact on ground water quality at Chinnalapatti. J.Ecotoxicol Envron.Monit. 22: 395-399.
17. 15. Davis SN, De Wiest RJ (1966) Hydrogeology. New York: Wiley.
18. Deepali S, Sapna P, Srivastava VS (2001) Groundwater quality at tribal town; Nandurbar (Maharashtra). Indian J. Environ. Ecoplan. 5: 475-479.
19. Jain CK, Kumar CP, Sharma MK (2003) Ground water qualities of Ghataprabа command Area, Karnataka. Indian J. Environ. Ecoplan. 7: 251-262.
20. Subhadra Devi G, Barbaddha SB, Hazel D, Dolly C (2003) Physico-chemical characteristics of drinking water at Velsao Goa. J. Environ. Monit. 13: 203-209.
21. Basavarajan D, Narayana J, Puttaiah (2008) Irrigation water quality assessment of Anjanapura reservoir. Indian J. Environ. Protection. 29: 145-151.
22. Varadarajan N, Purandara BK (2003) Environment and conservation 9: 253-262.
23. Sawyer CN, McCarty PL (1967) Chemistry for sanitary engineers (2ndedn). New York: McGraw-Hill Education.

11

The Response of Macrophytes to Nutrients and Implications for the Control of Phytoplankton Blooms in East Taihu Lake, China

Liu Guo-feng[1,2,3], Zhang lu[3], Fan Cheng-xin[3], Han Shi-qun[2*], He Jun[4], Hans W. Paerl[5]

[1]Freshwater Fisheries Research Center of Chinese Academy of Fishery Sciences, Wuxi, 214081, China
[2]Institute of Agricultural Resource and Environment, Jiangsu Academy of Agricultural Science, Nanjing 210014, China
[3]State Key Laboratory of Lake and Environment, Nanjing Institute of Geography & Limnology, CAS, Nanjing 210008, China
[4]Fisheries Technical Guidance Station of Wuxi Municipal Bureau of Agriculture, Forestry, Wuxi 214023, China
[5]Institute of Marine Sciences, University of North Carolina, USA

Abstract

When macrophytes are growing in the eutrophicated aquatic ecosystem, the vegetation induces important effects to the water quality and phytoplankton concentrations in the water which affected by macroscopic physical, chemical and biological processes and the effects are the results of direct and indirect interactions of the aquatic plants and water body. The interactions between macrophytes, nutrients and phytoplankton blooms were examined in the water and sediments of a shallow, eutrophic and typical East Taihu, China. The importance of macrophytes as a sink for nutrients, and the inhibitory effect of macrophytes on phytoplankton bloom potential were assessed through three different seasons. Luxuriant aquatic plants growth in this system led to decrease available nutrients for phytoplankton and prevented bloom development. Uptake of N and P by aquatic plants accounted for a major portion of the observed N and P loss from the water column and sediments. Luxury uptake of N and P were indicated by high biomass and tissues N and P concentrations, indicating the capacity of macrophytes to act as a nutrient sink in midsummer. Dissolved inorganic N (DIN) and soluble reactive phosphorous (SRP) in water were reduced in midsummer in the presence of macrophytes. The use of macrophytes to reduce the nutrients in water system and thereby inhibit freshwater phytoplankton blooms should be considered as an effective management strategy in shallow eutrophicated lakes. As aquatic macrophytes also develops considerable indirect effects that could have a vital impact than the direct uptake the nutrients into the plant biomass.

Keywords: Eutrophication; Macrophytes; Nutrients; Algae blooms; East Taihu Lake

Introduction

Submerged macrophytes are an important biotic component in freshwater ecosystems worldwide [1]. They provide refuge and food for various animals [2] and exert a strong influence on the water physical and chemical properties [3]. Macrophytes alter (generally increase) biodiversity in aquatic habitats [4]. They also provide shelter for zooplankton and young fish, reduce nutrient levels, serve as a habitat for macro-invertebrates [1,4]. Some field studies have demonstrated that the "clearing effect" increased with the macrophytes density and spatial extension of stands, but it was only restricted to a short distance outside the vegetation [5,6].

Freshwater phytoplankton blooms (mainly cyanobacterial) have become an increasingly problematic water quality issue worldwide [7-9]. They represent a health threat to domestic animals and human consumers of affected waters [10,11]. Blooms are primarily caused by excessive loading of nutrients [8] and global warming appears to enhance bloom potentials [12]. In recent years, periodic and widespread phytoplankton blooms have proliferated in the Taihu, China and this phenomenon was particularly serious in Taihu Lake in 2007 [13,14].

Submerged macrophytes can play an important role in the control of phytoplankton within enclosed shallow water bodies [15,16], and a possible negative feedback exists between shading provided by macrophytes and phytoplankton bloom development [17]. Enhanced nitrogen (N) and phosphorus (P) uptake and accumulation by summer biomass buildup of macrophytes leads to N limitation of the phytoplankton [15,18]. We investigated nutrient and macrophyte interactions with respect to their impacts on phytoplankton bloom potentials and water quality in East Lake Taihu, China (referred to as East Taihu). The lake has distinct regions where macrophytes form an important fraction of primary producers. In the relatively clear waters of East Taihu, macrophytes tend to dominate. Conversely, primary production in relatively turbid North Taihu is dominated by phytoplankton. So the purpose of the study was to investigate the role macrophytes played in nutrient cycling in the system by examining the partitioning of N and P between the macrophytes, water and sediment in East Taihu from March to December in 2009.

Material and Methods

Study area

The Eastern portion of Taihu is largely comprised of a large (130 km^2), shallow (mean depth 1.0 m) Bay, which was called Eastern Taihu. Submerged macrophytes have historically flourished in the lake region [19]. This region is also an important water source for the City of Shanghai [19]. The entire Taihu is currently eutrophic, with total phosphorus (TP) and total nitrogen (TN) concentrations of the water column rarely falling below the OECD threshold value of 0.02 mg/L TP and 0.2 mg/L TN for eutrophic lakes [20].

Three sampling sites (1#, 2#, 3#) in eastern Taihu were selected for transects according to prior results from the Taihu Monitoring

***Corresponding author:** Han Shi-qun, Institute of Agricultural Resource and Environment, Jiangsu Academy of Agricultural Science. Nanjing 210014, China
E-mail: shqunh@126.com

Program Stations (Figure 1). The three sites of 1# (E:120.41428°, N:30.98694°), 2# (E:120.4342°, N:31.03294°) and 3# (E:120.49765°, N:31.08131°) were located at the east suburb channel, near commercial crab culture operations, and the main channel of open water region, respectively. Sites were sampled in the March, August and December of 2009. These months represented the germination, maturation and senescence periods for aquatic macrophytes.

Sampling and analytical methods

Six quadrats were located along each transect crossing the studying area. Every sampling site had two quadrats. Water quality data and water samples were collected from all quadrats before collection of submerged macrophytes and sediment samples were undertaken to prevent contamination of the water with sediment. Secchi depth, temperature and dissolved oxygen (DO) were measured on sites. The suspended solids (SS), pH and EC of the water were measured in the laboratory.

Water samples were obtained using a water sampler deployed three times at different depths. Sub-samples were collected for chemical analyses. Surface sediment per quadrat was sampled and then frozen immediately in dry ice. The samples were analyzed for sediment chemical composition on the pooled samples of the top 2 cm of sediments. This is considered to be the portion of sediment where nutrients, P in particular, are enriched [21]. Lastly, the submerged macrophytes were collected from each quadrat, with either a submerged 0.28 m^2 quadrat by hand or a circular rake for very small plants [22] or large plants, respectively. The plants were then dried at 60°C for 24 h and weighed.

Water samples were analyzed for total phosphorus (TP), total dissolved phosphorus (TDP), soluble reactive phosphorous (SRP), total nitrogen (TN), total dissolved nitrogen (TDN), ammonium (NH_4^+-N), and nitrogen oxides (NO_X-N). Concentrations of TDN, NH_4-N and NO_X-N were determined colorimetrically on a Skalar Autoanalyser (Skalar-SA 3000/5000, Netherlands). TN, TDN were pretreated by digestion with potassium sulphite. TDP and TP were determined after perchloric acid digest, and SRP was determined using the ascorbic acid:molybdate method [23]. Chlorophyll *a* was measured with hot-ethanol extraction method as the phytoplankton concentration in the sampling water [24,25]. Sediment TP was measured colorimetrically of Murphy and Riley [26].

Results

Environmental conditions in the East Taihu

Surface water temperature increased from a March mean of 13°C to an August maximum average of 32.0°C (Table 1), and then declined to a December mean of 9.0°C. Bottom temperatures were similar to the surface water in spring, increasing more gradually over the summer time. The trend in surface DO shows a decline from March to August, then a rise to December, with some variability among sampling sites. The water pH and EC varied relatively little, except for some decrease in the aquatic macrophytes maturation period. Water Secchi depth of the 1# site was 0.15 m in March, and at the same time it were 0.90 m and 0.95 m at 2# site and 3# site, respectively. But the Secchi depth of 1# site reached 1.60 m in August, it was higher than that the other two sites and the water Secchi depth changed along with the season and macrophytes growth.

Nutrients changes in water and sediment

The selected three sampling sites exhibited similar water depth profiles and fluctuations (Figure 2). However, the three sites showed some variability among different forms of nutrients in the water. They had the higher concentration in March at 1# site, and then dropped to the minimum in August, but then increased again in December. The other sites showed an inverse pattern. Dissolved nitrogen and phosphorus concentrations in water column were sufficiently high to be available to macrophytes and phytoplankton. Sites 2# and 3# had

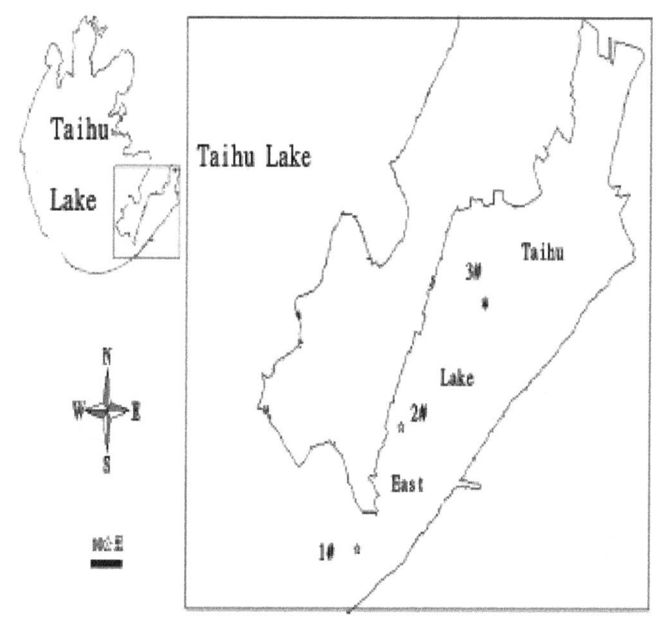

Figure 1: The sampling sites in the East Taihu Lake.
(The three sites of 1#, 2# and 3# were located at the east suburb channel, near the crab culture and the main channel of open water region, respectively).

Sampling site		1#			2#			3#	
Sampling time	Mar	Aug	Dec	Mar	Aug	Dec	Mar	Aug	Dec
Secchi depth (m)	0.15	1.60	1.00	0.95	1.22	0.90	0.45	0.85	1.30
Water depth (m)	1.00	1.60	1.50	0.95	1.22	0.90	1.00	1.60	1.40
Temperature (m)	13.0	32.0	9.0	13.0	32.0	9.0	13.0	32.0	9.0
pH	7.92	8.33	8.02	8.72	7.80	7.79	8.09	7.85	7.87
EC (µs/cm)	560.0	520.0	460.0	420	500.0	520.0	510.0	420.0	500.0
DO (mg/l)	10.10	9.85	4.50	7.56	4.87	9.70	8.30	7.65	9.60
SS (mg/l)	118.64	2.16	7.77	0.80	2.61	9.30	20.88	6.56	8.93

Table 1: The water environmental characteristic in East Lake Taihu in 2008
(Integrated samples of the water column were obtained using a water sampler deployed three times at different depths. Then water samples were brought back to the laboratory with the ice-bag protected and analyzed as soon as possible. The each sampling time was chosen at the same 9:00 am ; and all values presented are mean values).

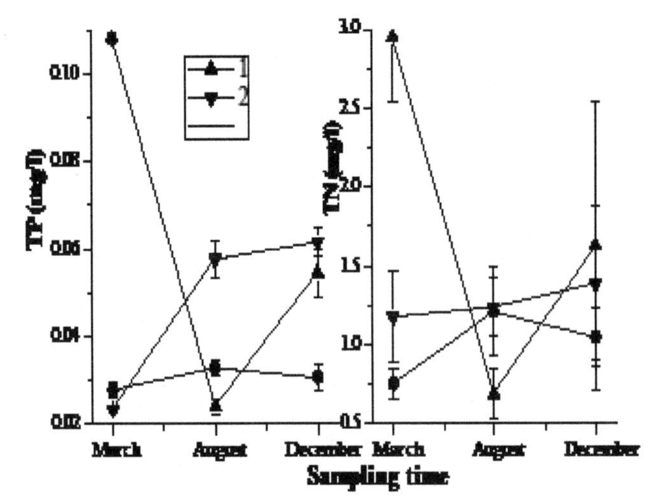

Figure 2: The concentration of TN and TP in water column.

the lower concentration of the TDN and TDP, and then increased from March to August. During the period from August to December, only the concentration of TDP in 1# sampling site rose (Figure 3). The PO_4^{3-}-P and NO_3^--N in water appeared to be directly available to macrophytes and phytoplankton. Though the NO_3^--N was higher in March, it decreased quickly in summer.

The TN and TP concentration in the sediments were similar to those in the water column. However, the 2# site had the higher concentration in March, and it was the time of the germination period of aquatic macrophytes, after which the nutrient concentrations declined during the summer (Figure 4). Nutrient accumulations appeared largely attributable to macrophytes decomposition.

Macrophytes and phytoplankton in the water column

The three sampling sites showed similar environmental conditions in the water and the macrophytes occurred in shallow regions to a maximum depth of 1.60 m in August and a maximum depth of 0.95 m in March. Macrophytes community composition also varied with sampling times and sites. The main species of the macrophytes included: *Potamogeton malaianus Miq.; Vallisneria natans L.; Elodea nuttallii; Hydrilla verticillata Royle; Ceratophyllum demersum* with other species comprising only a small percentage of the total composition (data not shown).

Site 1# is located at the east suburb channel, the flood discharge channel of Taihu and main water supply for the City of Shanghai. The macrophytes could obtain amounts of nutrients and grow well as the water velocity became lower at the 1# site (Figure 5). The 2# site had the highest biomass of aquatic plants during the selected sampling sites in March. But the 1# site had the highest biomass among three sampling sites in summer (fresh weight reaching 2,255.48 g/m^2). The 3# site had low biomass relatively except in August. Though the species composition and the macrophyte biomass showed little difference during the sampling time, in general there was an increase in aquatic plant biomass at sampling sites except some decreases at the 3# sampling site.

Phytoplankton biomass (dominated by cyanobacterial), estimated as Chl-a concentrations, varied among sampling times and sites (Figure 5). The concentration of phytoplankton biomass at 1# site was higher than that at the others sampling sites in March, but it was the lowest among the three sampling sites in August, only 5.58 μg Chla L^{-1}. The phytoplankton biomass generally increased in the summer, except site 3# showed a decrease. The biomass in 2# site became progressively higher over summer, reaching a peak in December (21.2 μg Chla L^{-1}).

Discussions

Conditions within the water column

The increase in water temperature in early summer at the germination period of aquatic macrophytes would favor growth of

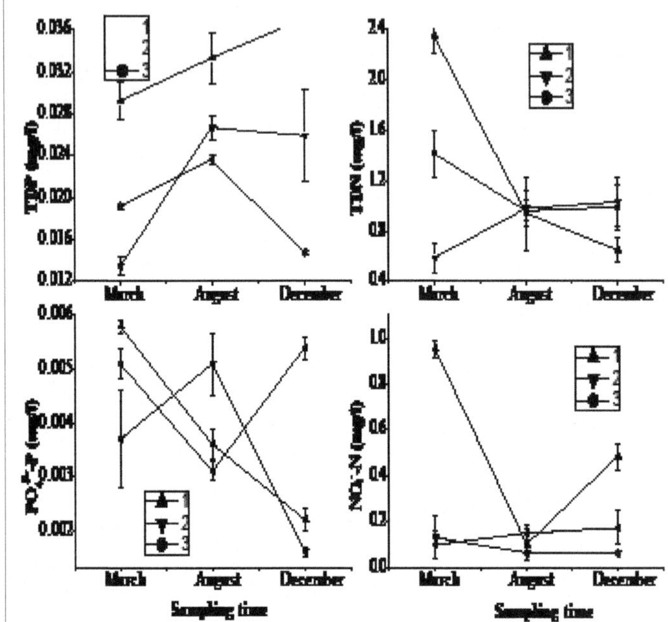

Figure 3: The dissolved nutrients in water (sampling time from March to December in 2008).

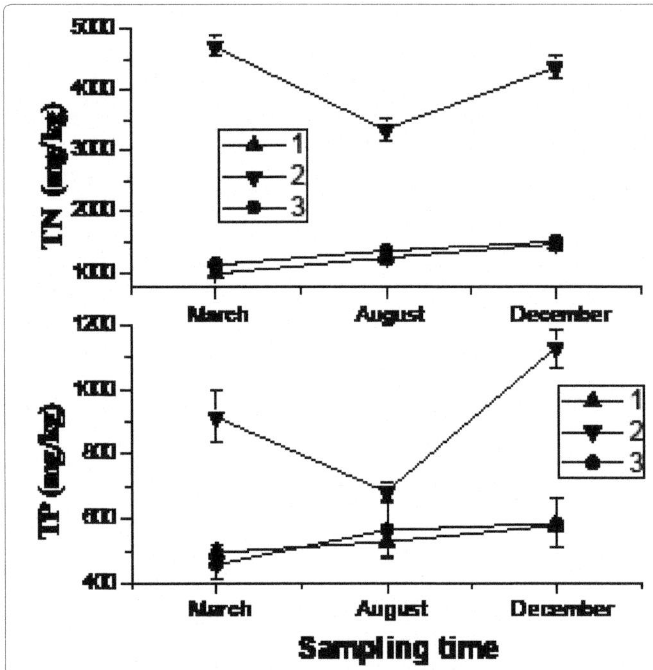

Figure 4: The concentration of TN and TP in sediment.

these plants. The early summer increase in biomass of photosynthetic plants would explain the increase DO concentrations in water in March. Differential photosynthetic use of carbon dioxide would also explain the fluctuation in pH at the three sampling sites. Higher water temperatures could also cause a decrease in the oxygen solubility in the water and therefore the August DO concentrations were reduced at all sampling sites (Table 1). As some macrophytes decomposed during this senescent period, DO consumption would also increase. As mentioned above, the macrophytes grew most vigorously from March to October, as temperature favored the aquatic plants and phytoplankton growth in this period. The results of Wang [27] indicate that water temperature and total phosphorus (TP) played dominant roles in controlling phytoplankton growth dynamics in most seasons; COD (chemical oxygen demand) and BOD (biological oxygen demand) presented significant positive relationships with phytoplankton biomass in spring, summer and autumn.

Macrophytes and nutrients

N and P values in the Taihu sediment are significantly different from those found in East Taihu sediments [28-30]. TN and TP concentration in the water was higher in 2# site in March and then decreased through the summer period. As in the water column, the TN in sediment also had higher content in December than that in August, although TP varied little during this period. The N and P concentration changes in the sediments indicate that nutrient losses from the top 2 cm of sediment were similar to the nutrient gains of the macrophyte community (Figure 5), of which the major species was the submerged macrophyte *Potamogeton malaianus*. The N data showed an increase in macrophyte N which was not matched by loss from the sediment, a phenomenon that has been observed by others [31,32]. TDN decreases in water column, although considerable (Figure 3). The additional N may have been derived from deeper sediments or from western Taihu water, although P data do not support this (Figure 3).

The macrophytes biomass progressively increased from March to August (Figure 5), a period when temperature and irradiance are favorable for macrophyte growth. Macrophyte species diversity was also enhanced during the same period (data not shown). Thriving aquatic plants during this period will optimally adsorb dissolved nutrients from the water, and increase the depth of the photic zone (Table 1). In a word, these effects would promote a positive-feedback, environment friendly ecosystem [33].

The Inhibition Effects of Macrophytes on Phytoplankton

The use of macrophytes for reducing N and P availability and thereby reducing phytoplankton blooms potential has been used in diverse aquatic ecosystems [18,34,35]. Possible inhibition of phytoplankton by allelochemicals released by submerged macrophytes has been proposed as one of the mechanisms that contribute to the maintenance of clear-water states in shallow lakes [36,37]. Although direct proof of allelopathy remains elusive, several authors have suggested possible involvement of allelopathy to explain phytoplankton successional patterns in whole-lake studies of vegetated, shallow lakes [38,39]. During this survey, the Chl-a concentration was low at the sampling sites except for 1# in March, when it was somewhat higher (Figure 5). The reason have been that this sample was taken from a site that was influenced by open Taihu water which passed through the 1# site, which was the main flood discharge channel of the Taihu. Following the summer macrophyte growth period, phytoplankton concentrations rapidly decreased. The macrophytes decayed at higher temperature in summer and caused oxygen deficiency in water column

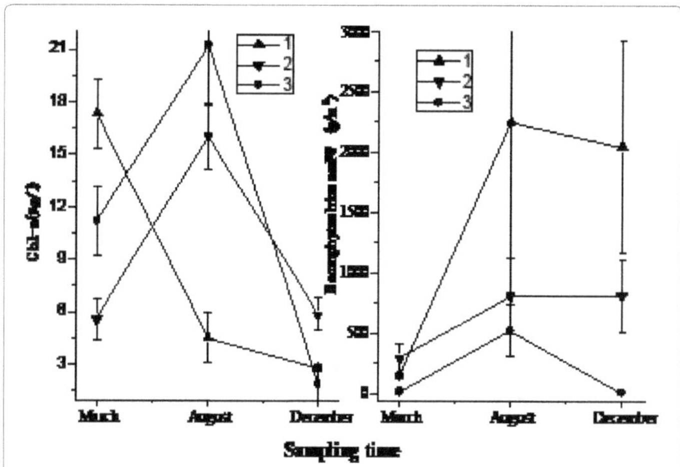

Figure 5: The biomass of macrophytes (fresh weight) and phytoplankton (Chla).

at the 2# site, as proven the nutrients concentration variety (e.g. higher concentration of PO_4^{3-}-P, TDP in water). The apparent inhibitory effects of macrophytes on phytoplankton are consistent with the other results [40-42]. Another explanation is salinity (conductivity) control of nutrient uptake by macrophytes [43], as the conductivity changed during surveying time (ranged from 560 to 42 μs/cm during the March to August time). Decreased salinity would promote uptake of N and P by these macrophytes, resulting in decreased availability of nutrients for phytoplankton.

Conclusions

Aquatic macrophytes have been shown to be a significant sink for nutrients and they increased clarity in East Taihu. Higher summer macrophyte biomass was responsible for uptaking large amounts of N and P. Macrophytes incorporation of N and P accounted for most of the observed nutrient loss from the sediment and water column. Surficial sediment N loss did not account for all the macrophyte N gain. Therefore other N sources, including cyanobacterial N_2 fixation and atmospheric N deposition, are postulated. Water column TDN and NO_3^--N significantly decreased during the *P. crispus* growth period. Decreased salinity (conductivity) maybe have also promoted N and P incorporation by macrophytes and hence the reduction of available nutrients for phytoplankton, explaining a reduction in phytoplankton biomass and blooms during the summer. The inhibitory effect of macrophytes on phytoplankton blooms has significant ramifications for management of phytoplankton blooms.

Acknowledgement

Thanks to the anonymous reviewer for his hard work in reviewing the article and he put forward many constructive suggestions and pointed out much minor mistakes in manuscript, his work is important for the authors to improve the paper quality.

This research was supported financially by the National Natural Science Foundation (NO. 41101525) and Independent Innovation Capital of Jiangsu province (NO. CX(12)5057).

Author Disclosure Statement

No competing financial interests exist.

References

1. Jeppesen E, Søndergaard M, Christoffersen K (1998) The structuring role of submerged macrophytes in lakes. Springer, New York, USA.
2. Proctor VW (1999) Charophytivorie, playas y papalotes, a local paradigm of

global relevance. Australian Journal of Botany 47: 399-406.

3. Sand-Jensen K, Frost-Christensen H (1998) Photosynthesis of amphibious and obligately submerged plants in CO_2-rich lowland streams. Oecologia 117: 31-39.

4. Scheffer M, Jeppesen E (1998) Alternative stable states. In: Jeppesen E, Søndergaard M, Christoffersen K (eds.) The Structuring Role of Submerged Macrophytes in Lakes. Ecological Studies 131: 397-406.

5. James WF, Barko JW, Butler MG (2004) Shear stress and sediment resuspension in relation to submersed macrophyte biomass. Hydrobiologia 515: 181-191.

6. Horppila J, Nurminen L (2005) Effects of different macrophyte growth forms on sediment and P resuspension in a shallow lake. Hydrobiologia 545: 167-175.

7. Reynolds CS (1987) Cyanobacterial water blooms. Adv Bot Res 13: 67-143.

8. Paerl HW (1988) Nuisance phytoplankton blooms in coastal, estuarine, and inland waters. Limnol Oceanogr. 33: 823-847.

9. Paerl HW, III Fulton RS (2006) Ecology of harmful cyanobacteria. In: Graneli E, Turner J (eds.). Ecology of Harmful Marine Algae. Springer-Verlag, Berlin.

10. Falconer IR, Burch MD, Steffensen DA, Choice M, Coverdale OR (1994) Toxicity of the blue-green alga Microcystis *aeruginosa* in drinking water to growing pigs, as an animal model for human injury and risk assessment. Environmental Toxicology and Water Quality 9: 131-139.

11. Chorus I, Bartram J (1999) Toxic cyanobacteria in water: A guide to their public health consequences, monitoring and management, London.

12. Paerl HW, Huisman J (2008) Blooms like it hot. Science 320: 57-58.

13. Guo L (2007) Doing Battle with the Green Monster of Taihu Lake. Science 317: 1166.

14. Song LR, Chen W, Peng L, Wan N, Gan N, et al. (2007) Distribution and bioaccumulation of microcystins in water columns: A systematic investigation into the environmental fate and the risks associated with microcystins in Meiliang Bay, Lake Taihu. Water Research 41: 2853-2864 .

15. Ozimek T, Gulati RD, Donk EV (1990) Can macrophytes be useful in biomanipulation of lakes? The Lake Zwemlust example. Hydrobiologia 61: 399-407.

16. Jackson HO, Starrett WC (1959) Turbidity and sedimentation as Lake Chautauqua, Illinois. J.Wild.Manage. 23: 157-168.

17. Scheffer M, Rinaldi S, Gragnani A, Mur LR, van Nes EH (1997) On the dominance of filamentous cyanobacteria in shallow, turbid lakes. Ecology 78: 272-282.

18. Donk EV, Gulati RD (1995) Transition of a lake to turbid state six years after biomanipulation: Mechanisms and pathways. Water Science and Technology 32: 197-206.

19. Huang YP (2001) The Water Enviromental and Contamination Control of Taihu Lake. Beijing, Science Press, China.

20. Ryding SO, Rast W (1989) The Control of Eutrophication of Lakes and Reservoirs. UNESCO. Paris and Parthenon.

21. Howard-Williams C (1981) Studies on the ability of a Potamogeton pectinatus community to remove dissolved nitrogen and phosphorus compounds from lake water. J. Appl. Ecol. 18: 619-637.

22. Marshall TR, Lee PF (1994) An inexpensive and lightweight sampler for the rapid collection of aquatic macrophytes. J. Aquat. Plant Manage. 32: 77-79.

23. Major GA, Dal Pont G, Klye J, Newell B (1972) Laboratory techniques in marine chemistry - a manual. CSIRO Division of Fisheries and Oceanography.

24. Jespersen AM, Christofersen K (1987) Measurements of chlorophyll-a from phytoplankton using ethanol as extraction solvent. Arch Hydrobiologia 109: 445-454.

25. Chen YW, Chen KN, Hu YH (2006) Discussion on possible error for phytoplankton Chl-a concentration analysis using hot-ethanol extraction method [J]. J.Lake Sci. 18: 550-552.

26. Murphy J, Riley JP (1962) A modified single solution method for the determination of phosphate in natural waters. Anal. Chim. Acta 27: 31-36.

27. Wang XL, Lu YL, He GZ, Han JY, Wang TY (2007) Exploration of relationships between phytoplankton biomass and related environmental variables using multivariate statistic analysis in a eutrophic shallow lake: A 5-year study. Journal of Environmental Sciences 19: 920-927.

28. Fan CX, Zhang L, Qin BQ, Hu W, Gao G, et al. (2004) Migration mechanism of biogenic elements and their quantification on the sediment-water interface of Lake Taihu:I.Spatial variation of the ammonium release rates and its source and sink fluxes. J. Lake Sciences 16: 10-20.

29. Qin BQ, Hu WP, Gao G, Luo L, Zhang J (2004) Dynamics of sediment resuspension and the conceptual schema of nutrient release in the large shallow Lake Taihu, China. Chinese Science Bulletin 49: 54-64.

30. Xu DL, Lei ZX, Wang HJ, Han BP, Liu ZW (2007) Distribution of Phosphorus in Sediments of Onshore Reed Areas of Lake Taihu. J.China Univ Mining & Technol 17: 557-561.

31. Chen RL, Barko JW (1988) Effects of freshwater macrophytes on sediment chemistry. Journal of Freshwater Ecology 4: 279-289.

32. Wilma JV (2001) Nutrient partitioning in the upper Canning River, Western Australia, and implications for the control of cyanobacterial blooms using salinity. Ecological Engineering 16: 359-371.

33. Scheffer M (2001) Ecology of Shallow Lakes. Kluwer Academic Publishers.

34. Stuart FM, Robert TW (1996) Grazing by black swans (*Cygnus atratus Latham*), physical factors, and the growth and loss of aquatic vegetation in a shallow lake. Aquatic Botany 55: 205-215.

35. Cassandra SJ, John WE, Keith H (2006) Responses of three invasive aquatic macrophytes to nutrient enrichment do not explain their observed field displacements. Aquatic Botany 84: 347-353.

36. Daniela E, Elisabeth MG (2006) Allelopathic activity of *Elodea canadensis* and *Elodea nuttallii* against epiphytes and phytoplankton. Aquatic Botany 85: 203-211.

37. Mulderij G, Mooij WM, Smolders AJP, Van Donk E (2005) Allelopathic inhibition of phytoplankton by exudates from *Stratiotes aloides*. Aquatic Botany 82: 284-296.

38. Blindow I, Hargeby A, Andersson G (2002) Seasonal changes of mechanisms maintaining clear water in a shallow lake with abundant Chara vegetation. Aquatic Botany 72: 315.

39. Lombardo P (2005) Applicability of littoral food-web biomanipulation for lake management purposes: Snails, macrophytes, and water transparency in northeast Ohio shallow lakes. Lake and Reservoir Management 21: 186-202.

40. Canfield DE, Shireman JV, Colle DE, Haller WT, Watkins II CE (1984) Prediction of chlorophyll a concentrations in Florida lakes: Importance of aquatic macrophytes. Canadian Journal of Fisheries and Aquatic Sciences 41: 497.

41. Meijer ML, Jeppesen E, Van Donk E, Moss B, Scheffer M, et al. (1994) Long-term responses to fish-stock reduction in small shallow lakes: Interpretation of five-year results of four biomanipulation cases in The Netherlands and Denmark. Hydrobiologia 275-276: 457-466.

42. Mjelde M, Faafeng B (1997) Ceratophyllum *demersum* hampers phytoplankton development in some small Norwegian lakes over a wide range of phosphorus concentrations and geographical latitude. Freshwater Biology 37: 355-365.

43. Moss B (1994) Brackish and freshwater shallow lakes-different systems or variations on the same theme? Hydrobiologia 275-276: 1-14.

Reducing Risks of Contamination Breaching the Drinking Water System through Catchment Sensitive Farming in Eastern Nigeria

Joachim Ibeziako Ezeji*
Rural Africa Water Development Project (RAWDP), Nigeria

Abstract

In Nigeria, over a score of urban water agencies depend on rivers and other surface water bodies for their raw water intakes. But many of these rivers are currently under unregulated threat from inadequate catchment management. The persistent failure to regulate and take advantage of the work that healthy watersheds perform naturally is currently overwhelming the ability of public water agencies to mitigate catchment sourced contaminants at a fraction of the cost of conventional treatment and guaranty safe drinking water to water consumers. This paper reviews a case study drawn from the Anambra-Imo River basin (one of the twelve River Basins in Nigeria) where intensive agriculture and poor land management activities, especially on catchments of key rivers serving water utilities, has increased pollution threats by non-point pollution; both organic and chemical. Human induced catchment activities, especially agriculture and deforestation has left sloping lands susceptible to rapid flood runoff and massive mudslides. However, an initiative by a Non-Governmental Organization (NGO) developed a collaborative liaison for land management leveraged by catchment sensitive farming, knowledge sharing and awareness creation amongst farmers and landowners etc. aimed at restoring quality on the fragile water bodies of the expansive river basin, particularly the Otammiri River, thus bolstering its resilience to diffuse pollution from agricultural nitrates and phosphates. The emerging result from the pilot was reduced risk of contamination breaching the drinking water system hence supporting the provision of the human population with safe drinking water, which is one of the most important issues in public health.

Keywords: Nigeria; Owerri; River; Agriculture; Pollution; Catchment; Runoff; Farming; Farmer; Water utility

Introduction

Otammiri River is the primary raw water source for the Imo State Water Corporation (ISWC), a public water supply agency. The Corporation supplies drinking water to the residents of Owerri, the capital city of the State. The Otammiri River is one of the many rivers in the larger Anambra-Imo River basin. However, upstream agriculture and deforestation activities have continued to diminish the ability of the river to continue to play this role in the future. The combination of intensive agriculture and deforestation, in addition to sand mining and residential activities continued to pose enormous threat to the over 2 million Owerri residents who depend on ISWC for their long term water needs.

Agriculture poses the greatest threat based on the enormous amount of fertilizer (organic and inorganic) used on commercial and private farms. The result is widespread Nitrates, a by-product of nitrogen-based farm fertilizer, animal manure, wastewater and leaky septic tanks. On top of these are wasteful and inefficient agricultural practices, including very low knowledge of fertilization application. For example, it was observed that many vegetable farmers can apply as much fertilizer as they want, within feet of the river.

Further observations and surveys in the course of this study showed that the leading source of nitrate pollution in many parts of the larger Anambra-Imo River basin-nitrogen fertilizer - is not regulated. For example, agriculture nitrate and phosphate has in combination to the region's high intensity sunshine and warmth, triggered blue-green algae bloom in the utility's reservoir [1-3]. The result has been incidents of toxic compounds released by the algae such as cyanotoxins gradually inundating the city's water-treatment plants. Already taste and smell has been reported by consumers. Such toxins especially Microcystins and nodularins are feared and can be dangerous for humans and animals [4]. As at the time of the pilot, it was already posing technical challenge for ISWC managers, for example, many purification methods used by the ISWC was incapable of removing the nitrate contaminants from the water. The result is that the quality of water supplied by ISWC became a public health challenge. An incident of poorly managed catchment has also been reported by [5] and the resultant expensive cost of water treatment which it posed.

The consequence is that the alternative reliance of city residents on waters provided by vendors sourcing from elsewhere, has come at the high cost of between US$0.92 -US$4.80 per 100 gallons thus taking clean drinking water above the reach of the average poor. Beyond drinking water concerns, eutrophication has also produced so much vegetation that recreational use of the Otammiri River and other neighbouring surface water bodies such as the Oguta Lake; a wetland of international importance is now constrained. Reference [6] notes that toxic blue-green algae in water used as drinking water or for recreational purposes poses a hazard to humans but has long been neglected or at most been treated on a local level. Scums of blue green algae or cyanobacteria accumulating along the shores of ponds and lakes also present hazards to wild and domestic animals [6]. Reference [4] notes that prevention of bloom formation is the most efficient method for avoiding cyanobacterial toxin contamination of drinking water because the presence of cyanobacterial toxins in drinking water supplies poses a serious problem to water treatment facilities, since not

***Corresponding author:** Joachim Ibeziako Ezeji, Rural Africa Water Development Project (RAWDP), Nigeria, E-mail: jo.ibeziakoezeji@gmail.com

all technical procedures are able to effectively remove these toxins to below acceptable levels.

Project location and background

The Project location lies within the tropical rain forest belt of Nigeria. It is found within the Owerri area (Figure 1) in Imo State, Nigeria and is bounded by latitudes 5°26' and 5°45' and longitudes 658' and 7°10' E. It covers an area of about 740 km. The Anambra-Imo River Basin is the largest river basin (Figure 2) in southern Nigeria and encompasses five states of Nigeria - Imo, Anambra, Enugu, Abia and Ebonyi states; with a whopping population of well over 15 million people. Another strategic importance of this basin is that it hosts the Oguta Lake, a wetland of international importance (Ramsar listed site) that measures 572 ha and can be located on coordinates 05°42' N 006°47' E.

All rivers in this region empty into the Imo River, including the Otammiri River. Agriculture and deforested lands account for more than half of the land use in the entire basin (Table 1). In this basin, agricultural activities, especially those on farmlands in both rural and urban area are located on critical watersheds and therefore use rivers as final discharge points for their non-point pollutions; both organic and chemical. Hundreds of indigenous plants and animal species such as large populations of antelopes, Buffon's kob (*kobus kob rob*), Bohor reedbuck (*Redunca redunca*) and the waterbuck (*kobus ellipsiprymnus defassa*) etc., are found in the basin, though eutrophication has joined the threat list of game hunting and bushfires to diminish their diversity.

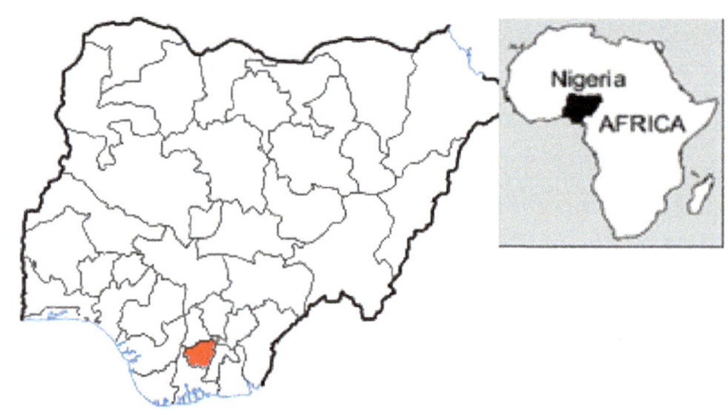

Figure 1: Map of Nigeria (showing Imo State in red).

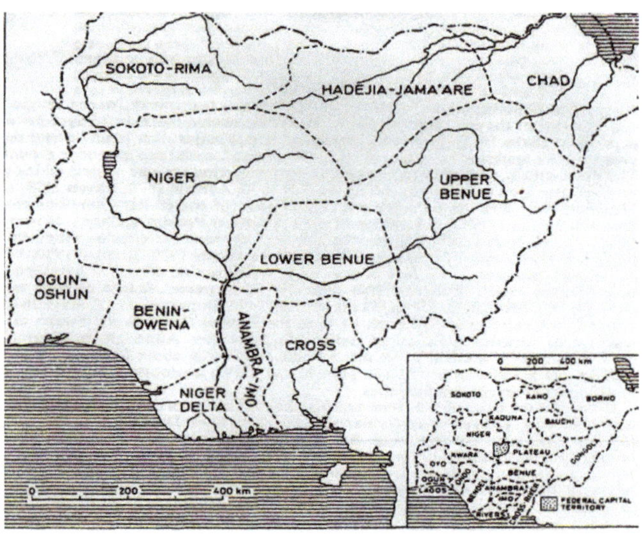

Figure 2: River Basin Authorities in Nigeria, 1979 (after Nwa 1978).

Land Use Characteristics	Northern	Southern	Central
Farmland/fallow areas	65%	55%	5%
Industrial, educational area	1%	5%	10%
Residential areas	25%	35%	80%
Nature conservation areas	8%	2%	1%
Erosion sites	0.1%	0.2%	0.3%
Others	0.9%	3.8%	4.7%
Relative population distribution	Moderate	Moderate	Dense

Table 1: Land Use characteristics of the Watershed.

Water pollution is getting critical in the study area. It is noted that upstream agricultural activities exacerbates non-point pollution through nitrates, while deforestation made sloping lands susceptible to rapid flood runoff and massive mudslides, in addition to increasing the economic and human toll of floods and other economic and social disasters, a toll almost certain to increase as climate change challenges the water cycle.

Another key problem posed by deforestation is the increased concentrations of soil particles washed into the stream by land disturbance, which sink to the bottom of the river and increase the bed loads while, depending on the stream velocity, smaller particles remain in suspension. Suspended matter in the Otammiri River obstructs the penetration of light and limits the photosynthetic zone to less than 1 m depth. Suspended sediments in the river in addition to nutrient loadings from nitrates and phosphates have become a serious concern for the water supply authorities in Owerri because they exert huge water treatment costs.

Studies by [7] shows that the Otamiri River has a maximum average flow of 10.7 m^3/s in the rainy season (September-October) and a minimum average flow of about 3.4m^3/s in the dry season (November-February). The total annual discharge of the Otamiri is about1.7 × 10^8 m^3/s, and 22 percent of this (3.74 x $10^7 m^3$) comes from direct runoff from rainwater and constitute the safe yield of the river. However, recent preliminary studies by the author shows that these are no longer the case as the average flow has shrunk by as much as 55% in both the raining and dry seasons.

Eutrophication by nutrients (especially nitrates and phosphates) has given rise to accelerated rise of algae and higher forms of plant life which has consequently led to the undesirable disturbance to the balance of organisms present in and dependent on the water and its quality. The resulting disturbed aquatic ecosystem is rapidly becoming an unsuitable habitat for other species such as fish and invertebrates. The threat to biodiversity of both the aquatic habitat and of other species in the food chain in the Otammiri ecosystem is not fully studied but it is believed that it poses very low oxygen levels for some species to tolerate, especially the local fish species. Threatened biodiversity includes those of fish, Cichlid and Tilapia etc. Small population of roan antelope (*Hippotragus equinus*) and the red-fronted gazelle (*Gazella rufifrons*) also exist in this watershed.

The blue-green algae bloom in ISWC reservoirs has been linked to the high intensity sunshine's and warmth of the watershed. [6] Posits that algal growth is influenced mainly by water composition, temperature and light intensity. Lowland water storage reservoirs, where the abstracted river water carries treated sewage effluent and agricultural run-off, contain relatively high levels of nitrogen and phosphorus compounds that promote the growth of algae. In many nutrient rich or so-called eutrophic waters, the growth potential of algae is determined by the dissolved phosphate concentration in the water [6]. Notes that a characteristic of algal blooms, which contributes to their nuisance value, is the tendency to form a surface scum. Cyanobacteria have a buoyancy control mechanism that maintains their vertical position in relation to the wind-induced circulation in the water column. Under calm conditions, when natural circulation within the water ceases, algae float and produce a scum that accumulates in the margins. This presents a possible health risk for recreational activities that involve water contact and for animals that drink from the margins of reservoirs. Another significant adverse effect of algal blooms is their propensity to cause de-oxygenation. During daytime, algae use carbon dioxide from the bicarbonate ions in the water and respire oxygen into the water. At night this process is reversed and the carbon dioxide respiration can cause localised oxygen depletion and consequent fish mortalities.

Some species of cyanobacteria produce toxins within their cell structure. These toxins can be released when the algae die and decay. Toxin release can also occur during the growth phase of young algae. The implications of toxic algal blooms are significant and likely to become more so if global warming results in reduced flow in rivers with correspondingly less dilution for effluents and agricultural run-off.

In its 2003 Background Document for the Development of WHO Guidelines for Drinking-Water Quality, WHO noted that recreational use of water contaminated with toxic algae had been linked to incidents of human illness in many countries. Symptoms ranged from sickness with diarrhoea to muscle and joint pains and irritation of the skin, eyes and throat. WHO also considered reports of illnesses associated with the consumption of contaminated drinking water reported in the USA, Australia, South Africa and China. In China incidence of liver cancer was related to consumption of contaminated water from ponds and ditches [6].

Project design and ownership

To meet the water challenges of the future in Owerri, there is the need to figure out how to deal with nitrates and one of the options is the collaborative project described in the paper.

The lead NGO responsible for the initiative was the Rural Africa Water Development Project (RAWDP). RAWDP is a community development program funded through open challenge funds. RAWDP's goal was to engage communities and its people in the protection of their water sources and the promotion of broad-based pro-poor growth in water and sanitation access delivery. To pursue this goal, RAWDP drives systemic change by mapping and delineating water resources pollution points, and widening the distribution and accessibility of appropriate clean water and sanitation infrastructure through active partnership with water resource management institutions, communities and entrepreneurs. Such partnerships aim to support innovative ideas, investment and regulatory reform that will bolster resource resilience, improve business performance, stimulate economic growth and ultimately provide benefits for the poor.

Other NGOs includes the Pan African Vision of the Environment (PAVE) and the Organised Centre for Women Empowerment (OCEAN) in addition to government institutions such as the Anambra-Imo River Basin Authority, the Imo State Water Corporation and other groups.

Methodology

The project used a six step process methodology that comprised of the following, with (Table 2) illustrating the steps, activities and timelines.

Stakeholders' mapping and engagement

The project used a 3 step approach to identify and bring key stakeholders on board. The first step was stakeholder mapping, which enabled the project to identify existing farmers, land owners and occupiers/tenants etc. The second step involved building awareness of these stakeholders on the need to address at source the poor land management challenges that impact water quality in the watershed [8]. The third step involved building-in support and systems for change in on-the-farm practices and other associated issues such as the shortlisting and cost analysis of possible interventions.

Identification of critical acres and designation of nitrate and phosphate vulnerable zones within the catchment

The identification of the sources of nitrogen was the first step towards remediation of nutrient pollution in aquatic ecosystem. Water samples were collected during both the wet and dry seasons from different parts of the Otamiri River and the Oguta Lake. This was followed by their analysis with focus on the seasonal and spatial variations of nitrate concentration and sources. This enabled the project to determine critical acres and designation of Nitrate and Phosphate Vulnerable Zones within the catchment. To effectively do that, the project employed the Oxygen Isotopic composition as an additional marker of nitrate sources due to the large variation of $\delta^{18}O$ ranges among different sources [9,10]. The key objective underpinning the use of the dual isotopic approach in tracing the nitrate dynamics was to investigate the spatial and seasonal variations of nitrate concentrations in the principal waters of the catchment. This approach is unlike the both the use of the Nitrogen Isotope and the traditional analysis of nitrate sources. While the later determines nitrate concentration in freshwater systems and potential nitrate sources of fresh water supplies. However, it suffers from the possibility of multiple nitrate sources and some biogeochemical processes. On the other hand, the former, specifically, $\delta^{18}O$ informs the sources of nitrate due to the distinct isotopic characteristics of main nitrate sources such as septic systems, animal waste, commercial fertilizer and decaying organic matter [11,12]. However, its efficacy is constrained by the difficulty associated with differentiating these sources because of some significant overlaps in $\delta^{15}N$-NO_3- ranges. Moreover, fractionation process may constrain the accuracy of NO_3- source identification. For example, kinetic isotope effects during microbial de-nitrification can cause the enrichment of the heavy isotopes in the remaining NO_3- [13]. The dual isotopic approach used in the project as described by the paper has successfully been used to investigate the nitrate sources of Mississippi Rivers and the Illinois River in the US [9,10] as well as the Yangtze [14] and the Yili Rivers [15], both in China, and the Han River in Korea [16]. Those studies reported high levels of sensitivity and reliability of using dual isotopic technique for tracking nitrate sources.

STEPS	Activities	Timeline
One	Stakeholders' mapping and engagement	May-December 2013
Two	Identification of critical acres and designation of Nitrate Vulnerable Zones	January-March 2014
Three	Formulation of intervention and cost	April-July 2014
Four	Implementation of selected interventions	Aug 2014-December 2014
Five	Monitoring and Evaluation activities	January 2015-March 2015
Six	Learning lessons	April 2015-Continuing

Table 2: Schedule of activities and timelines.

Formulation of intervention and cost

Considering the huge contribution of agriculture in the overall pollution of water resources, there was the great urge to curb its continued impact. But to do this effectively, the project needed to be sure of its shortlist. It therefore opted for interventions that are solution - effective, cost - efficient and ecologically green. Participatory approaches were adopted in the promotion of best practices in land use and farm infrastructure design and development. These were shortlisted in tandem with potency in overall soil, livestock, pesticide, fertilizer and organic manure management to control diffuse water pollution from agriculture in the catchment. Apart for the technical infrastructures that were paid for; all of the advisory services to the farmers was free of charge and were delivered through workshops, demonstration events and farm visits.

Implementation of selected interventions

Some of the interventions included the establishment and maintenance of artificial wetlands and use of riparian buffer strips. Others were in-field grass buffer strips, use of drills across the slopes, reduction of dietary Nitrate and Phosphate intakes by livestock, phase feeding of livestock, use of fertilizer recommendation system, non-application of fertilizer on high risk areas as well as the use of concrete surfaces in storing solid manure and simultaneously collecting the resulting effluent as well as supporting large farms to install and use Biogas plants etc. The project also provided at least 10,000 local farmers and residents within the watershed with adequate skills and awareness relevant in pollution prevention especially those emanating from local farms in this intensively farmed watershed.

Monitoring and evaluation activities

The Monitoring and Evaluation process chiefly used data collected on water quality indicators at their current (benchmark) levels (both ecological and social), and tracked changes in these indicators over time. The project had followed a systematic process of monitoring the achievements of the project operations, collecting and analysing farm activities to track progress toward planned results; using output information and evaluation to influence decision- making and resource allocation; and communicating results achieved, or not attained, to advance learning. In a nutshell this translates to learning from the on-the-farm real world experiences to improve effectiveness, efficiency and knowledge critical in achieving results. Results were often derived from rigorous investigation and analysis of farm activities; applying multi-disciplinary expertise and relying upon well-informed participatory insights from target farmers and land owners etc., The hierarchy of results captured within this iterative cycle occurred at the Activity Level, Project Level, Intermediate Level and the highest Level Strategic Level. In other words these are also categorized as Input, Output, Outcome and Impact Results. For example, the artificial wetlands built in the catchment had fences and channels that were sufficient to capture run-off and sediment from a field or group of fields that regularly discharge run-off water and sediment. The result at this level is that the wetlands are used for the "treatment" of wastewater generated from hard-standing areas and to intercept run-off water from a field or group of fields. These were useful in trapping sediment and through the retention of run-off, reduce Nitrate and Phosphate (soluble and particulate) and free inorganic loads in water exiting the wetland. At the utility level, effective monitoring indicates that nitrate concentration limits of 10 mgN/L in drinking water is no longer breached as was the case on the baseline in utter contravention of both the National Drinking Water Quality Standard (NDWQS) as well as those of the World Health Organisation (WHO).

Lessons learned

Effective mobilization and engagement of farmers and land owners is key: The active involvement of key stakeholders-farmers and land owners; at all stages of the project ensured that water quality deterioration was checked and mitigated. This requires adroit skills in social engagement. Excellence in managing relationship with stakeholders is essential to achieving success. In doing this, the project was able to effectively identify key stakeholder groups and designed appropriate communication strategies for each group. It created a favorable ground for the promotion of the desired change.

To maintain freshwater biodiversity, it is necessary to manage the physical processes in rivers properly: These include water quality, sediment dynamics, and, of course, flow. Flow is the main driver of biodiversity in rivers–it creates the aquatic habitats, it brings the food down from upstream, it covers the floodplain with water during high flows, and it flushes the sediments and poor water quality through the system. It has been observed that a fish which is sensitive to flow conditions, insects which lay eggs in the river, or water chemistry will give the first indication that the river is changing. Ecological components of a river can be used to indicate incipient problems, before they reach the stage of endangering people and their livelihood.

Restoring watersheds restores "Ability to Produce Clean Water" (APCW): The "Ability to Produce Clean Water" - an index used in measuring severe pressure on watersheds [17] is a desired input in water utility water safety plans. The project is in tandem with the multi-barrier approach for achieving drinking water safety as argued by [18]. The framework offers source protection to keep the raw water as clean as possible in order to reduce the risk of contamination breaching the drinking water system. Forested watersheds reduce storm runoff, stabilize stream banks, shade surface water, cycle nutrients, and filter pollutants [19]. For example, private companies that operate the water concessions in Manila are investing in catchment management to preserve the quality and quantity of the water on which the city depends [20]. Also, in countries with degraded water catchments, development of water storage infrastructure needs to be accompanied by improved protection of watersheds to sustain the investment [20].

Use of novel appropriate technology to advance knowledge: Through the sharing of experiences and the creation of networks across geographic regions and disciplines in Nigeria and overseas, the project hopes to access more tools, models and motivation to continue its activities. It acknowledges that emerging technologies offer unprecedented opportunities to advance water resource knowledge and information, using some of the most promising technologies such as the Thermal infrared (TIR) imagery; Digital Temperature Sensing (DTS); advanced automated samplers with telemetry; high–resolution digital topography; and spatially explicit, process–based hydro-ecological models. These new technologies could help to enormously reduce present costs for soil and water resource inventories and monitoring, and provide broad geographic and temporal coverage. Expanded application of these technologies and collaboration with partners will enable the project to improve the characterization and monitoring of water resources and better understand the processes controlling them. Such data could significantly improve the ability to forecast stream flows, predict changes in vegetation, assess threats to forest health, quantify wildfire risks, identify vulnerable and resistant landscapes, and provide early warning systems for threats to people, fish, and wildfire. It could greatly facilitate the implementation of more strategic, effective and robust watershed management practices.

Conclusion

By adopting the catchment sensitive farming approach to improve River water quality the paper has illustrated a concerted local effort to improve the management of the freshwater resource by forging a more effective connection among people, nature and the government. Particularly, the adoption of agricultural practices that are sensitive to the water ecosystem, the change in paradigm will empower hundreds of farmers to grow more crops without exacerbating river pollution or environmental damage. The paper recognises agriculture as the *dues ex machina* that triggers shared prosperity and inclusive growth. Land and water resources management in this context therefore are related first and foremost to the integrity, vitality, and resilience of ecosystem structures and processes. It is acknowledged that the key to ecosystem management is the goal of ecological sustainability–protecting and restoring critical ecological components, functions, and structures in perpetuity so that future as well as current generations will have their needs met, in this case–clean drinking water.

References

1. Kratz William A, Myers Jack (1955) Nutrition and growth of several blue green algae. American Journal of Botany Vol 42: 282-287.
2. Sivonen K (1990) Effects of light, temperature, nitrate and hepatotoxin production by oscillatonia agardhii strains. Appl Environ Microbiol Vol 56: 2658-2666.
3. Smith Val H (2011) Light and nutrient effects on the relative biomass of blue-green algae in Lake Phytoplankton Vol 43: 148-153.
4. Hitzfeld BC, Hoger SJ, Dietrich DR (2000) Cyanobacterial toxins: Removal during drinking water treatment and human risk assessment. Environmental Toxicology, Environ Health Perspect Vol 108: 113-122.
5. Ezeji JI, Smout I, Bosher L (2013) Coupling utility performance targets with catchment management. Proceedings of the Institution of Civil Engineers-Municipal Engineer 166: 130-138.
6. Foundation for Water Research, FWR (2015) Toxic Algal Blooms in Drinking Water Reservoirs. Accessed 05 November 2015.
7. Egboka BCE, Uma KO (1985) Water resources of Owerri and its environs, Imo State Nigeria. Journ Min Geol 22: 57-64.
8. O'Keeffe Jay, Quesne Tom Le (2009) Keeping Rivers Alive; A primer on environmental flows. World Wide Fund Water Security Series 2. WWF UK.
9. Panno SV, Hackley KC, Kelly WR, Hwang HH (2006) Isotopic evidence of nitrate sources and denitrification in the Mississippi River, Illinois. J Environ Qual 35: 495-504.
10. Panno SV, Hackley KC, Kelly WR, Hwang HH, Martinsek AT (2008) Sources and fate of nitrate in the Illinois river basin, Illinois. J Hydrol 359: 174-188.
11. Durka W, Schulze ED, Gebauer G, Voerkelius S (1994) Effects of forest decline on uptake and leaching of deposited nitrate determined from ^{15}N and ^{18}O measurements. Nature 372: 765-767.
12. Mayer B, Boyer EW, Goodale C, Jaworski NA, Breemen NV, et al. (2002) Sources of nitrate in rivers draining sixteen watersheds in the northeastern U.S. Isotopic constraints. Biogeochemistry 57: 171 -197.
13. Xue D, Botte J, De Baets B, Accoe F, Nestle A et al. (2009),Present limitations and future prospects of stable isotope methods for nitrate source identification in surface and groundwater. Water Res. 43: 1159-1170.
14. Lee KS, Bong YS, Lee DH, Kim KJ (2008) Tracking the sources of nitrate in the Han River watershed in Korea, using $\delta^{15}N\text{-}NO_3^-$ and $\delta^{18}N\text{-}NO_3^-$ values. Sci Total Environ 395: 117-124.
15. Zeng Haiao, Wu Jinglu (2015) Tracing the nitrate sources of the Yili River in the Taihu Lake Watershed: A Dual Isotope Approach. Water 7: 188-201.
16. Li S, Liu C, Li J, Liu X, Chetelat B, Wang F (2010) Assessment of the sources of nitrate in the Changjian River , China using a nitrogen and oxygen isotopic approach. Environ Sci Technol 144: 1573-1578.
17. Barnes M, Todd A, Whitney Lilja R, Barten P (2009) Forests, water and people: drinking water supply and forest lands in the Northeast and Midwest United States. NA-FR-01-08. Newton Square, PA: US. Department of Agriculture, Forest Service, North-eastern Area State and Private Forestry 71.
18. Hrudey Steve E, Hrudey Elizabeth J (2002) Walkerton and North Battleford -Key Lessons for Public Health Professionals: Commentary. Revue Canadienne De Sante Publique Volume 93: 332-333.
19. Furniss MJ, Staab BP, Hazelhurst S, Clifton CF, Roby KB, et al. (2010) Gen. Tech. Rep.PNW-GTR-812. Portland, OR: US. Department of Agriculture, Forest Service, Pacific Northwest Research Station 75.
20. Foster, Vivien, Briceno-Garmendia, Cecilia (2010) Africa's Infrastructure: a time for transformation. A co-publication of the Agence Francaise de Developpement and the World Bank. The World Bank, Washington.

Impact of Palm Kernel Oil (PKO) Biodiesel-Contaminated Catfish on Hepatic Function of Rat

Adeyemi Olalekan*

Department of Environmental Science, Federal University of Petroleum Resources, PMB 1221, Effurun, Nigeria

Abstract

The purpose of the present study was to assess the effects of consumption of catfish (*Clarias gariepinus*) exposed to palm kernel oil (PKO) biodiesel on functions and oxidative status of liver of albino rat. The author measured the responses of growth performance (average daily weight gain, feed conversion ratio, body weight, and relative liver weight), oxidative stress, and indices of liver functions of rat placed on feed formulated with catfish exposed to borehole water mixed with different concentrations of PKO biodiesel (0% w/v, 0.1% w/v, 0.25% w/v). The results showed that fish exposed to PKO biodiesel had no effect on relative liver weight, but average body weight was significantly reduced, serum levels of total and direct bilirubin were significantly elevated while those of bilirubin and globulin were significantly reduced. However, specific activities of catalase (CAT), superoxide dismutase (SOD) and glutathione-s-transferase (GST) of serum of rats fed with catfish exposed to lower concentration of PKO biodiesel (BD 0.1) were significantly higher than those of rats fed with catfish exposed to higher concentration of PKO biodiesel (BD 0.25) ($p<0.05$) while both are significantly lower than that of the control rats ($p<0.05$). Serum level of MDA in test rats increased significantly ($p<0.05$) as the concentration of PKO biodiesel increased. The data clearly demonstrate that consumption of catfish exposed to PKO biodiesel can result in liver dysfunction at a degree not dependent on concentration of PKO biodiesel by a mechanism of oxidative stress which is dependent on the concentration of PKO biodiesel.

Keywords: Palm kernel oil; Biodiesel; Oxidative status; Hepatic; Catfish

Introduction

Biodiesel is defined by ASTM International as a fuel composed of monoalkyl esters of long-chain fatty acids derived from renewable vegetable oils or animal fats meeting the requirements of ASTM D6751 [1]. Vegetable oils and animal fats are principally composed of triacylglycerols (TAG) consisting of longchain fatty acids chemically bound to a glycerol (1,2,3-propanetriol) backbone. The chemical process by which biodiesel is prepared is known as the transesterification reaction, which involves a TAG reaction with a short-chain monohydric alcohol normally in the presence of a catalyst at elevated temperature to form fatty acid alkyl esters (FAAE) and glycerol. The conversion of TAG to biodiesel is a stepwise process whereby the alcohol initially reacts with TAG as the alkoxide anion to produce FAAE and diacylglycerols (DAG, reaction [2].

Free radicals are atoms or molecules with unpaired electrons. Free radicals in the organism are of endogenous and exogenous origin. Endogenously, they occur in mammals during mitochondrial electron transport chain, oxidative reactions in phagocytic and endothelial cells, redox cycles, arachidonic acid metabolism, and autooxidation reactions [3]. Industrial pollutants, drugs, diet, ionizing radiation, ultraviolet light, cigarette smoke, and xenobiotics are exogenous sources of free radicals [3]. Free radicals are classified as Reactive Oxygen Species (ROS) and reactive nitrogen species (RNS). Superoxide radical, hydrogen peroxide, hydroxyl radical, hypochlorous acid, singlet oxygen, alkyl radical, hydroperoxyl radical, and alkoxyl radical are effective as ROS whereas nitric oxide is the most well known RNS [4,5]. Reactive oxygen species cause protein denaturation, damage in nucleic acids, and lipid peroxidation. As a result, impairment occurs in the membrane structure and cellular functions. Reactive oxygen species are involved in pathogenesis of many diseases [6]. When reactive oxygen species start to increase, phospholipids, glycolipids, glycerides, and sterols present in the cell membrane are decomposed. Peroxides, alcohols of these, ethane and pentane, nitric oxide and malondialdehyde are formed. Ultimately, oxidative damage is shaped [7].

Liver is the primary organ for the detoxification, as well as being the target of many of environmental chemicals including biodiesel. Today, we are all exposed to mixtures of different chemicals at a time because of industrialization and fast life style. Hepatic toxicity of chemical and/ or biological mixtures may have serious effects on the functions of liver, which is the non-stopping machine for biotransformation as well as several very important physiological processes.

McCormick [8] reviewed biodiesel emission data published through 2001 for heavy-duty engines. It was clear there were substantial reductions in the emissions of particulate matter (PM), carbon monoxide (CO), and hydrocarbon (HC), although there was an increase in nitrogen oxides (NOx) obtained by using B20 (blends of 20% biodiesel with 80% conventional diesel). A more detailed analysis of blends of biodiesel emissions data has indicated that the solid carbon fraction of the PM was reduced and the soluble organic fraction (SOF) was increased [9]. Some studies have compared biodiesel and conventional diesel extracts using mutagenicity assays [10]. The acute toxicity of biodiesel was assayed by the Microtox test, which demonstrated an increased toxicity in the semivolatile fraction compared with the particulates [11], and was shown to be a potent inflammatory agent [12] compared with exposure of diesel PM extracts. Moreover, the type of motor, fuel,

***Corresponding author:** Adeyemi Olalekan, Department of Environmental Science, Federal University of Petroleum Resources, PMB 1221, Effurun, Nigeria
E-mail: Adeyemi.olalekan@fupre.edu.ng

particle size, and chemical composition has all been shown to influence the mechanisms and magnitude of adverse manifestations [13].

Most biodiesel studies have focused on the analysis of exhaust emission material, motor efficiency, and the different processes for obtaining biodiesel with oleaginous microorganisms and palm, soybean, kitchen oil, palm oil, or animal fat [13,14]. Due to leaks in storage, fuels pollute underground waters. In that case, humans and animals are exposed to biodiesel cutaneously and orally. There are numerous studies investigating the effects of inhaled diesel and biodiesel fuels. However, there are very few studies on the fuels taken orally, directly [15,16] or indirectly through the food chain [17]. As a result of paucity of information, the present investigation was embarked upon to study the effects of catfish exposed to palm kernel oil (PKO) biodiesel on the functions and oxidative status of liver of rats.

Materials and Methods

Reagents and solvents were of analytical grade and are products of British Drug House, Poole, England.

Palm Kernel Oil (PKO)

Palm kernel oil was purchased at the local market in Effurun, Nigeria. 100 g PKO was used for the transesterification process. The ethanol used (99% pure) is an analytical grade with boiling point of 78°C; while the NaOH used was also an analytical grade product of Aldrich Chemicals, England. The blender used was a Dry and Wet mill Blender with a clear glass (1,250 cc capacity) containers and stainless steel cutting blades. Other major materials used include scales, translucent white plastic container with bung and screw-on cap, funnels, PET bottles and thermometer.

Preparation of bio-diesel from PKO

Biodiesel was prepared from PKO in accordance with the method described by Aremu et al. [18].

Experimental water and fish treatment: The Biodiesel from PKO was diluted with borehole water to obtain 0.25 and 0.1% v/v. Twenty-four healthy juvenile catfish (*Clarias gariepinus*) were obtained from a commercial fish pond at Ekpan in Delta State, Nigeria and acclimatized for ten days prior to the commencement of the experiment. The catfish were grouped into three (3) of eight catfish and were kept in 30 L plastic aquaria. Group A served as control and the catfish here were cultured in borehole water while those in Groups B and C were exposed to the different mixtures (0.1% v/v and 0.25% v/v respectively) of Biodiesel from PKO. The catfish were fed *ad libitum* with commercial fish meal for 30 hrs during which the experiment lasted. Each of these treatments had three replicates.

At the end of the 30 h experimental period, the catfish were harvested, oven dried at 40°C and used as a source of protein (25%) to

NUTRIENT	Experimental diet		
	CONTROL	BD0.1	BD0.25
PROTEIN (%)	20.2 ± 2.2^{a}	18.39 ± 2.1^{a}	18.19 ± 1.9^{a}
FAT (%)	7.7 ± 1.2^{a}	4.88 ± 1.2^{b}	3.96 ± 1.1^{b}
ASH (%)	6.2 ± 1.1^{a}	7.38 ± 0.7^{a}	7.30 ± 0.7^{a}
FIBRE (%)	5.3 ± 0.5^{a}	9.22 ± 1.1^{b}	5.56 ± 0.9^{a}
WATER (%)	6.3 ± 0.7^{a}	8.83 ± 1.2^{b}	7.83 ± 1.1^{ab}
CARBOHYDRATE (%)	54.3 ± 2.4^{ab}	51.30 ± 2.4^{a}	57.16 ± 2.5^{b}

Values are means ± SEM for 3 determinations. ^{a,b,c} Row values with different superscripts are significantly different ($p<0.05$).

Table 1: Proximate composition of experimental diet.

formulate diet for albino rats. The diet for each group was formulated by mixing known quantities of sources of each food class comprising corn starch (52%), oil (4%), maize cob (4%), sucrose (10%) and vitamin/mineral mixture (5%). The food items were mixed together and manually made into pellets to feed albino rats. Proximate analysis of the experimental diet is shown in Table 1.

Experimental rats and treatment: Thirty albino rats of (51 ± 2.0 g) were procured from the animal house of Department of Anatomy, University of Benin, Benin-City, Nigeria. All animals were housed under standard laboratory conditions with free access to water *ad-libitum* and balanced pellets food. The housing temperature was (25 ± 1°C) with 12 h dark /light cycle and 50% humidity. All ethical guidelines on the use of animals for investigational purposes were followed and the experiment protocol was approved by Federal University of Petroleum Resources, Effurun (FUPRE), Nigeria ethics committee.

The animals were grouped into three with each group containing ten rats. The rats in Group 1 served as the control and they were fed on the control diet, which was formulated with catfish cultured in borehole water. Animals in Groups 2 and 3 were fed on diet formulated with catfish exposed to the different mixtures of biodiesel (0.1 and 0.25% v/v respectively). The feeding lasted for a period of thirty (30) days (after an acclimatization period of ten days) during which the weight and feed intake were monitored.Each of these treatments had three replicates.

At the end of the feeding exercise, the rats were anaesthetized by placing them in a jar containing cotton wool soaked with chloroform before being sacrificed by jugular puncture. The liver of experimental rat was removed into a beaker containing ice cold 0.25 M sucrose solution. The blood was obtained through their jugular veins. Each blood sample was thereafter centrifuged at 3500 rpm for about 15 min using refrigerated centrifuge RC650s and the serum obtained was preserved at -8°C until required for use. A portion of the liver was homogenized for biochemical studies and enzyme assays. MDA determination was based on method described by Bird et al. [19]. The method described by Jollow et al. [20] was used to determine reduced glutathione (GSH) concentration. Catalase activity was determined according to the method of Sinha [21]. The activity of superoxide dismutase (SOD) was determined by the method of Misra and Fridovich [22]. The cytosolic glutathione s-transferase activity was determined spectrophotometrically at 37°C (340 nm) by the procedure described by Habig et al. [23]. The activity of Gamma glutamyltranspeptidase (GGT) was determined following the method described by Tietz [24]. Concentrations ofserum bilirubin and albumin were determined by a method earlier reported [24]. The determination of serum globulin concentration was carried out using the method described by Tietz [25] by subtracting the concentration serum albumin from the total protein content determined using Biuret method as described by Henry et al. [26]. Activities of alanine transaminase (ALT) and aspartate transaminase (AST) in the liver of experimental animals were determined following the procedure reported by Reitman and Frankel [27] modified by Schmidt and Schmidt [28].

Statistical Analysis

All numerical results were obtained from the three (3) groups (control and treated). Data obtained were presented as mean ± SEM and subjected to statistical analysis using a one way analysis of variance (ANOVA) by employing the method of Steel and Torrie [29]. Significant difference between the treatment means was determined at 95% confidence level using Duncan's Multiple range test [30]. Post-hoc comparison using Duncan's New Multiple Range Test (DNMRT)

Group of Rats	Average daily weight gain/loss (g/rat/day)	Average daily feed intake (g/rat/day)	Feed Conversion Ratio (FCR)
Control	0.20 ± 0.003^{a}	6.39 ± 0.18^{a}	10.65 ± 0.23^{a}
BD0.1	0.012 ± 0.002^{b}	6.33 ± 0.27^{a}	15.85 ± 0.19^{b}
BD0.25	0.007 ± 0.001^{c}	6.36 ± 0.16^{a}	15.90 ± 0.21^{b}

Values are means ± SEM for 10 rats. ^{a,b,c} Column values with different superscripts are significantly different (p<0.05).

Table 2: Performance of rats placed on diet formulated with catfish exposed to biodiesel polluted water.

Group of Rats	Bilirubin		Albumin (g/dL)	Globulin (g/dL)
	Direct (mM/L)	Total (mM/L)		
CONTROL	1.04 ± 0.02^{a}	5.45 ± 0.21^{a}	16.89 ± 1.35^{a}	10.54 ± 1.09^{a}
BD0.1	1.26 ± 0.02^{b}	6.02 ± 0.52^{ab}	14.66 ± 1.64^{b}	7.72 ± 1.30^{b}
BD0.25	1.94 ± 0.02^{c}	6.56 ± 0.64^{b}	12.31 ± 1.55^{b}	6.48 ± 0.97^{b}

Values are means ± SEM for 10 rats. ^{a,b,c} Column values with different superscripts are significantly different (p<0.05).

Table 3: Serum concentration of selected liver function indicators of rats placed on diet formulated with catfish exposed to biodiesel polluted water over a period of 30 days.

Group of Rats	AST	ALT	GGT	CAT	SOD	GST
CONTROL	24.9 ± 1.19^{a}	14.6 ± 1.64^{a}	67.9 ± 4.36^{a}	0.82 ± 0.03^{a}	1.21 ± 0.02^{a}	1.32 ± 0.02^{a}
BD0.1	18.8 ± 1.94^{b}	10.4 ± 1.35^{b}	53.2 ± 4.26^{b}	0.61 ± 0.02^{b}	0.92 ± 0.01^{b}	0.88 ± 0.02^{b}
BD0.25	15.2 ± 1.69^{b}	10.0 ± 1.89^{b}	47.3 ± 4.90^{b}	0.55 ± 0.03^{c}	0.81 ± 0.02^{c}	0.56 ± 0.04^{c}

Values are means ± SEM for 10 rats. ^{a,b,c} Column values with different superscripts are significantly different (p<0.05).

Table 4: Specific activity (U/mg protein) of selected enzymes of the liver of rats placed on diet formulated with catfish exposed to biodiesel polluted water over a period of 30 days.

was employed. DNMRT is a One-Way Analysis of Variance (ANOVA) that allows multiple comparisons. The error rate of each pairwise comparison is considered rather than an overall rate, and allows a higher rate for pairs of sample averages that are further apart when ordered by size.

Results

Rats were placed on diets formulated with catfish exposed to biodiesel contaminated water as source of protein. The performance of rats placed on the diets formulated with catfish exposed to biodiesel (test diet) is presented in Table 2. Average daily weight gain of rats placed on the BD 0.25 diets is 1/3rd that of rats placed on the control diet. In contrast, there was no significant difference in the average daily feed intake between the control diet and the test diet (p>0.05). The feed conversion ratio (FCR) of test rats was significantly higher than that of the rats placed on control diet (p<0.05).

The growth of rats placed on diets formulated with catfish exposed to biodiesel was monitored over a period of 30 days and the growth response curve is presented in Figure 1. Generally, both test and control rats showed positive growth response. The growth of controls was found to be significantly higher than that of the rats placed on diets formulated with catfish exposed to biodiesel.

Figure 2 presents the relative liver weight of rats placed on diet formulated with catfish exposed to biodiesel. Although the value of relative liver weight of test rats was numerically lower than that of control rats, there was still no significant difference between the control and the test rats (p>0.05).

The serum levels of some indicators of liver function were monitored. Table 3 presents the results of serum concentrations of bilirubin, albumin and globulin of rats placed on feed formulated with catfish exposed to biodiesel over a period of 30 days. Generally, it was observed that concentrations of direct and total bilirubin of rats increased significantly (p<0.05) as the level of biodiesel increased while the serum concentrations of albumin and globulin decreased significantly (p<0.05) with increasing level of biodiesel. The concentration of direct bilirubin in BD 0.25 rats was about 2 folds that of the control. There was no significant difference (p>0.05) between the concentration of serum total bilirubin of BD 0.1 rats and the control, while that of BD 0.25 rats was significantly higher than that of control (p<0.05). The serum globulin concentration of control rats was about 1½ fold that of the BD 0.25 and BD 0.1 rats.

Figure 3 presents the level of serum malondialdehyde (MDA) of rats placed feed formulated with catfish exposed to biodiesel. The general trend observed was that serum level of MDA increases significantly (p<0.05) with increasing level of biodiesel. The serum MDA levels of test rats were about 1 ½ fold that of the control.

Results of concentration of serum reduced glutathione (GSH) of rats placed on diet formulated with catfish exposed to biodiesel over a period of 30 days is presented in Figure 4. Here, concentration of serum GSH of test rats was significantly lower than that of the control (p<0.05). GSH concentration of serum of control rats was twice that of the rats placed on diet formulated with catfish exposed to biodiesel.

Specific activities of selected liver enzymes of rats exposed to biodiesel were assayed and the result is presented in Table 4. Specific activities of alanine transaminase (ALT), asparatate transaminase (AST) and Gama glutamyltranspeptidase (GGT) of serum of BD 0.1 and BD0.25 rats were not significantly different (p>0.05) but significantly lower than those of control rats. Conversely, specific activities of catalase (CAT), superoxide dismutase (SOD) and glutathione-s-transferase (GST) of serum of BD 0.1 is significantly higher than that of BD 0.25 (p<0.05) while both were significantly lower than that of the control rats (p<0.05). Generally, specific enzyme activities of the selected enzymes decreased as the level of biodiesel increased.

Discussion

The current study demonstrated that catfish exposed to biodiesel from PKO significantly affects function and oxidative status of liver of albino rats. This is the first investigation to report that catfish exposed to biodiesel from PKO decreased liver antioxidant status of albino rats. In addition, these data are the first to demonstrate liver dysfunction of albino rats fed with catfish exposed to environmentally relevant concentration of biodiesel from PKO.

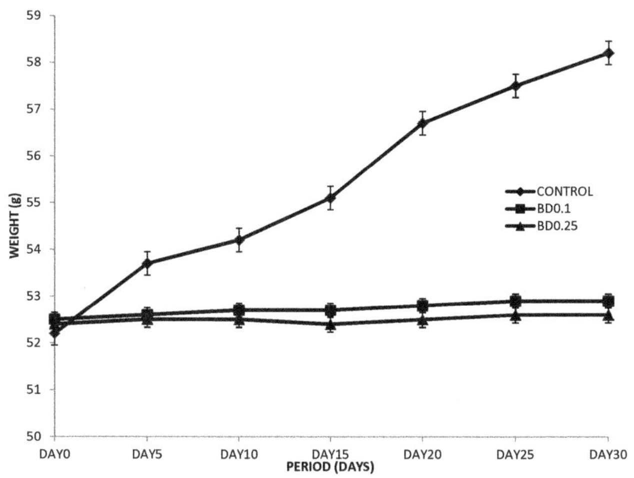

Figure 1: Effects of catfish exposed to PKO biodiesel on growth of rats. Plotted values are means ± SEM for 10 rats.

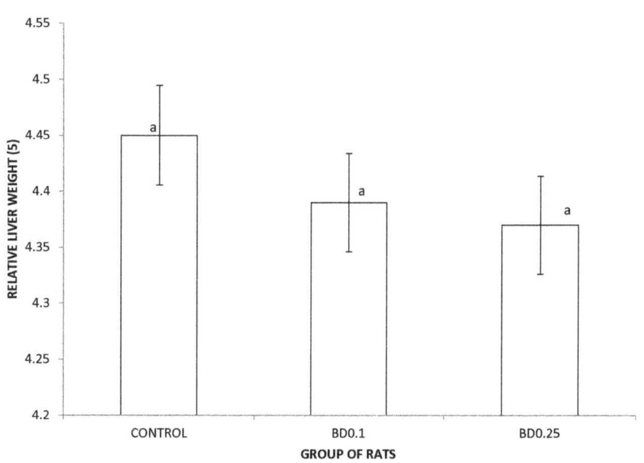

Figure 2: Effects of catfish exposed to PKO biodiesel on relative liver weight. Plotted values are means ± SEM for 10 rats. [a,b,c]Column values with different superscripts are significantly different (p<0.05).

The performance of rats placed on the formulated diets as reported in Table 1 suggested that the biodiesel from PKO decreased the nutrient in the catfish exposed to it thereby increasing the FCR. It could also be that the albino rats could not convert the catfish exposed to biodiesel from PKO into useful nutrients for body growth and development creating a situation where quantity of food consumed by the animals was not commensurate with body weight eventhough positive growth response was observed (Figure 1). A number of studies have demonstrated positive growth response to consumption of contaminated water and food from contaminated water bodies [31-33]. Data from this study are in agreement with these earlier findings.

Since organ weight evaluation is an essential part of the toxicologic and risk assessment of drugs, chemicals, biologics, food additives, and medical devices, the relative liver weights of the experimental rats were determined (Figure 2). These results revealed that consumption of fish exposed to biodiesel form PKO did not significantly affect relative liver weights of albino rats. However, Haley et al. [34] reported that organ weight changes themselves were not necessarily toxic effects and emphasized that the organ weight data should be assessed in the context of the entire study. Therefore, the data on relative liver weight may not imply non-toxic effect.

Bilirubin is an endogenous compound that can be toxic [35], especially in neonates. Bilirubin is the ultimate breakdown product of haemoglobin and serves as a diagnostic marker of liver and blood disorders. Elevated serum bilirubin of rats observed in this study (Table 2) suggested liver dysfunction which may not be unconnected with catfish exposed to biodiesel from PKO. The liver is responsible for clearing the blood of bilirubin. It does this by the following mechanism: bilirubin is taken up into hepatocytes, conjugated (modified to make it water soluble) and secreted into the bile, which is excreted into the intestine. Increased total bilirubin causes jaundice and can signal a number of problems. Studies had shown that if direct (that is, conjugated) bilirubin is normal, then the problem is an excess of unconjugated bilirubin and the location of the problem is upstream of bilirubin excretion. Anemia, viral hepatitis, or cirrhosis can be suspected. If direct bilirubin is elevated, then the liver is conjugating bilirubin normally, but is not able to excrete it. Bile duct obstruction by gallstones or cancer should be suspected [17]. In this study, rats fed with catfish exposed to PKO biodiesel presented elevated serum concentrations of both total and direct bilirubin suggesting that the liver is not able to excrete bilirubin which is an evidence of liver dysfunction.

Serum albumin and globulin concentrations are some biochemical indices for monitoring liver function in the blood (Table 2). Abnormal levels of these proteins have been reported to be associated with haemolysis or increased breakdown of RBC and/ or liver damage [36]. Decreased serum concentrations of albumin and globulin as observed in this study lend credence to the submission that the liver function may be impaired. Both globulin and albumin are produced by the liver. If the liver is damaged, it may no longer be able to produce these proteins. The results presented on serum proteins are consistent and all pointing to the fact that the liver may have been damaged by consumption of the contaminated fish.

Hepatic ALT, AST, GGT, GST, CAT, SOD, GSH and lipid peroxidation product (malondialdehyde, MDA) are commonly used as biomarkers of hepatotoxicity [37]. Abnormal values of serum GGT, ALT and AST have been observed for different diseased conditions, but most especially in hepatic diseases (Adeyemi et al., 2009). The experimental data (Tables 3) revealed that the catfish exposed to PKO biodiesel might have altered protein metabolism, among others, at the subcellular level and this may be indicative of impairment of the function of the tissues.

Ecological stressor such as biodiesel generate reactive oxygen species (ROS) such as O_2^-, H_2O_2 and OH radicals have potential to interact with many cellular components, causing significant damage to membrane and other cellular structures [38]. Some of the ROS are highly toxic and must be detoxified by cellular responses [39]. The ROS scavenging depends on the detoxification mechanism, which may occur as a result of sequential and simultaneous action of a number of antioxidant enzymes, including superoxide dismutase (SOD), catalase (CAT), and glutathione-S-transferase (GST). The observed decreased activity of SOD, CAT and GST in the liver of rats fed with catfish exposed to PKO biodiesel was indicative of oxidative stress.

Malondialdehyde (MDA) is a lipid peroxidation product generated from reactive oxygen species (ROS), and as such is assayed in vivo as a bio-marker of oxidative stress [40]. Malondialdehyde reacts with deoxyadenosine and deoxyguanosine in DNA, forming DNA adducts, malondialdehyde is reactive and potentially mutagenic. It has been found in heated edible oils such as sunflower and palm oils. The

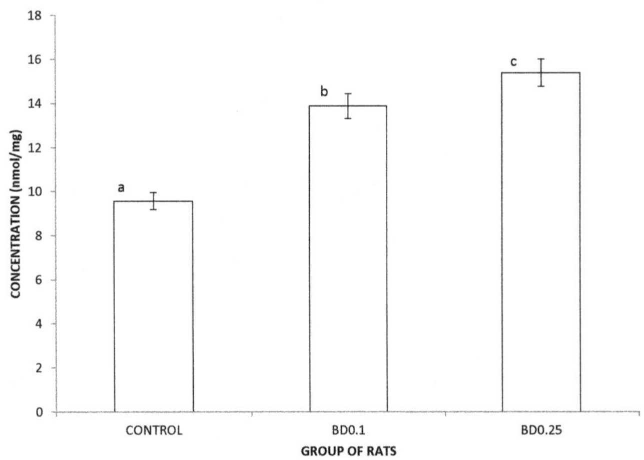

Figure 3: Effects of catfish exposed to PKO biodiesel on liver MDA. Plotted values are means ± SEM for 10 rats. [a,b,c] Column values with different superscripts are significantly different ($p<0.05$).

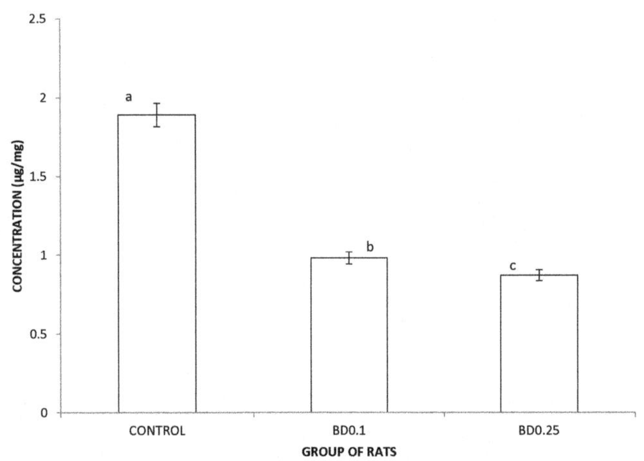

Figure 4: Effects of catfish exposed to PKO biodiesel on relative liver GSH. Plotted values are means ± SEM for 10 rats. [a,b,c] Column values with different superscripts are significantly different ($p<0.05$).

significant increase in the levels of MDA of liver of experimental rats (Figure 3) lend credence to the view that biodiesel caused a reduction in the total antioxidant status by reactive oxygen species.

GSH plays a very important role in the detoxification of xenobiotics. In vitro examinations proved that the free thiol group of glutathione reacts with xenobiotics to form conjugates. These conjugates reveal toxic properties [41]. In this study, catfish exposed to PKO biodiesel quickly depletes hepatocyte glutathione levels (Figure 4). It is therefore, a potential agent to which can lead to further lipid peroxidation.

In conclusion, the author has shown evidence that consumption of catfish exposure to PKO biodiesel alters growth and performance of albino rats. The indices of liver function observed, including serum levels of bilirubin, albumin, globulin, and activities of AST, ALT, GGT have not been reported previously in investigations of the effects of catfish exposed to PKO biodiesel on albino rats. The data on MDA, GSH and antioxidant enzymes (CAT, SOD, GST) of the liver of experimental rats suggests that consumption of catfish exposed to PKO biodiesel may predispose the organ to a condition of oxidative stress. Future study should consider light micrograph of the liver.

Conflict of Interest

The author declares that there are no conflicts of interest.

Acknowledgement

The author acknowledges Prof O.B Oloyede and Prof (Mrs) A.T. Oladiji, Department of Biochemistry, University of Ilorin, for their assistance.

References

1. ASTM Standard specification for biodiesel fuel (B100) blend stock for distillate fuels. In: Annual Book of ASTM Standards, ASTM International, West Conshohocken, Method D6751-08; 2008a.
2. Ajav EA, Singh B, Bhattacharya TK (2000) Thermal balance of a single cylinder diesel engine operating on alternative fuels. Energy Conversion Management, 41: 1533–1541.
3. Forde RC, Fitzgerald DJ (1997) Reactive Oxygen Species and platelet activation in reperfusion injury. Circulation, 95: 787-789.
4. Babior BM (2000) Phagocytes and oxidative stress. Am J Med 109: 33-44.
5. Halliwell B, Gutteridge JMC (1989) Free radicals in biology and medicine. In Halliwell B, and Gutteridge JMC (eds) (2ndedn), Clarendon Press, Oxford.
6. Cheeseman KH, Slater TF (1993) An introduction to free radical biochemistry. Br. Med. Bull, 49: 481-493.
7. Serarslan G, Aaltug E, Kontas T, Atik E, Avci G (2007) Caffeic acid phenetyl ester accelerates cutaneous wound healing in a rat model and decreases oxidative stress. Clin. Exp. Dermatol, 32: 709-715.
8. McCormick RL (2007) The impact of biodiesel on pollutant emissions and public health. Inhal. Toxicol, 19: 1033–1039.
9. Graboski MS, McCormick RL, Alleman TL, Herring AM (2003) The Effect of Biodiesel Composition on Engine Emissions from a DDC Series 60 Diesel Engine.NREL/SR-510–31461. Colorado Institute for Fuels and Engine Research, Colorado School of Mines, Golden, CO.
10. Bunger J, Krahl J, Franke HU, Munack A, Hallier E (1998) Mutagenic and cytotoxic effects of exhaust particulate matter of biodiesel compared to fossil diesel fuel. Mutat. Res, 415: 13–23.
11. Liu YY, Lin TC, Wang YJ, Ho WL (2008) Biological toxicities of emissions from an unmodified engine fueled with diesel and biodiesel blend. J Environ Sci Health A Tox Hazard Subst Environ Eng. 43: 1735-1743.
12. Swanson KJ, Madden MC, Ghio AJ (2007) Biodiesel exhaust: the need for health effects research. Environ. Health Perspect 115: 496-499.
13. Di Y, Cheung CS, Huang Z (2009) Experimental investigation on regulated and unregulated emissions of a diesel engine fueled with ultra-low sulfur diesel fuel blended with biodiesel from waste cooking oil. Sci. Total Environ 407: 835-846.
14. Tiyapongpattana W, Wilairat P, Marriott PJ (2008) Characterization of biodiesel and biodiesel blends using comprehensive two-dimensional gas chromatography. J Sep Sci 31: 2640–2649.
15. Poon R, Chu I, Valli VE, Graham L, Yagminas A, et al., (2007) Effects of three biodiesels and a low sulfur diesel in male rats-A pilot 4-week oral study. Food and Chem. Toxicol 45: 1830-1837.
16. Poon R, Valli VE, Rigden M, Rideout G, Pelletier G (2009) Short-term oral toxicity of three biodiesels and an ultra-low sulfur diesel in male rats. Food Chem. Toxicol 47: 1416-1424.
17. Adeyemi O, Osubor CC, Adeyemi O (2010) Toxicological evaluation of the effect of Clarias gariepinus (African catfish) cultivated in water contaminated with phthalate, benzene and cyclohexane on liver of albino rats. Afri J Fd Sci 4: 026-031.
18. Alamu OJ, Akintola TA, Enweremadu CC, Adeleke AE (2008) Characterization of palm-kernel oil biodiesel produced through NaOH-catalysed transesterification process. Sci Res and Essay 3: 308-311.
19. Bird RP, Drapper HH, Valli VE (1982) Toxicological evaluation of Malondialdehyde: a 12-month study of mice. J. Toxicol. Environ. Health, 10: 897-905.

20. Jollow DJ, Mitchell JR, Gillette JR (1974) Bromobenzene induced liver necrosis: Protective role of glutathione and evidence for 3,4-Bromobenzene oxide as the hepatotoxic metabolite. Pharmacol, 11: 151-169.

21. Sinha KA (1971) Colorimetric assay of catalase. Anal. Biochem, 47: 389-394.

22. Misra HP, Fridovich I (1972) The role of superoxide anion in the antioxidation of epinephrine and a simple assay of superoxide dismutase. J Bio1 Chem 247: 2417-3170.

23. Habig WH, Pabst MJ, Jakoby WB (1974) Glutathione-s-transferase: the First Step in Mercapturic Acid Formation. J Biochem 249: 7130- 7139.

24. Tietz NW (1990) Clinical guide to laboratory tests, (2ndedn) Philadelphia, USA: W.B. Saunders Company.

25. Tietz NW (1995) Clinical guide to laboratory tests, (3rdedn) W.B. Saunders Company, Philadelphia.

26. Henry RJ, Cannon DC, Winkleman JW (1974) Clinical chemistry, principles and techniques, Harper and Row.

27. Reitman S, Frankel S (1957) A colorimetric method for the determination of serum glutamate – oxaloacetate and pyruvate transaminases. Amer. J. Clin. Pathol. 28: 56-63.

28. Schmidt E, Schmidt FW (1963) Determination of serum GOT and GPT. Enzyme. Biol. Clin, 3:1.

29. Steel RGD, Torrie JH (1960) Principles and procedures of statistics. McGraw Hill Book Company Inc. London.

30. Duncan DB (1955) Multiple range and multiple F test. Biometrics, 11: 1-10.

31. Bakare AA, Mosuro AA, Osibanjo O (1999) Effect of simulated leachate on chromosomes and mitosis on roots of Allium cepa. J Environ Biol 21: 263-271.

32. Oladiji AT, Adeyemi O, Abiola OO (2004) Toxicological Evaluation of the Surface Water of Amilegbe River Using Rats. BIOKEMISTRI 16: 94-101.

33. Adeyemi O, Oloyede OB, Oladiji AT (2007) Effect of Leachate-Contaminated Groundwater on the Growth and Blood of Albino Rats. Int. J. Hem, 3: 1-6.

34. Michael B, Yano B, Sellers RS, Perry R, Morton D, et al., (2005) Evaluation of Organ Weights for Rodent and Non-Rodent Toxicity Studies: A Review of Regulatory Guidelines and a Survey of Current Practices. Toxicol Pathol, 33: 404-7.

35. Tiribelli C, Ostrow JD (2005) The molecular basis of bilirubin encephalopathy and toxicity: report of an EASL Single Topic Conference. J Hepatol 43: 156-160.

36. Islam MS, Lucky NS, Islam MR, Ahad A, Das BR, et al., (2004) Haematological parameters of Fayoumi, Assil and local chickens reared in sylhet region in Bangladesh. Int J Poult Sci 3: 144-147.

37. Saura CS, Philip PS, Robert LS, Thomas FXC, Ivan AR, et al., (2005) Hepatotoxicity of androstenedione in pregnant rats. Food Chem. Toxicol 43: 341-34.

38. Adeyemi O, Adeyemi O, Osubor CC (2009) Effect of Clarias gariepinus cultured in contaminated water on rats. *Hum* Exptal Toxicol 28: 759-764.

39. Adeyemi O, Oloyede OB, Oladiji AT (2008) Biochemical evaluation of oxidative damage induced by leachate contaminated groundwater on selected tissues of rats. Int J Toxicol, 4: 1-7.

40. Day BJ, Batinic-Haberle I, Crapo JD (1999) Metalloporphyrins are potent inhibitors of lipid peroxidation. Free Radic. Biol Med, 26: 730-736.

41. Flora SJS (2009) Structural chemical and biological aspects of antioxidants for strategies against metal and metalloid exposure. Oxid Med Cel Longev 2: 191-206.

Assessment of Potential Risks from Trihalomethanes in Water Supply at Alexandria Governorate

Ali M Abdullah*

Holding Company for Water and Wastewater, Cairo, Egypt

Abstract

Chlorination is used in the worldwide to produce drinking water in the most developing world. Frequently lack of proper sanitation and pollution control increases the organic content in water sources thereby increasing the potential of trihalomethanes (THMs) formation. This paper evaluated the lifetime cancer risk and the hazard index caused by THMs contained in drinking water from four areas of Alexandria Governorate, northern of Egypt. Oral exposure and the health risk were estimated using a probabilistic approach. This was done by monitoring the free residual chlorine and the THMs content in drinking water and by obtaining the actual population characteristics. Population characteristics considered, among other variables, the water intake rate, the body weight and the exposure time, and were expressed as empirical frequency distribution curves. Results showed that the 95th percentile of the carcinogenic risk estimated for bromodichloromethane (BDCM) and dibromochloromethane (DBCM) were above the acceptable level of one in a million (10-6) even though in 26% of the cases tap water did not meet the minimum free residual chlorine content required by the Egyptian drinking water standard (0.35 mg/l). Until proper sanitation is implemented and water is managed integrally (in quantity and quality), the Egyptian government need to consider alternate disinfection systems otherwise may be review integrally its water supply policy in these areas.

Keywords: Trihalomethanes; Drinking water; Risk assessment

Introduction

A water service is currently supplied to 99% of the Egyptian population. It is estimated that 99% of the distributed water is chlorinated. Chlorination of water has certainly contributed to reducing the incidence of gastrointestinal diseases such as cholera, typhoid fever, hepatitis, etc. Disinfection with chlorine has been recognized as one of the major public health achievements. However, appropriate levels of free residual chlorine in supply systems are needed to protect health and chlorine disinfectant is added to sources that are becoming polluted. The addition of chlorine reduces microbial risk but poses chemical risks when disinfection by-products (DBPs) are formed. DBPs occur when chlorine reacts with natural organic matter (humic and fulvic acids) presents naturally in water [1].

Natural organic matter, commonly measured as total organic carbon (TOC) is the organic precursor, while bromide ion is the inorganic one.

DBPs found in chlorinated water, trihalomethanes (THMs), which include chloroform ($CHCl_3$), bromodichloromethane ($CHCl_2Br$), dibromochloromethane ($CHClBr_2$) and bromoform ($CHBr_3$), has been widely studied because they are consid ered potentially carcinogenic [2]. In addition, recent studies suggest that they also produce reproductive disorders [3] if ingested during pregnancy [4-6]. Therefore, water utility managers try to reduce their formation while maintaining a free residual chlorine content that should be enough to inactivate microorganisms and prevent their re-growth in the distribution system. To control health risks caused by THMs, several countries have established a maximum content in drinking water. In the USA, the US EPA has set a value of 0.08 mg/l, in the United Kingdom and Canada [7] the limit is 0.10 mg/l, and also in Egypt it is 0.10 mg/l [8]. However, in Egypt, it is evidence that; water sources are becoming polluted due to a lack of sanitation raises concerns about the possible presence of DBPs in drinking water, so the risks are bigger because most of the population consumes tap water instead of bottled water.

Chlorination, the most commonly used method to disinfect tap water, has led to a sharp decrease in both mortality and morbidity from many diseases known to be waterborne [9]. However, the presence of chlorinated disinfection by-products (DBP) in tap water is of concern from a public health aspect because they are suspected to be carcinogenic [10]. The most significant group of DBP formed during chlorination is the THM such as $CHCl_3$, $CHBrCl_2$, $CHBr_2Cl$, and $CHBr_3$. $CHCl_3$ is classified in Group 2B as a possibly carcinogenic to humans, based on limited evidence of carcinogenicity in humans but sufficient evidence of carcinogenicity in experimental animals [9]. $CHBrCl_2$ is a weekly mutagenic and it has been classified as probably carcinogenic to humans, with sufficient evidence in animals and inadequate evidence in humans [9]. Between the four THMs found in drinking water, $CHBrCl_2$ appears to be the most potent rodent carcinogen. $CHBr_2Cl$ and $CHBr_3$ are classified in group 3 due to the inconclusive genotoxicity [9]. The second prevalent DBP group is haloacetic acids (HAAs). Aside from THMs and HAAs, many other compounds that comprising the DBP group have been found in treated waters, which include haloacetonitriles, haloketones, haloaldehydes, halopicrin, cyanogen chloride, halophenols and chloral hydrate and many others [9]. The main THM effects are cancer and adverse reproduction problems such as abortion, miscarriage, and retarded fetal development [11].

Chloroform concentrations measured in breath or blood after swimming and showering have been correlated with the activity time, and the concentration of this compound found in water and air [12]. Rafael et al [13] have reported that the THM concentration increases in blood as compared with their pre-activity blood levels in individuals after water-consuming.

***Corresponding author:** Ali Mahmoud Ahmed Abdullah, Ph. D, Technical Manager at Holding Company for Water and Wastewater, Cairo, Egypt
E-mail: dr2252000@dr.com

The aims of this research to evaluate the life cancer risk and the hazard index caused by THMs in drinking water for four areas communities of Alexandria using field data to characterize the local population and the quality of the drinking water.

Materials and Method

For this research, four communities located in the Alexandria governorates were considered: Abou Qir, Sidi Gaber (Siouf WTP and Mamoura WTP), Agimy and Amireya (K40 WTP and Noubaria WTP) with less than 2,000,000 total residents. Each community has its own source to supply water where chlorine is automatically added before water enters the distribution network.

The field study was divided into two parts. In the first one, an environmental study was performed to determine the quality of drinking water. In the second part, data to characterize the population of each community was gathered to use it for the human health risk assessment. Field data was gathered during a 4-week sampling program in May 2012.

Water samples were collected from each area of household water intake points and from tap water inside the houses, after passage through individual storage tanks. Sampling sites were selected and some criteria were taken in consideration: (a) the population size; (b) the direct supply from the municipal network and (c) the household's agreement to participate in the monitoring of their water and in the survey to gather information for the risk assessment analysis.

Parameters	Value
Weight of Population, $C = \frac{C_i \times P_i}{P_{total}}$	
C_i : Concentration of i region P_i : Population of water supply in the i region P_{total} : Total population of water supply in the i region	
Exposure pathway of Ingestion, $CDI = \frac{(CW \times 0.8 \times IR \times EF \times ED)}{(AT \times BW)}$	
Chronic daily intake, CDI (mg (Kg-day)$^{-1}$)	
THMs concentration of drinking water, CW	
Intake quantity, IR	2.5(liters day^{-1})
Average exposure time, AT	70(year)×365(day/year)
Exposure during, ED	70(year)
Exposure frequency, EF	365(day year^{-1})
Body Weight, BW	Male : 64.8 ± 10(Kg)
	Female : 56.3 ± 9.09(Kg)
Absorptivity of body	100%

Table 1: References data and formula for exposure assessment [16].

Chlorine Residual (mg/l)	Abou Qir	SidiGaber	Agimy	Amireya
<0.1	3	0	9	14
0.1-0.3	6	3	23	18
0.35-0.6	12	9	22	16
0.65-1.0	30	16	10	10
>1.0	21	44	8	14
Average (mg/l)	0.74	1.21	0.56	0.76
Standard Deviation	12.6	16.4	19.3	11.3
Skewness	0.57	0.87	0.82	0.64

Table 2: Free residual chlorine at household water in tap water sample.

Chloroform (μg/l)	Abou Qir	SidiGaber	Agimy	Amireya
<5	2	0	0	0
5-15	42	36	6	8
16-40	20	26	31	34
>40	8	10	35	30
Average	26.4	23.3	29.8	32.4
Standard deviation	31.3	44.5	34.5	18.5

Table 3: Chloroform at household water in tap water samples.

The environmental study consisted of 24 sampling points in different households (tap water). The parameters measured were free-residual chlorine (Cl_2) and THMs.

The THM extraction and analysis was carried out according to EPA 551 method [14]. Free residual chlorine was analyzed using a friendly colorimetric kit (DPD4).

The monitoring of residual chlorine in water carried out in the morning and in the afternoon, at the household's water intake point and tap water.

To characterize population exposure scenarios questionnaires were applied to 100 adults from the four areas to determine their: tap water ingestion habits, time of residence at the site and body weight, and were expressed as empiric distribution functions (EDFs) to describe the variability of those parameters. To evaluate the lifetime cancer risk (R_{ci}) caused by the exposure to halogen compounds in drinking water, a risk assessment model was applied (equation 1) based on United States Environmental Protection Agency guidelines [15]. The non-cancer hazard quotient (HI_i) was used (equation 2) to estimate the risk caused by chloroform as a secondary carcinogen.

$$R_{ci} \equiv \frac{EF \times ED}{BW \times AT} \times C_{ai} \times IR_a \times Sf_i \quad (1)$$

$$HI_i \equiv \frac{EF \times ED}{BW \times AT} \times \frac{C_{ai} \times IR_a}{RfD_i} \quad (2)$$

Where C_{ai} is the concentration of the trihalomethanes measured (mg/l) in drinking water for each community, IR_a is the water ingestion rate (l/d), Sf_i is the specific cancer slope factor (mg/kg.d)$^{-1}$ for bromide compounds and RfD_i is the reference dose for chloroform (mg/kg.d). The other exposure variables are EF, exposure frequency (d/y); ED, exposure duration (y); BW, body weight (kg) and AT, average exposure time (d).

Results and Discussion

Free residual chlorine

A total of 288 measurements were done at households' in tap water after passing through the individual storing tanks. Only 73.6% of the measurements (Table 1) [16] in the tap water fulfilled the free residual chlorine content established by the Egyptian drinking water standard of 0.35 [8]. This is a serious concern about the safeness of water, particularly from a microbial point of view.

According to Table 2, Amireya has the worse water quality supply and the major impact on water quality due to storage followed by Agimy, where all the values were less than or equal to the minimum enough level free residual chlorine of 0.35 mg/l. Less adverse conditions were observed in Abou Qir and Sidi Gaber.

THMs exposure levels

The total THMs concentration varied from 12 to 74 μg/l, always meeting the Egyptian drinking water standard of 100 μg/l. THMs concentration ranges for the different chemical species were as follows:

$CHCl_3$, 8 – 46.5 µg/l; $CHCl_2Br$, 4 – 21.3 µg/l; $CHClBr_2$, 0.0 – 6.9 µg/l and $CHBr_3$ 0.0-2.0 µg/l. Chloroform ($CHCl_3$) was then the major specie, and was notably present in Agimy and Amireya as shown in Table 3. $CHCl_2Br$ was the most important brominates compound, $CHBr_3$ being the least one. This is consistent with other studies [8]:

(a) Chloroform (Table 3 and Figure 1).

(b) Dichlorobromomethane (DCBM) (Table 4 and Figure 2).

(c) Dibromochloromethane (DBCM) (Table 5 and Figure 3).

(d) Bromoform (Table 6 and Figure 4).

Risk estimation

The probabilistic health risk estimation [17] was estimated for drinking water ingestion. The cancer risk equation (1) was used for bromide compounds while the chloroform hazard quotient was estimated using equation (2). The empirical distribution functions developed for the ED, BW, C_{ai} and IR_a variables were used as inputs to the risk equations in order to obtain the risks as probabilistic functions.

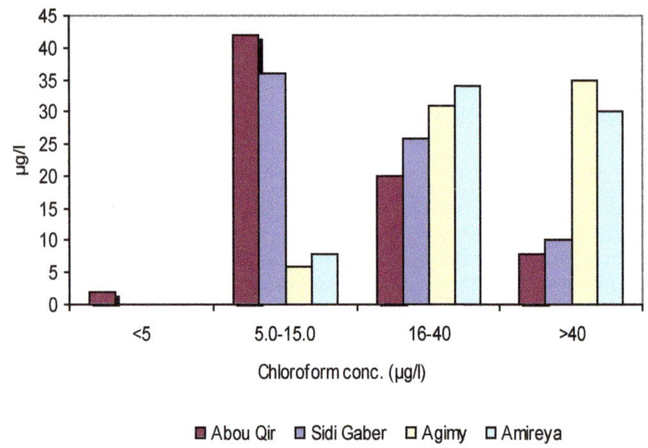

Figure 1: Chloroform at household water in tap water samples.

Dichlorobromomethane(µg/l)	Abou Qir	SidiGaber	Agimy	Amireya
<5	48	44	0	0
5-15	16	18	44	52
16-40	8	10	28	20
>40	0	0	0	0
Average	18.3	17.4	21.2	19.8
Standard Deviation	34.5	39.4	36.7	33.9

Table 4: Dichlorobromomethane at household water in tap water samples.

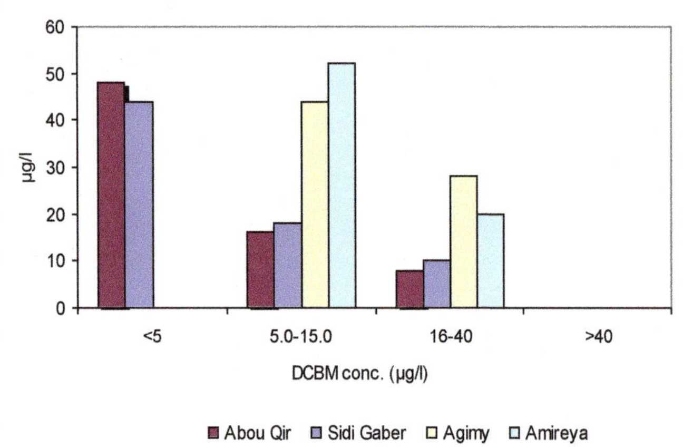

Figure 2: Dichlorobromomethane at household water in tap water samples.

Dibromochloromethane(µg/l)	Abou Qir	SidiGaber	Agimy	Amireya
<5	62	58	47	51
5-15	10	14	14	8
16-40	0	0	11	13
>40	0	0	0	0
Average	11.3	12.3	14.8	15.2
Standard Deviation	44.2	50.3	55.6	59.6

Table 5: Dibromochloromethane at household water in tap water samples.

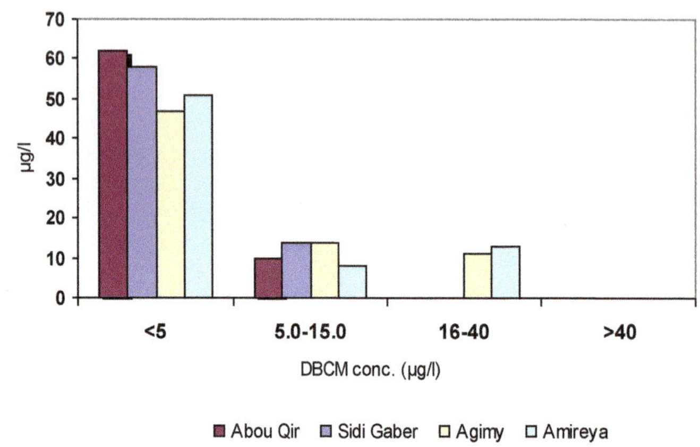

Figure 3: Dibromochloromethane at household water in tap water samples.

Bromoform (µg/l)	Abou Qir	SidiGaber	Agimy	Amireya
<5	72	72	66	65
5-15	0	0	6	7
16-40	0	0	0	0
>40	0	0	0	0
Average	1.3	1.2	3.9	4.2
Standard Deviation	14.3	16.7	22.3	24.7

Table 6: Bromoform at household water in tap water sample.

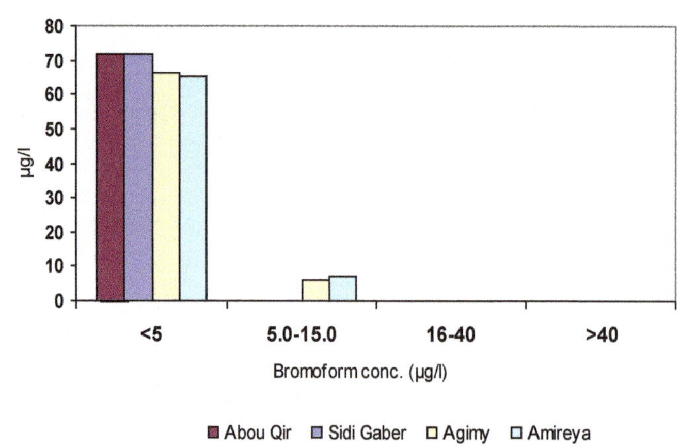

Figure 4: Bromoform at household water in tap water samples.

Bromoform (μg/l)	Abou Qir	SidiGaber	Agimy	Amireya
Chloroform	6.82E-07	6.08E-07	6.07E-07	5.89E-07
Bromodichloromethane	1.27E-05	1.18E-05	1.22E-05	1.39E-05
Chlorodibromomethane	1.59E-05	1.51E-05	1.24E-05	1.33E-05
Bromoform	1.80E-05	1.80E-05	1.68E-05	1.65E-05

Table 7: The 95th percentile from the risk distribution functions.

The hazard quotient for chloroform was estimated using the reference dose value given by IRIS [18], which is considered as protective to carcinogenic effects.

The 95th percentile of the probabilistic distribution functions (Table 7) calculated for the four areas communities showed in all cases that risk caused by chloroform and bromoform are acceptable (< than 1 for the first compound and < than 10-6 for the second one). However, the 95th percentile for bromodichloromethane and dibromochloromethane for tap water was greater than 10^{-6} for all communities. These results show that oral exposure to these compounds is higher than international acceptable levels, although content in water fulfill the Egyptian regulation.

Conclusions

This investigation included statistical analysis, epidemiology data and cancer risk analysis and assessment of THMs species in drinking water in Alexandria. It is more significant to establish an assessment procedure for the decision-making in policy of drinking water safety predominantly.

Specification the derive conclusions lie above from this study:

- The chloroform concentration is the major DBP species in the local regions of Cairo.
- The Southern region presented a high cancer risk (Amireya and Agimy).
- Residents of some districts were found to have a higher cancer risk through the oral ingestion of THMs. hazard indexes of THMs in different districts were found to be lower than unity, which did not indicate the noncancerous effects of THMs.
- Quantifying the risk factors is important for population and decision-making policy for drinking water safety. Fortunately, the Benchmark model and MCS and Risk supply the methodology were used for risk calculation. The standard for the total THMs species in Taiwan was 100 ug/l presently.
- We suggest that the standard be separated using separate dibromochloromethane, bromoform, chloroform, and bromodichloromethane standards. This may establish a control management for individual material to reduce the harmful risk. It displays the legislation limit values for different counties for DBPs levels.
- A methodology for decision-makers in formulating a procedure considering the economic, political, and feasible technology to reduce the standard value limits is necessary. An acceptable policy for safe drinking water and optimum social cost is the next objective of our study.

The techniques can be used for removal of THM compounds are activated carbon, enhancement coagulation, alternatives of disinfectants such as ozone, chloramine and chlorine dioxide.

References

1. Gallard H, Gunten U (2002) Chlorination of natural organic matter: kinetics of chlorination and of THM formation. Water Res. 336: 65-74.
2. McGeehin MA, Reif JS, Becher JC, Mangione EJ (1993) Case-control study of bladder cancer and water disinfection methods in Colorado. Am. J. Epidemiol. 138: 492-501.
3. Graves CG, Matanoski GM, Tardiff RG (2002) Weight of evidence for an association between adverse reproductive and developmental effects and exposures to disinfection by-products: a critical review. Regulatory Toxicology and Pharmacology 34: 103-124.
4. Bove FJ, Fulcomer MC, Klotz JB, Esmart J, Dufficy EM, et al. (1995) Public drinking water contamination and birth outcomes. Amer. J. Epidemiol. 141: 850-862.
5. King WD, Marrett LD (1996) Case-control study of bladder cancer and chlorination by-products in treated water (Ontario, Canada). Cancer Causes Control 7: 596-604.
6. Dodds L, King WD, Woolcott C, Pole J (1999) Trihalomethanes in public water supplies and adverse birth outcomes. Epidemiology 10: 233-237.
7. Health Canada (2001) Summary of guidelines for drinking water quality, Federal-Provincial subcommittee.
8. Ali M. Abdullah (2008) Minimizing the Formation of DBPs (Trihalomethanes and Haloacetic acids) in drinking water: 1st international conference and exhibition on sustainable water supply and sanitation (SWSSC'2010), 25-27 July, 2010, Cairo, Egypt.
9. Chowdhury S, Rodriguez MJ, Sadiq R (2011) Disinfection byproducts in Canadian provinces: associated cancer risks and medical expenses. Journal of Hazardous Materials 187: 574-584.
10. Hsu CH, Jeng WL, Chang RM, Chien LC, Han BC (2001) Estimation of potential lifetime cancer risks for trihalomethanes from consuming chlorinated drinking water in Taiwan. Environ. Res. 85: 77-82.
11. Gottlieb MS, Carr JK, Clarkson JR (1982) Drinking water and cancer in Louisiana: A retrospective mortality study. Am. J. Epidemiol. 116: 652-667.
12. Hossein P, Stevens AA (1995) Relationship between trihalomethanes and haloacetic acids with total organic halogen during chlorination. Water Res. 29: 2059-2062.
13. Rafael J, Villanova G, Garcia C, Gomez JA, Paz Garcia M, et al. (1999) Formation, Evaluation and Modelling Rats and B6C3F1 Mice (gavage studies). National Toxicology Program Technical Report Series No. 350. DHHS Publications No. (NIH) 89-2805.
14. U.S. Environmental Protection Agency (USEPA) (1995) The Use of the Benchmark Dose (BMD) Approach in Health Risk Assessment. Final report. EPA/630/R-94/007. Risk Assessment Forum, USEPA, Washington DC.
15. US EPA (1989) Risk Assessment Guidance for Superfund. Volume 1. Human Health Evaluation Manual (Part A). Interim Final. Office of Emergency and Remedial Response, Washington DC, EPA/540/1-89/002.
16. USEPA (US Environmental Protection Agency) (2003) Draft final guidelines for carcinogen risk assessment. Risk Assessment Forum EPA/630/P-03/001A, NCEA-F-0644A.
17. US EPA (2001) Risk Assessment Guidance for Superfund: Volume III - Part A. Process for Conducting Probabilistic Risk Assessment. Office of Emergency and Remedial Response USEPA, Washington DC, EPA 540-R-02-002, OSWER 9265.7-45, PB2002 963302.
18. IRIS (2005) Integrated Risk Information System.

Optimization of Water Reuse Network Using Water Pinch Technology (WPT) by Considering Single Contaminant for Kaduna Refinery and Petrochemical Company, Nigeria

Abdullahi SA*

Faculty of Engineering, Ahmadu Bello University, Zaria, Nigeria

Abstract

Conventionally, freshwater has been the only source of water for processes used in the industries due to the fact that it is cheap, and the wastewater generated in these processes has been treated in a central common facility in order to remove contaminants and to meet regulatory specifications for wastewater disposal. This research identify an optimum design of water reuse network for Kaduna refinery and petrochemical company (KRPC), while considering suspended solid (SS), hardness (H, Chemical oxygen demand (COD) and Free hydrocarbons (FH) wastewater contaminants and analysing the feasibilities of reuse and recycling for water and wastewater minimization in the water network of the company by using graphical method known as Water Pinch Technology (WPT). The analysis indicated the possibility of reducing the fresh water use by 76.8 m^3/h (32.3%) if Free Hydrocarbons (FH) was considered as the reference contaminant.

Keywords: Freshwater; Optimization; Wastewater; Environment; Contaminant; Pinch

Introduction

Large quantity of water is required in many industries, which is used in many processes for washing or as part of the product. The resulting wastewater is usually treated in a central common facility in order to remove contaminants and to meet regulatory discharge requirements. In some cases, the treated wastewater may still contain certain level of pollutants. Consequently, the environmental pollution increases due to this disposal trend. As opposed to this conventional approach, reusing and re-routing the water streams in an integral water network helps in reducing the consumption of freshwater, and minimizes the amount of wastewater to be treated and disposed of to the environment. This is very important, especially in developing countries where the relatively high cost and limited manpower pose additional challenges to effective wastewater treatment. One of the methods that can be used to minimise wastewater production is by Graphical method known as Water Pinch Technology (WPT) or Water Pinch Analysis (WPA).

WPT or WPA originates from the concept of heat pinch technique. It is a systematic technique for reducing water consumption and wastewater generation through integration of water-using activities or processes. It was first introduced by Wang and Smith. They used a limiting composite curve to solve water minimization problems. Since then, it has been widely used as a tool for water conservation in industrial process plants. It was extended in 1998 by Nick Hallale at the University of Cape Town, who developed it as a special case of mass exchange networks for capital cost targeting. Manan et al. [1] applied it successfully in urban/domestic buildings. Most of the methods used in WPT/WPA are based on the mass exchange of one or several contaminants. If the mass exchange is based on mass transferring of one contaminant, the problem will be solved as a single contaminant [2-4]. Nevertheless, if it includes mass transferring of two or more key contaminants, the problem will be solved as multiple contaminants [5]. Graphical methods are so practical to solve single contaminant problems. However, they are complicated and sometimes impossible for multiple contaminant problems [6].

This paper focuses on wastewater minimisation using single-contaminant technique using WPT/WPA based on the composite curves introduced by Wang and Smith [7] and the work of Mohammadnejad et al. [6]. Furthermore, it will address the problem of an optimum design of water minimization by using a combination of WPA, simple mathematical calculation and possible forbidden or forced water network interconnections restrictions. The proposed method tries to addresses shortcomings in the conventional WPA by considering the possibility of interconnections of different processes in any industry [8,9]. The method thus represents an improvement over existing state-of-the-art approaches. This paper will also test the feasibility of the method using data obtained from the Nigerian Kaduna petrochemical refinery. The feasibilities of reuse and recycling of water and wastewater minimization in the water network of the refinery will be analysed by considering suspended solid (SS), hardness (H), chemical oxygen demand (COD) and Free Hydrocarbons (FH) wastewater contaminants [10,11].

This will eventually lead to providing a means of minimizing the potential dangers facing the aquatic ecosystem as a result of the reduction in the discharge volume of industrial effluents containing harmful contaminants in concentrations above the maximum permissible levels and facilitate efficient and cost effective water usage planning and management.

Materials and Methods

This research is on the Kaduna Refining and Petrochemical Company Limited (KRPC) (Figure 1), a Subsidiary of Nigerian National Petroleum Corporation (NNPC).

***Corresponding author:** Abdullahi SA, Faculty of Engineering, Ahmadu Bello University, Zaria, Nigeria, E-mail: abdummu1@gmail.com

The Refinery project was completed and the Fuels Plant was commissioned in 1980. While the two other subsections of the refinery, the Lubes Plant and Petrochemical Plant were commissioned in 1983 and 1988 respectively. Kaduna Refinery occupies an area of 2.89 square kilometres. The plot plan is designed with safety and ease of maintenance in mind. The design also attempts to minimize operating costs. The refinery was designed to process both imported paraffinic and Nigerian crude oils into fuels and lubes products and was constructed by Chiyoda Chemical Engineering and Construction Company (now Chiyoda Corporation) of Japan (http://www.nnpcgroup.ng/KRPC.aspx.htm, 2015). The total installed capacity of the refinery is 110,000 BPSD. The water usage network existing in KRPC shows that fresh water from the river is the only source of water for process and other utilities use. The typical characteristics of this water are high suspended solids (with seasonal variation), high dissolved solids (metals), and variable turbidity. From the information obtained, the flow rate of the fresh water use is 650 m^3/h. The existing water usage network by various units is shown in Figure 2.

Table 1 shows the flow rate and the maximum allowable concentration at inlet (C_{in}) and outlet (C_{out}) of major water using units. However the value for fire unit indicated as 60.0 m^3/h may reach 150.0 m^3/h during fire fighting.

Water pinch analysis steps

Water Pinch Analysis (WPA) was carried out using source and sink composite curves, which is a graphical tool for setting water recovery targets as well as for design of water recovery networks [12]. It considers each water-using operation as being described by the mass transfer of a certain contaminant(s) from the process itself to the water stream (Figure 3a) and represented by the Loading and Concentration Figure (Figure 3b). First of all, in order to achieve a better understanding of the composite curves, a single water using unit is analysed in this section.

Figure 1: The picture of Kaduna Refining and Petrochemical Company Limited (KRPC).

WPA involves five key steps as stated by Wan and Manan [13], they are:

1. Analysis of water network: The first step is to analyse the existing or the base case water network through plant auditing as shown in Figure 2. In general, water-using operations can be classified into two main categories, i.e., mass-transfer-based (MTB) and non-mass-transfer-based (NMTB) operations [1,14].
2. Data extraction: The second step is to identify water sources and water sinks having potential for reuse and recycling and to extract the limiting water flow rate and limiting concentration data. Non-limiting concentrations in the water system are situated at a concentration below their maximum limits and do not affect the water sink for that operation.
3. Setting the minimum utility targets: The third step is to establish the minimum possible quantity of fresh water requirement and wastewater generation, or the minimum water targets using a targeting method.
4. Water network design/retrofit: The fourth step is to design a water recovery network to realise the minimum water targets.
5. Economic evaluation: The final step is to evaluate the economics of the water network.

The processes that use large amount of water rate such as cooling tower, Ion exchangers, and Service water used in desalter, fuel carbon cracking, Lube plant, caustic treating and laboratory in the refinery were selected [15-19]. They use total rate of 565 m^3/h which is equivalent to 82.5% of the total water rate used by the refinery. While the contaminants selected are Suspended solids (SS), Hardness (H), Chemical oxygen demand (COD) and free hydrocarbons (FH).

These contaminants are analysed separately using single contaminant method of WPA/WPT, by establishing the pinch point using mass load which is calculated as follows (Equation 1):

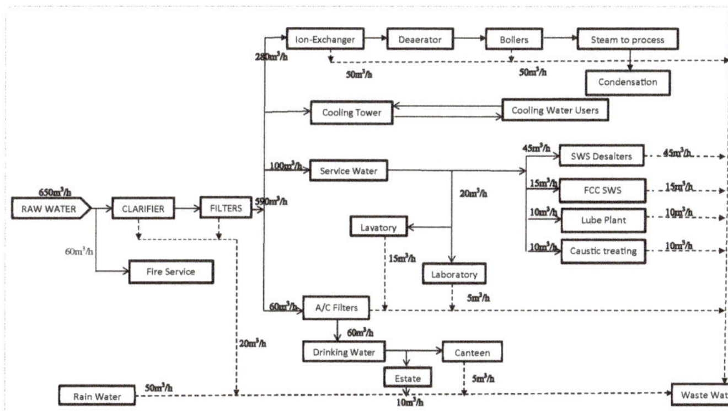

Figure 2: The existing water usage network by various units.

Unit No.	Unit name	Flow Rate of Fresh Water (m^3/h)	Suspended Solid		Hardness		Chemical Oxygen Demand (COD)		Free Hydrocarbons	
			C_{in} (ppm)	C_{out} (ppm)	C_{in} (ppm)	C_{out} (ppm)	C_{in} (ppm)	C_{out} (ppm)	C_{in} (ppm)	C_{out} (ppm)
1	Cooling Tower	150	40	500	5	10	0	150	0	5
2	Ion Exchangers	280	10	20	20	100	0	150	0	5
3	Laboratory	20	25	250	50	200	400	1000	0	5
4	SWS Desalters	45	30	300	20	100	400	1000	0	1000
5	FCC SWS	15	15	30	20	100	600	1200	0	10
6	Caustic treating	10	15	20	20	100	600	1200	0	10
7	Drinking	60	20	150	10	200	0	0.05	0	0
8	Fire	60	-	-	-	-	-	-	-	-

Table 1: The maximum allowable value of major water using units.

$$\Delta m_{opi} = \frac{(C_{out} - C_{in}) f_{opi}}{1000} \qquad (1)$$

Where Δm_{opi} is the mass-load for operation i, C_{out} and C_{in} are the outlet and inlet concentration respectively, and f_{opi} is the fresh water flowrate.

The Schematic representation of water balance superstructure model of KRPC water using operations in all processes is illustrated in Figure 4. This can be used as the bases for analysis of water using single contaminant method.

Results

Water Pinch Analysis (WAP) using single contaminant method

The WAP analysis using single contaminant method is done by constructing limiting composite curve (LCC). To construct the LCC, limiting water profiles of individual processes are plotted on an impurity concentration vs load graph. The limiting water profiles are plotted according to their limiting inlet and outlet concentrations which define the concentration intervals. The impurity load is then added within each concentration interval to form the limiting composite curve [20-25].

The limiting water use data for suspended solids (SS), Hardness (H), Chemical oxygen demand (COD) and free hydrocarbons (FH) contaminants and their calculated cumulative mass load (ppm) are shown in Tables 2-5 respectively. The data were established by referring to the water consuming units as sink and water generating units as source and it was used to construct the limiting composite curves which indicate the pinch point for the particular contaminant considered as a single contaminant to all the processes [26-28]. The sink and source were arranged in ascending order from lower contaminant value process to the higher one. The curves are shown in Figures 5-8.

Figure 5 is the graphical representation of limiting water use data for suspended solids (SS) in Table 2. The graph indicates a pinch point at 20 ppm concentration of SS, and 3.07 ppm cumulative mass load. This means that there is possibility of reduction of fresh water source of 365 m³/h if SS was considered as the only contaminant in the refinery water supply, which is 56% of the total fresh water use.

Figure 6 is the graphical representation of limiting water use data for hardness (H). The graph indicates a pinch point at 100 ppm concentration of H, and 23.15 ppm cumulative mass load. This means that there is possibility of reduction of fresh water source of 430 m³/h if H was considered as the only contaminant in the refinery water supply, which is 66% of the total fresh water use.

Figure 7 is the graphical representation of limiting water use data for Chemical oxygen Demand (COD). The graph indicates a pinch point at 150 ppm concentration of COD, and 64.5 ppm cumulative mass load. This means that there is possibility of reduction of fresh water source of 490 m³/h if COD was considered as the only contaminant in the refinery water supply, which is 75.4% of the total fresh water use.

Figure 8 is the graphical representation of limiting water use data for Free Hydrocarbon (FH). The graph indicates a pinch point at 10 ppm concentration of FH, and 2.49 ppm cumulative mass load. This means that there is possibility of reduction of fresh water source of 534 m³/h if FH was considered as the only contaminant in the refinery water supply, which is 82% of the total fresh water use, because of the range of acceptance of pollution level of FH by different processes.

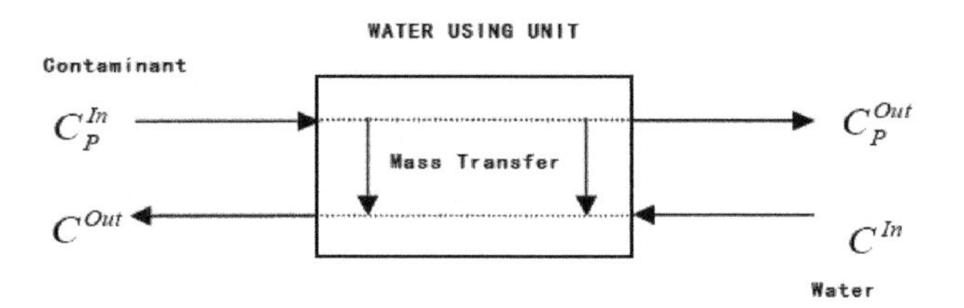

Figure 3a: Model of mass transfer in water using unit.

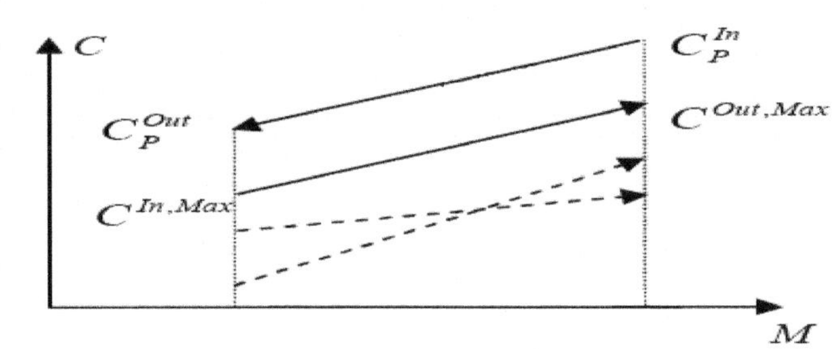

Figure 3b: Loading and Concentration Figure.

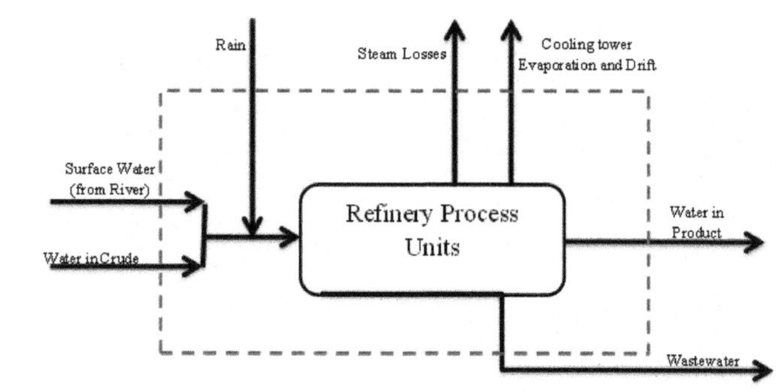

Figure 4: Schematic representation of water balance in Kaduna Refinery.

Process number	Sink	Sink Flow Rate m³/h	C_{in} (ppm)	Source	Source Flow Rate m³/h	C_{out} (ppm)	Mass Load (ppm)	Cummulative Mass Load (ppm)	Cumulative Flow m³/h
8	SK8	60	-	SR8	60	-	-	-	60
2	SK2	280	10	SR2	280	20	2.8	2.8	340
5	SK5	15	15	SR5	15	30	0.225	3.025	355
6	SK6	10	15	SR6	10	20	0.05	3.075	365
7	SK7	60	20	SR7	60	150	7.8	10.875	425
3	SK3	20	25	SR3	20	250	4.5	15.375	445
4	SK4	50	30	SR4	45	300	12.15	27.525	490
1	SK1	150	40	SR1	150	500	69	96.525	640

Table 2: Limiting water data for suspended solid (SS) contaminant.

Process number	Sink	Sink Feed Flow Rate m³/h	C_{in} (ppm)	source	Source Flow Rate m³/h	C_{out} (ppm)	Mass load (ppm)	Cummulative massload (ppm)	cumulative flow m³/h
1	SK1	**150**	5	SR1	**150**	10	0.75	0.75	150
2	SK2	280	20	SR2	280	100	22.4	23.15	430
4	SK4	45	20	SR4	45	100	3.6	26.75	475
5	SK5	15	20	SR5	15	100	1.2	27.95	490
6	SK6	10	20	SR6	10	100	0.8	28.75	500
7	SK7	60	10	SR7	60	200	11.4	40.15	560
3	SK3	20	50	SR3	20	200	3	43.15	580
8	SK8	60	-	SR8	60	-	-	-	640

Table 3: Limiting water data for hardness (H).

Process number	Sink	Sink Feed Flow Rate m³/h	C_{in} (ppm)	Source	Source Flow Rate m³/h	C_{out} (ppm)	Mass load (ppm)	Cumulative massload (ppm)	Cumulative flow m³/h
7	SK7	60	0	SR7	60	0.05	0.003	0.003	60
1	SK1	**150**	0	SR1	**150**	150	22.5	22.503	210
2	SK2	280	0	SR2	280	150	42	64.503	490
3	SK3	20	400	SR3	20	1000	12	76.503	510
4	SK4	45	400	SR4	45	1000	27	103.503	555
5	SK5	15	600	SR5	15	1200	9	112.503	570
5	SK6	10	600	SR6	10	1200	6	118.503	580
8	SK8	60	-	SR8	60	-	-		640

Table 4: Limiting water data for chemical oxygen demand (COD).

Process number	Sink	Sink Feed Flow Rate m³/h	C_{in} (ppm)	Source	Source Flow Rate m³/h	C_{out} (ppm)	Mass load (ppm)	Cumulative massload (ppm)	Cumulative flow m³/h
7	SK7	60	0	TSR7	60	0	0	0	60
1	SK1	150	0	SR1	150	5	0.75	0.75	210
2	SK2	**280**	0	SR2	**280**	5	1.4	2.15	490
3	SK3	20	0	SR3	20	5	0.1	2.25	510
5	SK5	15	0	SR5	15	10	0.15	2.4	525
6	SK6	10	0	SR6	10	10	0.1	2.5	535
4	SK4	45	0	SR4	45	1000	45	47.5	580
8	SK8	60	-	SR8	60	-	-	-	640

Table 5: Limiting water data for free hydrocarbons (FH).

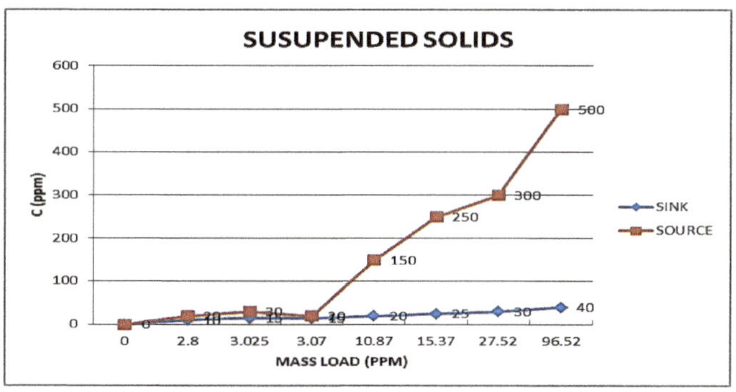

Figure 5: The limiting composite curve for Suspended Solids (SS).

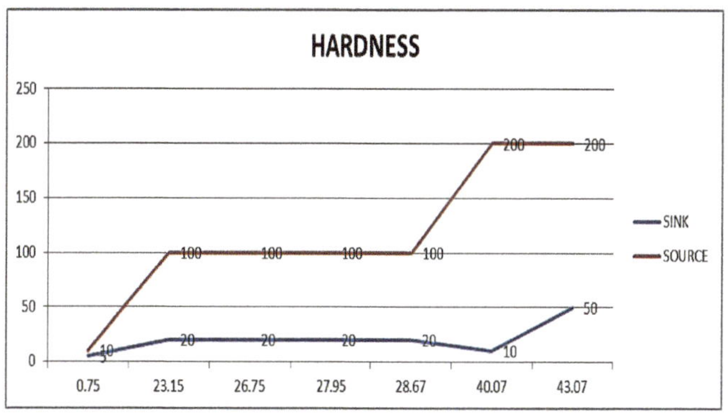

Figure 6: The limiting composite curve for Hardness (H).

The result obtained by this graphical analysis of pinch point can be justified by analysing the concentration grid of each contaminant and considering that a water generating process with output concentration lower than the minimum input concentration of another process can be re-used if the process allow re-use of water from another process [26].

The concentration grid for suspended solids (SS) contaminant is shown in Figure 9. The result indicate that water from three processes namely; Ion Exchangers, FCC SWS, Caustic treating and Drinking can be re-used in the refinery which gives a total saving of fresh water source of 365 m³/h if SS is considered as the only contaminant as shown by the pinch graphical analysis. This agrees with the result obtained from WPT analysis [29-31].

The concentration grid for Hardness (H) contaminant is shown in Figure 10. The result indicate that only water from Cooling Tower and Drinking can be re-used in the refinery which gives a total saving of fresh water source of 210 m³/h if H is considered as the only contaminant, which is 32.3% of the total fresh water use. This contradicts the value

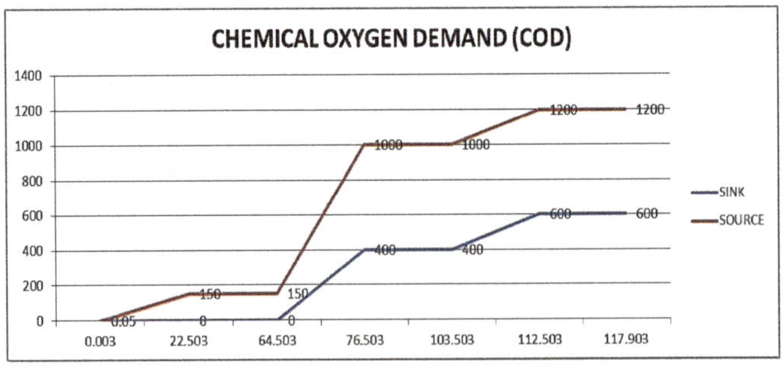

Figure 7: The limiting composite curve for Chemical Oxygen Demand (COD).

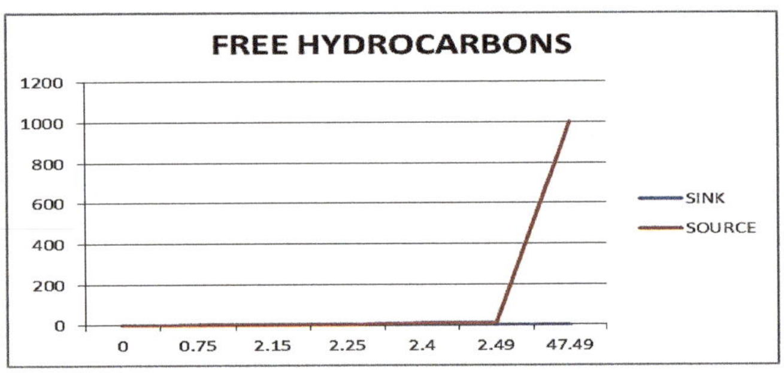

Figure 8: The limiting composite curve for Free Hydrocarbons (FH).

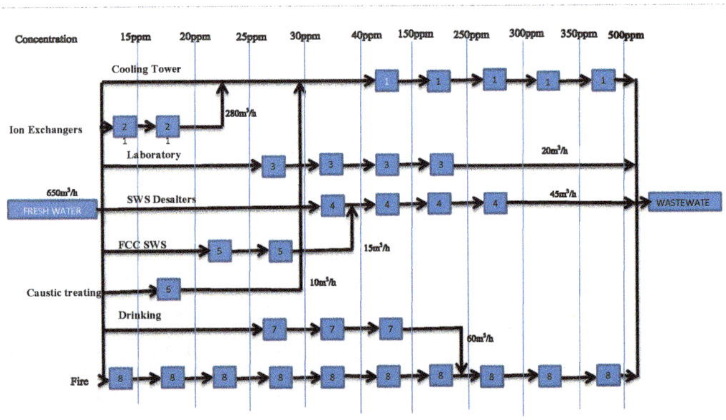

Figure 9: The concentration grid for Suspended Solids (SS).

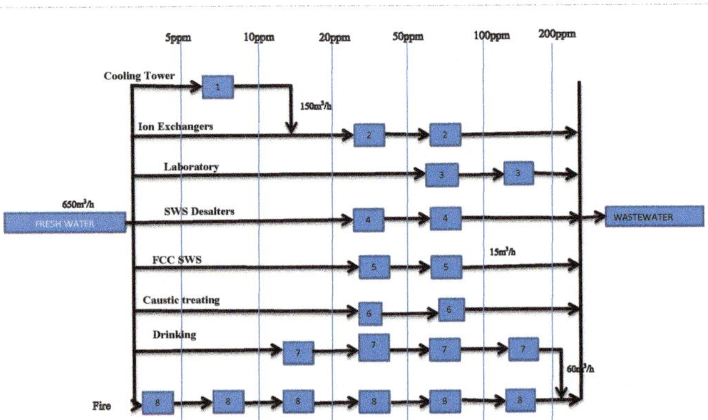

Figure 10: The concentration grid for Hardness (H).

shown by the pinch graphical analysis. Possibly due to the presence of four processes (Ion Exchangers, SWS Desalter, FCC SWS, and caustic treating) that has equal concentration ranges of 20 ppm to 100 ppm.

Furthermore, the concentration grid for Chemical Oxygen Demand (COD) contaminant is shown in Figure 11. The result indicate that water from three processes namely; Cooling Tower, Ion Exchangers, and Drinking can be re-used in the refinery which gives a total saving of fresh water source of 490 m^3/h if COD is considered as the only contaminant as shown by the pinch graphical analysis.

The concentration grid for Free Hydrocarbons (FH) contaminant is shown in Figure 12. The result indicate that only water from Drinking can be re-used in the refinery which gives a total saving of fresh water source of 60 m^3/h if FH is considered as the only contaminant, which is 9.2% of the total fresh water use. This contradicts the value shown by the pinch graphical analysis. Possibly due to the fact that all the processes have equal input concentration ranges of 0 ppm.

Therefore, the pinch analysis for a single contaminant should be used only in the situation where all the processes have different concentration ranges. Otherwise equal concentration ranges should be eliminated before carrying out the analysis.

Based on the above analysis of single contaminant, the amount

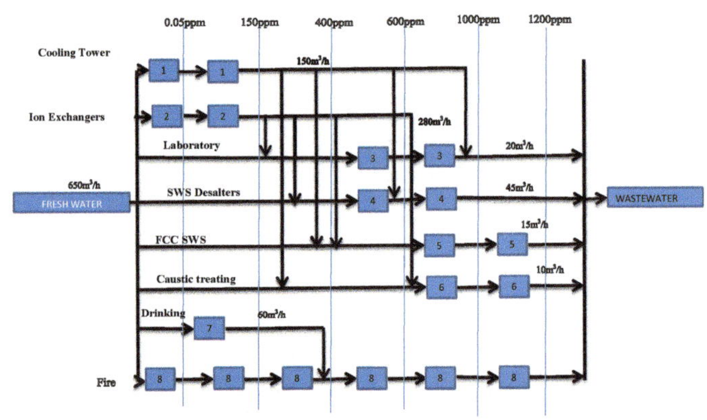

Figure 11: The concentration grid for Chemical Oxygen Demand (COD).

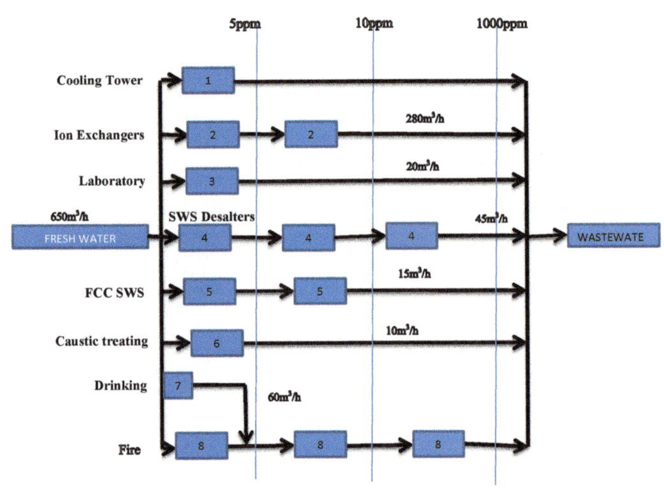

Figure 12: The concentration grid for Free Hydrocarbons (FH).

of fresh water can be reduced by re-using 365 m^3/h (56%) if SS was considered as the only contaminant, in the refinery water supply, by 210 m^3/h (32.3%) if H was considered as the only contaminant, by 490 m^3/h (75.4%) if COD was considered as the only contaminant, and by 60 m^3/h (9.2%) if FH was considered as the only contaminant.

Conclusion

In this paper, an optimum design of water reuse network for Kaduna refinery by considering suspended solid (SS), hardness (H), chemical oxygen demand (COD) and Free Hydrocarbons (FH) wastewater contaminants was achieved. The analysis of the feasibilities of reuse and recycling for water and wastewater minimization in the water network of the refinery was also carried out. The result indicate that there are possibilities of reduction of fresh water use of 365 m^3/h, 430 m^3/h, 490 m^3/h, and 534 m^3/h if SS, H, COD, and FH are considered as the single-contaminant in the refinery water supply, respectively. This will leads to 56%, 66%, 75.4%, and 82% reduction of the total fresh water use if SS, H, COD, and FH are considered as the single-contaminant in the refinery water supply, respectively. However, these results were further analysed by using the concentration grid of each contaminant. The values obtained for SS and COD agree with the result obtained from WPT analysis, but give the values of 210 m^3/h and 60 m^3/h, which correspond to the reduction of 32.3% and 9.2% for H and FH respectively.

References

1. Manan ZA, Wan Alwi SR, Ujang Z (2006) Water pinch analysis for urban system: a case study on the Sultan Ismail Mosque at Universiti Teknologi Malaysia (UTM) Desalin 194: 52-68.
2. Alnouri SY, Linke P, El-Halwagi M (2014) Water integration in industrial zones: a spatial representation with direct recycle applications. Clean Techn Environ Policy 16: 1637-1659.
3. Bandyopadhyay S (2015) Mathematical foundation of Pinch Analysis.
4. Boix M, Montastruc L, Pibouleau L, Azzaro-Pantel C, Domenech S (2010) Multiobjective optimization of industrial water Networks with contaminants. Comput Aided Chem Eng 28: 859-864.
5. Bagajewicz M, Savelski M (2001) On the use of linear models for the design of water utilization systems in process plants with a single contaminant. Chem Eng Res Des 79: 600-610.
6. Mohammadnejad S, Nabi Bidhendi GR, Mehrdadi N (2011) Water and wastewater minimization in petroleum refinery through water pinch analysis-single and double contaminants approach. Res J Environ Sci 5: 88-104.
7. Wang YP, Smith R (1994) Wastewater Minimisation. Chem Eng Sci 49: 981-1006.
8. Chaturvedi ND, Manan ZA, Wan Alwi SR, Bandyopadhyay S (2016) Effect of multiple water resources in a flexible-schedule batch water network. J Clean Prod 125: 245-252.
9. Dakwala M, Mohanty B, Bhargwa R (2009) A process integration approach to industrial water conservation: A case study for a starch industry. J Clean prod 17: 1654-1662.
10. Deng C, Feng X, Bai J (2008) Graphically based analysis of water system with zero liquid discharge. Chem Eng Res Des 86: 165-171.
11. Doyle SJ, Smith R (1997) Targeting water reuse with multiple contaminants. Proc Saf Environ Prot 75: 181-189.
12. Kim KJ, Smith R (2004) Automated design of discontinuous water system. J Proc Saf Environ Prot 82: 238-248.
13. Wan Alwi SR, Manan ZA (2013) Water pinch analysis for water management and minimisation: an introduction. Klemeš (Ed.), Process Integration Handbook, Woodhead Publishing, Cambridge, UK pp: 353-382.
14. Chew KH, Klemeš JJ, Alwi SRW, Manan ZA (2013) Industrial implementation issues of total site heat integration. Appl Therm Eng 61: 17-25.
15. Feng X, Bai J, Zheng X (2007) On the use of graphical method to determine the targets of single-contaminant regeneration recycling water systems. Chem Eng Sci 62: 2127-2138.
16. IPIECA (2010) The global oil and gas industry association for environmental and social issues. 5th Floor, 209-215 Blackfriars Road, London SE1 8NL, United Kingdom.
17. Jeżowski J (2010) Review of water network design methods with literature annotations. Ind Eng Chem Res 49: 4475-4516.
18. Khezri SM, Lotfi F, Tabibian S, Erfani Z (2010) Application of water pinch technology for water and wastewater minimization in aluminum anodizing industries. Int J Environ Sci Tech 7: 281-290.
19. Dunn RF, Ristau JS (2016) Using Process Integration Technology to Retrofit Chemical Plants for Energy Conservation and Wastewater Minimization. Chem Proc Retrofitt Revamp: Tech Appl 167.
20. Majozi T, Brouckaert CJ, Buckley CA (2006) A graphical technique for wastewater minimisation in batch processes. J Environ Manage 78: 317-329.
21. Nick H (2002) A New Graphical Targeting Method for Water Minimisation. Adv Environ Res 6: 377-390.
22. Olesen SG, Polley GT (1997) A Simple Methodology for the Design of Water Networks Handling Single Contaminants. Chem Eng Res Des 75: 420-426.
23. Savelski JM, Bagajewicz JM (2001) Algorithmic procedure to design water utilization systems featuring a single contaminant in process plants. Chem Eng Sci 56: 1897-1911.
24. Sharma S, Rangaiah GP (2016) Designing, Retrofitting, and Revamping Water Networks in Petroleum Refineries Using Multiobjective Optimization. Ind Eng Chem Res 55: 226-236.
25. Sorin M, Be´dard S (1999) The global pinch point in water reuse networks. Proc Saf Environ Prot 77: 305-308.
26. Li B, Zhang G, Ye M, Du J, Xiang X, et al. (2016) Network optimization and performance evaluation of the water-use system in China's straw pulp and paper industry: a case study. Clean Tech Environ Pol 18: 257-268.
27. Savelski JM, Bagajewicz JM (2000) Design of water utilization systems in process plants with a single contaminant. Was Manage 20: 659-664.
28. Savelski JM, Bagajewicz JM (2000) On the optimality conditions of water utilization systems in process plants with single contaminants. Chem Eng Sci 55: 5035-5048.
29. Savelski JM, Bagajewicz JM (2003) On the necessary conditions of optimality of water utilization systems in process plants with multiple contaminants. Chem Eng Sci 58: 5349-5362.
30. Takama N, Kuriyama T, Shiroko K, Umeda T (1980) Optimal water allocation in a petroleum refinery. Comput Chem Eng 4: 251-258.
31. Ting L, Castro PM, Zhimin L (2016) Models and relaxations for the wastewater

Aluminum Pollution Removal from Water Using a Natural Zeolite

Ali M Abdullah*
Holding Company for Water and Wastewater, Cairo, Egypt

Abstract

Aluminum ion adsorption capacity of zeolite from polluted tap water was investigated in this study. For the experimental studies, zeolite mineral which was 20-200 mesh size was used by activating with HCl and aluminum ion removal efficiencies were examined. Experiments were studied under laboratory, batch conditions were run at different pH values, temperatures. The highest removal efficiency was found as about 84.7% at pH 6.5. Similarly, experiments were carried out at different temperature values, and the maximum efficiency was obtained at 30°C. The efficiency obtained under the same conditions was 93%. The highest aluminum removal efficiency was obtained with 180 rpm shaking speed.

Keywords: Aluminum removal; Water; Adsorption; Zeolite

Introduction

Aluminum can occur in a number of different forms in water. It can form monomeric and polymeric hydroxy species, colloidal polymeric solutions and gels, and precipitates, all based on aquated positive ions or hydroxylated aluminates. In addition, it can form complexes with various organic compounds (e.g. humic or fulvic acids) and inorganic ligands (e.g. fluoride, chloride, and sulfate), most but not all of which are soluble. The chemistry of aluminum in water is complex, and many chemical parameters, including pH, determine which aluminum species are present in aqueous solutions. In pure water, aluminum has a minimum solubility in the pH range 5.5-6.0; concentrations of total dissolved aluminum increase at higher and lower pH values.

The concentration of aluminum in natural waters can vary significantly depending on various physicochemical and mineralogical factors. Dissolved aluminum concentrations in waters with near-neutral pH values usually range from 0.001 to 0.05 mg/l but rise to 0.5-1 mg/l in more acidic waters or water rich in organic matter. At the extreme acidity of waters affected by acid mine drainage, dissolved aluminum concentrations of up to 90 mg/l have been measured [1].

The presence of Al in drinking water has given rise to discussions on possible of risks, because of its suspected connection with Alzheimer's diseases or dialysis encephalopathy [1]. Driscoll and Letterman [2] reported that dialysis patients exposed to elevated Al may exhibit dialysis enchepholopathy, and/or bone mineralization disorders such as dialysis osteodystrophy. Martyn et al. [3] based on a survey of 88 county districts in England and Wales reported that rate of Alzheimer's disease was 1.5 times higher in districts where the mean Al concentration exceeded 0.11 mg/l than in districts where concentrations were less than 0.01 mg/l. Al shows low acute toxicity. Chronic exposure data are limited, but indicate that Al likely interferes with phosphorus absorption, and causes weakness, bone pain and anorexia.

The most widely used methods for removing metal ions are: precipitation with chemical and electrochemical methods [4]. A major problem with this type of treatment is the disposal of the precipitated wastes. Ion exchange technique which the second most widely used method for metal ion removal does not present a sludge disposal problem and has the advantage of reclamation of Al (III) [5]. This method can reduce heavy metals to very low levels. However, ion exchange treatment does not appear to be economical. Activated carbon is also efficient for removal of trace elements from the waste water, but its high cost when usage. The adsorption phenomenon has still been found economically appealing for the removal of toxic metals from water by choosing some adsorbents under optimum operation conditions.

It has been reported that some aquatic plants [6], agricultural by-products [7,8], sawdust [9], clay [10], zeolite, turba [11] and microorganisms have the capacity to adsorb and accumulate heavy metals. Cost comparisons are difficult to make due to the scarcity of consistent cost information. Although many experiments have been accomplished in the area of low-cost sorbents, a great deal of work is necessary to understand better low-cost adsorption processes and to demonstrate the technology.

Natural zeolites have ion exchange and removal capacity. zeolites has high removal capacity of metal ions Pb^{+2}, Cu^{+2}, Zn^{+2}, Cd^{+2}, Ni^{+2}, Al^{+3}, Fe^{+2} and Mn^{+2}. It is an economical ion exchange material comparing with the synthetic resin.

Natural zeolites have been used as adsorbent and ion exchanger with discovering crystal feature and some chemical structure. It is physical structure look like a selective sieve and named as molecular sieve by Mc Bain.

General formula of the Zeolites is $X[(M^{+}_{1}, M^{++}_{1/2}).(AlO_2)].ySiO_2.zH_2O$. M is a cation ($Na^+$, K^+, Ca^{++}, Pb etc.) which ratio (y/x) changes from 1 to 5. M^{++} is a cation (Ca^{++}, Mg^{++} and Ba^{++} etc.) change with M^{+}_{1}. SiO_2/AlO_2 molar ratio or (y/x) zeolite changes with zeolite types from 1 to 5. SiO_4 and AlO_4 octahedrals are the monomers of the zeolites that Si or Al was centre of the structure and oxygen was on the corners. Natural zeolites were also uses at industrial fish ponds [12]. Chelischev [13] used zeolites as ion exchange for heavy metals. Similarly Sato and Fukagawa, used zeolites for nitrogen removal at detergent industry waste water.

The aim of this study was removal of the ionic aluminum pollution

***Corresponding author:** Ali Mahmoud Ahmed Abdullah, Ph. D, Technical Manager at Holding Company for Water and Wastewater, Cairo, Egypt
E-mail: dr2252000@dr.com

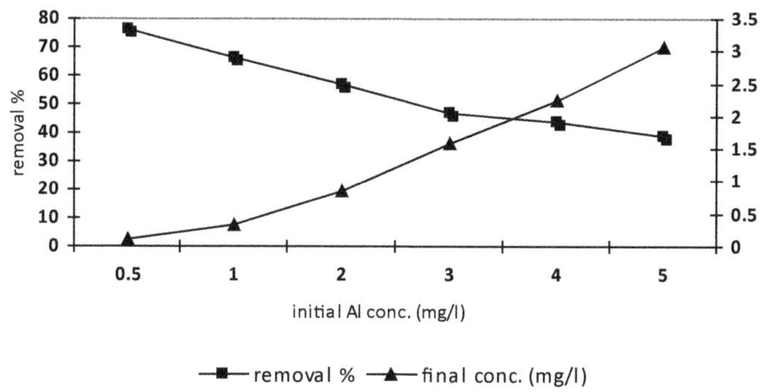

Figure 1: Al^{+3} removal efficiency of zeolite from aqueous solutions containing different Al^{+3} levels.

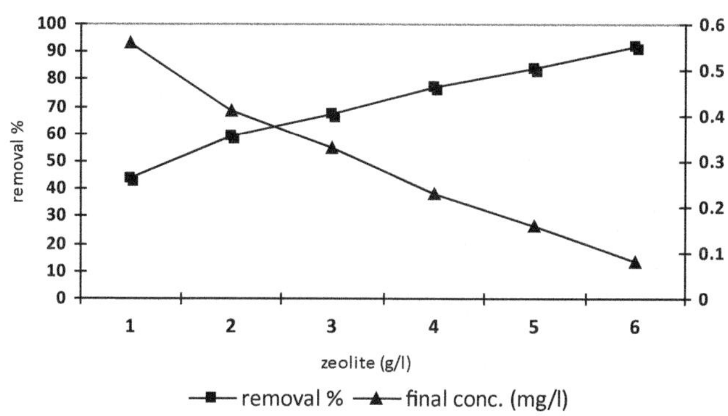

Figure 2: Effect of different adsorbent dose of zeolite for Al^{+3} adsorption.

from the preliminary treated water using zeolite as a natural and economical material.

Materials and Method

- Zeolite used provided from Iran (Chem-lab, Co., Cairo, Egypt).
- The chemical analysis showed that, the zeolite was in hydrated alumina-silicate mineral containing mostly alkali and soil alkali metals originated in holes of volcanic rocks. They contain hydrate in holes connected each other. Particle size is 50-2000 µm of 82.56% materials and 2-50 µm others. Porosity ratio was about 50% and natural hydrate capacity and ion exchange capacity were 25.68% and 71.73 meq/100 g respectively.
- Materials used in this study were between 20-200 mesh sizes.
- *Gallenkap* (U.K.) thermostatic shaker incubator was use for batch experiments.
- The concentration of Al used in the present studies varies from 0.5 to 5 mg/l.
- Al concentrations residues were measured by Dr. Lange Cadas-200 Spectrophotometer (Germany).
- The pH was adjusted to required values using *Jenway 3010* pH meter.
- *Gec-Avery VA/WA* Analytical Balances was used to weighing procedure for preparation of solution.
- Zeolite in 20-200 mesh size was treated with acid or base solution in different concentration to improve its adsorption capacity. For this process, 2.0 grams of zeolite in 100 ml distilled water containing different amounts of acid or base was shaken at different shaking seeped and temperature, and finally 1 h 200 rpm and 23 h 100 rpm at 30°C stable temperature.
- Zeolite samples were washed twice 2 with distilled water after treating aside or alkali solution. With this procedure, the best activation media was determined 2.0 ml 1 M HCl additions. For this reason, activated zeolite samples with 2.0 ml 1 M HCl were used in following experiments. After designing the activation media, to optimum adsorbent amount, each 2 grams of activated zeolite was added in 100 ml tap water containing different concentration of Al^{+3}.
- After 1 h contact time in batch rector, zeolite was filtered with fine nylon filter and Al^{+3} concentrations were measured using Cuvette test via Cadas 200 UV -Vis spectrophotometer.
- Effect of pH on the adsorption capacity of 2 grams of zeolite was tested adjusting pH of the 100 ml sample solution (containing 40 mg-Al/l) via adding different amount of acid or base. Effect of ambience temperature was also investigated.
- Experiments were run at the temperatures of 10, 20, 30, 40 and 50°C the conditions designated earlier experiments.

Results

Zeolite samples were activated using 1 M HCl or NaOH to increase the adsorption capacity. For this purpose, firstly 1, 2 and 3 ml 1M HCl were added in three conical flask containing 2 grams of zeolite and 100 ml distilled water separately, secondly 1, 2 and 4 ml NaOH were added in an other conical flask containing 2 grams of zeolite separately, and one conical flask containing 2 grams of zeolite was kept without acid or alkali addition. Results showed that the best activation condition was addition of 2.0 ml 1 M HCl in100 ml distilled water containing 2 grams of zeolite.

Effect of initial Al (III) concentration by the zeolite adsorption

Determination of effective Al^{+3} concentrations on removal efficiency of zeolite was investigated with addition of 2.0 grams of activated adsorbent in different concentration of Al^{+3} in 100 ml water solutions. Figure 1 shows that the maximum removal efficiency was obtained with 1.6 mg/L Al^{+3} concentrations, a little decrease was found below from this values. Later on, decreasing removal efficiency was detected with increasing initial heavy metal concentrations in the solutions. It is likely that a given mass of adsorbent material has a finite number of adsorption sites, and that as metal concentrations increase, these sites become saturated. That is, there is some metal concentration that produces the maximum adsorption for a given adsorbent mass, and there after, adding more metal cannot increase adsorption because no more sites are available: all are occupied [7].

Effect of adsorbent doses on the Al^{+3} adsorption by the zeolite was investigated with changed adsorbent doses from 1.0 to 4.0 g that maximum removal efficiencies have been achieved to be 67% with 2.0 g zeolites /100 ml solution (Figure 2). Residual Al^{+3} concentration seemed until to 2.0 g a clear increasing and than a smaller increasing after this value.

pH effect on adsorption capacity

Effect of pH on the adsorption of Al^{+3} by the zeolite was also investigated with the changing pH values from 3.0 to 9.0. The results are depicting the dependence of the Al^{+3} removal as a function of pH

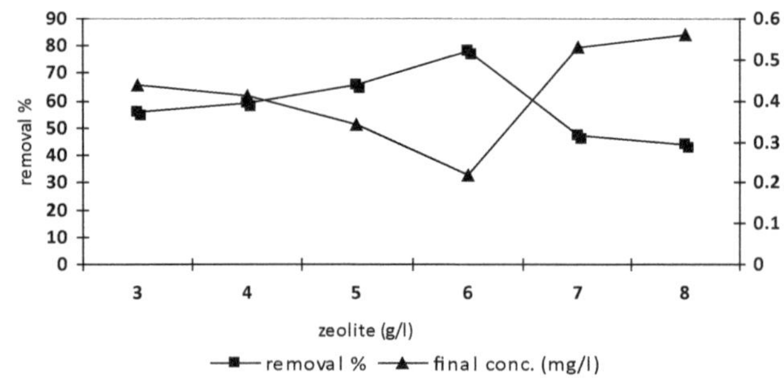

Figure 3: Effect of solution pH on removal efficiency of Al^{+3} from water.

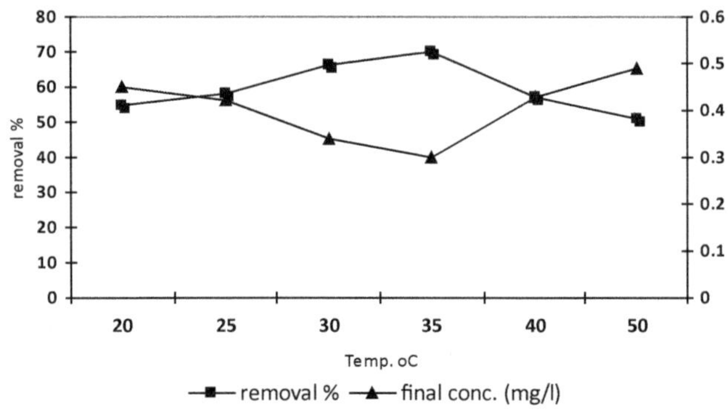

Figure 4: Effect of sample temperature on the zeolite removal efficiency of Al^{+3} from the water.

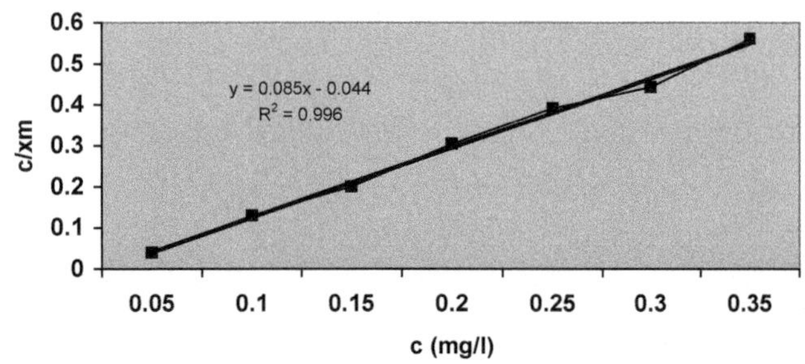

Figure 5: Langmuir adsorption isotherm plot for Al^{+3} adsorption.

showed that maximum removal efficiencies have been achieved to be 85% at pH 6.0 (Figure 3). According to different pH values Al ions may present in different forms. For this reason, Al^{+3} ions dominates at pH<7 and $Al(OH)_3$ dominates at pH>7 [8].

Effect of sample temperature on Al^{+3} removal capacity of zeolites

Activated 2 grams of zeolite in 100 ml water at pH 6.0 was tested for adsorption of Al^{+3} at different sample temperature between 10 to 50°C. Results showed that optimum working temperature was about 30°C (Figure 4). Removal of Al^{+3} from the water was 88% at this temperature; it was lowered at lower and at higher temperatures.

The equilibrium adsorption isotherm for Al^{+3} on activated zeolite was plotted for varying adsorbent doses changing from 0.25 to 3.0 g L^{-1}) and fixed initial Al^{+3} concentration (1.6 mg L^{-1}) at pH 6.0. The percentage of Al^{+3} adsorption decreased with increasing adsorbate amount, which indicates that the adsorption depends upon the availability of the binding sites for Al^{+3}. In order to determine the adsorption capacity of the activated zeolites, the equilibrium data for the adsorption of Al^{+3} were analyzed in the light of adsorption isotherm models. Mostly used adsorption isotherms are Langmuir and Freundlich equations which were given below respectively (Equation 1 and 2):

$$(x/m) = (1/a) + (1/ab).(1/c) \quad (1)$$

$$\mathrm{Ln}\,(x/m) = \mathrm{Ln}\,(k) + (1/n).\mathrm{Ln}\,(c) \quad (2)$$

where x/m is the amount of Al^{+3} adsorbed from per unit weight of adsorbent (mg g^{-1}), c is the concentration of Al^{+3} at equilibrium (mg L^{-1}), $1/ab$ is slope for plot with $1/a$ contact point on y axis for Lanmuir adsorption isotherm, and $1/n$ is slope for plot with contact point of *ln* k constant on crossing point of y axis for Freundlich adsorption isotherm. The straight line nature of the graph indicates that the adsorption confirms the Langmuir model. The experimental data points were fitted to the Langmuir equation (Figure 5), but it was not fit well to the Freundlich equation. Chemical kinetics of Al^{+3} removal was also investigated in batch reactor for 60 min. contact time that checked with °0, °1, °2 and °3 order reaction equations. Only our date is quite well fit to 2nd order reaction kinetic (Figure 6). Square of correlation coefficient was R^2=0.9905 which confirms good relation between contact time and concentration change by experimental period. Reaction kinetic of Al^{+3} removal from the aqueous environment by zeoliteis 2nd order that adsorption of zeolite was physical character and it is suitable for reuse.

Discussions

This paper presents the results of Al^{+3} ions removal from aqueous solution using common, naturally occurring zeolites. Operational parameters such as initial Al^{+3} concentrations, adsorbent doses, pH of the solution, temperature, and contact time clearly affect the removal efficiency. The optimum Al^{+3} removals by the zeolite were obtained at pH 6. At pH higher than 7, Al^{+3} removal probably resulted primarily from precipitation rather than adsorption. According to results reported in this study, Al^{+3} adsorption mechanisms by zeolite may be explained as follows:

- Vander-walls and hydrogen binding of Al^{+3} with hydroxyl group of zeolite surfaces.
- Ion exchange between Al^{+3} ions and exchangeable cations (Na, K etc.).
- Diffusions of Al^{+3} through selective sieve of zeolite named as molecular sieve (Mc Bain, 1932).

According to these results, natural zeolite appeared to be a promising

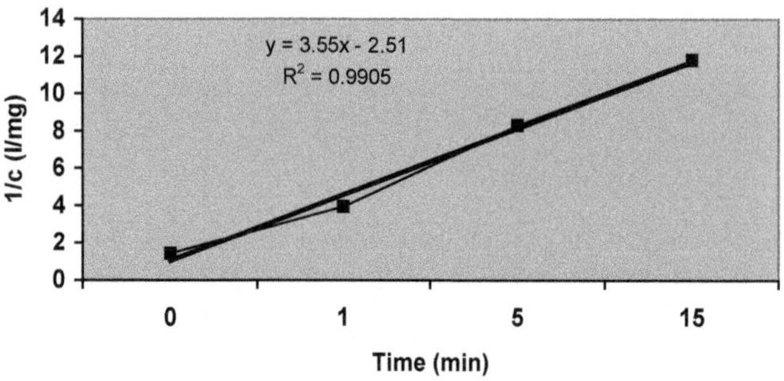

Figure 6: Reaction kinetic (2nd order) plot for the Al^{+3} adsorption of activated zeolite (1.6 mg L Al^{+3} concentration and 15 min contact time period).

adsorbent for the removal of the Al^{+3} from the polluted water. A process using the zeolite for the removal of heavy metal seemed to be potentially more economical than current process technology. However, further studies would suggest, since different metals and different adsorbents are known to have different properties structures.

Heavy metals affect to human health because of their harmful effect. Nevertheless, a lot of heavy metals were using in the industry and they join to food chain. In this respect, wastewaters coming from pollution sources must be treated to decrease their concentration down to limits given in water regulations before to give in the environment. Main principle of removal processes base to chemical precipitation of metal ions as hydroxide. Adsorption, ion exchange, membrane filtration and reverse osmosis etc. were used for advanced treatment. Removal of the metal pollution from the water with the classical methods was not sufficient, practical and economical. Ion exchange method may be used after pre-treatment operation. Most of the heavy metals recovery was not easy from the precipitate and economic. For this reason, adsorption processes by the natural zeolites may be preferred in the application.

In this investigation, a natural zeolite, as an adsorbent was used for removal of Al^{+3} from the water and very high removal efficiencies were found with different working conditions. Zeolite was activated treating with different concentrations of acid and alkali solution in the first step. Tests showed that optimum activation condition was addition of 2 ml 1 M HCl in 100 ml water containing 2 grams of zeolites.

After activation process, other experimental conditions were investigated. The second test was determination of usage of concentration of Al^{+3} which was optimum at 1.6 mg L^{-1}. Removal rate was reached about 66% with this Al^{+3} concentrations. In a separate investigation, with activated zeolite at same condition given above and 40 mg L^{-1} Al^{+3} concentrations adsorption solution was optimum condition.

pH of the adsorption solution was other important condition which was tested. A maximum removal efficiency of 85% was found at pH 6. This showed that zeolite was activated in acidic condition changing cations with H^{+} ions and Al^{+3} ions form in the working solution. After this experiment temperature and contact time were determined. Removal efficiency was increased until 30°C, and then decreased.

All these experimental results showed that zeolite is suitable adsorbent for removal of heavy metals. Zeolite is available in economical price. It is suitable material for heavy metal removal from the industrial waste water.

References

1. Jekel MR (2005) Aluminum in water: How it can be removed? Use of aluminum salts in treatment. Proc. of the Int. Water Supply Ass., Copenhagen, Denmark, May 25-31.
2. Driscoll CT, Letterman RD (2005) Chemistry and fate of Al III in treated drinking water. J. Environ. Eng. Div., ACSE 114: 21-37.
3. Martyn CN, Osmand C, Edwardson JA, Barker DJ, Harris EC, et al. (2008) Geographical relation between Alzheimer's disease and aluminum in drinking water. The Lancet. 1: 59-62.
4. Chen G (2004) Electrochemical technologies in wastewater treatment Separation and Purification Technology 38: 11-41.
5. Petrus R, Warchol J (2003) Ion exchange equilibria between clinoptilolite and aqueous solutions of Na^{+}/Cu^{2+}, Na^{+}/Cd^{2+} and Na^{+}/Pb^{2+}. Microporous and Mesoporous Materials.
6. Axtell NR, Sternberg SP, Claussen K. (2003) Lead and nickel removal using Microspora and Lemna minor. Bioresour. Technol. 89: 41-48.
7. Argun ME, Dursun Ş, Özdemir C, Karataş M (2007a) Heavy metal adsorption by modified oak sawdust: thermodynamics and kinetics. J Hazard. Materials. In Press.
8. Argun ME, Dursun S, Gür K, Özdemir C, Karatas M, et al. (2005) Adsorption of copper on the modified wood (pine) materials. Cell. Chem. Technol. 39: 581-593.
9. Özdemir C (2005) Chromium removal from the wastewater on raw material of saw dust. Cell Chem. Technol. 39: 325-329.
10. Márquez GE, Ribeiro MJP, Ventura JM, Labrincha JA (2004) Removal of nickel from aqueous solutions by clay-based beds. Ceramics International 30: 111-119.
11. Ho YS, Wase DAJ, Forster CF (1995) Batch nickel removal from aqueous solution by sphagnum moss peat. Water Res. 29: 1327-1332.
12. Ames LL (1967) Zeolitic Removal of Ammonium Ions from Agricultural Waste Waters, Proc. 13th Pasific Northwest Ind. Waste Conf, Washington State Univ, 135-152.
13. Chelischev NF (1974) Ion Exchange of Heavy Metals on Clinoptilolite, Doc. Acad Nauk. S.S.S.R. 217: 1140-1141.

Modeling Total Ammonia Nitrogen Concentration: A Case Study in Houston Ship Channel and Upper & Lower Galveston Bay

Qing Li*, Fengxiang Qiao and Lei Yu

Innovative Transportation Research Institute, Texas Southern University, Houston, Texas, USA

Abstract

Nitrogen is an essential nutrient to form amino acids for fauna and flora. Its compound ammonia, however, is also one of the most important contaminants in an aquatic environment for its highly toxic nature and ubiquity in the surface water. Monitoring and controlling the total ammonia nitrogen are vital for human health and sustainable economic development. This paper attempts to develop an optimal model to monitor and predict the development of total ammonia nitrogen in a water body. A case study was carried out in the Houston Ship Channel and Galveston Bay in Texas, which aquatic environment is a nursery and spawning ground for diverse types of marine life. Meanwhile, the Bay also assimilates ammonia pollutants from Texas's municipal and industrial wastewater. The toxic threat of total ammonia nitrogen in the bay was assessed, based on observed samples and forecasted values. Forty years of samples were collected from the Texas Commission on Environmental Quality. Correlations analysis was conducted between all physical, biological and chemical parameters and the total ammonia nitrogen. The trends of total ammonia nitrogen were modeled through a multivariate regression and an auto-regression, followed by an estimation of hazard quotient. The outcome shows that the total ammonia nitrogen in this study area is spatially correlated. For upstream flow, most measured parameters were highly correlated with the total ammonia nitrogen, whereas for downstream flow, weak correlations were noticed. Modeling results indicate that the auto-regression models can better fit the observed data than the multivariate regression models. Meanwhile, the predicted total ammonia nitrogen and the hazard quotient will remain at a lower level, meeting the ambient water criteria continuous concentration.

Keywords: Water pollution; Hazard quotient analysis; Time series model; Multivariate regression; Correlation analysis

Abbreviation

TAN: Total Ammonia Nitrogen

Introduction

Fauna and flora require nitrogen as an essential nutrient to form amino acids. These amino acids perform critical roles in processes, such as neurotransmitter transport and biosynthesis [1]. Even though 75% of the atmosphere is nitrogen in its diatomic form (N_2 gas), few living creatures can utilize this type of nitrogen directly. Nitrogen needs to be converted into other forms before most aquatic plants can utilize it.

Ammonia (NH_3), or azane, is one of the important sources of nitrogen for the living system. It contributes significantly to the nutritional needs of terrestrial organisms, serving as a precursor to food and fertilizers. Ammonia is widely used in many commercial cleaning products and is considered as one of the most important contaminants in an aquatic environment for its highly toxic nature and ubiquity [2]. Fish and amphibians lack a mechanism to prevent ammonia's build-up in the bloodstream, like humans and other mammals do. They usually eliminate ammonia from their bodies by direct excretion. Therefore, ammonia is toxic to the aquatic environment, even at low concentrations.

NH_3 is an unionized ammonia that can react with water to form ionized ammonia (NH_4^+) in a weak base; it is represented by the chemical equilibrium in Equation 1.

$$NH_4^+ + OH^- \leftrightarrow NH_3 \cdot H_2O \leftrightarrow NH_3 + H_2O \quad (1)$$

The unionized ammonia and ionized ammonia exist simultaneously in the water; it can be measured as the total ammonia nitrogen (TAN). The proportion of NH_3 in TAN is subject to the values of pH, temperature, and salinity in the aquatic environment [3-7]. The toxicity of TAN increases as the pH decreases, and as temperature decreases. The higher base leads to higher ionized ammonia, thereby increasing toxicity [8].

The toxic concentrations of NH_3 range from 0.53 to 22.8 mg/L for freshwater organisms [8]. Though plants are more tolerant of ammonia than animals, the hatching and growth rates of fish may result in changes in the tissue of gill, liver, and kidney. For human beings, an excessive concentration of ammonia may cause a loss of equilibrium, convulsions, coma, and death. Therefore, to prevent any chronic and acute aquifer toxicity, a sufficient monitoring approach is required. A monitoring approach is usually referred to biological measurements at regular sites or random sites throughout an area and state, which are time-consuming work and cannot predict seasonal change in water quality. Modeling could complement these imperfections in the approach of biological measurements.

This paper attempts to develop an optimal model to monitor the dynamic TAN concentration in a water body and assess the toxic threat of TAN to the aquatic environment, based on a case study in the Houston ship channel and Galveston Bay (HSC&GB) in Texas (USA) was selected. The Bay assimilates ammonia pollutants from Texas's wastewater discharge for more than 4.5 million populations;

***Corresponding author:** Qing Li, Innovative Transportation Research Institute, Texas Southern University, Houston, Texas
E-mail: liq@tsu.edu

this includes raw or partially treated industrial wastes, and non-point source storm water runoff [9]. The overload of nutrients, fecal coliform bacteria, and suspended solids are attributed to deleterious effects on the HSC&GB's water quality and aquatic life.

Since the late 1960's and earlier 1970's, many remediation plans have been implemented in the HSC&GB, such as wastewater treatment at the point sources and a federal/state permitting process [10]. Though the concentration of pollutants has been controlled, the monitoring of the nutrient dynamics is still required for sustainable economic development.

Materials and Methods

Sample collection

The data samples used in this paper were collected in the HSC&GB, in the State of Texas (USA). The Upper & Lower Galveston Bay is a shallow bay with an average depth of 7 feet (2.1 meters), a length of 35 miles (56 km), and a width of up to 19 miles (31 km). It is the largest estuary in Texas and the 7th largest estuary in the US. The Houston ship channel provides deep-water access to both the Gulf of Mexico and Houston [11]. The unique and complex mixing of water in this area provides important nursery and spawning grounds for diverse types of marine life, including crabs, shrimp, oysters, and fish. Consequently, this spawning ground contributes to Texas' economy by 4.2 billion dollars annually [9].

The TAN observed concentrations for this research were obtained from the water database of the Texas Commission on Environmental Quality [12] from the early 1970's until 2010. A total of 206 samples were collected from the Houston Ship Channel; 1978 samples were collected from the Upper & Lower Galveston Bay. The entire study area is divided into six groups, as illustrated in Figure 1.

Analytical methods

Analytical procedure: The method to analyze and model the TAN in the HSC&GB is to follow a five-step procedure:

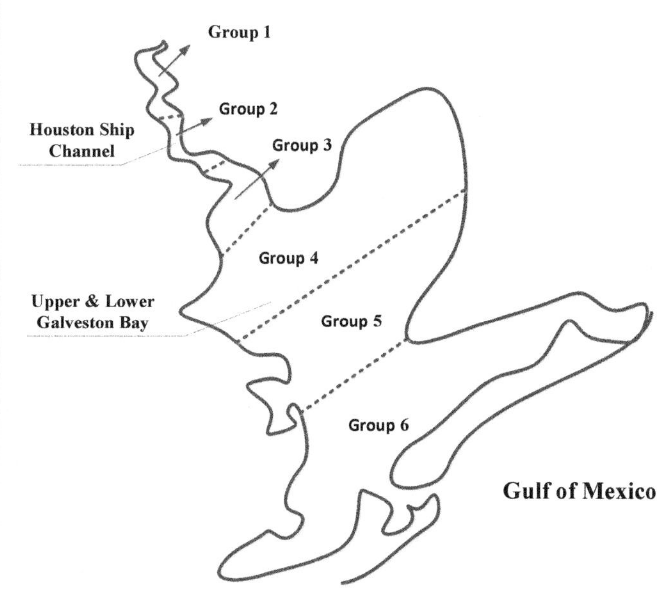

Figure 1: Six divided groups in the study area.

dividing the entire study area into six groups, as shown in Figure 1, and calculating the mean concentration for each group;

conducting time/spatial correlation analyses among the observation groups and a dependency correlation analyses to identify potential dependent parameters for modeling;

conducting multivariate regression modeling between the identified dependent variables and the TAN from relevant groups;

conducting time series modeling for the TAN from all groups; and

evaluating the toxic threat of ammonia to the aquatic environment by calculating the Hazard Quotient (HQ) of the TAN.

Correlation analyses: Due to the connected water body in the study area, the TAN concentration may be spatially correlated to each other. The spatial correlation analyses of TAN concentrations among observation groups will be conducted.

Besides, the un-ionized ammonia concentration is very sensitive to pH and temperature, which vary from time to time and season to season. The un-ionized ammonia concentration is often estimated through TAN. The dependency correlations between the TAN and the parameters indicating water and ecological quality were studied as well, including pH, temperature, Total Suspended Solids (TSS), Total Organic (TO), Chlorophyll a, salinity, Dissolved Oxygen (DO), ortho-phosphorus (OP), Biochemical Oxygen Demand (BOD), Specific Conductance (SC), fecal, and E coli, Enterococci. The cross analysis of dependency correlation attempts to identify potential dependent parameters for the TAN from each group.

Multivariate regression modeling: The multivariate regression attempts to determine a formula that can describe how elements in a vector of parameter (e.g., TAN) respond simultaneously to the changes in others (e.g., the parameters that may have higher correlation with TAN in the same group). The multivariate regression is distinct from the multivariable regression, which has only one dependent variable [13]. A multivariate normal regression is the regression of a multi-dimensional response on a design matrix of predictor variables, with normally distributed errors.

$$y_i(t) = \sum_{j=1}^{n_i} \alpha_{ij} x_{ij}(t) + e_i \tag{2}$$

where: y_i (t) is the TAN for group i in year t, x_{ij} (t) is the jth parameters for the TAN of group i in year t. n_i is the total number of parameters for group i. $\{\alpha_{ij}\}$ is the regression coefficients and e_i is the error term following a normal distribution.

Time series modeling: No matter whether the dependent parameters can be successfully identified in step 1, the TAN from all groups can be treated as chronic datasets called "time series" with "no consideration of" the dependent parameters. The "raw" datasets should be pre-processed by removing the trend components from the TAN observations, while the residuals with a zero mean should be used for non-dependent parameter modeling [14].

The candidate "best fit" models include Autoregressive (AR) models and Moving Average (MA) models, or their combination ARMA model [14]. As the Moving Average (MA) part needs error information from real observations, which is not feasible during the forecasting stage, only the Autoregressive part as AR models are considered in this research.

Assuming x_t is the TAN observation for year t, the estimate of x_t can be expressed in Equation 2.

$$\hat{x}_t = \hat{\varnothing}_1 x_{t-1} + \hat{\varnothing}_2 x_{t-2} + \ldots + \hat{\varnothing}_n x_{t-n} \quad (3)$$

In Equation (3), x_{t-i} means the TAN observation in the ith year before year t. $\{\varnothing_i\}$ are the coefficients of $\{x_t\}$, i=1, 2, ...n, n is the order of the model. $\{\varnothing_i\}$ can be calibrated using algorithms such as the Forward-Backward algorithm (FB), the Least Squares algorithm (LS), the Yule-Walker algorithm (YW), the Burg's algorithm (BURG), and the geometric lattice method (GL) [14]. The goodness of the models is based on indexes such as the Akaike Information Criterion (AIC) (please refer to Akaike (1974) [15] for more AIC detail). The modeling error α follows an independent normal distribution $a \sim \text{NID}\left(0, \sigma_a^2\right)$.

Risk assessment: The toxic threat of ammonia to the aquatic environment in HSC&GB will be assessed by the Hazard Quotient (HQ) of TAN for each group, which is expressed as the concentration ratio between the observed and reference values (Equation (4)).

$$HQ = \frac{C_{Observed}}{C_{Reference}} \quad (4)$$

The ambient water Criteria Continuous Concentration (CCC) for saltwater was adopted as the reference concentration. The CCC's should be selected based on the water salinity level, pH values, and temperature in the different groups.

Results and Discussion

Characteristics of the TAN in the HSC&GB

The time spatial distribution of TAN concentrations in the HSC&GB is shown in Figure 2. The primary TAN pollutants sources come from the Houston Ship Channel, as the TAN concentrations in groups 1 and 2 (in the Ship Channel and up to 5 mg/l in mid-1970) are significantly are higher than others. The TAN concentrations in groups 4 and 5 (within the Galveston Bay) decrease to much less than 0.5 mg/l, which may be the result of the dilution of the seawater from the Gulf of Mexico. The TAN values in group 3 are in the middle.

Spatial correlation

The correlations of the TAN concentration among all six groups are listed in Table 1. Strong correlations are found between groups 1 & 2, 2 & 3, and 3 & 4 with correlation coefficients of 0.91, 0.86, and 0.80, respectively. The correlations between the first four groups (1-4) and the last two groups (5 and 6) are apparently weaker. The weakest correlation is between group 5 and the other five groups, ranging from 0.10 to 0.30.

Dependency correlation

Ammonia in an aquatic environment is sensitive to many factors. Therefore, it is necessary to conduct a correlation analyses between the TAN and related physical, biological and chemical parameters. Their correlation results are listed in Table 1.

In Table 1, apparently almost all correlation coefficients for groups 4 to 6 (all in Galveston Bay area) are less than 0.5, which implies that, in the Galveston Bay area, the TAN is not well correlated with all of the listed parameters. It is more likely that Galveston Bay pays a crucial role in the dilution and degradation of contaminants coming from the Houston Ship Channel.

By excluding the parameters with continuously missing data, the dependent variables for TAN are: TSS, TO, DO, Ortho-Phosphorus, and TN for group 1; TSS, Chlorophylla, DO, and Ortho-Phosphorus for group 2; TSS, Chlorophylla, Ortho-Phosphorus, and TN for group 3; and Chlorophylla and Ortho-Phosphorus for group 4.

From the modeling perspective for groups 1 to 4, multivariate regression models could be an option to depict the relationships between TAN and its correlated parameters. For groups 5 to 6, since there is no significant independent variable identified, the time series model, especially the auto-regression (AR) model, could be a possible choice. The AR models could also be applied to TANs for groups 1 to 3, as long as the modeling errors are acceptable.

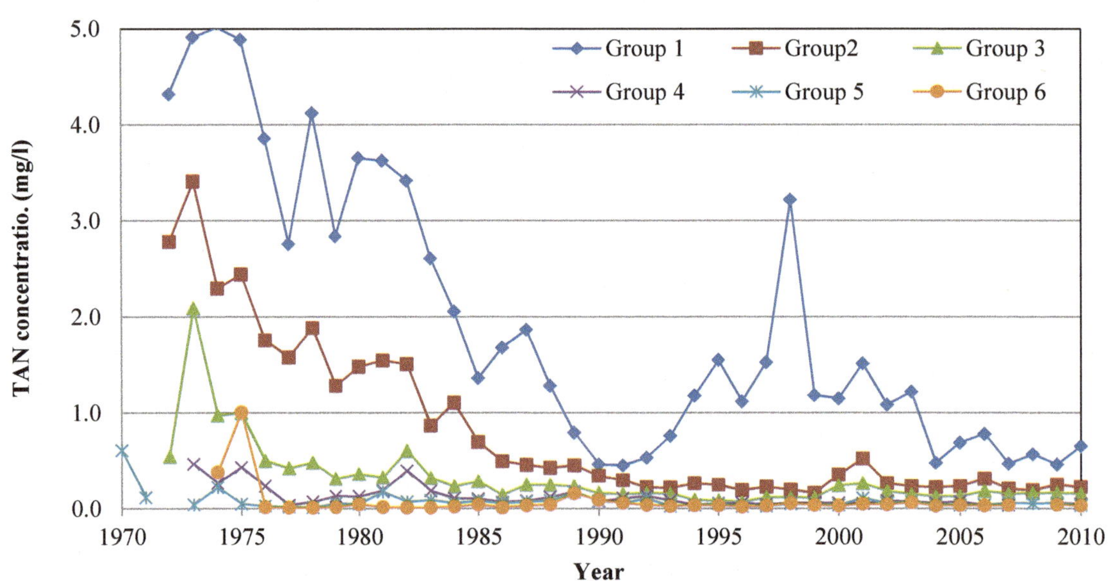

Figure 2: Concentrations of TAN in HSC&GB from 1970 to 2010.

	Group/ Parameter	Group 1	Group 2	Group 3	Group 4	Group 5	Group 6
Spatial Correlation	1		**0.91**	**0.73**	**0.63**	0.10	0.36
	2			**0.86**	**0.66**	0.06	0.33
	3				**0.80**	0.10	0.34
	4					0.30	*0.51*
	5						0.22
Dependency Correlation	HSC1_pH	-0.04	0.02	0.37	0.30	0.01	0.24
	Total Suspended Solids (TSS)	**0.71**	**0.53**	**0.50**	0.15	-0.06	0.08
	Total Organic (TO)	**0.71**	0.06	-0.11	0.32	0.06	0.21
	Chlorophyll a	0.21	**0.59**	**0.69**	0.54	-0.06	-0.46
	Salinity	-0.20	0.21	0.27	-0.04	-0.05	-0.15
	Temperature	-0.13	MD	0.05	-0.03	0.07	-0.03
	Dissolved Oxy (DO)	**-0.80**	**-0.71**	-0.14	0.09	-0.19	0.22
	Ortho-Phosphorus (OP)	**0.61**	**0.91**	**0.75**	**0.59**	-0.26	*0.65*
	Biochemical Oxygen Demand (BOD)	**0.65***	0.11	0.67*	0.26	0.03	0.40
	Specific Conductance (SC)	0.02	-0.09	-0.28	-0.19	-0.15	-0.34
	Fecal	MD	MD	MD	-0.02	0.36	0.01
	E coli	MD	MD	MD	MD	MD	MD
	Enterococci	MD	MD	MD	**0.51***	-0.14	-0.09
	Total Nitrate Nitrite Nitrogen (TN)	**-0.61**	-0.26	**0.69**	0.09	0.35	-0.10

Note: MD: Significantly missing data; *Excluded in the modeling, since the data are not fully continuous; The bold numbers are those with higher correlations.

Table 1: Spatial correlation and the correlation between the TAN concentrations and the other physical, biological and chemical parameters.

Multivariate regression modeling for groups 1 to 4

Figures 3a-3d show the observed and MR predicted TANs for Groups 1 to 4, respectively. At first glance, the MR modeled values can basically catch the trends of the observed TANs. The calibrated coefficients of all MR models are listed in Table 2.

Time series modeling for all groups

The chronic TAN for each group from 1970 to 2010 is a time series. Even though the MR models have been applied to the modeling of the TANs for groups 1 to 3, time series could be an ideal alternative to model them and compare the modeling errors. Therefore, the time series modeling process is applied to the TANs for all groups (*i.e.*, groups 1 to 6).

In this sense, the time series models will be built up with no dependent variables during the modeling process, which follows a four-step procedure: (1) spatial correlation analyses among observation groups; (2) dependency correlation analyses for potential dependent variables; (3) removal of trend components from the TAN observations; and (4) the time series modeling for the residuals of the TAN observations with the trend components removed.

The first two steps are to understand the correlations of the TAN spatially and with other possible dependent factors; while the last two steps are the standard processes of time series modeling [16].

Removing the determinative trends from the TAN sets: One pre-

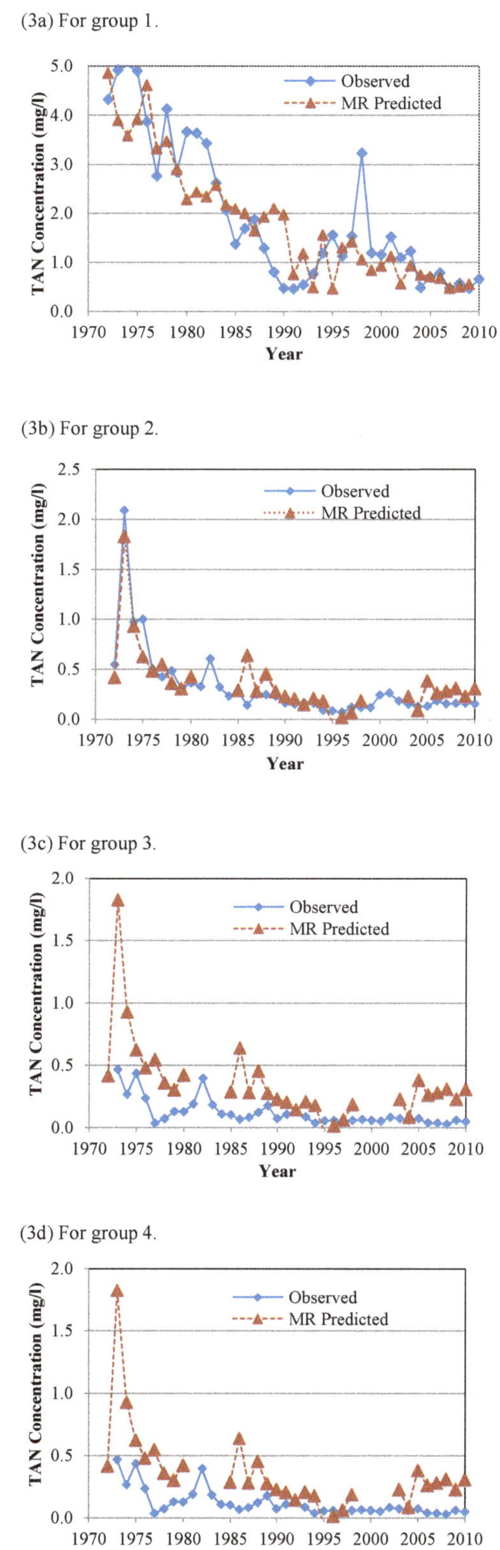

Figure 3: Observed and MR predicted TANs for groups 1-4.

Parameters	Group 1	Group 2	Group 3	Group 4
Total Suspended Solids (TSS)	0.035976672	0.016241028	-0.005974875	N/A
Total Organic (TO)	0.065535998	N/A	N/A	N/A
Chlorophylla	N/A	0.006295538	0.009072259	0.000118654
Dissolved Oxy (DO)	-0.161808029	-0.098797239	N/A	N/A
Ortho-Phosphorus (OP)	0.528104801	0.95821269	0.4602417	0.00872595
Total Nitrate Nitrite Nitrogen (TN)	0.080082469	N/A	0.243803128	N/A

Table 2: Calibrated coefficients α_ of all MR models.

Group	Component	Equations
1	Exponential	$y_{11} = 7 * 1047 * e^{-0.055t}$ $R^2 = 0.6434$
	Linear	$y_{12} = 0.0153t - 30.855$ $R^2 = 0.0649$
	De-mean	$\bar{y}_{13} = 0.007961527$
	Time Series	$\hat{y}_{14,t} = 0.4642y_{14,t-1}$
2	Exponential	$y_{21} = 4 * 10^{63} * e^{-0.074t}$ $R^2 = 0.7987$
	Linear	$y_{22} = -0.0241t + 48.166$ $R^2 = 0.4033$
	De-mean	$\bar{y}_{23} = 0.058382543$
	Time Series	$\hat{y}_{24,t} = 0.4823y_{24,t-1} + 0.25y_{24,t-2} + 0.3221y_{t-3} - 0.1769y_{24,t-4} - 0.0009422y_{24',t-5}$
3	Exponential	$y_{31} = 2 * 10^{41} * e^{-0.049t}$ $R^2 = 0.5621$
	Linear	$y_{32} = 0.0148t + 29.662$ $R^2 = 0.271$
	De-mean	$\bar{y}_{33} = 0.020847395$
	Time Series	$\hat{y}_{34,t} = 0.5844y_{34,t-1} + 0.2618y_{34,t-2} + 0.1373y_{34,t-3} - 0.1552y_{34,t-4} + 0.03495y_{34,t-5}$
4	Linear	$y_{42} = -0.0065t + 13.15$ $R^2 = 0.4377$
	De-mean	$\bar{\ }_{43} = 0.08926$
	Time Series	$\hat{y}_{44,t} = 0.4695y_{44,t-1} - 0.08773y_{44,t-2} + 0.07929y_{44,t-3} - 0.1595y_{44,t-4} - 0.2242y_{44,t-5} + 0.259y_{44,t-6}$
5	Exponential	$y_{51} = 4 * 10^{17} * e^{-0.022t}$ $R^2 = 0.1106$
	Linear	$y_{52} = -0.0023t + 4.6912$ $R^2 = 0.0873$
	De-mean	$\bar{y}_{53} = -0.07657$
	Time Series	$\hat{y}_{54,t} = 0.3155y_{54,t-1} + 0.3158y_{54,t-2} + 0.04417y_{54,t-3} + 0.02185y_{54,t-4} - 0.05981y_{54,t-5} - 0.1181y_{54,t-6}$
6	Exponential	$y_{61} = 6 * 10^{9} * e^{0.0077t}$ $R^2 = 0.0054$
	Linear	$y_{62} = -0.0052t + 10.489$ $R^2 = 0.111$
	De-mean	$\bar{y}_{63} = -0.0916592$
	Time Series	$\hat{y}_{64,t} = 0.8513y_{64,t-1} - 0.012y_{64,t-2} - 0.01424y_{64,t-3}$

Table 3: Identified determinative trends and time series models for groups 1-6.

requisite of time series modeling is to remove the known determinative trends, as well as the mean values, from the datasets. The typical determinative trends to consider include exponential functions, linear functions, and Sine functions. The de-trended residuals should go through a de-mean process to make sure the residual datasets are with the zero mean. The identified determinative trends for all groups are listed in Table 3.

Time series modeling results: The de-trended and time series modeling equations are listed in Table 3. The observed and AR predicted TAN for groups 1 to 6 are illustrated in Figures 4a-4f, respectively,

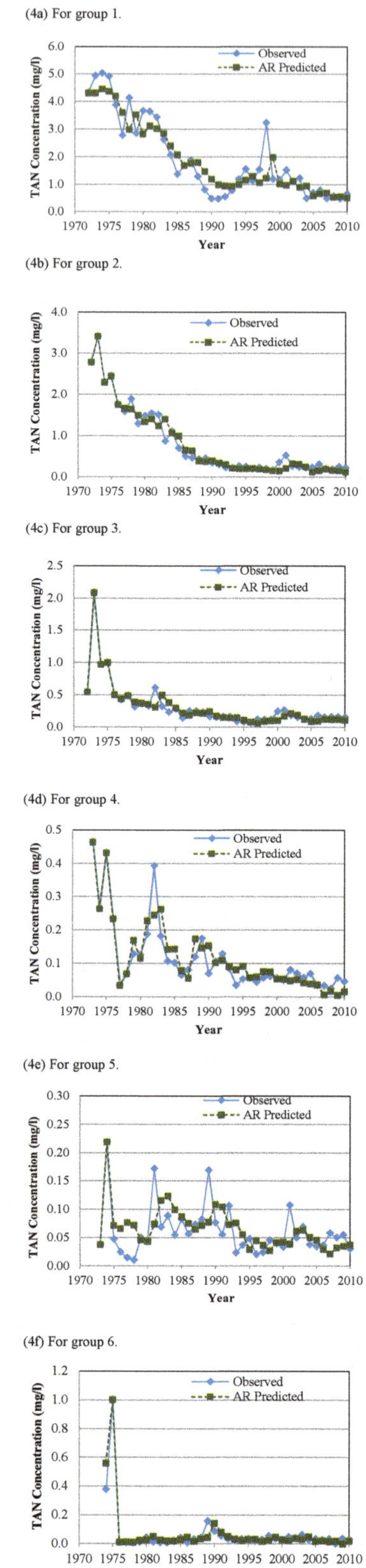

Figure 4: Observed and predicted TAN for groups 1-6.

Group	MR Models	AR Models		
	RMSD	Order	AIC	RMSD
1	0.794	AR(1)	-1.013	0.581
2	0.528	AR(5)	-3.344	0.153
3	0.198	AR(5)	-4.748	0.076
4	0.284	AR(6)	-5.931	0.039
5		AR(6)	-6.241	0.036
6		AR(3)	-6.961	0.040

Table 4: Comparison of the modeling RMSDs.

which shows that the predicted values were very close to the observed one. Their modeling errors are discussed in the following section.

Discussion of the modeling: The orders of time series modeling, AIC values, and the Root Mean Squared Deviation (RMSD) of AR modeling errors for all six groups, plus the RMSD for all MR models, are listed in Table 4.

For groups 1 to 4 in Table 4, the RMSDs of AR are all smaller than their corresponding ones for MR. This means that the time series modeling errors are all less than the multivariate regression ones. Therefore, time series models (the AR models) are adopted for modeling the TANs of all groups. The equations in Table 3 can be used to model the TANs from 1970 until 2010, and even forecast the future year's TAN.

Characteristics of the water body in HSC& GB

Over forty years, there is trend of decline in TAN, TSS, chlorophyll a, ortho-P, and TO, while DO increased. These are the sign of better water quality.

Table 5 lists the mean values and standard deviations of all observed parameters. From Group 1 to 6, the pH and temperature were stable, whereas the salinity, SC, and TSS increased visibly. The closer to the Gulf of Mexico, the water becomes saltier, more dissolved solids (such as salt), and more organic and inorganic particles dispersed in water. The increased DO indicates less contaminant that induces oxygen consumption in the group closely the Gulf. On the contrary, the TO, TAN, Orthor-P, BOD, TN declined obviously, which indicate the decrease in total organic and inorganic matters, nutrients, and biological organisms in the water body closer to the Gulf. Besides, Chlorophll a in group 3-5 were apparently higher than those in other groups, which indicates higher phytoplankton abundance and biomass as well as poor water quality.

Further, the larger standard deviation of TSS, TO, and SC depict that the organic and inorganic particles, total organic, and dissolved solids in the water varied from year to year greatly.

Risk assessment of the TAN for the aquatic environment in HSC&GB

The total ammonia nitrogen Criteria Continuous Concentration (CCC) for each group was identified, according to the water salinity level, pH values, and temperature. In the last four decades, the salinity concentrations ranged between 4 g/kg and 12 g/kg in Group 1 to 3, and 13 g/kg and 17 g/kg in Group 4 to 6, respectively. Saltwater with a salinity of 10 g/kg and 20 g/kg was adopted in the estimation of the Hazard Quotient (HQ) for group 1-3, and group 4-6, respectively. The overview of the HQ is illustrated in Figure 5.

In Figure 5, the samples displayed above the threshold line indicate that the opportunity of toxic effects caused by ammonia to the aquatic environment is relatively high, while below the line represents a lower possibility. In the 1970's, there was higher risk taking place in groups 1 and 2 (Houston Ship Channel). This implies that the TAN seriously contaminated the water body. In fact, in that period, the water quality had raised a great concern, which motivates the implementation of various plans to monitor and remedy the freshwater in HSC&GB.

In the 1990's, the HQ increased dramatically, crossing the 6 groups. In the last two decades, the HQ in group 1 fluctuated slightly, but was still below the threshold line.

This substantial decline may be attributed to the great success of the remediation plans. Moreover, the plans also contribute to the maintenance and monitoring of the water quality in the last two decades. The HQ in the rest of the groups remained stable, as less than 0.25. In other words, since the 1990's, the water quality in the HSC&GB has improved significantly, and therefore, the toxic threat of the TAN to the aquatic environment was lessened.

The identified models were further used to predict the TAN concentrations for the year of 2011, which is still not available in the public. The predicted TAN in 2011, together with the compared 2010 observations and the related HQ, are listed in Table 6.

Table 6 shows that the TAN observations in the both the years of 2010 (observed) and 2011 (forecasted) for all groups were lower than the corresponding CCC. Their Hazard Quotient (HQ) values were also relatively lower, between 2.27E-04 and 2.71E-01. This means that the toxic threat of ammonia to the aquatic environment is low. The

Parameter	Group 1		Group 2		Group 3		Group 4		Group 5		Group 6	
	mean	std	mean	std	mean	std	mean	std	mean	std	mean	std
pH	7,33	0,18	7,55	0,17	7,96	0,21	8,17	0,19	8,18	0,17	8,12	0,22
TSS	26,45	16,65	23,20	8,13	28,52	10,67	26,96	10,11	33,50	24,11	34,38	21,60
TO	11,36	5,84	16,08	2,83	16,69	3,13	6,62	4,41	5,30	4,13	5,04	3,99
Chlorophyll a	6,03	3,65	8,75	7,33	13,72	10,94	18,18	18,70	12,21	14,51	6,80	5,45
Salinity	3,76	1,65	8,69	2,90	11,61	3,21	12,86	4,07	15,78	3,87	16,73	4,75
Temperature	23,58	1,61	23,00	1,59	22,80	1,78	22,24	1,75	22,39	1,58	22,40	2,24
DO	3,76	1,56	5,13	1,61	7,38	1,32	8,48	0,76	8,35	0,69	8,24	0,57
Ortho-P	1,32	0,63	0,95	0,62	0,53	0,31	0,39	0,23	0,18	0,28	0,20	0,24
BOD	5,96	2,92	3,83	1,52	4,26	1,55	4,17	1,71	3,88	1,95	4,31	2,26
SC	6909,11	2393,74	13911,72	4344,22	18089,14	4691,22	20815,20	5318,22	24535,18	5784,32	25846,60	6766,73
TAN	1,95	1,45	0,81	0,84	0,32	0,36	0,12	0,11	0,08	0,10	0,07	0,17
Total Nitrogen	1,40	1,01	1,29	2,37	0,59	0,87	0,18	0,08	0,11	0,07	0,06	0,07

Table 5: Mean values of all parameters over forty years in HSC& GB.

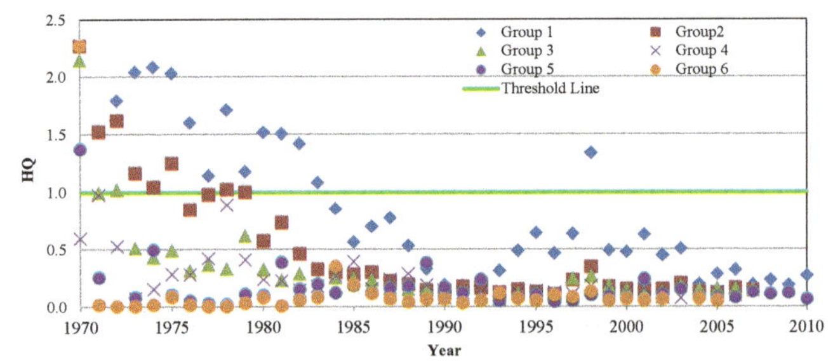

Figure 5: HQ of TAN for all six groups.

Parameters	Group 1	Group 2	Group 3	Group 4	Group 5	Group 6
Salinity concentrations (g/kg)	4-12			13-17		
Saltwater salinity criteria (g/kg)	10			20		
T °C (2010)	20	20	20	20	20	20
pH (2010)	7.6	7.8	8	8.4	8.4	8.4
CCC* (mg/l)	2.4	1.5	0.97	0.44	0.44	0.44
TAN (mg/l) (2010)	0.65	0.22	0.15	0.05	0.03	0.03
HQ (2010)	2.71E-01	1.47E-01	1.55E-01	1.14E-01	6.82E-02	6.82E-02
TAN (mg/l) (2011)	0.52	0.06	0.07	0.00	0.02	0.01
HQ (2011)	2.17E-01	4.00E-02	7.22E-02	2.27E-04	4.55E-02	2.27E-02

Note:*Published in the Environmental Protection Agency (1989) for ambient aquatic life water quality criteria for ammonia [16].

Table 6: Identification of toxicity criteria for each group and HQ estimation for2010 and 2011.

remediation plans for wastewater treatment proposed in the late 1960's and early 1970's continue to contribute to the water quality in the Houston ship channel and Galveston Bay in Texas (USA).

Conclusions

In this investigation, the TAN concentration in the divided six groups of the HSC&GB were analyzed and modeled. Since the implementation of remediation plans in the late 1960's and earlier 1970's, the water quality has significantly improved. The treatment at the point sources has made a great contribution. The correlation results show that the TAN concentration is highly correlated to the different geo-locations. This phenomenon could be the result of the higher capacity of dilution and degradation that the upper and lower Galveston Bay possesses. For the upstream flow, there were 2-4 parameters that were highly correlated with TAN (R>0.50), while for the downstream flow, weak correlations were noticed between the TAN and the 14 candidate parameters. By eliminating the parameters with the continuously missing data, the highly correlated parameters were chosen as dependent variables to model the chronic TAN concentrations using a multivariate regression (MR) in groups 1 to 4. On the other hand, the time series AR models were applied to all groups. The results show that the AR is able to model and predict the trends of the TAN more accurately than the MR. The predicted TAN and the HQ will remain at a lower level, meeting the ambient water Criteria Continuous Concentration (CCC).

Acknowledgment

The author sincerely appreciates the detailed instructions from Dr. Maruthi Sridhar B. Bhaskar on data processing. The author also wishes to acknowledge the supports in part by the US. Tier 1 University Transportation Center TranLIVE # DTRT12GUTC17/KLK900-SB-003, and the U.S. National Science Foundation (NSF) under grant #1137732.

References

1. Wagner I, Musso H (1983) New Naturally Occurring Amino Acids. Angew Chem Int Ed Engl 22: 816-828.
2. Russo RC (1985) Ammonia, nitrite, and nitrate. In: Fundamentals of aquatic toxicology and chemistry. Hemisphere Publishing Corp., Washington, D.C. 455-471.
3. Emerson K, Russo RC, Lund RE, Thurston RV (1975) Aqueous ammonia equilibrium calculations: Effect of pH and temperature. Journal of the Fisheries Research Board of Canada. 32: 2379-2383.
4. Erickson RJ (1985) An evaluation of mathematical models for the effects of pH and temperature on ammonia toxicity to aquatic organisms. Water Research. 19: 1047-1058.
5. Thurston RV (1988) Ammonia toxicity to fishes. In: Fish physiology, fish toxicology, and fisheries management: Proceedings of an International Symposium, Guangzhou, PRC. 183-196.
6. Whitfield M (1974) The hydrolysis of ammonium ions in seawater-a theoretical study. J. Mar. Biol. Assoc. U.K. 54: 565-580.
7. Wood CM (1993) In: Evans DH (Ed.) Ammonia and urea metabolism and excretion. The physiology of fishes. CRC. Press, Ann Arbor, MI. 379-424.
8. Oram B. Ammonia in Groundwater, Runoff, and Streams. Water Research Center, na.
9. Galveston Bay Estuary Program 2003-2013.
10. The Galveston Bay Plan. Point Sources of Pollution.
11. Encyclopædia Britannica. IOP Publishing Physics.
12. Texas Commission on Environmental Quality (TCEQ).
13. Hidalgo B, Goodman M (2013) Multivariate or multivariable regression? American journal of public health 103: 39-40.
14. Pandit SM, Wu SM (1983) Time Series and System Analysis with Applications. John Willey and Sons, Inc. New York.
15. Akaike H (1974) A new look at the statistical model identification. IEEE Transactions on Automatic Control 19: 716-723.
16. Environmental Protection Agency (EPA) (1989). Ambient Water Quality Criteria for Ammonia (Saltwater)-1989.

Determination of the Sea Waters Quality of Arzew-Algeria Gulf

Redouane F* and Mourad L
LAAR University of Science and Technology, Mohammed Boudiaf, BP 1505, Oran, Algeria

Abstract

This work consists of assessing the degree of pollution caused by wastewater near the gulf of Arzew-Algeria. This study is based on the analysis of the physicochemical parameters, the parameters indicating pollution and the determination of heavy metals (Pb, Zn, Cd, Fr, Cr, Ni) on seawater samples taken from the site. Four companions of seasonal sampling were carried out during the year 2015. The results obtained show significant values in physicochemical parameters, pollution indicator parameters and heavy metal content in seawater samples. The results of the pollution indicator parameters show the presence of important pollution at the site studied. The results obtained are far superior to the Algerian standards defined.

It is clear from the determination of heavy metals that the site under investigation shows a slight contamination by metals. The absence of treatment and the long-term exposure to these discharges will lead to severe metal pollution in seawater.

Keywords: Arzew; Marine pollution; Physicochemical parameters; Pollution indicators; Heavy metals

Introduction

Every year, pollution is gaining ground. Even beaches allowed for swimming are not totally clean. Water is one of the essential elements for most large industrial enterprises. After being used, most of this wastewater is returned to the environment. As it is usually loaded with organic matter, it becomes a major source of pollution for the environment that receives it [1] . The study of the content of heavy metals (Pb, Zn, Cd, Fe, Cr and Ni) in polluted superficial sea waters and sediments in the Gulf of Arzew shows that these levels vary with harvest and season.

The Algerian coast would undoubtedly deserve a little more attention from public authorities and citizens or from the 487 beaches on the Algerian coast, 213 are prohibited to bathing, the majority of which are closed because of pollution. In recent years, several industries have been active in Arzew. There are mainly chemical and petrochemical industries. These industries, like all human activities, produce various effluents and waste. Untreated effluent is discharged into the sea. The large slabs of petrochemicals presented along the coast off the industrial poles of Arzew and Bethioua, causing adverse effects on marine fauna and flora. Estimated quantities of more than 20,000 tons of petrochemicals are illegally dumped by foreign oil tankers and LNG tankers into the waters of these two ports linked to the industrial complexes of Arzew and Bethioua. In this study, we will show the impact of effluent discharges on aquatic receptor. For this, we measured the main parameters of this pollution.

Several authors Boutiba et al. [2], Samir [3] have shown a growing pollutionthroughout the Algerian coast, especially in the coastal areas of large cities like the Oran coast. The pollution of marine ecosystems by hydrocarbons has been a concern for national governments and the international community for many years. According to the results obtained by Laama [4], Sahnoun et al. [5], Remili and Kerfouf [6] concerning the analysis of sea water samples taken at different locations are summarized in the following Table 1.

Selection and Description of Study Area

The industrial zone of Arzew was realized in 1970 east of Oran. By virtue of its strategic geographical position, its existing infrastructure accentuates more by the important urban-industrial development of the establishment of the industrial zone which was originally conceived as a factor of spatial planning policy and unfortunately, not ensure socio-economic development in a harmonious way. The cities of gulf (Arzew, Bethioua and Ain-bia) are characterized by a problem specific to the presence of complexes that gives it an industrial function. The industrial zone of Arzew is one of the three most important industrial of the country.

This industrial zone covers a total area of 961.69 ha (Figure 1) and allows the establishment of 15 production units and 17 service units.

Materials and Methods of Analysis

The coastline of the Arzew gulf is increasingly attacked today by various forms of nuisance due to industrial activity and massive urbanization, resulting in an ever-increasing extent of pollution from domestic and Industrial development. The coastal fringe is increasing wastewater discharge without any treatment, causing biological and physicochemical contamination of marine waters [6,7]. This coastal fringe is experiencing a galloping population expansion accompanied by anarchic urbanization and intense port associated with various industrial activities, which have consequences for the coastal environment [8].

In our study, we carried out 4 season sampling companions (Figure 2), the 1st on winter (January), the 2nd during the spring (April), the 3rd on summer (July) and the last on autumn (October).

To take a sample, I used a glass bottle, to keep it sterilized, to introduce it under the surface of the water to 15-30 cm below the level

***Corresponding author:** Fares Redouane, LAAR University of Science and Technology, Mohammed Boudiaf, BP 1505, Oran, Algeria
E-mail: fares-redouane@hotmail.com

LAAMA. 2009			SAHNOUN. 2010				REMILI. 2009	
Port of Algiers			Industrial area of Arzew				Wastewater discharged at the level of the coast of the city of Oran	
	S1	S2		S1	S2	S3	Cd (mg/L)	0.52
MES (mg/L)	83	152	DCO (mg/L)	530	900	585	Pb (mg/L)	0.23
DBO5 (mg/L)	23.3	16.7	DBO5 (mg/mL)	220	360	250	Zn (mg/L)	0.61

Table 1: Various studies on the analysis of sea water samples.

Figure 1: View of the Arzew golf with the industrial zone and the various ports.

Figure 2: Different sampling stations.

of the water then to remove the cap to fill the Bottle to avoid any kind of contamination of the sample.

Seawater samples taken at different sites are preserved in polyethylene bottles with capacities ranging from 250 ml to 1 liter, previously washed with distilled water and transported at low temperatures 4°C in portable coolers. The conservation of the water samples was done according to the guide for the preservation and handling of samples according to ISO 5667/3. The temperature of the water is measured in situ using a two-way oximetry (type WTW OXI 96), pH and electrical conductivity are measured using a pH meter of type (WTW 525). For the determination of the BOD_5, the OxiTop is used; COD (COD-type meter WTW) was used. The MES was quantified by filtration (circular WHATMAN filters with a 47 mm diameter and a porosity of 0.45 µm) is determined according to the AFNOR standard (T90-105). The determination of the traces of metals is carried out by atomic emission spectroscopy with plasma coupled by induction according to standard NF EN ISO 11885.

The analytical methods used are described by [9]. The analyzes were carried out at the Physico-chemical, Materials, Catalysis and Environment laboratory within the USTO's university.

Results and Discussion

All results are grouped in Table 2.

Physical parameters

Temperature (T): The mean values of the temperature recorded in the sea water of the studied sites vary between 12°C and 21°C. According to the results, we note that the sea water temperature at the site level is within the norm (below 30°C) which is considered as the limit value of direct rejection in the receiving environment [10].

pH: The pH is an important parameter which represents the degree of ionization of the medium studied; it gives us an indication on the level of the pollution of the water. It should be strictly monitored during the sampling period. The average of pH values recorded in sea water ranged from 6.8 to 8.3. The results obtained remain in the global standards which recommend a pH of the discharges comprised between 6.5 and 8.5.

Dissolved oxygen: Dissolved oxygen is a very useful parameter to indicate the quality of the water; it is considered one of the parameters most sensitive to pollution. Its value gives information on the state of water pollution. The dissolved oxygen analyzes performed on seawater samples ranged from 4.3 mg/L to 10.5 mg/L.

Conductivity: The recorded salinity of the raw effluent oscillates at an average of 36000 µS/cm. This salinity is mainly linked to the discharge of wastewater. The values recorded are relatively high, which can cause aggressive problems in the pipes. Beyond 3000 µS/cm, the conditions are unfavorable for a normal ecological balance. For the samples of seawater analyzed, the results obtained reveal averages varying between 34933,33 µS/cm and 38433,33 µS/cm. The results obtained show that the electrical conductivity is very high in seawater. The conductivity of seawater is very high due to the high concentration of dissolved salts because it makes it possible to evaluate the total load in electrolytes of water. Conductivity gives an idea of the mineralization of water and is therefore a good marker of the origin of water. Indeed, the measurement of the conductivity makes it possible to appreciate the quantity of the dissolved salts in the water, and therefore of its mineralization. The electrical conductivity depends on the loads of endogenous and exogenous organic matter generating salts after decomposition and mineralization and also with the phenomenon of evaporation which concentrates these salts in water; it also varies according to the geological substrate traversed.

Suspended matter: The suspended matter represents all the mineral and organic particles contained in the waters. High levels of suspended solids can be considered a form of pollution [7]. The results obtained during our study reveal very high concentrations. For the samples obtained, the mean values recorded ranged from 294 mg/L to 3062 mg/L. The results obtained during our study show that the concentrations of the suspended solids in all the samples analyzed far exceed the national standards defined (35 mg/L). These concentrations are probably caused by the boats emptying, sewage and urban domestic waste that are rich in colloidal matter of mineral or organic origin.

Chemical parameters

Chemical oxygen demand (COD): The COD monitoring of seawater shows very high values which far exceed the limits allowed by the national standard (120 mg/L) and international of direct discharges. Indeed, the COD values range between 460.8 and 4944 mg/L with an average of 1600 mg/L.

Biochemical oxygen demand for 5 days (BOD_5): Monitoring of BOD_5 follows that of COD with lower values. It oscillates between 207.57 mg/L and 2001.62 mg/L with an average of 582.93 mg/L. These

Parameter	Norms	Range of results	
		MIN	MAX
Temperature °C	30	12	21
pH	6.5-8.5	6.8	8.29
Dissolved Oxygen (mg/L)	-	4.3	10.5
Conductivity (µS/cm)	3 000	34933.33	38433.33
Suspended Matter SM (mg/L)	35	294	3062
Chemical Oxygen Demand COD (mg/L)	120	460.8	4944
Biochemical Oxygen Demand BOD5 (mg/L)	40	207.57	2001.62
Total Hydrocarbons (HCT) (mg/L)	10	9.2	11.3
Organic Matter (mg/L)	-	15.83	71
Total Phosphorus (mg/L)	2	0.15	3.46
Nitrogen Kjeldahl Total (NTK) (mg/L)	30	16	21.4
Nitrate (mg/L)	50	1.66	2.66
Nitrite (mg/L)	-	0.02	0.05
Ammonium (mg/L)	-	0.21	4.54
Lead	1(mg/l)	0.003 µg/L	0.24 µg/L
Zinc	5 (mg/l)		0.66 µg/L
Cadmium	0.2(mg/l)	0.14 µg/L	8.58 µg/L
Iron (mg/L)	1	0.056	0.28
chrome (mg/L)	0.1	0.005	0.024
Nickel (mg/L)	5	0.019	0.033

Table 2: Results and norms of seawater physicochemical parameters of the Arzew gulf.

values also exceed 2 to 20 the permissible standard (40 mg/L) for direct discharge.

Biodegradability ratio COD/BOD_5: The ratio of biodegradability allows us to assess the derivability of organic matter. The ratio of biodegradability concerns the COD/BOD_5 ratio which serves to specify the degree of biodegradability of the organic matter at the sites studied. The results of this ratio are all between 2.5 and 4 so the sites studied are highly biodegradable.

Total hydrocarbons (HCT): The values obtained for HCT stay near to the norm (10 mg/L). Values fluctuated from 9.2 mg/L to11.3 mg/L with an average of 10.22 mg/L.

Organic matter: Average values of organic matter in seawater range from 15.83 mg/L to 71 mg/L. The results obtained allow us to observe the variations of the organic matter during the period of study. MO concentrations increased throughout the winter season and decreased in the spring. The increase in organic matter content at study sites is due to the high levels of organic matter discharged into the environment. This high organic matter content is explained by the pollutant load carried by the anthropogenic coastal discharges of the waste water discharged at the sites studied.

Total phosphorus: Analysis of total phosphorus in the samples collected revealed very high concentrations. The mean values obtained are between 0.15 mg/L and 3.46 mg/L. Compared to the Algerian limits set at 2 mg/L, the phosphorus contents found are well above the defined standard. Much of the organic phosphorus also comes from wastes of protein metabolism and its elimination as phosphates in the urine by man and the rest comes mainly from detergents. Phosphorus has an important role in the eutrophication of the marine environment because too much phosphorus can accelerate it. An excessive phosphorus intake can also create favorable conditions for the development of cyanobacteria water flows, especially when the surface water temperature is relatively high and there is a lack of dissolved oxygen.

Nitrogen kjeldahl total (NTK): In practice, Nitrogen Kjeldhal is an indicator of environmental pollution and its monitoring makes it possible to monitor the evolution of contamination. In our case, the amplitude of NTK variations during the study remains low. Values range from 16 mg/L to 21.40 mg/L and are more or less within the permitted standard (30 mg/L).

Nutrient Salts

The nitrate: Nitrates represent only one of the many forms of nitrogen present in water, while in general constituting the most abundant form of mineral nitrogen. Analysis of the nitrate content at sea level revealed concentrations between 2.66 mg/L and 1.66 mg/L, compared to the Algerian limits set at 50 mg/L, the nitrate contents recorded remain Very inferior. As a result, the waters studied are not subject to a risk of nitrate pollution.

The nitrite: Average concentrations of nitrite in seawater range from 0.02 mg/L to 0.05 mg/L. The low concentrations of nitrites found in the waters studied could be explained by the fact that the nitrite ion (NO_2^-) is an intermediate between ammonium and nitrates and is unstable in the presence of oxygen, whose concentration is generally much lower than that of the two forms which are related to it, nitrate ions and ammonium. For the seawater samples analyzed, the results show a continuous decrease of nitrite at the site studied throughout the four study seasons, a decrease in winter and then a stabilization of the concentration in spring, we note that despite the High levels of nitrite; The nitrates always remain in surplus. This is obviously the reflection of a massive and continuous supply of nitrates or a possible nitrification of ammoniacal nitrogen.

Ammonium: Results from ammonium analysis in seawater samples show concentrations ranging from 0.21 mg/L to 4.54 mg/L. The ammonium ion, NH_4^+, is the reduced form of nitrogen. It comes mainly from the decomposition of natural proteins contained in phytoplankton and microorganisms. It may also come from the supply of purified urban effluents or industrial discharges. It is found in natural waters at concentrations ranging from 0.1mg/L to more than 10 mg/L. The amount of ammonium in the sea water increased in winter and decreased in the spring. The high levels of ammonium can be explained by the enrichment of these sites in this nitrogen compound, which results from the discharges of domestic and industrial wastewater, besides the bacterial decomposition of organic nitrogen compounds, the bacterial mechanism called ammonification. Urban ammonium releases can cause a significant increase in the concentrations of this element in the receiving environment, this reduced form of mineral nitrogen, which can lead to the consumption of dissolved oxygen in this medium during its oxidation. Prove to be toxic.

Determination of heavy metals

The lead: The determination of lead in water samples gives information on the content of this metal in the sea. Average concentrations range from 0.003 µg/L to 0.24 µg/L. By comparing our results with the Algerian standard of 1 mg/L defined by the OJ [10], we find that the site studied is far from being polluted by lead.

The zinc: The mean values of zinc found in samples of seawater range from 0.14 µg/L to 0.66 µg/L. The sets the zinc content in liquid effluents at 5 mg/L, so our results are still below the norm [10].

The cadmium: Cadmium is a relatively rare metal in nature. Its stability in water is a function of pH and oxidation-reduction potential. Its origin, from an industrial point of view, is related to plastics, engine oils, batteries and in products of thermal stability. For pH above 8, cadmium precipitates with carbonates. When mixing sewage with seawater, cadmium forms very stable complexes with chlorides. Cadmium does not degrade in the environment, but physical and chemical processes can alter its mobility, bioavailability and residence time in different environments. In aquatic environments, the mobility and bioavailability of cadmium increases when pH, hardness, suspended solids concentration and salinity are low and oxidation-reduction potential is high [11]. The mean values of cadmium recorded in the studied water fluctuate between 0.24 µg/L and 8.58 µg/L. According to the standards defined by the OJ [10] which fix the cadmium content in liquid effluents at 5 mg/L, our results are much lower than this standard, so there is no cadmium pollution at the D study.

Iron, chrome and nickel: The results show that the values of total metal pollution (iron, chromium, nickel) show negligible variations between the different sampling campaigns. The iron concentrations of the analyzed water ranged from 0.056 mg/L to 0.28 mg/L with an average of 0.122 mg/L. The chromium concentration shows that the extreme values of 0.005 mg/L and 0.024 mg/L with an average concentration of 0.012 mg/L. The recorded nickel values range from 0.019 mg/L to 0.033 mg/L with an average value of 0.026 mg/L [12].

Conclusion

Marine pollution from sewage is a serious problem in the Mediterranean Sea, especially in riparian countries in the process of industrialization like Algeria. With the absence of stations and sewage treatment systems in these countries, wastewater is discharged into the sea without any prior treatment, leading to degradation of the sanitary quality of coastal waters and disruption of marine ecosystems. This work deals with an assessment of the state of marine pollution along the Arzew golf course.

Space-temporal monitoring of several physicochemical tracers has given us the image of a relatively intense pollution which results in a large organic and mineral load. The results of the pollution indicator parameters reveal that the site studied is over the month polluted by the wastewater discharged. Analysis of nutrient salts shows the richness of the site studied in nutrients imported by wastewater discharges.

The determination of heavy metals indicates the presence of slight pollution by lead, zinc, cadmium, iron, chromium and nickel. The long-term exposure to these releases will increase the metal content that can alter aquatic life. However, we find that the impact of the releases studied is relatively moderate. However, continuous exposure can degrade the marine ecosystem and affect wildlife and marine life. These results show the need for prior treatment of raw wastewater, biological treatment should be considered to improve their quality to the required standards and to meet the expectations of the public authorities with regard to protection of the environment and human health.

References

1. Belghyti D, Guamri YEL, Ztit G, Uahidi ML, Joti MB (2009) Physicochemical characterization of abattoir wastewater for the implementation of adequate treatment: Kenitra cases in Morocco. Afri Sci 5: 199-216.
2. Boutiba Z, Zoheïr TM, Abi-Ayad SMEI (2003) State of marine pollution of the Oranan coast. In : Gharb DE, (ed.) Oran, pp: 69.
3. Samir G (2010) Benthic populations of soft substrates on the Algerian coast: Taxonomy, structure and ecological status. PhD Thesis, Oran, Department of Biology, pp : 360.
4. Laama C (2009) Contribution to the study of the physico-chemical and bacteriological parameters of the waters of the port of Algiers. Same. DEUA in marine sciences; Higher National School of Marine Sciences and Coastal Planning; Algeria, pp: 45.
5. Sahnoun F, Bendraoua A, Hadjel H M (2010) Control of marine pollution on the coast of Oran. Communi Sci Technol No. 8. Oran.
6. Remili S, Kerfouf A (2006) Impact of faecal pollution in the recreational areas of the Oran coast (Algeria, northwestern France). Proceedings of the 1st edition of the Francophone Days of Microbiology of Hydric Environments (Agadir, Morocco), pp: 134-141.
7. Redouane F, Mourad L (2016) Pollution characterization of liquid wastes of the factory complex fertial (Arzew, Algeria): J Air Wast Manage Assoc 66: 260-266.
8. Remili S, Kerfouf A (2009) Environmental monitoring and coastal sustainable development (case of the Oran littoral). Proceedings of the 5th International Colloquium on "Energy, Climate Change and Sustainable Development" (Hammamet, Tunisia).
9. Rodier J (1975) An analysis of water, natural water, waste water, sea water: Chemistry, bacteriology, biology.
10. OJ (2002) Official Journal of Algeria. Algerian regulations of liquid effluent discharges.
11. DSPR (2005) Directorate of surveillance and risk prevention. Report on lead and cadmium. Ministry of territorial development, Water and Environment; Kingdom of Morocco.
12. BS EN ISO 5667-3:2003 (2003) Water quality, Sampling. Guidance on the preservation and handling of water samples.

Microplastics in the Marine Environment: Current Status, Assessment Methodologies, Impacts and Solutions

Ogunola Oluniyi Solomon[1]* and Thava Palanisami[2]

[1]International Studies in Aquatic Tropical Ecology, University of Bremen, Germany
[2]Global Centre for Environmental Remediation, University of Newcastle, Adelaide, Australia

Abstract

Microplastics exhibit a global distribution and have been detected in all levels of the marine environment. The most visible and disturbing impact of microplastics are their ingestion and consequent suffocation of hundreds of marine species. Microplastics can contribute considerably to the transport of non-indigenous marine species to a new area thereby threatening the marine biodiversity and the food web. They can also accumulate toxic substances on their surface therefore represent a potential concentrated source of environmental pollution or act as a vector of toxic pollutants in the food web with some severe health implications. Therefore, assessment of the status, effect and mitigations are needed. The current status, effects and solutions to the occurrence of microplastics in the marine environments were conducted by reviewing most of the literatures on the subject matter. Here, we particularly discussed old and new techniques for extracting and quantifying microplastics from marine matrices. Although, it is an intricate and herculean task to address the problems of microplastics in the ocean, some of the approaches to mitigate them are discussed.

Keywords: Microplastics; Marine environment; Impact; Mitigate

Introduction and Historical Background

Plastic pollution has become a global environmental issue as plastic debris has reached and is found in all oceans of the world with adverse impacts on marine biota, biodiversity as well as human health [1]. Although, the threats posed by plastics to the marine environment were initially ignored for a long time, they recently gained attraction and recognition [2]. Since the development of the first plastic "Bakelite" in 1907, a number of cheap production techniques have been developed to mass produce the modern day plastics [3]. Plastics started to enter the ocean in increasing quantities from the 1950s from a wide variety of land and sea based sources; river, run-off, beach-goers and tourists, ship [4]. Although, there are no reliable estimates of the inputs at a regional or global scale, the trend assumed that the total quantities have increased over the years as in Figure 1 [5]. In 1975 alone, it was estimated that the world's fishing fleet dumped into the ocean approximately 135,400 tons of plastic fishing gears and 23,600 tons of packaging materials [6,7]. A survey carried out in South African beaches in 5-year intervals revealed that the quantities of plastic debris dumped increased tremendously over the years. Plastics, which are synthetic polymers, are still one of the most widely used and versatile materials in the world [8]. The global production of plastic in 2012 was 288 million tonnes [9]. They are light-weight, highly durable, strong and cheap [10]. These properties make them suitable to produce a wide range of products, persistent and hazardous in the environment [10]. Their buoyancy makes them to be dispersed over a long distance and finally settle in the sediment where they persist for centuries [11]. One major aspect of plastic pollution is the occurrence of microplastics in the aquatic ecosystems. Through some physical, chemical and biological processes such as UV-light, wave action, ocean-current, suspension and resuspension of plastics, large plastic debris fragments can degrades into micro-sized plastic commonly referred to as Microplastics. Microplastics is defined as the plastic particles in the size range of 1 nm to <5 mm [5]. The scientific community is currently focusing on microplastics more than plastics because they tend to pose a greater threat to marine biota and increasing changes in the integrity of the habitats at alarming rate globally [12]. For instance, microplastic ingestion has been recorded in a wide variety of marine biota [13,14] resulting in physiological disorders [15]. Therefore, since the issue of microplastics in the environment is a global intricate subject matter with environmental, economic and health implications and is a new rapidly evolving area of marine research that requires local and international attentions. This prompts and arouses interest and necessitates writing of this review article whose objectives are;

1. To review the various microplastics assessment methodologies for different marine matrices (Sampling and Analytical Methods).
2. To summarize the biological and ecological impacts of microplastics on marine biota and the environment.
3. To review the solutions to mitigate the disposal, occurrence and curb the menace of microplastics in the ocean.

Distribution of Microplastics in the Marine Environment

Microplastics have been found in most marine habitats of the world and throughout the water column. Understanding the distribution of microplastics both vertically and horizontally in the oceans, between geographic regions, between open and enclosed seas, between various compartments which can be surface, mid-water and benthic sediments, is a requirement for assessing the potential impacts on the marine biota and the environment [5]. Although, the major sources of microplastics are land based (urban and storm run-off, beach-visitors, industrial activities, construction and illegal dumping, inadequate waste disposal [16,17], in general, it is extremely difficult to identify and point out the ultimate sources of microplastics due to their fragmentation and

***Corresponding author:** Ogunola Oluniyi Solomon, International Studies in Aquatic Tropical Ecology, University of Bremen, Germany
E-mail: solomonunilag@yahoo.com

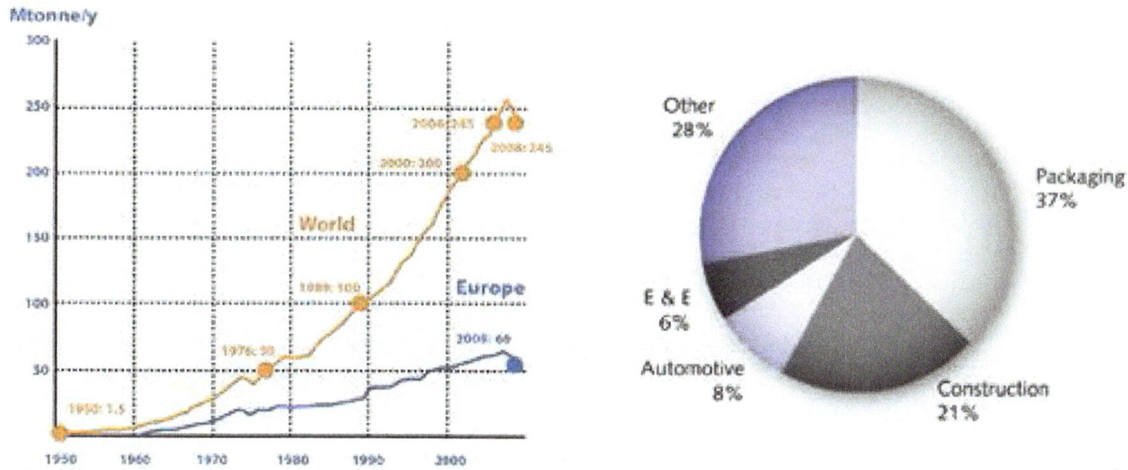

Figure 1: World plastics production 1950-2008. From the compelling facts about plastic, Plastics-Europe [141].

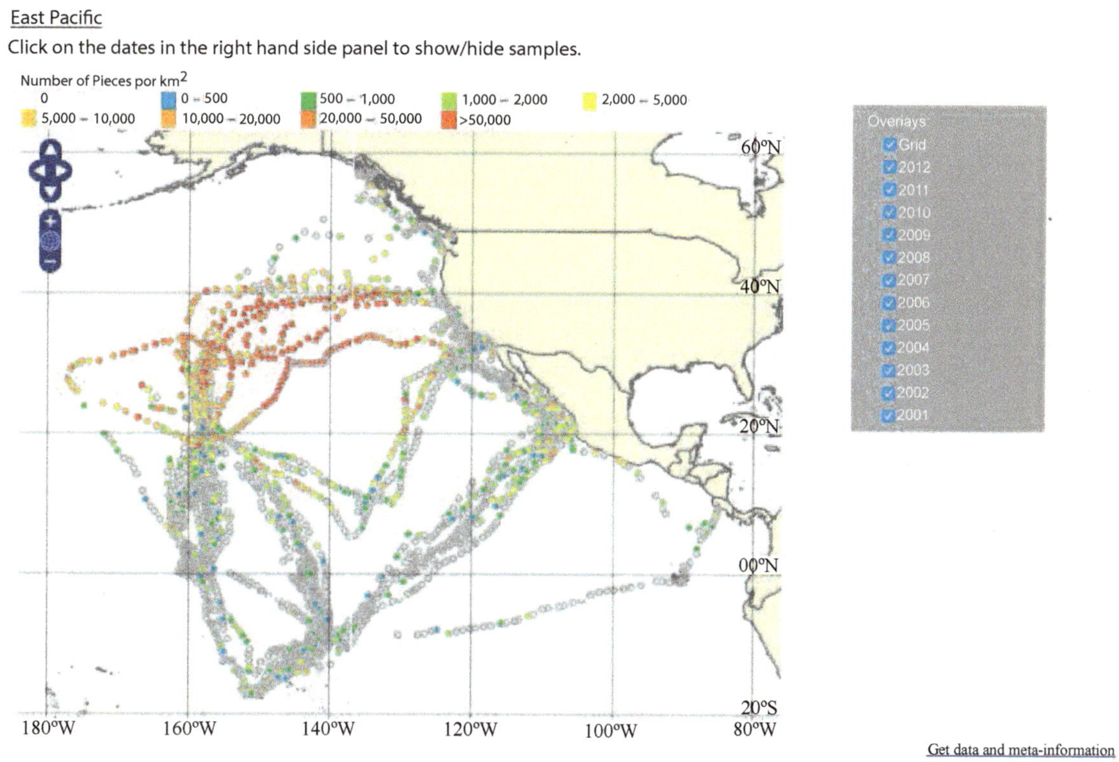

Figure 2: Distribution of microplastics in the western North Pacific, 2001-2012. Sea Education Centre, Woods Hole, MA (downloaded from: http://onesharedocean.org/open_ocean/pollution/floating_plastics).

degradation nature of the debris occurring in small and heterogeneous assemblages. It is not possible to observe the microplastic particles that float below the surface of the sea-water by flight observations or satellite and there is no accurate data estimating the global plastic inputs into the ocean and the parts that sink to the ocean floor [1]. The geographical coverage of microplastics is growing on a yearly basis. Most of the microplastic surveys have been conducted in the Northern Hemisphere and were conducted on the sea-surface. There is strong evidence that low-density microplastics such as Poly-propylene (PP), Polyethylene (PP) and Poly-Styrene (PS) predominate the oceanic surface [18,19]. Microplastics at the sea surface have also been reported in the coastal ocean [20,21], the open ocean [22-24] and also in enclosed and semi-closed seas [25,26] and South China [27]. Floating microplastics are best sampled in the North Atlantic and Pacific oceans and in the Mediterranean Seas due to their widespread distribution and concentration in those regions as shown in Figure 2 [5]. Recently, efforts have been made to estimate and quantify surface plastic on a global scale [28,29]. Furthermore, microplastics have also been reported on beaches at numerous locations globally [4,30]. Most studies conducted were on the surface sediments, some have also taken samples below the surface sediments [31]. Van Cauwenberghe et al. [32] conducted and reported microplastics occurrence in the deep sea sediments. High density microplastics such as Poly-Vinyl Chloride (PVC), Poly-Esters (PES) and Poly-Amide (PA) are likely to be found in largest quantities in the benthos and determining their magnitude is hindered by high cost of sampling [33].

Based on the spatial and temporal trends, the concentration of floating microplastics in the ocean is increasing over time as a result of increased plastic production, use and inefficient waste disposal management since the 1980s. For-instance, Goldstein et al. [34] reported two-order of magnitude increase of floating microplastics in the Eastern North Pacific between 1999 and 2010.

Attributes of Microplastics in the Marine Environment

Although, many different types of plastic are produced worldwide, the market is mostly dominated by 6 classes of plastics which include: Polyethylene (PE), Polypropylene (PP), Polyvinyl Chloride (PVC), Polystyrene (PS), Poly-Urethane (PUR), and Polyethylene Terephthalate (PET) [5]. Plastics originate from fossil fuels but biomass can also be used as feedback. Figures 3 and 4 show the production and sources of most artificial and natural polymers while Table 1 shows their common applications:

Microplastics encounter in the marine matrices are broadly classified into 2 types; 1) Primary Microplastics and 2) Secondary Microplastics.

1. Primary microplastics are originally and intentionally manufactured in the size range of 1 nm to 5 mm and have applications in personal care products like toothpaste, shower gel, scrubs, cosmetics, air-blasting [35].

2. Secondary microplastics result from the breakdown of large plastic items, e.g from fishing gears, ships, aquaculture, recreational activities, transport of plastic products into finer particles [5,36,37]. The formation of microplatics in the ocean

Resin Types	Common Applications	Specific Gravity
Polyethylene	Plastic bags, Storage containers	0.91-0.95
Polypropylene	Rope, bottle caps, gear, Strapping	0.90-0.92
Polystyrene (expanded)	Cool boxes, floats, cups	0.01-1.05
Polystyrene	Utensils, Containers	1.04-1.09
Polyvinyl Chloride	Film pipe, Containers	1.16-1.30
Polyamide or Nylon	Fishing Nets, Rope	1.15-1.15
Poly (ethylene-terephthalate)	Bottles, Strapping	1.34-1.39
Polyester Resin + glass-fibre	Textiles, boats	>1.35
Cellulose Acetate	Cigarette-fibre	1.22-1.24

Table 1: Densities and common applications of plastics in the marine environment [41].

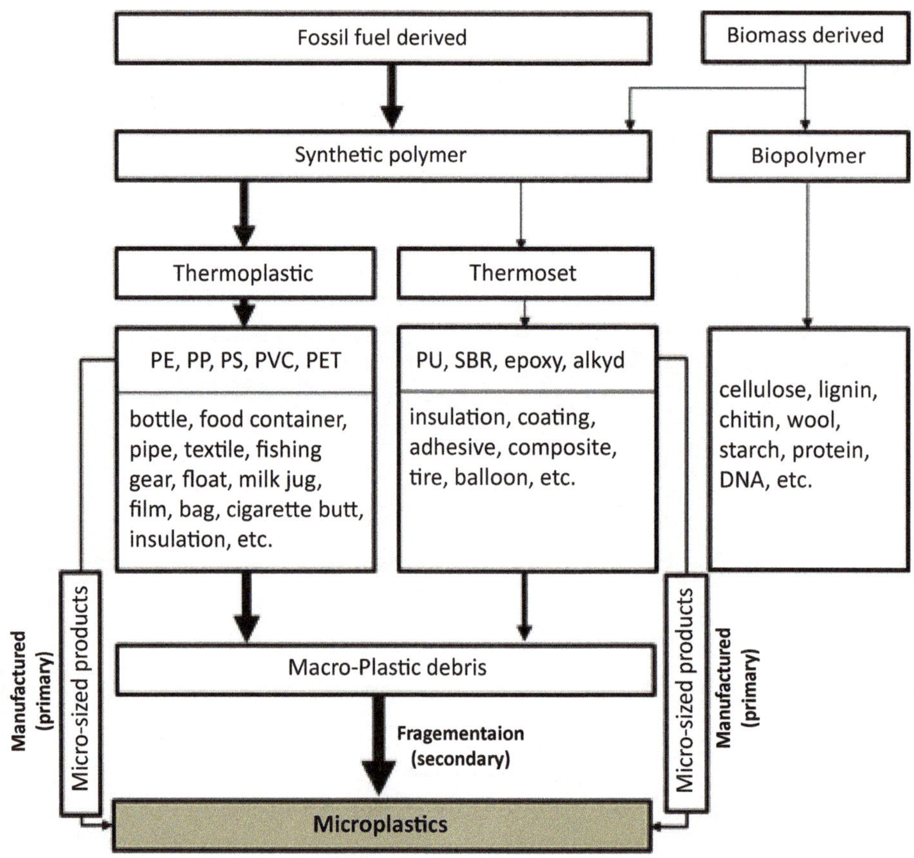

Figure 3: Production of the most common artificial (plastic) and natural polymers, including some typical applications. Microplastics are manufactured for particular applications, such as industrial scrubbers or in personal cleaning products such as toothpaste. All plastics can be subject to fragmentation on environmental exposure and degradation into (secondary) microplastics. The proportion of plastic reaching the ocean to become plastic litter depends on the effectiveness of the re-use, recycle and waste management chain [5].

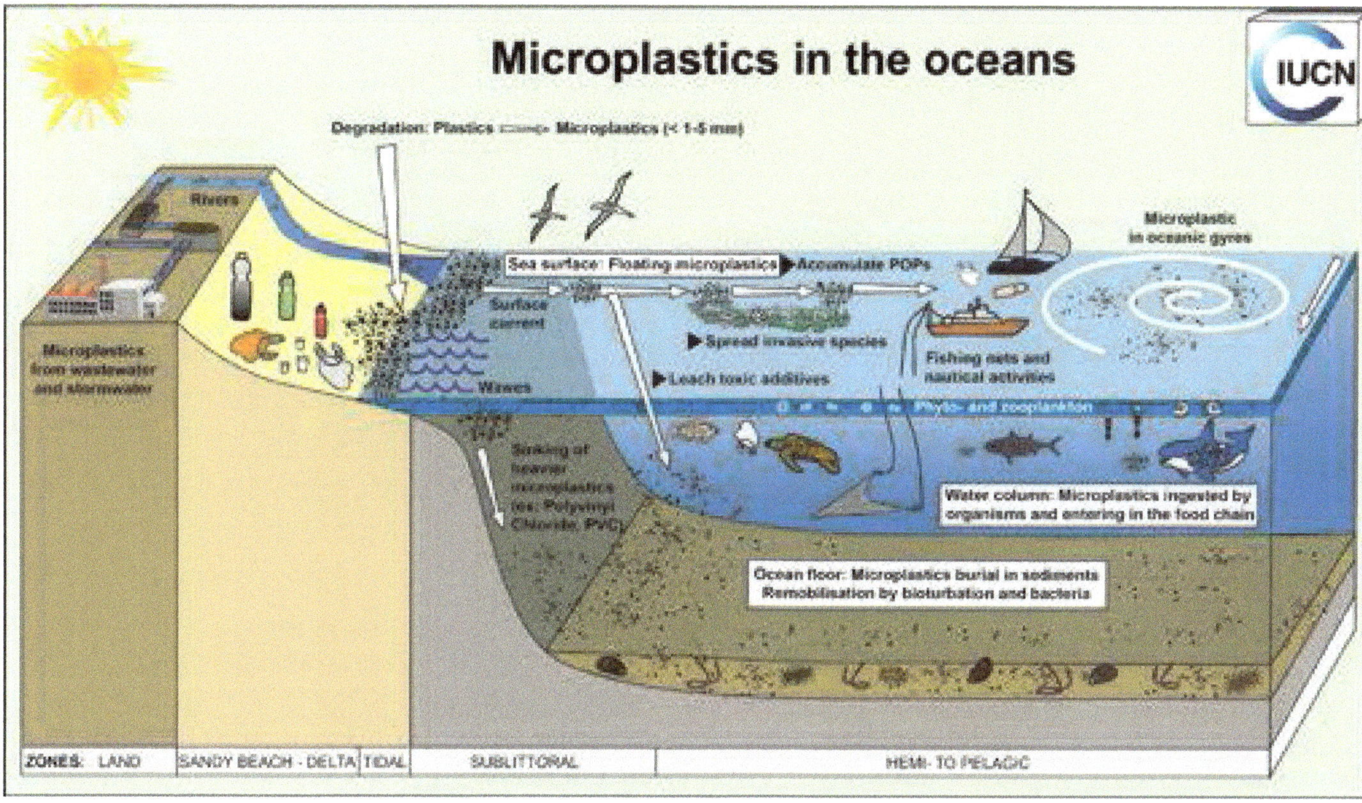

Figure 4: Schematic drawing showing the main sources and movement pathways for plastics debris in the oceans (Source: Florian Thevenon) [1].

is greatly influenced by a combination of environmental factors such as (1) solar ultra-violet radiation that facilitates oxidative degradation of polymers [38] and causes it to loose mechanical strength (2) Mechanical abrasion such as wind, wave, ocean current, animal bite, human activity that can break the polymer further into smaller fragments. This process is called "Weathering" and tends to occur in decreasing order of plastics float in water, in the mid-water column and in the sediment.

Few plastics have the tendency to undergo complete degradation or mineralization (into carbon dioxide or methane) in the marine environment. These include plastics of aliphatic polyester, bio-polymer and some bio-derived polymer origins but the cost of production of these polymers is very high [5]. These bioplastics are synthetic polymers made from plant biomass used as feedstock and are not chemically different from synthetic polymers of fossil fuel origin. They tend to degrade under certain conditions of oxygen, light levels and microbial species [39]. Microplastics could be bio-fouled and then eventually sink to the bottom where it becomes part of the sediment or may be eaten by the benthos. If plastic does become part of the sediment, whether on the shore or sea bottom, then it is no longer available for degradation by micro-organisms as it is in the water, especially if it is buried [40]. Most commodity plastics are not bio-degraded or mineralized and tend to persist for hundreds to thousands of years (decades to centuries) in the marine environment, even longer especially in the deep sea and polar region due to their high resistance to natural degradation as seen in Figure 5 [1,33,41].

Sampling and Analytical Assessment of Microplastics

Although, the presence of plastic litter is ubiquitous in all the oceans of the world, there are still no standardized sampling and quantification methods and common units. These impede an accurate evaluation of the fate and the effects of microplastics in the marine columns. Microplastic is widely considered as under-researched components of marine debris [20]. The problem that is yet to be resolved is how to find and recommend different reproducible analytical methods with low technology and cost effective to improve characterization, identification and quantification of plastic fragments [1]. Nevertheless, a number of sampling techniques have been developed and designed to assess the presence of microplastics in the marine matrices (water, sediment and biota) [42].

Sampling of water

Microplastics in the water column can be collected using various plankton nets of different mesh sizes to harness both the pelagic and benthic zones. Neuston or manta nets of mesh size approximately 300 micro-meters are usually towed using boats or pulled along the shores for a defined distance as seen in Figure 6. Afterwards, the net is thoroughly washed or flushed with seawater and empty into a sample jar. The quantity of microplastics trapped in the net is divided by the towed area or volume of sampled water. Alternatively, microplastic particles are separated from the plankton using conventional gravity separation techniques [43]. Bongo net and benthic trawl are another sampling devices that have been explored to sample microplastics in the mid-water and sea-bed regions respectively as shown in Table 2 [44-46]. Sample bottles and rotating drum samplers have also been utilized to sample microplastics at different depths in the water column [1]. It is expedient to note that using meshes of different apertures have tendency to produce large variations in the quantity of microplastics collected. KIMO (Kommunenes Internasjonale Miljoorganisasjon),

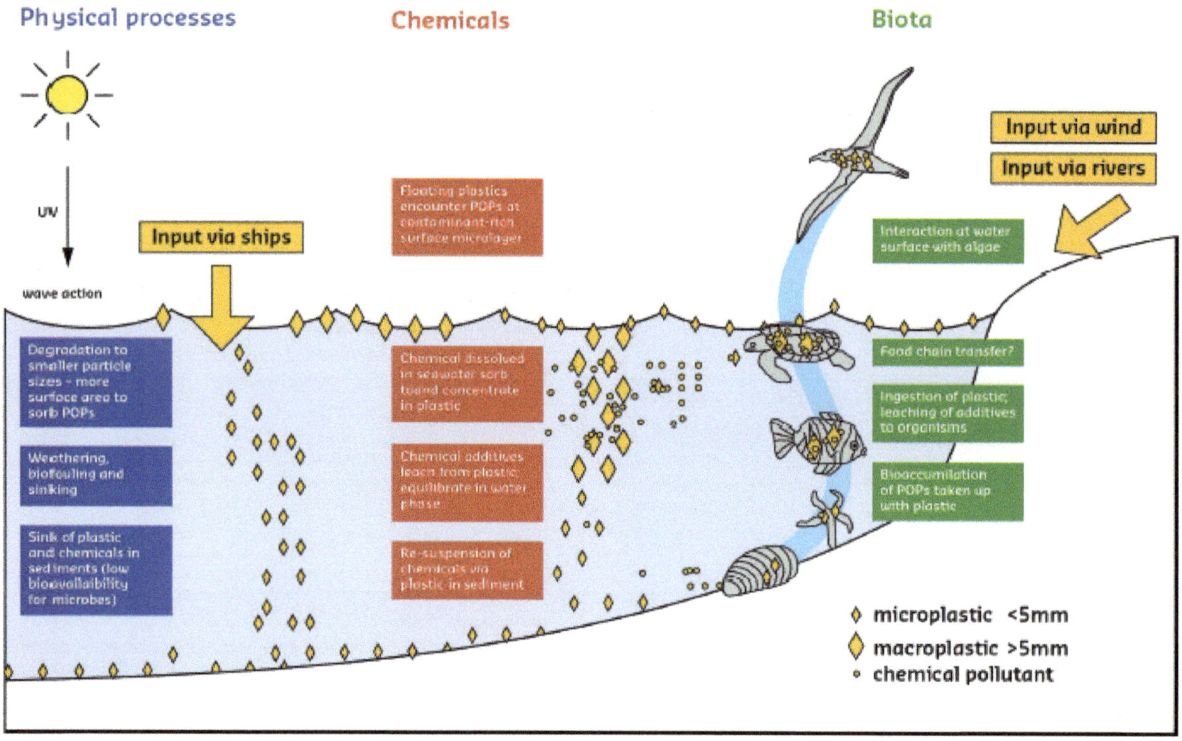

Figure 5: Sources of marine microplastics and the various physical, chemical and biological processes affecting microplastics in the marine environment [39].

Figure 6: Manta trawl with flow meter (left); Manta trawl in action (right). Samples in the nets are collected in glass containers, and quantitatively transferred from the net to the container with clean drinking water (not seawater). Onboard ship, seawater microplastics samples may be treated with preservatives. To rid the sample of organic matter, a H_2O_2 step is sometimes applied. Ridding samples of organic matter is useful when visual inspection is applied to separate polymer material from other materials [142]. Photos Leslie HA.

Type of Sampler	Lower Size limit(µm)	Water Sampled	Reference
Mazur	330	Samples surface water with flow meter	NOAA
Regular Plankton or neuston net (Continuous plankton recorder)	330	Sample surface water at 10m depth	U. Plymouth (UK)
Algalita Manta trawl	333	Sample Surface water, Approx. 500-3000m^2 per trawl	Algalita (USA), Cefas (UK)
Bongo Plankton Net	333	Samples mid-depth water column sample	Lattin et al. [129]
Epibenthic Sled	333	Sample water column near sea bottom	Lattin et al. [129]

Source: Adapted from (Leslie et al. [39]).

Table 2: Methods of sampling microplastics from seawater [39].

Sweden investigated and reported that 80 micron-meter meshes is more efficient than 450 micron-meter meshes for sampling microplastics which is 100,000 times higher in concentration of the microplastics sieved [47]. Continuous Plankton Recorder is another tool used to determine the abundance of microplastics in the open ocean and from which long term data could be collected [45].

Sampling of sediment

This procedure allows the assessment of microplastics in the benthic sediments from beaches, estuaries and the sea-floor [31]. Grab samplers have been widely used to collect surface sediments in large quantities (several kilograms). Noren [48] collected sediments from the Swedish coastal area with the Eckman grab. The demerit of this device is that the sediments are disturbed, layers mixed up and information regarding the historical deposition could not be obtained [1]. These problems are solved using sediment corers to sample sediments without mixing up the layers [1]. Sediments are collected manually from beaches [49] and can also undergo hand sieving before further processing in the laboratory (Figure 7) [5].

Sampling of biota

Biological assessment involves examining microplastic particles consumed by the organisms especially during feeding. Stomach content analysis has been observed in several species of fish, bivalves, crustaceans and birds as seen in Figure 8 [5]. Sampling techniques of the biota depend on the type of movement performed by the animal in question i.e whether it is sedentary/sessile, mobile, fast or slow runners. Vandermeersch et al. [50] collected Hotspot mussel (*Mytilus galloprovincialis*) by hand from three European estuaries in the coastal areas of Portugal, Sacca di Goro, Italy and Fangar Bay in Spain prior to microplastic analysis. Sampling of birds depends on the recovery of dead organisms (carcass) at shorelines or coastal nestling sites. Analysis of recovered carcasses of the northern Fulmar operates in the North Sea for monitoring and providing ecological quality based on the assessment of its stomach contents [5].

Figure 7: Illustration of the amount of visible microplastics found in beach sand. Photo Vethaak AD.

Microplastic analysis

Microplastic Analysis involves an array of steps to arrive at the identification and quantification of microplastics in the marine matrices. The lack of uniform standardized approach among regions or studies in the same region make comparisons difficult [44]. Precautions involved before analyzing microplastics are; to avoid background contamination when sampling and during extraction procedures and to remove non-plastic materials [51].

Sample preparation: This involves sieving the sediment at 150 or 500 micron-metres depending on the size range to obtain homogeneity of the material.

Separation: Microplastics in the sediment need to be separated from both organic and inorganic non-plastic particles prior to counting, weighing and identification of the polymer type. Separation is effected by density principles using salt solutions of varying densities (sodium chloride, sodium iodide, sodium polytungstate). Sediment is mixed with concentrated NaCl solution and vigorously shaken for specific period of time. The supernatant is filtered with a filtration unit and a vacuum pump. The filter (usually nitro-cellulose 0.45 micron-meter) is then dried and sealed in a petri-dish for further analysis [30,45]. The use of NaCl to separate microplastics by floatation is highly recommended as it is very cheap and eco-friendly. The setback in its usage is that it is only suitable for separating low-density plastic polymers like Polyethylene, Polypropylene and Polystyrene from non-plastic particles but unsuitable for high density plastic polymers like Polyvinyl chloride and Polyethylene terephthalate. Therefore, this problem is overcome by using Sodium Iodide (NaI) of higher density (1.6 g/cubic cm) in a subsequent step. This approach has been found to be highly economical and reduces analytical cost [52]. There have been recent advances to separate microplastics from natural organic materials in surface water sample using enzymatic digest to remove planktonic organisms [53]. In a more advanced technological approach, microplastics have been more efficiently extracted from sediment based on the principle of elutriation

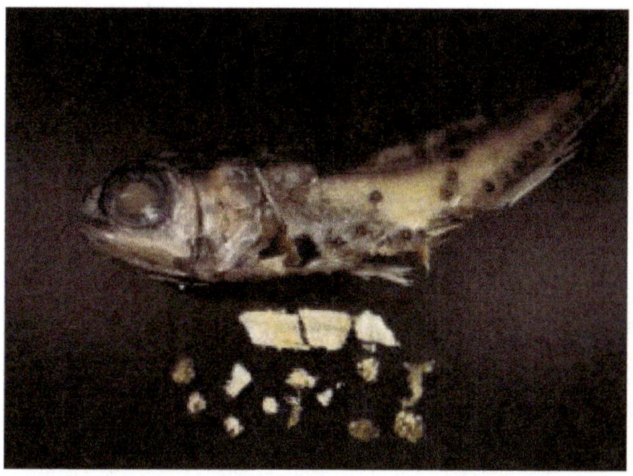

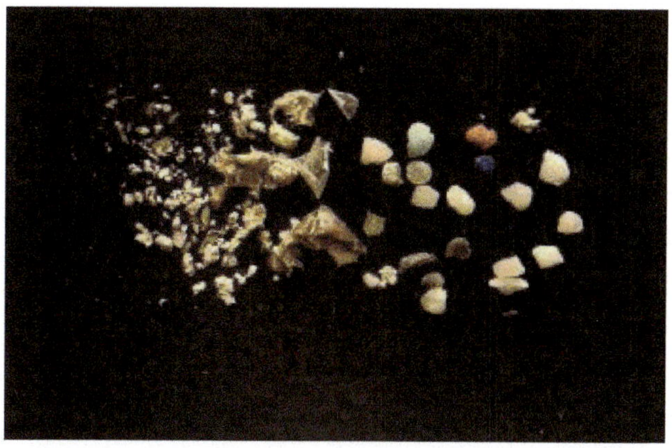

Figure 8: Lanternfish with large piece of plastic (unpassable) which broke into three pieces (left); Stomach contents – plankton on left, plastic on right (right). Reprinted with permission of Christina Boerger.

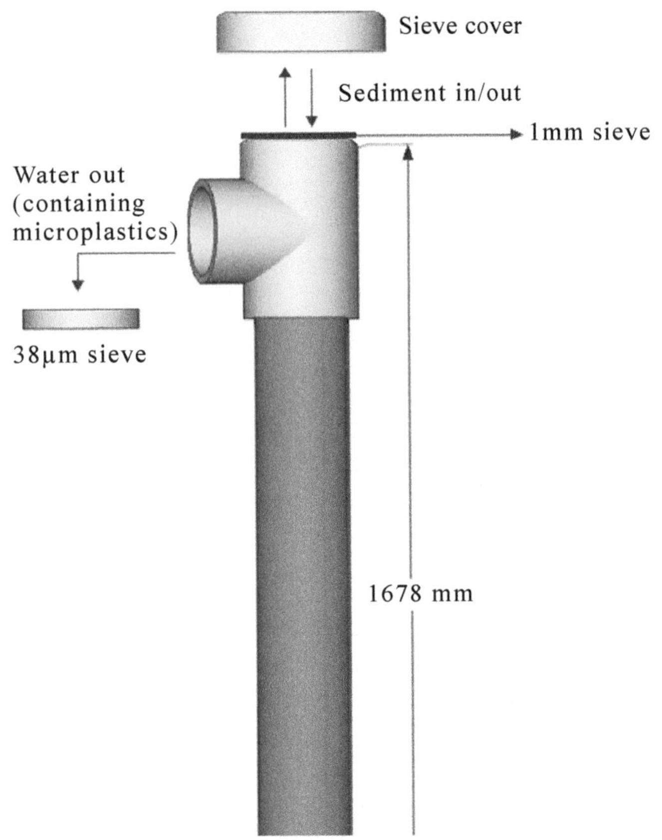

Figure 9: Elutriation: schematic representation of the elutriation column, used for separating microplastics from the heavier sand particles [54].

which is a process of separating lighter particles from heavier ones using an upward stream of gas or liquid phase (water) [54] as shown in Figure 9.

In biological assessment, the principle of wet digestion of tissues is employed. Nitric acid or Acid mix method (combination of nitric acid and perchloric acid in the ratio of 4:1 v:v) is used separately. Both methods are highly effective to destroy or digest the tissues, reduce greasy tissue fractions and enhance extraction. After complete digestion at a regulated temperature of 60 degree centigrade in the 250 ml conical flask, the digested tissue is diluted with warm deionized water and filtered over a 5 micron-metre cellulose nitrate membrane (Whatman AE 98), the filter is dried at 40 degree Celsius for 24 hours in a petri-dish and subsequently quantified under a powerful microscope [1,50,54]. Hydrogen peroxide and sodium hydroxide could also be used for digestion and extraction.

Visual sorting under microscope is necessary to separate the microplastic fragments from non-plastic particles. Each plastic piece could be picked up with forceps and placed in graduated dish to be counted, measured, photographed and classified [21]. Digital photography enhances the morphological parameters and surface of substantial plastic particles to be computed automatically. This technique was used to measure microplastics in the sea surface of the North-east Pacific ocean [55] and the North Sea [52].

Sometimes, it might be difficult and impossible to distinguish microplastics from non-plastic particles due to decreased size observed under microscope. This problem is solved with use of spectroscopic approach. Microscopic FT-IR (Fourier Transformed – Infra Red) and Raman Spectroscopy are the most widely used to confirm the identification of plastic and their synthetic polymers for particles in the size range<1 mm [5]. This device provides reliable results and datasets and does not give room for prior pre-sorting by hand [56].

The Impacts of Microplastics on the Marine Environment

Microplastics have both direct and indirect harmful effects on the marine biota. Although, the threats posed by plastic pollution on the living resources continue to gain recognition worldwide, the effects on marine organisms, food webs, community structure and ecosystems are still poorly understood [1,57]. A study by Goldberg [58] argues that the deleterious effect of microplastics on marine life is on the rise. The threat to marine life is due to mechanical or chemical effects after the ingestion of microplastic particles. The size of ingested plastic materials is related to the body size of the organism [44,59] and ontogenetic phase [60,61]. The degree of impacts is proportional to the size, shape and amount of the ingested items and other factors which are behavioral, physiological and geographical, foraging strategies and diet [62-64].

The physical effects of microplastics ingestion

A number of field studies have revealed that microplastic particles are ingested by a wide variety of biota from all oceanic regions; planktivorous fish from North Pacific Gyre [65], mussel from Belgian coastal waters [66], harbor seals from the North Sea [67], Fulmar from the North Sea [68], Whales [69]. It was discovered that a good number of these animals tend to mistake plastic wastes for their potential prey items which they ingest accidentally or voluntarily by feeding on lower trophic organisms that have themselves consumed microplastics [35,70]. Albatrosses may take up red plastic residues for squids and ingest them unknowingly [1]. As a consequence, some obstructions/blockage/clogging of intestinal tract or digestive organs occur and impede the organisms from taking in more food resources or diminish their feeding stimulus (pseudo-satiation) [18,71]. This can lead to starvation. For instance, Eriksson and Burton [72] reported that Fur seals tend to ingest and bio-accumulate plastic debris by eating pelagic fish species that have already ingested plastic particles during feeding. Other harmful effects reported are; adsorption of the microplastics on the organism's surface , lowered steroid hormone level, delayed ovulation, reproductive failure, internal injuries like gut perforations, toxin introduction, delayed growth and death [15,18,73-78]. Spear et al. [79] provided evidence that the higher the number of plastic particles ingested; the worse the physical condition (body weight) is in seabirds from the tropical Pacific. On the other hand, microplastics can also alter the physical conditions of the marine habitats. On sandy beaches, they can affect the permeability and temperature of the sediments when they settle on the matrix with consequential effect for fauna which are temperature-dependent sex determination like the reptiles [80].

The chemical effects of microplastics ingestion

Microplastics contain some potentially harmful plasticizers called 'Additives' such as Bisphenol A (BPA) and di-butyl phthalates which are incorporated into the parent plastics during the manufacturing process [81] to give tensile strength or flexibility. These plastic particles also tend to adsorb and accumulate toxic materials from the surrounding seawater such as polychlorinated biphenyls (PCBs), polycyclic aromatic hydrocarbons (PAHs), persistent organic pollutants like DDT and heavy metals [82-86]. These chemicals are not chemically bound to the surface of the microplastics and eventually tend to leach into the surrounding environment or animal tissues when ingested [1]. Tagatz et al. [87] showed that high concentration of dibutyl phthalates can adversely affect the macrobenthic communities that ingest the microplastic particles. Among the reported impacts include; reproductive dysfunctionality during developmental stage [88], hepatic stress in fish [89], decreased steroid level and delayed ovulation [90], endocrine disruption and mortality [68,91].

Transport of invasive species

Microplastics floating in the sea have the tendency of providing raft substrates for various epifauna and microbes such as bacteria, algae, diatoms, barnacles, hydroids, tunicates [92-94] and transporting them to areas where they were not existing before. This widens the range of introduced species into an environment where they were previously absent [95]. Zettler et al. [96] and Harshvardhan and Jha [97] reported that certain microbes called 'Plastisphere' colonized microplastic fragments collected in the North Subtropical Gyre and Arabian Sea in India which were detected using Scanning Electron Microscopy (SEM) and gene sequencing analyses. The invasion of these species can be detrimental to littoral, intertidal and shoreline ecosystems [98,99]. Gregory [100] argued that not all the species that are transported will necessarily prove harmful to their new host environment.

Social and economic effects

It has been shown that when the condition of natural asset like the ocean is altered by plastic pollution, it would influence the socio-economic systems by changing the environmental quality for future generations as seen in Table 3 [101] and results in health implications [88]. The ingestion of microplastics by marine species (fish) from the North Pacific ocean could impact the fishing industries in the future [102]. The accumulation of microplastics along the food chain as in fish and shell-fish like mollusks, crustaceans could pose a threat to the food safety and generate serious health implications on final human consumers through their diet [103]. Exposure of human population to Bisphenol A is associated with heart diseases, diabetes, alterations in circulating hormonal levels [104,105].

Possible Management Strategies to the Problems of Microplastics

The intricacy and rising scale of the marine plastic pollution issue is too large for any single agency or country to resolve [106]. Therefore, it is imperative to integrate and harmonize several measures (management, regulatory, technical and operational tools) at local, national, regional and international levels to address and tackle the problem [18].

Legislation

The most important legislation that is in place addressing the problem of marine pollution is the 1978 Protocol to the International Convention for the Preservation of Pollution from Ships (MARPOL) which forbids or bans all vessels to dispose their waste of plastic origin into the marine environment [107]. The Annex V of MARPOL is the key international authority controlling ships at sea source of marine plastic debris [108]. Although, 79 countries have so far ratified the Annex V [109], the signatory countries are implored to fully implement the convention. Furthermore, in 2012, 64 countries and the European Union adopted the Manila Declaration with the aim of implementing United Nations Global Program of Action for the Protection of the Marine Environment from land-s-based sources. The signatories of this declaration also agreed to formulate Global Partnership on Marine Litter (GPML) whose main goal is to include sea-based sources of marine debris [1]. The most challenging aspect confronting legislation as a mitigation tool is how to enforce it in an area as vast as the world's oceans and non-compliance of the member states to the convention [18].

Efficient waste disposal systems

Sufficient litter and recycling bins must be provided on beaches and in coastal areas. There must be regular collection and evacuation of commercial, municipal and agricultural wastes from the residential areas, streets, parks and dumping sites. Large number of incinerators must be put in place to burn plastic wastes [1].

Recycling/valorization of plastic materials

This method is regarded as the most important to reduce the negative environmental impacts of open landfills and open air burning that are commonly practiced in the developing countries to manage domestic wastes and step down the spread of ocean plastic pollution [1].

Education and public enlightenment/engagement/marine litter clean-ups

Education is a powerful tool to address the issue if it is discussed in schools. This exists in two forms; formal and informal. Formal education involves incorporating the environmental topics into the country's educational curricular systems. Informal education involves the people or the general public engaging in volunteering projects and learning such as beach clean-ups, campaigns, seminars, rally [5]. Harvesting of plastic litters and microplastics from the marine environment are costly operations but engaging volunteers in clean-up activities can help reduce costs (although the time of volunteers also has an economic value) and improve awareness [110]. For instance, Ocean-Conservancy [111] reported that over 650,000 participants worldwide engaged in beach clean-up day in 2013 across 92 countries.

Integration and harmonization of trans-disciplinary

To actualize the implementation of action plans to reduce the input of plastics into the marine environment around the world, there is urgent need to involve different stake-holders involving plastic industries, fishing industries, food industries, tourism, shipping companies, NGOs, community, local authorities, national, regional

Sector	Impacts of marine litter	Estimated Cost
Fishing	Marine litter can lead to the loss of output or loss of value in the sales of certain types of seafood or fish	The loss of marketable lobster due to abandoned of lost fishing gear is estimated to lead to a global loss of US$250 million per year. Microplastics are estimated to lead to a loss of up to 0.7% of annual income every year for the UK Aquaculture sector
Shipping	Plastic debris can foul ship propulsion equipment, disrupting operations, requiring clean-up, repair and rescue efforts, loss of life injury	In 2008, 286 rescues of vessels with fouled propellers in UK waters were carried out at a cost of between €830,000 and €2,189,000. Cost of removing litter and addressing damage in the Scottish aquaculture industry is estimated at €155,549/year
Tourism	Polluted beaches can discourage visitors from certain beaches, leading to loss of revenues for the tourism sector.	In Goeje Island (Republic of Korea), marine debris led to loss of revenue from tourist of between €29-37 million in 2011. In the Asia Pacific Economic Community(APEC) region, marine debris Is estimated to cost the tourism sector approximately US$622 million/year

Table 3: Potential economic impacts and costs of marine litter Source: Watkins et al. [110].

and international parastatals, in addressing socio-economic and environmental problems associated with plastic pollution and also in decision making process [1].

Innovating bio-degradable plastics and other alternatives

Advanced technology needs to be developed to mass produce eco-friendly biodegradable and photo-degradable plastics [112,113] at low cost that can be affordable by the public. Although, the effects of final degradation of the products are not yet known, this may portend more dangers [57,112]. Another way of reducing plastic is to use products made from other alternative materials such as stainless steel (drink containers), aluminium, glass e.g straws or cotton e.g shopping bags but their effectiveness are yet to be assessed [114].

Discussion

For decades, scientific interest has grown and expanded knowledge for microplastics but some fundamental questions and issues still remain pending and unresolved [42]. There are discrepancies in the sample strategies, equipment, procedures and assessors that might influence the results of microplastics quantification unknowingly [50]. These differences include; the use of filters of different pore sizes for retaining particles, this could lead to results obtained being underestimated and biased. Diverse digestion methods are described in the literature as shown in Figure 10 and choosing one technique singly could influence the estimation. Claessens et al. [54] and the International Council for the Exploration of the Sea (ICES, 2015) reported that the Acid mix method of digestion tends to be highly detrimental and destructive to microplastics of Nylon origin. Similarly, Cole et al. [53] was of the opinion that alkaline treatment (the use of NaOH, KOH which is an alternative approach) causes damage and discoloration of microplastics beyond recognition. Nuelle et al. [52] supported the idea by demonstrating that using oxidizing agent (Hydrogen peroxide) treatment of 35% could result in the size reduction of the plastic particles of polyethylene and polypropylene and also unable to remove all the biological materials from completely from the mixture. The use of spectrometry for direct identification of microplastics is not reliable as it can be hindered by the presence of pigments [103]. To date, there is no unified and agreed scientific method for estimating and quantifying microplastics. Therefore, there is a need for a standardized approach to be developed for quantifying microplastic iin a reproducible and science-based manner [50].

A multi-disciplinary approach can help to some extent to provide solutions to the problems of microplastics because knowledge and ideas of multiple experts from different fields are harnessed to arrive at a joint research question and focus on the area of concern for the conservation of biota and habitats [115-118]. There is need to involve disciplines that can understand the economic and social barriers and effect change behavior and markets [119-121] and evaluate the benefits.

Conclusion and Recommendation

To solve the problems posed by plastic pollution, there is an urgent need to identify the various sources and transport mechanisms involved which will guide and help us to close the tap, develop mitigation strategies and clean-up programs that are highly efficient on a long term basis [1,122-127]. At present, the environmental fate, bioaccumulation and quantification of microplastics still remain vague. Therefore, it is mandatory and useful to develop a common standardized and harmonized methods and units so as to make comparisons feasible [50]. There is a critical need to understand the impacts of chemicals associated with plastic waste on the environment and human health and what cost-effective measures should be put in place to address the issue [128-133]. For example, there is little information on the actual health risks associated with ingestion of microplastics – both to marine species and to human consumers [110]. All sectors of the community (the general public, the scientific organizations and NGOs) have a key and crucial role to play [134-139]. There is no one perfect solution

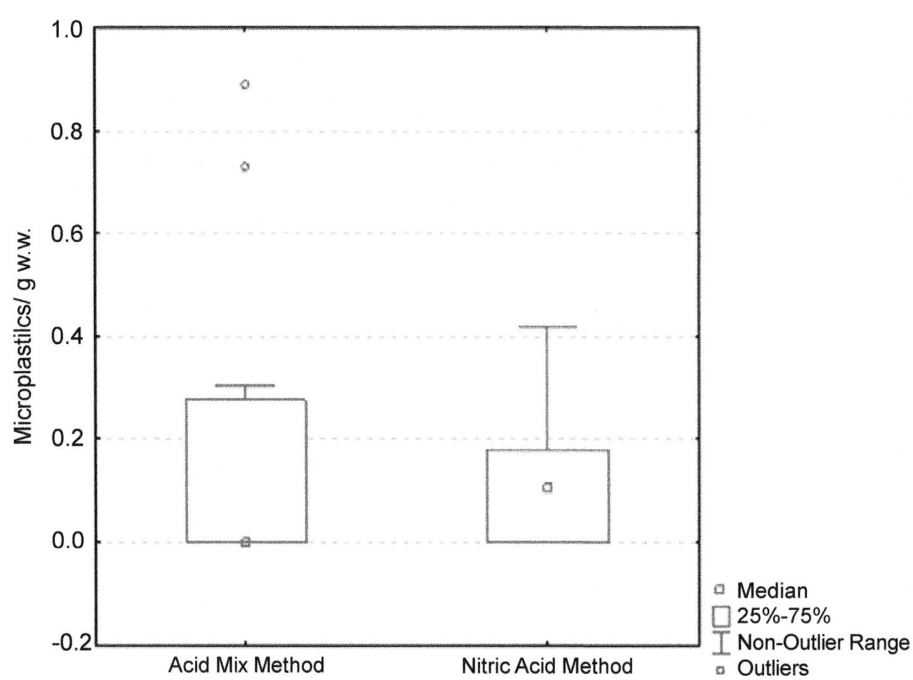

Figure 10: Total number of microplastics in hotspot mussels (three estuaries together) evaluated by the Acid mix Method and the Nitric acid [50].

to the issue [5]. All hands must be on deck (policy formulation and implementation, surveillance and monitoring operations, adherence to the existing laws and regulation) and the combination of different strategies (change in producer's product design, provision of waste management infrastructure and wastewater treatment facilities, economic incentives, bans on plastic bags, plastic blasting in shipyards, awareness raising activities among consumers to help them reduce their consumption of plastic bags and cosmetic products containing microbeads, fishing for plastic litter) must be harnessed to resolve the issue [110]. It is time to take local, regional and global level actions and work synergistically to prevent or control the entry of plastics into the oceans [1]. Above all, understanding risk and benefit perception of the people would go a long way to influence their behaviors and acceptability of regulatory measures [140]. Thinking globally and acting locally would go a long way to reduce the environmental threats drastically [18].

Acknowledgements

I would like to thank Professor (Dr) Tilmann Harder of the Marine Chemistry Department, University of Bremen, Germany for proof-reading, reviewing, criticisms, insights, inputs, corrections/adjustments and suggestions made for the successful write up of this paper. Special thanks to German Academic Exchange Service (DAAD) for their financial support.

References

1. Thevenon F, Carroll C, Sousa J (2014) Plastic Debris in the Ocean: Characterization of Marine Plastics and their Environmental Impacts, Situation Analysis Report. Gland, Switzerland, IUCN 52.
2. Stefatos A, Charalampakis M, Papatheodorou G, Ferentinos G (1999) Marine debris on the seafloor of the Mediterranean Sea: examples from two enclosed gulfs in Western Greece. Marine Pollution Bulletin 36: 389-393.
3. Plastics-Europe (2010) Plastics-The Facts 2010.
4. Browne MA, Crump P, Niven SJ, Teuten EL, Tonkin A, et al. (2011) Accumulations of microplastic on shorelines worldwide: sources and sinks. Environ SciTechnol 45: 9175-9179.
5. Gesamp (2015) Sources, Fate and effects of Microplastics in the Marine Environment: A global Assessment. (Kershaw PJ (ed.) (IMO7FAO7UNESCO-IOC7UNIDO7WMO7IAEA7UN7UNEP7UNDP Joint Group Experts on the Scientific Aspects of Marine Envtal. Protection) 96.
6. Cawthorn M (1989) Impacts of marine debris on wildlife in New Zealand coastal waters. In: Proceedings of Marine Debris in New Zealand's Coastal Waters Workshop, 9 March 1989, Wellington, New Zealand. Department of Conservation, Wellington, New Zealand pp: 5-6.
7. DOC (1990) Department of Conservation. Marine Debris. Wellington, New Zealand.
8. Shehada A, Land M (2014) What are the effects of plastic particles on growth and mortality of marine organism?. PS5 Pilot Study.
9. Plastics-Europe (2009) The Compelling Facts About Plastics, Brussels.
10. Laist DW (1987) Overview of the biological effects of lost and discarded plastic debris in the marine environment. Marine Pollution Bulletin 18: 319-326.
11. Goldberg ED (1997) Plasticizing the seafloor: an overview. Environmental Technology 18: 195-202.
12. Ivar do Sul JA, Costa MF, Barletta M, Cysneiros FJA (2013) 'Pelagic microplastics around an archipelago of the Equatorial Atlantic'. Mar Pollut Bull 75: 305-309.
13. Wright SL, Rowe D, Thompson RC, Galloway TS (2013a) Microplastic ingestion decreases energy reserves in marine worms. Current Biology 23: R1031-R1033.
14. Wright SL, Thompson RC, Galloway TS (2013b) The physical impacts of microplastics on marine organisms: A review. Environmental Pollution 178: 483-492.
15. Tanaka K, Takada H, Yamashita R, Mizukawa K, Fukuwaka MA, et al. (2013) Accumulation of plastic-derived chemicals in tissues of seabirds ingesting marine plastics. Mar Pollut Bull 69: 219-222.
16. Gordon M (2006) Eliminating land-based discharges of marine debris in California: A plan of action from the Plastic Debris Project. Prepared by the California Coastal Commission.
17. Jayasiri HB, Purushothaman CS, Vennila A (2013) Plastic litter accumulation on high-water strandline of urban beaches in Mumbai, India. Environ Monit Assess 185: 7709-7719.
18. Derraik JGB (2002) The pollution of the marine environment by plastic debris: a review. Marine Pollution Bulletin 44: 842-852.
19. Gregory MR (1996) Plastic "scrubbers" in hand cleansers: a further (and minor) source for marine pollution identified. Marine Pollution Bulletin 32: 867-871.
20. Doyle MJ, Watson W, Bowlin NM, Sheavly SB (2011) Plastic particles in coastal pelagic ecosystems of the Northeast Pacific ocean. Marine Environmental Research 71: 41-52.
21. Reisser J, Shaw J, Wilcox C, Hardesty BD, Proietti M, et al. (2013) Marine Plastic Pollution in Waters around Australia: Characteristics, Concentrations, and Pathways. PLoS ONE 8: e80466.
22. Law KL, Moret-Ferguson S, Maximenko NA, Proskurowski G, Peacock EE, et al. (2010) Plastic accumulation in the North Atlantic subtropical gyre. Science 1185-1188.
23. Goldstein MC, Rosenberg M, Cheng L (2012) Increased oceanic microplastic debris enhances oviposition in anendemic pelagic insect. Biol Lett 8: 817-820.
24. Eriksen M, Maximenko N, Thiel M, Cummins A, Lattin G, et al. (2013) Plastic pollution in the South Pacific subtropical gyre. Marine Pollution Bulletin 68: 71-76.
25. Collignon A, Hecq JH, Galgani F, Voisin P, Collard F, et al. (2012) Neustonic microplastic and zooplankton in the North Western Mediterranean Sea. Marine Pollution Bulletin 64: 861-864.
26. Yamashita R, Tanimura A (2007) 'Floating plastic in the Kuroshio Current area, western North Pacific Ocean'. Marine Pollution Bulletin 54: 485-488.
27. Zhou P, Huang C, Fang H, Cai W, Li D, et al. (2011) 'The abundance, composition and sources of marine debris in coastal seawaters or beaches around the northern South China Sea (China)'. Mar Pollut Bull 62: 1998-2007.
28. Cózar A, Echevarría F, González-Gordillo JI, Irigoien X, Úbeda B, et al. (2014) Plastic debris in the open ocean. Proceedings of the National Academy of Sciences of the United States of America 111: 10239-10244.
29. Yamashita R, Takada H, Fukuwaka MA, Watanuki Y (2011) Physical and chemical effects of ingested plastic debris on short-tailed shearwaters, Puffinus tenuirostris, in the North Pacific Ocean. Mar Pollut Bull 62: 2845-2849.
30. Hidalgo-Ruz V, Gutow L, Thompson RC, Thiel M (2012) Microplastics in the marine environment: a review of the methods used for identification and quantification. Environ Sci Technol 46: 3060-3075.
31. Claessens M, Meester SD, Landuyt LV, Clerck KD, Janssen CR (2011) Occurrence and distribution of microplastics in marine sediments along the Belgian coast. Mar Pollut Bull 62: 2199-2204.
32. Van Cauwenberghe L, Vanreusel A, Mees J, Janssen CR (2013c) 'Microplastic pollution in deep-sea sediments'. Environmental Pollution 182: 495-499.
33. Barnes DKA, Galgani F, Thompson RC, Barlaz M (2009) Accumulation and fragmentation of plastic debris in global environments. Philosophical Transactions of the Royal Society B: Biological Sciences 364: 1985-1998.
34. Goldstein MC, Titmus AJ, Ford M (2013) Scales of spatial heterogeneity of plastic marine debris in the northeast Pacific Ocean. PLOS ONE 8: e80020.
35. Fendall LS, Sewell MA (2009) Contributing to marine pollution by washing your face: Microplastics in facial cleansers. Mar Pollut Bull 58: 1225-1228.
36. Astudillo JC, Bravo M, Dumont CP, Thiel M (2009) Detached aquaculture buoys in the SE Pacific: potential dispersal vehicles for associated organisms. Aquat Biol 5: 219-223.
37. Bowmer T, Kershaw P (eds.) (2010) Proceedings of the GESA microplastic international workshop on micro-plastic particles as a vector in transporting persistent, bioaccumulating and toxic substances in the oceans. UNESCO-IOC, Paris 69.
38. Tourinho PS, do Sul JAI, Fillmann G (2010) Is marine debris ingestion still a problem for the coastal marine biota of southern Brazil? Mar Pollut Bull 60: 396-401.

39. Leslie HA, van Velzen MJM, Vethaak AD (2013) Microplastic survey of the Dutch environment. Novel data set of microplastics in North Sea sediments, treated wastewater effluents and marine biots. Amsterdam, The Netherlands: IVM Institute for Environmental Studies, Final report R-13/11.

40. Teuten EL, Rowland SJ, Galloway TS, Thompson RC (2007) Potential for plastics to transport hydrophobic contaminants. Environ SciTechnol 41: 7759-7764.

41. Andrandy AL (2011) Microplastics in the marine environment. Marine Pollution Bulletin 62: 1596-1605.

42. Cole M, Lindeque P, Halsband C, Galloway TS (2011) Microplastics as contaminants in the marine environment: A review. Marine Pollution Bulletin 62: 2588-2597.

43. Collignon A, Hecq JH, Galgani F, Collard F, Goffart A (2014) Annual variation in neustonic micro- and meso-plastic particles and zooplankton in the Bay of Calvi (Mediterranean–Corsica). Marine Pollution Bulletin 79: 293-298.

44. Ryan PG, Moore CJ, van Franeker JA, Moloney CL (2009) Monitoring the abundance of plastic debris in the marine environment. Philos Trans R Soc B Biol Sci 364: 1999-2012.

45. Thompson RC, Olsen Y, Mitchell RP, Davis A, Rowland SJ, et al. (2004) Lost at sea: where is all the plastic? Science 838.

46. Browne MA, Galloway TS, Thompson RC (2010) Spatial patterns of plastic debris along estuarine shorelines. Environ Sci Technol 44: 3404-3409.

47. Lozano RL, Mouat J (2009) Marine Litter in the North-East Atlantic Region: Assessment and Priorities for Response. KIMO International.

48. Norén F (2008) Small plastic particles in coastal Swedish waters. N-Research report, commissioned by KIMO Sweden 11.

49. Dekiff JH, Remy D, Klasmeier J, Fries E (2014) Occurrence and spatial distribution of microplastics in sediments from Norderney. Environ Pollut 186: 248-256.

50. Vandermeersch G, Van Cauwenberghe L, Janssen CR, Marques A, Granby K, et al. (2015) A critical Review on Microplastic quantification in aquatic organisms. Envt Res 143: 46-55.

51. Fries E, Dekiff JH, Willmeyer J, Nuelle MT, Ebert M, et al. (2013) Identification of polymer types and additives in marine microplastic particles using pyrolysis-GC/MS and scanning electron microscopy. Environmental Science: Processes & Impacts 15: 1949-1956.

52. Nuelle MT, Dekiff JH, Remy D, Fries E (2014) A new analytical approach for monitoring microplastics in marine sediments. Environmental Pollution 184: 161-169.

53. Cole M, Lindeque P, Fileman E, Halsband C, Goodhead R, et al. (2013) Microplastic Ingestion by Zooplankton. Environ SciTechnol 47: 6646-6655.

54. Claessens M, van Cauwenberghe L, Vandegehuchte MB, Janssen CR (2013) New Techniques for the detection of microplastics in sediments and field organisms. Marine Pollution Bulletin 70: 227-233.

55. Goldstein MC, Goodwin DS (2013) Gooseneck barnacles (Lepas spp.) ingest microplastic debris in the North Pacific Subtropical Gyre. Peer J 1: e184.

56. Löder M, Gerdts G (2013) FT-IR analysis for monitoring marine microplastics, YOUMARES 4, Oldenburg.

57. Quayle DV (1992) Plastics in the marine environment: problems and solutions. Chemical Ecology 6: 69-78.

58. Goldberg ED (1995) The health of the oceans-a 1994 update. Chemical Ecology 10: 3-8.

59. Furness RW (1985) Plastic particle pollution: accumulation by Procellariiform seabirds at Scottish colonies. Marine Pollution Bulletin 16: 103-106.

60. Ramos J, Barletta M, Costa M (2012) Ingestion of nylon threads by Gerreidae while using a tropical estuary as foraging grounds. Aquat Biol 17: 29-34.

61. Dantas D, Barletta M, Ramos J, Lima A, Costa M (2013) Seasonal diet shifts and overlap between two sympatric catfishes in an estuarine nursery. Estuaries Coasts 36: 237-256.

62. Moser ML, Lee DS (1992) A fourteen-year survey of plastic ingestion by Western North Atlantic seabirds. Colonial Waterbirds 15: 83-94.

63. Laist DW (1997) Impacts of marine debris: Entanglement of marine life in marine debris including a comprehensive list of species with entanglement and ingestion records. In: Coe JM, Rogers DB (eds.) Marine Debris. Sources, Impacts, Solutions. Springer-Verlag New York, Inc 99-140.

64. Vegter AC, Barletta M, Beck C, Borrero J, Burton H, et al. (2014) Global research priorities to mitigate plastic pollution impacts on marine wildlife. Endangered Species Research 25: 225-247.

65. Boerger CM, Lattin GL, Moore SL, Moore CJ (2010) Plastic ingestion by planktivorous fishes in the North Pacific Central Gyre. Marine Pollution Bulletin 60: 2275-2278.

66. Van Cauwenberghe L, Claessens M, Vandegehuchte M, Janssen C (2012a) 'Occurrence of microplastics in mussels (Mytilus edulis) and lugworms (Arenicola marina) collected along the French-Belgian-Dutch coast'. VLIZ Special Publication 55.

67. Rebolledo ELB, van Franeker JA, Jansen OE, Brasseur SMJM (2013) 'Plastic ingestion by harbour seals (Phoca vitulina) in The Netherlands'. Mar Pollut Bull 67: 200-202.

68. Van Franeker JA, Blaize C, Danielsen J, Fairclough K, Gollan J, et al. (2011) Monitoring plastic ingestion by the northern fulmar Fulmarus glacialis in the North Sea. Environmental Pollution 159: 2609-2615.

69. De Stephanis RJ, Gimenez E, Carpinelli C, Gutierrez-Exposito, Canadas A (2013) 'As main meal for spermwhales: Plastics debris'. Marine Pollution Bulletin 69: 206-214.

70. Browne MA, Dissanayake A, Galloway TS, Lowe DM, Thompson RC (2008) Ingested microscopic plastic translocates to the circulatory system of the mussel, Mytilus edulis (L.). Environ Sci Technol 42: 5026-5031.

71. Thompson RC (2006) Plastic debris in the marine environment: consequences and solutions. In: Krause JC, Nordheim H, Bräger S (eds.), Marine Nature Conservation in Europe. Federal Agency for Nature Conservation, Stralsund, Germany 107-115.

72. Eriksson C, Burton H (2003) Origins and biological accumulation of small plastic particles in fur seals from Macquarie Island. Ambio 32: 380-384.

73. Zitko V, Hanlon M (1991) Another source of pollution by plastics: skin cleansers with plastic scrubbers. Marine Pollution Bulletin 22: 41-42.

74. McCauley SJ, Bjorndal KA (1999) Conservation implications of dietary dilution from debris ingestion: sublethal effects in post-hatchling loggerhead sea turtles. Conserv Biol 13: 925-929.

75. Mader D (2006) Reptile medicine and surgery. Elsevier Inc., St Louis, MO.

76. Teuten EL, Saquing JM, Knappe DRU, Barlaz MA, Jonsson S, et al. (2009) Transport and release of chemicals from plastics to the environment and to wildlife. Philosophical Transactions of the Royal Society B: Biological Sciences 364: 2027-2045.

77. Gray H, Lattin GL, Moore CJ (2012) Incidence, mass and variety of plastics ingested by laysan (Phoebastria im -mutabilis) and black-footed albatrosses (P. nigripes) recovered as by-catch in the North Pacific Ocean. Mar Pollut Bull 64: 2190-2192.

78. Bhattacharya P, Lin S, Turner JP, Ke PC (2010) Physical adsorption of charged plastic nanoparticles affects algal photosynthesis. J Phys Chem C 114: 16556-16561.

79. Spear LB, Ainley DG, Ribic CA (1995) Incidence of plastic in seabirds from the tropical Pacific, 1984–91 –relation with distribution of species, sex, age, season, year and body-weight. Marine Environmental Research 40: 123-146.

80. Carson HS, Colbert SL, Kaylor MJ, McDermid KJ (2011) Small plastic debris changes water movement and heat transfer through beach sediments. Marine Pollution Bulletin 62: 1708-1713.

81. Meeker JD, Sathyanarayana S, Swan SH (2009) Phthalates and other additives in plastics: human exposure and associated health outcomes. Philos Trans R Soc Lond B Biol Sci 364: 2097-2113.

82. Mato Y, Isobe T, Takada H, Kanehiro H, Ohtake C, et al. (2001) Plastic resin pellets as a transport medium for toxic chemicals in the marine environment. Environ Sci Technol 35: 318-324.

83. Ashton K, Holmes L, Turner A (2010) Association of metals with plastic production pellets in the marine environment. Mar Pollut Bull 60: 2050-2055.

84. Holmes LA, Turner A, Thompson RC (2012) Adsorption of trace metals toplastic resin pellets in the marine environment. Environ Pollut 160: 42-48.

85. Rochman CM, Hentschel BT, Teh SJ (2014a) Long-term sorption of metals is similar among plastic types: Implications for plastic debris in aquatic environments. PloS ONE 9: e85433.

86. Rochman CM, Lewison RL, Eriksen M, Allen H, Cook AM, et al. (2014b) Polybrominated diphenyl ethers (PBDEs) in fish tissue may be an indicator of plastic contamination in marine habitats. Sci Total Environ 476-477: 622-633.

87. Tagatz ME, Plaia GR, Deans CH (1986) Toxicity of dibutyl phthalate-contaminated sediment to laboratorycolonized and field-colonized estuarine benthic communities. Bull Environ Contam Toxicol 37: 141-150.

88. Talsness CE, Andrade AJM, Kuriyama SN, Taylor JA, vom Saal FS (2009) Components of plastic: experimental studies in animals and relevance for human health. Philos Trans R Soc Lond B Biol Sci 364: 2079-2096.

89. Rochman CM, Browne MA, Halpern BS, Hentschel BT, Hoh E, et al. (2013) Policy: Classify plastic waste as hazardous. Nature 494: 169-171.

90. Azzarello MY, Van Vleet ES (1987) Marine birds and plastic pollution. Marine Ecology Progress Series 37: 295-303.

91. Bouland AJ, White AE, Lonabaugh KP, Varian-Ramos CW, Cristol DA (2012) Female-biased offspring sex ratios in birds at a mercury-contaminated river. J Avian Biol 43: 244-251.

92. Carpenter EJ, Smith KL (1972) Plastics on the Sargasso sea surface. Science 175: 1240-1241.

93. Minchin D (1996) Tar pellets and plastics as attachment surfaces for Lepadid cirripedes in the North Atlantic Ocean. Marine Pollution Bulletin 32: 855-859.

94. Carpenter EJ, Anderson SJ, Harvey GR, Miklas HP, Peck BB (1972) Polystyrene spherules in coastal waters. Science 178: 749-750.

95. Winston JE (1982) Drift plastic-an expanding niche for a marine invertebrate. Marine Pollution Bulletin 13: 348-351.

96. Zettler ER, Mincer TJ, Amaral-Zettler LA (2013) Life in the "Plastisphere": Microbial Communities on Plastic Marine Debris. Environ Sci Technol 47: 7137-7146.

97. Harshvardhan K, Jha B (2013) Biodegradation of low-density polyethylene by marine bacteria from pelagic waters, Arabian Sea, India. Marine Pollution Bulletin 77: 100-106.

98. Gregory MR (1991) The hazards of persistent marine pollution: drift plastics and conservation islands. Journal of the Royal Society of New Zealand 21: 83-100.

99. Gregory MR (1999) Plastics and South Pacific Island shores: environmental implications. Ocean and Coastal Management 42: 603-615.

100. Gregory MR (2009) Environmental implications of plastic debris in marine settings-entanglement, ingestion, smothering, hangers-on, hitch-hiking and alien invasions. Philos Trans R Soc Lond B Biol Sci 364: 2013-2025.

101. Balance A, Ryan PG, Turpie JK (2000) How much is a clean beach worth? The impact of litter on beach users in the Cape Peninsula, South Africa. S Afr J Sci 96: 210-213.

102. Choy CA, Drazen JC (2013) Plastic for dinner? Observations of frequent debris ingestion by pelagic predatory fishes from the central North Pacific. Mar Ecol Prog Ser 485: 155-163.

103. Van Cauwenberghe L, Janssen CR (2014) 'Microplastics in bivalves cultured for human consumption'. Environ Pollut 193: 65-70.

104. Galloway T, Cipelli R, Guralnick J, Ferrucci L, Bandinelli S, et al. (2010) Daily bisphenol A excretion and associations with sex hormone concentrations: results from the In CHIANTI adult population study. Environ Health Perspect 118: 1603-1608.

105. Lang IA, Galloway TS, Scarlett A, Henley WE, Depledge M, et al. (2008) Association of urinary Bisphenol A concentration with medical disorders and laboratory abnormalities in adults 300: 1303-1310.

106. Donohue MJ (2003) How multiagency partnerships can successfully address large-scale pollution problems: a Hawaii case study. Mar Pollut Bull 46: 700-702.

107. Lentz SA (1987) Plastics in the marine environment: legal approaches for international action. Marine Pollution Bulletin 18: 361-365.

108. Ninaber E (1997) MARPOL Annex V, commercial ships, and port reception facilites: makingit work. In: Coe JM, Rogers DB (eds.), Marine Debris-Sources, Impacts and Solutions. Springer- Verlag, New York 239-243.

109. CMC-Center for Marine Conservation (2002) MARPOL/MPPRCA.

110. Watkins E, Ten Brink P, Withana S, Mutafoglu K, Schweitzer JP, et al. (2015) Marine litter: socio-economic study. Scoping report. London, Brussels.

111. Ocean Conservancy (2014) Turning the tide on trash. International Coastal Cleanup 28.

112. Wolf N, Feldman E (1991) Plastics-America's packaging dilemma. Environmental Action Coalition. Island Press, Washington DC.

113. Gorman M (1993) Environmental Hazards-Marine Pollution. ABCCLIO Inc, Santa Barbara.

114. Barlow CY, Morgan DC (2013) Polymer film packaging for food: an environmental assessment. Resour Conserv Recycling 78: 74-80.

115. Hamann M, Godfrey MH, Seminoff JA, Arthur K (2010) Global research priorities for sea turtles: informing management and conservation in the 21st century. Endang Species Res 11: 245-269.

116. Sutherland WJ, Clout M, Côté IM, Daszak P, Depledge MH, et al. (2010) A horizon scan of global conservation issues for 2010. Trends in Ecology & Evolution 25: 1-7.

117. Laurance WF, Dell B, Turton SM, Lawes MJ (2011) The 10 Australian ecosystems most vulnerable to tipping points. Biol Conserv 144: 1472-1480.

118. Lewison R, Oro D, Godley BJ, Underhill L (2012) Research priorities for seabirds: improving conservation and management in the 21st century. Endang Species Res 17: 93-121.

119. Stern PC (2000) New environmental theories: toward a coherent theory of environmentally significant behavior. J Soc Issues 56: 407-424.

120. Brulle RJ (2010) From environmental campaigns to advancing the public dialog: environmental communication for civic engagement. Environ Commun 4: 82-98.

121. Ham S (2013) Interpretation, making a difference on purpose. Fulcrum Publishing, Golden, CO.

122. Cefas (2008) Literature review on particle assimilation by molluscs and crustaceans. Cefas Environ Rep RL 10: 17.

123. Colton JB, Knapp FD, Burns BR (1974) Plastic particles in surface waters of the Northwestern Atlantic. Science 185: 491-497.

124. Day RH, Shaw DG (1987) Patterns in the abundance of pelagic plastic and tar in the North Pacific Ocean, 1976-1985. Mar Pollut Bull 18: 311-316.

125. Eriksson C, Burton H, Fitch S, Schulz M, Van Den Hoff J (2013) Daily accumulation rates of marine debris on sub- Antarctic island beaches. Mar Pollut Bull 66: 199-208.

126. Fossi MC, Panti C, Guerranti C, Coppola D, Giannetti M, et al. (2012) Are baleen whales exposed to the threat of microplastics? A case study of the Mediterranean fin whale (Balaenoptera physalus). Marine Pollution Bulletin 64: 2374-2379.

127. Frias JPGL, Sobral P, Ferreira AM (2010) 'Organic pollutants in microplastics from two beaches of thePortuguese coast'. Marine Pollution Bulletin 60: 1988-1992.

128. Karapanagioti HK, Endo S, Ogata Y, Takada H (2011) Diffuse pollution by persistent organic pollutants as measured in plastic pellets sampled from various beaches in Greece. Mar Pollut Bull 62: 312-317.

129. Lattin GL, Moore CJ, Zellers AF, Moore SL, Weisberg SB (2004) A comparison of neustonic plastic and zooplankton at different depths near the southern California shore. Mar Pollut Bull 49: 291-294.

130. Lusher AL, McHugh M, Thompson RC (2013) 'Occurrence of microplastics in the gastrointestinal tract of pelagicand demersal fish from the English Channel'. Marine Pollution Bulletin 67: 94-99.

131. Moore CJ, Moore SL, Leecaster MK, Weisberg SB (2001) 'A Comparison of Plastic and Plankton in the North Pacific Central Gyre'. Marine Pollution Bulletin 42: 1297-1300.

132. Ng KL, Obbard JP (2006) Prevalence of microplastics in Singapore's coastal marine environment. Mar Pollut Bull 52: 761-767.

133. Noren F, Naustvoll LJ (2011) Survey of microplastic anthropogenic particles in Skagerrak. Flodevigen, Norway: Institute of Marine Research 21.

134. Ocean Conservacy (2010) International Coastal Cleanup (ICC), Trash travels: From our hands to the sea, around the globe, and through time (ISBN: 978D-615-34820-9), Washington, DC.

135. Pruter AT (1987) Sources, quantities and distribution of persistent plastics in the marine environment. Marine Pollution Bulletin 18: 305-310.

136. Rochman CM, Hoh E, Kurobe T, Teh SJ (2013) Ingested plastic transfers contaminants to fish and induces hepatic stress. Scientific Reports.

137. Song YK, Hong SH, Jang M, Kang JH, Kwon OY, et al. (2014) 'Large Accumulation of Micro-sized Synthetic Polymer Particles in the Sea Surface Microlayer'. Environ Sci Technol 48: 9014-9021.

138. Vianello A, Boldrin A, Guerriero P, Moschino V, Rella R, et al. (2013) 'Microplastic particles in sediments of Lagoon of Venice, Italy: First observations on occurrence, spatial patterns and identification'. Estuarine, Coastal and Shelf Science 130: 54-61.

139. Wong CS, Green DR, Cretney WJ (1974) 'Quantitative tar and plastic waste distributions in Pacific Ocean'. Nature 247: 30-32.

140. Zlatev M, Pahl S, White M (2010) 'Perceived risk and benefit for self and others as predictors of smokers' attitudes towards smoking restrictions'. Psychol Health 25: 167-182.

141. Plastics-Europe (2008) The compelling facts about plastics 2007: An analysis of plastics production, demand and recovery in Europe 24.

142. Arthur C, Baker J, Bamford H (2009) In: Proceedings of the International Research Workshop on the Occurence, Effects and Fate of Microplastic Marine Debris. NOAA Technical, Memorandum NOS-OR&R-30.

A Review on Western Yamuna Canal Water Quality Upstream and Downstream of Yamunanagar, Industrial Town

Sakhuja N[1*] and Sharma V[2]

[1]Department of Applied Sciences, JMIT Radaur, District Yamunanagar, India
[2]PG Department of Industrial Chemistry, Guru Nanak Khalsa College, Yamunanagar, India

Abstract

Yamuna Nagar is a city in the Indian state of Haryana. It is located south-east of the state capital Chandigarh. This town is known for the cluster of plywood units. It is also known for providing the country's finest timber to even larger industries. The older town is called Jagadhri. It used to be a green, clean and prosperous industrial city. However, due to recent spur in Industrial units in and around the city have resulted in severe air, water and soil pollution issues. Yamuna Nagar has the river Yamuna running through the district, and forming the eastern boundary with the neighbouring Saharanpur district. This boundary is also a state boundary, as Saharanpur is in the state of Uttar Pradesh. The district also separates the Yamuna system from the Satluj river system. Yamuna's pollution starts from Tajewala in the upper segment. Here two canals, the Western Yamuna Canal (WYC) and the Eastern Yamuna Canal (EYC), divert river waters into Haryana and Uttar Pradesh (UP). The WYC crosses Yamuna Nagar, Karnal and Panipat before reaching the Haiderpur treatment plant (which supplies part of Delhi's water), receiving wastewater from Yamuna Nagar and Panipat. The WYC at Yamuna Nagar, and rejoins the canal about 80 km downstream at Karnal. All domestic and industrial discharges from the effluents of paper mill, sugar mill, metal industries, distilleries, starch mill etc. are let out into this canal located at Yamunanagar. The water has a bad odour and contains a high amount of metal content, beside a high content of B.O.D. The water samples were collected at regular intervals from various strategic points at the exit of effluents from various industries. The parameters studied were temperature, pH, turbidity, solids, hardness, chloride content, dissolved oxygen, biochemical oxygen demand, chemical oxygen demand, turbidity, conductivity, total *E. coli*, fungi, etc. These contaminants have contributed the water unfit for drinking.

Keywords: Yamuna canal; Effluents; Contaminants; Seepage; Yamunanagar; WYC

Introduction

Water is one of the prime necessities of life. We can hardly live for a few days without water. In a man's body, 70-80% is water. Cell, blood, and bones contain 90%, 75%, and 22% water, respectively. The general survey reveals that the total surface area of earth is 51 crore km^2 out of which 36.1 crore km^2 is covered sea. In addition to this, we get water from rivers, lakes, tanks, and now on hills. In spite of such abundance, there is a shortage of soft water in the world. Physicochemical parameter of any water body plays a very important role in maintaining the fragile ecosystem that maintains various life forms. Potable water is the water that is safe to drink. The quality of water is of serious concern as it is seriously linked with human welfare. It must also meat very high standards of hygiene. In 2012, 89% of people had access to water suitable for drinking. The largest source of water pollution in India is untreated sewage [1]. Other sources of pollution include agricultural runoff and unregulated small scale industry. Rivers, canals, lakes, streams etc. are the various natural sources of water which supply abundant supply of water containing large amount of impurities in most of the cases. In 2010 the water quality monitoring found almost all rivers with high levels of BOD. The worst pollution, in decreasing order, were found in river Markanda (490 mg O/l), followed by river Kali (364), river Amlakhadi (353), Yamuna canal (247), river Yamuna at the Delhi (70) and river Betwa (58) [2].

The total length of the River Yamuna from its origin near Yamunotri to its confluence with Ganga River at Allahabad is 1376 Kilometers. It flows through various industrial towns. Western Yamuna Canal is originating from Tajewala, downstream from Tajewala it passes through industrial town of Yamunanagar and ultimately reaches Delhi under the name of Delhi Feeder Canal. During its passage through Yamunanagar, several drains and Nalahs containing effluents , sewage and domestic water merge into water of WYC, contaminating its water and making it unfit for human consumption. Disposes of carcases of cattle and other animals, immersion of statues, flowers, religious ceremonies in and around the rivers add to the pollution load. The river flow is not uniform throughout the year. The discharge peaks are very high in monsoon and post monsoon periods, but in summer the discharge rates are very low. Yamunanagar is the second biggest industrial town of Haryana. Many industries like paper mill, sugar mill, distillery, cement, metal industries etc. passes their waste water into WYC [3]. The various contaminants in waste water are a complex mixture of organic and inorganic compounds which make the natural water unfit for human consumption. The water of nearby colonies also gets polluted due to seepage of contaminated water of WYC.

Experimental

Nine strategic points were selected from Chitta Mandir to Garhi Gujran at a stretch of 16 km through which WYC passes through the industrial town of Yamunanagar. Samples were collected from these nine strategic points at regular intervals during Pre monsoon (April to June) and Post monsoon (July-Sept) twice a year. Water samples

***Corresponding author:** Sakhuja N, Department of Applied Sciences, JMIT Radaur, District Yamunanagar, India, E-mail: nitisakhuja22@gmail.com

were also collected from various strategic points at the exit of effluents from paper mill, sugar mill, metal industries, distilleries, starch mill, etc. Samples were also collected from the hand pump water of different colonies alongside Western Yamuna Canal in order to analyse the seepage effect of contaminated water of WYC. The various parameters analysed include pH, alkalinity, hardness, solids, chloride content, turbidity, conductance, Chemical oxygen demand, bio-chemical oxygen demand, chemical oxygen demand, total *E. coli*, total fungal count etc. using the standard methods for the examination of waste water [4-7] and water. The samples were analysed within few hours after collection to get the accurate results. These samplings sites have been shown in Figures 1 and 2.

Result and Discussion

Free carbon dioxide, phenolphthalein alkalinity, acidity

Acidity is a measure of an aggregate property of water and can be interpreted in terms of specific substances only when the chemical composition of the sample is known. Alkalinity measures the amount of alkaline compounds in the water, such as carbonates, bicarbonates and hydroxides. These compounds are natural buffers that can remove excess hydrogen, or H^+, ions. The value of alkalinity in the pollution infested zone is maximum 480 ppm and the seepage effected zone is 340 ppm. Though the alkalinity itself is not harmful to human beings, water supplies with less than 100 ppm are desirable for domestic use. Higher value for free carbon dioxide and acidity was observed in both pollution infested and seepage zone (Figure 3).

PH, conductance, turbidity, hardness

PH is an important limiting chemical factor for aquatic life. If the water in a stream is too acidic or basic, the H^+ or OH^- ion activity may disrupt aquatic organism's biochemical reactions by either harming or killing the stream organisms. Conductivity is a measure of how well water can pass an electrical current. It is an indirect measure of the presence of inorganic dissolved solids such as chloride, nitrate, sulphate, phosphate, sodium, magnesium, calcium, iron and aluminium. Turbidity is a measure of the cloudiness of water. Cloudiness is caused by suspended solids (mainly soil particles) and plankton (microscopic plants and animals) that are suspended in the water column. Moderately low levels of turbidity may indicate a healthy, well-functioning ecosystem, with moderate amounts of plankton present to fuel the fuel the food chain. However, higher levels of turbidity pose several problems for stream systems. The hardness of a water is governed by the content of calcium and magnesium salts (temporary hardness), largely combined with bicarbonate and carbonate and with sulfates, chlorides, and other

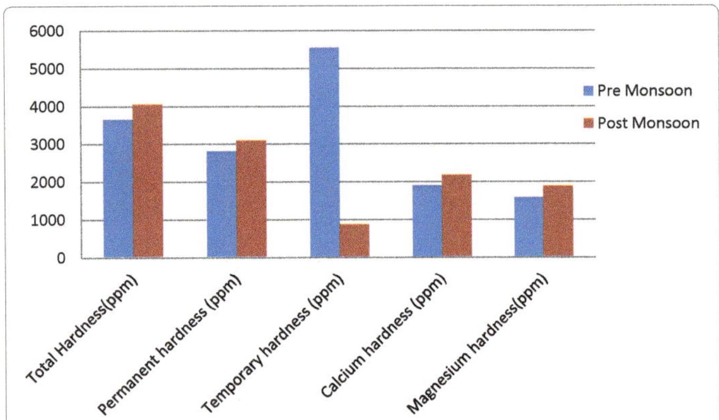

Figure 1: Water quality of WYC during pre and post monsoon.

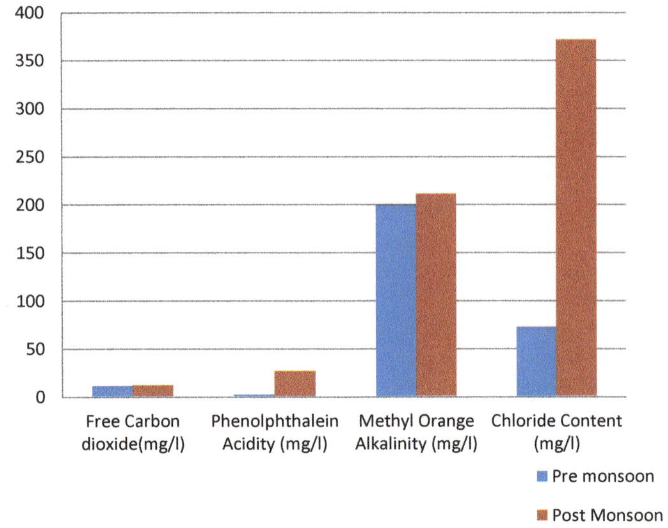

Figure 2: Water quality of WYC during pre and post monsoon.

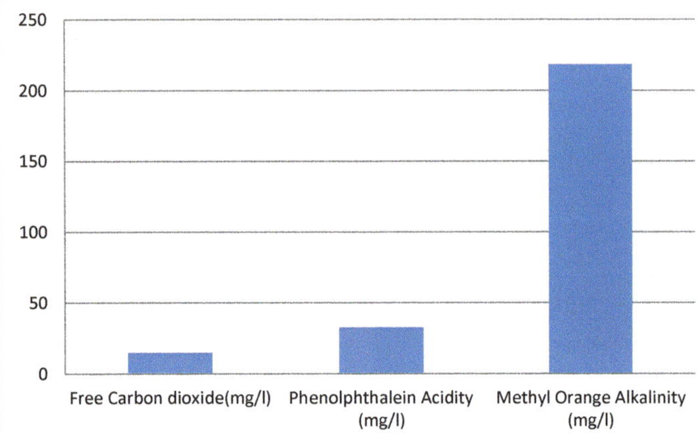

Figure 3: Water quality of nearby colonies alongside WYC.

anions of mineral acids (permanent hardness). Both temporary and permanent hardness show high value in polluted zone (Figure 1). Even the quality of underground water of nearby colonies is effected seriously, which is the cause of water borne diseases.

DO, BOD, COD and total dissolved solids

Dissolved oxygen gets into the water by diffusion from the atmosphere, aeration of the water as it tumbles over falls and rapids, and as a waste product of photosynthesis. Dissolved oxygen criteria for drinking water should be 5 mg/L minimum for aquatic life. Dissolved oxygen (DO) levels between 6.5 and 8.5 are acceptable for most drinking water. The DO level at site no.7 is much below the optimum level i.e. 7. The chemical oxygen demand, or COD, is used as a measure of the oxygen equivalent of the organic matter content of a sample that is susceptible to oxidation by a strong chemical oxidant. For samples from a specific source, COD can be related empirically to BOD, organic carbon, or organic matter. COD values are beyond the permissible limit (Tables 1-3). The Biological Oxygen Demand, or BOD, is the amount of oxygen consumed by bacteria in the decomposition of organic material. It also includes the oxygen required for the oxidation of various chemical in the water, such as sulphides, ferrous iron and ammonia. While a dissolved oxygen test tells you how much oxygen is available, a BOD

Parameters	site 1	Site 2	Site 3	Site 4	site 5	Site 6	Site 7	Site 8	Mean
Free carbon dioxide(mg/l)	2.09	2.78	3.1	30.01	7.32	28.32	10.8	5.2	11.21
Phenolphthalein acidity (mg/l)	2.45	3.12	13.23	50.1	15	25	55	26	2.08
Methyl orange alkalinity (mg/l)	42.54	55.12	132	130	142	375	453	265	199.33
Chloride content (mg/l)	18.2	18.38	18.08	38.14	21.35	70.14	130	267.2	72.73
D.O (mg/l)	8.2	7.8	5.2	6.1	4.7	3.9	nil	nil	5.98
B.O.D (mg/l)	23	26	38	65	20	62	Nil	Nil	29.25
C.O.D (mg/l)	54.25	57.05	62.6	64.4	132	190	1990	2256	600.78
Total hardness (ppm)	560.12	690.54	1766	2000.12	1990	3854.5	9486	8888	3650.9
Permanent hardness (ppm)	390.78	440.8	1126	1546.7	1298.8	2289	8765	6550	2800.9
Temporary hardness (ppm)	181.34	245.87	567.9	578.9	767.89	1887.9	991	2558	5540.4
Calcium hardness (ppm)	450.9	490.89	790	1234.67	1125.5	2870	3981	4211	1894
Magnesium hardness (ppm)	180	230.8	805.6	889.9	940.8	1090.9	5909	2555	1575.2
Total solid content (ppm)	500	490	550	590	600	660	3410	2540	1167.5
Total suspended solids (ppm)	270	230	250	280	290	300	1500	600	465
Total dissolved solids (ppm)	230	260	300	310	310	360	1910	1940	3922.5
pH	7.6	8.2	6.8	7	7.4	8	8.9	9.2	7.9
Temperature (°C)	28	26	28	29	30	29.5	31.5	32	29.25
Conductance (mhos)	5.8×10^{-2}	5.7×10^{-2}	5.9×10^{-2}	7.2×10^{-2}	7.9×10^{-2}	8.3×10^{-2}	2.24×10^{-2}	2.89×10^{-2}	5.74×10^{-2}
Sulphate Content (% age)	1.13	1.15	6.987	5.772	7.987	8.786	1.9	2.456	4.52
Nitrogen Content (% age)	0.023	0.0256	0.034	0.043	0.045	0.029	0.4	0.421	0.65
Total *E. coli* (Mpn)	Nil	Nil	42	>300	>300	>300	>300	>300	>300
Total fungal count	Nil	Nil	Nil	6	4	5	6	5	5.2
Total bacterial count	12	13	46	>300	>300	>300	>300	>300	>300
Turbidity (NTU)	Nil	12	22	38	Above100 NTU	45	Above100 NTU	60	5.98

Table 1: Characteristics of Western Jamuna canal at different monitoring stations (Pre monsoon).

Parameters	site 1	Site 2	Site 3	Site 4	site 5	Site 6	Site 7	Site 8	Mean
Free carbon dioxide (mg/l)	2.6	3.18	3.58	31.43	7.98	29.31	10.9	5.78	11.84
Phenolphthalein Acidity (mg/l)	2.55	3.44	14.23	54.1	18.5	27.6	60	30	26.3
Methyl Orange Alkalinity (mg/l)	45.64	58.12	139	140	152	390	480	280	210.59
Chloride Content (mg/l)	18.8	19.68	20.08	40.14	23.35	78.14	136.34	276.22	371.05
D.O (mg/l)	8.6	7.9	5.8	6.9	4.9	4.2	nil	nil	6.38
B.O.D (mg/l)	25	28	42	68	25	65	Nil	Nil	42.16
C.O.D (mg/l)	57.25	60.15	65.6	69.4	137	200	2180	2345	639.3
Total hardness (ppm)	580.12	700.54	2765.5	2200.12	2200	3990.5	9985.5	9877.8	4037.51
Permanent hardness (ppm)	400.78	480.8	1439.8	1656.7	1499.8	2348	8995.3	7750	3071.39
Temporary hardness (ppm)	179.34	219.74	560.2	543.42	700.2	1552.5	990.2	2127.8	859.17
Calcium hardness (ppm)	460.9	480.89	820	1554.67	1625.5	2980	4890.89	4510.9	2165.4
Magnesium hardness (ppm)	119.22	219.65	1945.5	645.45	574.5	1010.5	5094.61	5366.9	1872.04
Total solid content (ppm)	700	500	650	650	700	760	4450	2940	1418.75
Total suspended solids (ppm)	390	350	450	480	390	400	1800	700	620
Total dissolved solids (ppm)	310	150	200	170	310	360	2650	2240	798.75
pH	7.2	7.8	5.23	5.8	6.47	7.54	8	8.67	7.08
Temperature(°C)	26	27	26	28	28.5	28	31.5	31	28.25
Conductance (mhos)	5.3×10^{-2}	5.5×10^{-2}	5.8×10^{-2}	7.1×10^{-2}	7.7×10^{-2}	8.2×10^{-2}	2.10×10^{-2}	2.24×10^{-2}	5.49×10^{-2}
Sulphate content (% age)	1.23	1.25	5.887	5.882	6.887	7.186	1.657	1.856	3.979
Nitrogen content (% age)	0.027	0.0286	0.0356	0.053	0.065	0.039	0.418	0.431	0.1371
Total *E. coli* (Mpn)	Nil	Nil	42	>300	>300	>300	>300	>300	>300
Total fungal count	Nil	Nil	Nil	2	3	4	4	6	4.8
Total bacterial count	8	9	37	>300	>300	>300	>300	>300	>300
Turbidity (NTU)	Nil	12	22	38	Above100 NTU	45	Above100 NTU	60	33

Table 2: Characteristics of Western Yamuna canal at different monitoring stations (Post monsoon).

test tells you how much oxygen is being consumed. BOD is determined by measuring the dissolved oxygen level in a freshly collected sample and comparing it to the dissolved oxygen level in a sample that was collected at the same time but incubated under specific conditions for a certain number of days. The difference in the oxygen readings between the two samples in the BOD is recorded in units of mg/L. Unpolluted; natural waters should have a BOD of 5 mg/L or less. Raw sewage may have BOD levels ranging from 150-300 mg/L. Dissolved solids refer to any minerals, salts, metals, cations or anions dissolved in water. Total dissolved solids (TDS) comprise inorganic salts (principally calcium, magnesium, potassium, sodium, bicarbonates, chlorides, and sulphates) and some small amounts of organic matter As per Bureau

Parameters	site 9	Site 10	Site 11	Site 12	site 13	Site 14	Site 15	Mean
Free carbon dioxide (mg/l)	29.1	2	4.89	4.98	22.35	18.98	19.88	14.6
Phenolphthalein acidity (mg/l)	68	0.9	12	10	53	41	42	32.41
Methyl orange alkalinity (mg/l)	240	284	340	100	107	214	243	218.3
Chloride content (mg/l)	17.22	70.58	8.3	9.87	10.34	45.66	24.99	26.71
D.O (mg/l)	6.2	7.8	4.5	4.7	5.9	5.2	6.2	5.786
B.O.D (mg/l)	22	43	55	67	53	88	74	57.43
C.O.D (mg/l)	120	99	310	488	340	330	370	293.9
Total hardness (ppm)	2245	2178	1000	1225	2324	1900	1995	1838
Permanent hardness (ppm)	1478	1450	550	655	1445	1170	1167	1131
Temporary hardness (ppm)	767	728	450	570	879	730	828	707.4
Calcium hardness (ppm)	1355	1124	554.5	635.22	744.5	1666.5	109	884
Magnesium hardness (ppm)	890.2	1054	445.6	589.78	1580	233.5	1886	954.1
Total suspended solids (ppm)	260	220	130	190	324	199	187	215.7
Total dissolved solids (ppm)	2010	1775	360	470	560	570	550	899.3
pH	7.6	7.9	8.5	7.8	8.2	7.8	7.77	7.939
Temperature(°C)	34	33.2	33.8	33	32.7	32.8	33.6	33.3
Conductance (mhos)	480	590	180	240	440	590	550	438.6
Sulphate content (% age)	3.43	2.8	4.21	4.256	3.301	2.52	2.877	3.342
Nitrogen content (% age)	0.015	0.014	0.019	0.0192	0.031	0.0321	0.0267	0.022
Total *E. coli* (Mpn)	Nil	Nil	Nil	Nil	Nil	9	8	8.5
Total fungal count	Nil	Nil	Nil	Nil	Nil	4	4	4
Total bacterial count	19	17	13	26	22	45	41	44.86
Turbidity (NTU)	8	3	1	2		4	11	7.137

Table 3: Water quality of various colonies alongside western Yamuna canal.

of Indian Standards (IS: 10500) guidelines, 500 mg/L is the desirable limit and 2000 mg/L is the permissible limit, which means drinking water having TDS in excess of 2000 mg/L must be rejected. Presence of TDS beyond 500 mg/L in drinking water decreases palatability and may cause gastrointestinal irritation. The TDS values at site no. 7 and 8 are 1910 ppm and 1940 ppm (pre monsoon) which is much above the permissible limit.

The contaminated water samples have also shown the presence of a number of viruses and bacteria like *Salmonella* group (typhoid) *Shigella* (Bacillary dysentary) *Mycobacterium* (tuberculosis and virrio cholera which causes diseases like gastroenteritis, diarrhea, respiratory illness, heart diseases, liver diseases etc.). These diseases are severely affecting the health of residents of these colonies. The bacteriological examination has shown the MPN index of coli form bacteria exceeding 70 per 100 ml. The permissible level of MPN index of coli form is zero or less than1.0 in the treated water and 20 per 100 ml in raw water. Toxic effects of non-biotic contaminants also affect the health. The high MPN values and BOD values also indicates organic pollution in the river.

Conclusion

Every drop of water is essential. Drinking purified water does a lot more than quench your thirst...it keeps your body in great working order. There are many benefits to drinking clean, fresh water. We are comprised of mostly water; over 70% of our bodies are water. To break it down even further our brains are over 75% water, blood 80%, and the human liver which is our main filter is 96% water. Water pollution throughout the world is affecting food chains and food webs. Water such as lakes, rivers, streams, creeks, and oceans become polluted in many different ways. One main way is the dumping of trash, or littering. Many creeks, rivers, and even oceans have been polluted by manmade items such as trash. Today there is widespread realisation that there can be no readymade solutions that technology cannot alone solve problems. Urbanisation, industrial growth, transportation system, agriculture, housing all have to stop imitating practices that are already being phased out in the developed world. The few ways to prevent water pollution [8] are sewage treatments, prevent river water to get polluted, treatment of wastes before discharge, strict enforcement of water laws, treatment of drainage water, treatment plants, keep the pond water clean and safe, routine cleaning, sanitation, public awareness. Common public should be aware about the effect of water pollution. Voluntary organization should go door-to-door to educate the people about environmental problems. They should perform street plays for creating awareness about the environment. They should run environmental education centres. The lives of plants, animals, and human beings are not possible without water. It is duty of all of us to try over utmost to make proper use of water as drop of water speaks "Be A part of the solution, not the pollution". It is soul and hope of future.

References

1. Evaluation of Operation and Maintenance of Sewage Treatment Plants in India-2007 (2008) Central Pollution Control Board, Ministry of Environment & Forests.
2. Bansal S (1989) Physicochemical studies of the water of river Betwa in U.P.I.J.E.P. 2: 899-903.
3. Vashisht HS (1981) The ecology of polluted water of Yamunanagar. J Ecol 1: 134-135.
4. Kudesia VP (1990) Water Pollution, Pragati Parakashan, Meerut.
5. Chatwal GR, Mehra MC, Satake M, Katyal T, Katyal M, et al. (1989) Environment Analysis. Anmol Publications, New Delhi.
6. Lenove SC, Arnold E, Greenberg R, Rhodes T (1990) Standard methods for the examination of water and waste water. American Public Health Association 1015 fifteenth street N.W. Washington DC.
7. De AK (1993) Environmental Chemistry. Wiley Eastern Limited Publication.
8. Goel PK (2006) Water Pollution-Causes, Effects and Control. New Delhi: New Age International.

Implication of Roasting Goats with Tyre on Human Health and the Environment in Abakaliki, Ebonyi State, Nigeria

Aya FC[1]* and Nwite JN[2]

[1]*Department of Soil Science and Environmental Management, faculty of agriculture and natural resources management, Ebonyi State University, P.M.B 053, Abakaliki, Nigeria*
[2]*Ebonyi State Ministry of Environment, Abakaliki, Nigeria*

Abstract

Implication of roasting goats with tyre on human health and the environment in Abakaliki metropolis, Ebonyi State was reviewed in this study. Tyre has 45.49, 40, 0.23 and 1.17% composition for carbon black, zinc oxide, wax and sulphur toxic materials which could pose health threats to human being and the environment. Most of the butchers in Abakaliki and its environs roast slaughtered goats with scrap car tyres. Scraped soot from the roasted goats have significant ($P<0.05$) high levels of 5.40, 96.10, 38.03 and 61.00 mg kg^{-1} heavy metal residues for pb, Zn, Cu, and Fe which are washed into the soil, water bodies and air that cause their contamination. These heavy metals concentrations are higher than maximum permissible limit (MPL). Appropriate laws should be made and enforced in Ebonyi State to stop local butchers from using tyres as roasting materials. Scraped soot should also be disposed off safely as is done elsewhere in order to reduce the risk of contamination of soils, water bodies and air with heavy metals in Abakaliki metropolis of Ebonyi State.

Keywords: Abakaliki metropolis; Environment; Goats; Human; Implication; Roasting; Tyre

Introduction

In Abakaliki, Ebonyi State of Nigeria and in several other African countries, roasting in an open fire is the major process by which hair on the skin of slaughtered goats is removed [1]. This is because roasting maintains the carcass hide for consumption and also evokes flavours in the meat which are acceptable to the consumer [2]. Roasting is done mainly by the use of firewood as fuel, but the relative scarcity of firewood in recent times has resulted in Abakaliki local butchers using scrap car tyre in place of firewood [1].

Tyres are believed to contain toxic materials which could contaminate the environment. US Agency for Toxic Substances and Disease Registry revealed that "tyre derived fuel" (TDF) contained several heavy metals such as lead (Pd), zinc (Zn), and Copper (Cu) that could be carcinogenic when exposed to consumers over a long period. Research by Obiri- Danso et al., [1] also revealed high levels of heavy metal residues in goat and cattle hides roasted with scrap tyres, which made them unsafe for consumption.

Furthermore, tyre being non-biodegradable combustible material, when burnt under open and uncontrolled fire conditions as low temperatures produces black plumes of smoke as well as toxic volatilization products such as soot, which become incorporated into the ambient environment cause air and water pollution [3] as well as result in human and environmental hazard. In addition, large amounts of greenhouse gases and becomes absorbed. Dioxin does not breakdown; it accumulates in the fatty tissues of animals and humans that consume the contaminated vegetation, meats, chickens and dairy products. In humans, these Dioxins can lead to productive impairments, development of injuries and an increase in the risk of diabetes. It is against this background of health implication and environmental concern as a result of contamination of soil, water bodies and air by deposits from roasting goats with tyre that this review becomes imperative. For instance, deterioration in water quality has adverse effect on human health as well as aquatic ecosystem either directly or indirectly [4-6]. The objective of this review was to highlight implication of roasting goats with on human health and the environment and therefore create awareness for people within the Abakaliki environment to be cautious.

Methodology

The review was carried out at Abakaliki metropolis, the capital of Ebonyi State, Nigeria using extrapolation of information from previous studies. Abakaliki metropolis is located by latitude 06°4'N and longitude 08°65'E in the derived savannah of the Southeast agro-ecological zone of Nigeria. Ebonyi State is one of the last six states created in 1996 by the then military regime in the country. It is characterized by bimodal pattern of rainfall which spreads from April – July and September – November with break in August. Minimum and maximum animal rainfall range from 1700mm to 2000mm, respectively. Temperatures are 27°C and 31°C for rainy and dry seasons [7]. Ebonyi State is an agrarian state which grows most common food crops such as yam, cassava, maize, rice and vegetables and rears native livestock for their livelihood. Nevertheless, part of the population engages in civil service work.

Results and Discussion

Composition of tyre

A review of tyre composition shows that styrene butadiene had highest percentage composition of 46.78% followed by carbon black with 45.49%. Others include Aromatic coil, Zinc oxide, stearic acid, Antioxidant 6C, wax, Sulphur, and accelerator Cz with 1.74%, 1.40%, 0.94%, 1.40%, 0.23%, 1.17% and 0.75% compositions of tyre, respectively.

***Corresponding author:** Aya FC, Department of Soil Science and Environmental Management, faculty of agriculture and natural resources management Ebonyi State University, P.M.B 053, Abakaliki
E-mail: ayason4real@yahoo.co.uk

Tyre as a product is environmentally friendly. However, the additives, as well as their products of chemical reactions and degradation products incorporated into the polymeric material have the potential to be released into the environment where they cause significant health and environmental problems [8].

Statistics of disposal of scrap tyre

Nigeria does not have statistical data on scrap tyre generation and disposal. However, a number of organisations, most prominently the US Rubber Manufacturers' Association and the European Tyre Recycling Association (ETRA) have data on scrap tyres as given in Table 1.

The Table 2 illustrates the disposal methods of scrap tyre by EU and US as well as their diverse uses. It also indicates the percentages of their usage. For instance, energy recovery by EU is at 22.40% while US is 40.90%. The amount of scrap tyres that go to landfill by EU countries is estimated to be 30.8% while that of US is 22.40%. Rubber recycling of scrap tyre by EU and US is estimated to be between 15.70% and 14.60%, respectively. Scrap tyre export is seen to be higher in EU countries by 7.50% than US by 5.30%. United States (U.S.) uses scrap tyres for engineering purpose and this is 14.20% higher than EU countries which are 8.0%.

Nigeria does not have statistical data on scrap car tyres but it is believed that Nigeria which statistical data is not known has even more number of scrap tyres being disposed off without recycling. One of the disposal methods in Nigeria is by using tyre to roast goats or hides of animals. This is attributed to the influx of imported fairly used tyres from western countries into Nigeria. This situation has made Nigeria a "dumping ground" of fairly used tyres. As a result, the ubiquity of scrap tyres necessitates their use for roasting goats and other slaughtered animals due to scarcity of firewoods.

Material	Percentage
Styrene butadiene	46.78%
Carbon black	45.49%
Aromatic oil	1.74%
Zinc oxide	1.40%
Stearic acid	0.94%
Antioxidant 6C	1.40%
Wax	0.23%
Sulphur	1.17%
Accelerator CZ	0.75%

Source: Diane Kihara (2004); http//www.calrecycle.ca Advisories.gov/LEA/

Table 1: Tyre composition.

Disposal method	EU		US	
	Tyres disposed (tonnes)	Percentage	Tyres disposed (tonnes)	Percentage
Energy recovery	563.69	22.40%	1,035,000	40.90%
Landfill/stockpiling	775,300	30.80%	567,000	22.40%
Rubber recycling	395,287	15.70%	369,000	14.60%
Used tyre export	189,509	7.50%	135,000	5.30%
Civil engineering	200,607	8.00%	360,000	14.20%
Miscellaneous	394,013	15.60%	63,000	2.50%
Total	2,518,406	-	2,529,000	-

Source: Kurt Reschner, (2003).

Table 2: Statistics of scrap tyre disposal for 2000-2001 in US and EU member states.

Emissions from open tyres fires

Emissions like sulphur (IV) Oxides (SO_2) and other Oxides of Nitrogen (NO_x) from open tyre fires are being suspected of having a serious impact on human health and the environment. Due to lack of data, it was uncertain to know exactly what was being emitted, how much was being emitted, and how dangerous these emissions are, especially to sensitive individuals such as the children and the elderly. These products can also contaminate soil, surface water and ground water.

When fire from tyre burns, it is believed that the smoke would contain hazardous substances. When the fire is cooling down it could release other poisonous substances, for example, the Benzene which is produced in this process once inhaled, ingested or touched will lead to symptoms such as dizziness, euphoria, giddiness, headache, nausea, weakness, drowsiness, respiratory irritation, pulmonary edema, pneumonia, skin and eyes irritation. The health conditions render the victim incapacitated and consequently reduces healthy work force.

Concentration of heavy metals in scrap soot from roasted slaughtered goats on soil, water bodies and air

Table 3 below shows concentrations of heavy metals in scrapped soot from roasted slaughtered goats. The scraped soot from the roasted animals has higher levels of heavy metal residues such as Pb, Zn, Cu and Fe with 5.40, 96.10, 38.03, 61.00 mg kg^{-1} concentrations respectively and their corresponding individual maximum permissible limits (MPL) of 0.10, 50.00, 20.00 and 50.00MPL, which when above the recommended level pose threats to soil, water bodies air and even to human beings who depend on crops grown in the soil or water from the ground or inhale air for sustenance of life.

The soot from the burnt tyre is often washed into drains, and when there is rainfall, the soots are transported into various water bodies within and around. This could be one of the sources of heavy metals in soils, water air and even plants as reported by Qui et al. [9]. Heavy metals dissolve in water and are taken up by plants during photosynthesis [10]. When animals feed on these plants over a period of time, the heavy metals accumulate in their tissues and skin, hence the source of heavy metals detected in the hides of the un-roasted slaughtered animals. Furthermore, some butchers fetch water from nearby streams and other water bodies for washing the roasted slaughtered goat [11]. This therefore, may explain why washing roasted slaughtered goat with water could result in increased heavy metal risks of ecotoxicity as reported by Obiri- Danso et al., [1].

Heavy metals contamination of soil reduces crop yields and eventually destroys plots of agricultural land. The side effects of human exposure to these toxins in our food chain can lead to serious health conditions. Lead poisoning destroys human nervous systems,

Heavy metal	Concentration (mg/kg)	SEM(+/-)	MPL (mg/kg)
Pb	5.40**	0.01	0.1
Zn	96.10**	0.012	50
Cu	38.03**	0.015	20
Fe	61.00**	0.006	50
Mn	ND	-	-
Cd	ND	-	0.05

**= ($p<0.01$); MPL= European Commission's Maximum Permissible Level of Product Acceptability, SEM = standard error of means, ND= Not detected.
Source: Adam et al., (2013).

Table 3: Concentration of heavy metals in scraped soot from roasted slaughtered goats.

causes retardation, learning difficulties, bone marrow deficiencies and stunted growth in children. Chromium and arsenic can also cause cancer [12].

In Abakaliki metropolis, Ebonyi state of Nigeria, some abattoir workers engage in combustion of old vehicle tyres as fuel to burn the fur on cattle and goats. This process releases dark sooty smoke into the atmosphere. Tyre is composed of natural and synthetic rubber, sulphur, filler, accelerators, antioxidants, fabrics and optionally steel wires. For the fact that majority of the components of tyre are of hydrocarbon origin; high temperatures are usually employed in the combustion processes. The emissions from such operations are most likely to contaminate the air [13]. Human beings are at health risks if they inhale such contaminated air.

According to Schwartz [14], air could also be polluted through emission of particulate matter (Black carbon) that occurs during the burning of fuels such as diesel or coal. The release of noxious gases such as SO_x, CO, NO_x and chemical vapours and particulate matter could lead to air pollution [15]. Hence, air pollutions are either particulates or gases. Gaseous and particulate pollutants may share some common sources however; they have different types of problems. Besides, some of these gases contribute immensely to climate change [16] which today has caused distortion in farming system, low crop and soil productivity with consequent famine in the country.

Health and environmental implication of roasting goats with tyres

Table 4 shows the permissible limits of CO, SO_2, NO_2, Pb, PM (2.5) as 9 ppm, 0.17 ppm, 10 ppm, 0.15 μg/m^3, 150 μg/m^3, and 15.0 μg/m^3 respectively with their corresponding average time of 8hours, 24hours, 1hour, 24hours and 24hours above which could pose serious health hazards. For instance, carbon monoxide (CO) is a harmful pollutant and it's injurious to human health at a level greater than the safe concentration limit of 9ppm. Carbon monoxide (CO) reacts with oxygen carrying blood tissue to form carboxy-hermoglobin. When SO_2 concentration exceeds the permissible limit, it could cause respiratory difficulty especially with asthmatic patient, children and the elderly. Sulphur dioxide (SO_2) can also form acid deposition that has the ability of disrupting plant growth and damage to building roof.

The Table 4 above is also in accordance with WHO Allowable limits. Nitrogen dioxide (NO_2) formed during high combustion temperature, reacts with water vapour to give HNO_3, an irritant and highly corrosive substance [17]. Nitrogen dioxide (NO_2) contributes to the formation of ground level (tropospheric) ozone and fine particle pollution. It is linked with a number of adverse effects on the respiratory system. This includes airway increased respiratory symptoms in people with asthma among others. The emission of SO_2 from combustion of tyre can be traced to the presence of sulphur in tyre. To increase the tensile strength and durability of tyre, sulphur is usually incorporated as raw material during the vulcanization process.

Pollution	Safe concentration zone	Averaging time
Carbon monoxide	9ppm (10mg/m^3)	8hours
Sulphur dioxide	0.017ppm(ug/m^3)	24hours
Nitrogen dioxide	10ppm	1hour
Lead	0.15ug/m^3	-
Particulate matter PM(10)	150ug/m^3	24hours
Particulate PM (2.5)	15.0ug/m^3	24hours

Table 4: US- EPA standards for six principal pollutions.

Use of figures

The figures below show the processes used in abattoir by butchers in Abakaliki metropolis to roast goats in open fire from scrap tyres at different parts of the town.

As shown in Figure 1, roasting in an open fire is the major process by which hair on the skin of slaughtered goats is removed [1]. This is because roasting maintains the carcass hide for consumption and also adds some flavours in the meat which are acceptable to the consumer [2].

Roasting is done mainly by the use of firewood as fuel as indicated in Figure 2, due to relative scarcity or high cost of firewood in recent times resulting in local butchers using scrap car tyres in place of firewood [1]. The use of car tyres as fuel for roasting carcasses is reported to impart some toxic heavy metals to the roasted skin which could diffuse into the meat, thus rendering them unsafe for human consumption, as it poses several health threats to the consumer [18-21] in Adam et al., [22]. US-Agency for Toxic substances and Disease Registry carried out research revealing that "tyre derived fuel" (TDF) contains several heavy metals such as lead (pb), zinc (Zn), copper (Cu) that could be carcinogenic when exposed to consumers over a long period. Research by Obiri-Danso et al, [1] also revealed high levels of heavy metal resides in goat

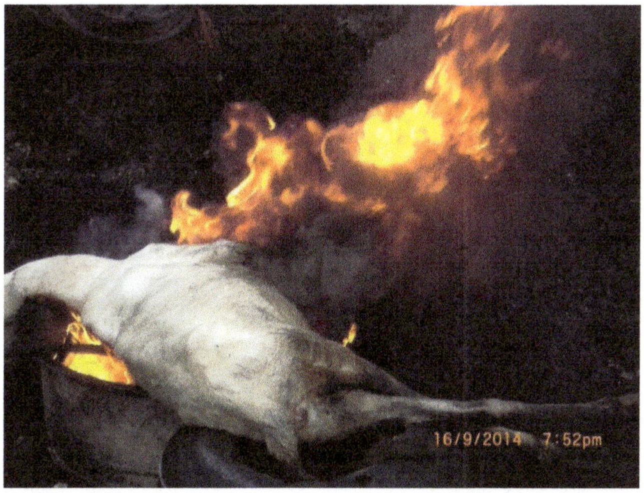

Figure 1: Goat being roasted with Tyre at number 63 Gunning Road, Abakaliki.

Figure 2: Roasting done with Firewood at Garki off Ogoja road, Abakaliki.

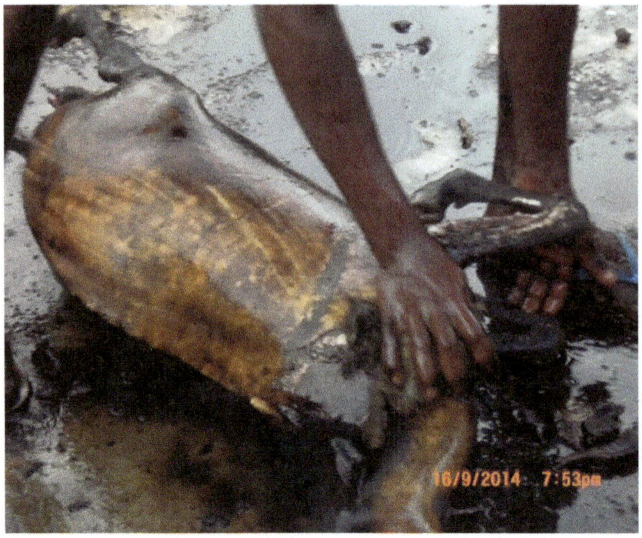

Figure 3: one of the carcasses being washed Gunning Road, Abakaliki.

and cattle hides roasted with scrap tyres, which made them unsafe for human consumption.

According to Adam et al., [22], boiling and washing of the slaughtered goats as in Figure 3 slightly reduced the heavy metal concentrations, but the heavy metal residues after boiling were still higher than the maximum permissible level (MPL), indicating that boiling cannot render tyre-roasted carcasses wholesome for consumption.

Conclusion

Tyres are believed to contain toxic materials which could pose health threats to human beings, soil, water, and air. The scrapings or soot of roasted slaughtered animals (goats) have very high heavy metal concentrations above maximum permissible level (MPL) and incidentally are potential sources of heavy metals in soils, surface water and in plants. Roasting of slaughtered goats with tyres increases the heavy metal concentration in the hides or skin of the slaughtered animal which diffuse into the body beyond acceptable limits. Boiling of these carcasses slightly reduces the heavy metal concentration, but not below the safe or acceptable limits.

Consumption of slaughtered goats roasted with tyres could pose health threats to consumers. More importantly, human beings who eat the crops grown in soot contaminated soil, drink water or inhale air are at risk of ecotoxicity. Therefore, regulatory agencies should checkmate the activities of these local butchers in Abakaliki metropolis and beyond to ensure total compliance and henceforth slaughtered goats should be disposed of appropriately in order to minimize the contamination of soils, water bodies, air and the surrounding environment with heavy metals.

References

1. Obiri-Danso K, hogarh JN, Antwi-Agyei p (2008) Accessemnt of contamination of singed hides from cattle and goats by heavy metals in Ghana. African journal of Environmental Science and Technology 2: 217-221.
2. FAO (1985) Animal by-product: Processing and Utilization. FAO Animal production and Health series No9 FAO.
3. Mohammad V, Seyyed MM, Reza V, Ehsan R (2012) Air, Water, and Soil Pollution Study in Industrial Units Using Environmental Flow Diagram. Journal of Basic and Applied Scientific Research 2: 12365-12372.
4. Dibofori-orji AN, Braide SA (2008) Emission of NO_x, SO_X and CO from the combustion of vehicle tyres in an abattoir. Journal of natural sciences research. 3.
5. Kurt R (2003) Scrap tyre recycling. Accessed online on February 26, 2015.
6. Scrap Tyre and Rubber Users Directory (2002). Recycling Research Institute.
7. US scrap tyre markets (2001) Rubber Manufacturers' Association.
8. Kihara D (2004) http//www.calrecycle.ca Advisories.gov/LEA/. Accessed online on February 20, 2015.
9. Qui CAI, Long M, Lui J, Zhu M, Zhou Q-zhen, et al. (2008) Correlation between Heavy Metals Concentration in Cattle Tissues and Rearing Environment. Chinese journal of Ecol 27: 202-207.
10. Leita L, Enne G, De Nobili M, Baldini M, Sequi P (1991) Heavy metal bioaccumulation in lamb and sheep bred in smelting and mining areas of S.W. Sardinia, Italy. Bull Environ Contam Toxical 46: 887-893.
11. Personal Communication (2014) Gunning Road Roasting Zone, field survey carried out in Abakaliki Metropolis, Ebonyi State.
12. Anon (2012) http://lexbirchviewsoftheworld.wordpress. Accessed online on February 20, 2015.
13. Sivacoumar R, Bhanarkar AD, Goyal SK, Gadkari SK, Aggarwal AL (2001) Air Pollution Modelling for an Industrial Complex and Model Performance Evaluation. Journal of Environmental Pollution 111: 471-477.
14. Schwartz ND (1994) Emissions of air pollutants ICI Ltd, Runicom.
15. Ramzan N, Degenkolbe S, Witt W (2008) Evaluating and Improving Environmental Performance of HC's Recovery System: A case Study of Distillation Unit. Journal of Chemical Engineering 140: 201-213.
16. Nwite JN, Nnabo PN, Nnoke FN (2007) Essentials of agricultural climatology and biogeography, SNAAPS Publishers Ltd, Enugu Nigeria. 1-25.
17. Wong TW (2002) Health Effects of Air Pollutants. Department of Community and Family Medicines, Chinese University Press, Hong Kong.
18. United States Fire Administration (USFA) (1999) Report on Tyre-fires submitted to Committee on Commerce, Science and Transportation, United States Senate and Committee on Science, United States. House of Representative 12 -14.
19. Okiel W, Ogunlesi M, Alabi F, Osiughwu B, sojinrin A (2009) Determination of toxic metal concentrations in flame treated meat products. African journal of Biochemistry Research 3: 332-339.
20. Jayasekara S, Samarajeewa U, Jayakody AN (1992) Trace metals in foods of animal origin in Srilanka. ASEFAN Journal of food 7: 105-107.
21. Adam I, Okyere D, Teye M (2013) Assessment of Heavy Metal Residues in hides of goats singed with tyres, and the effect of boiling on the heavy metal concentrations in the hides. Journal of veterinary advances 3: 165-169.
22. US-EPA (1985) Standard for six pollutants. Environmental Quality Washington DC.

Evaluation of TSS, BOD5, and TP in Sewage Effluent Receiving Sambul River

Rono AK*

Department of Biological Sciences, Moi University, Kenya

Abstract

Proper treatment of sewage to remove suspended solids, organic matter and nutrients is necessary before the effluent is discharged into aquatic bodies. The major environmental concern at Sambul river is the potential pollution by effluent received from sewage treatment plant in Moi University. In this study, evaluation of the Sambul river was analysed using key water quality parameters, and comparisons made with those of the National Environmental Management Authority (NEMA). Purposive sampling design was used to select three sampling points; outlet where sewage effluent from STP is discharged into Sambul river, downstream of Sambul river where bio-treated sewage has mixed with river Sambul waters, and upstream (control) of Sambul river where the river water does not mix with STP effluent. A total of 72 water samples were collected for this study. At each sampling point, triplicate water samples for analysis were collected bi-weekly in sterilized plastic bottles from May to August 2015. Total suspended solids (TSS), BOD_5 and total phosphorus (TP) were determined using standard laboratory methods. Results showed that levels of TSS, BOD_5 and TP were within the acceptable standards of NEMA. Upstream parameters showed no significant differences with those of downstream (BOD $F_{4,115}$=38965.46; p=0.9734; TSS $F_{4,115}$=708.50; p=0.9999; total phosphorus $F_{4,115}$=2107.17; p=0.9972), indicating that there is no change in water quality at river Sambul as a result of effluent received from STP.

Keywords: Pollution; Sewage effluents; Sambul river

Introduction

Aquatic ecosystems in developing countries are polluted through indiscriminate release of untreated or improperly treated wastewater. Sources of **wastewater include** surface runoff, domestic or sewage effluents, industrial discharges and agricultural activities. Wastewater must be treated effectively to prevent pollution of the receiving environment and to safeguard public health. Numerous nutrients, solids, and other contaminants exist in raw sewage hence treatment of sewage is thus vital before effluent is released to the ecosystem [1].

In many developing countries poor sanitation condition is widespread. In 2012, only 30% of the population in sub-Saharan Africa had access to improved sanitation [2]. The main improved sanitation systems such as pit latrines and toilets dominated in many countries whereas systems that safeguarded the collection, transportation and proper treatment of wastewater remained very low. In Kenya the available good quality water is presently projected at 650 m^3 per year per capita and could drop to about 350 m^3 per year per capita by the year 2020 due to pollution and drought factors [3]. Water scarcity in Kenya slows development hence the need for water saving and regulation policies. Sewage effluent discharges must meet environmental standards to ensure environmental protection. In Kenya, wastewater discharge permits are issued by the National Environmental Management Authority (NEMA) in accordance to the Environmental Management and Co-ordination [4]. NEMA coordinates and publishes regulations on all matters relating to the environment.

The major environmental concern of Sambul river is the potential pollution from effluent Moi received from Moi University sewage treatment plant. At the treatment plant wastewater undergoes biological treatment in stabilization ponds [5]. The receiving Sambul river could be contaminated through nutrient loading which is likely to lead to eutrophication and algal blooms. Furthermore possible pollution of Sambul river will reduce its water quality to harmful levels for aquatic life and riparian population downstream [6]. The physicochemical and biological characteristics of sewage effluent must be within the set environmental levels for discharge provided by NEMA for the effluent receiving environments and public health protection.

Total suspended solids are solids in water that can be trapped by a filter including decaying plants, silt, animal matter, industrial wastes, and sewage. In general, raw wastewater comprises of solids of different types and sizes. The presence of excess suspended solids is harmful to the health of a water body consequently affecting aquatic life. TSS values are used to monitor and assess efficacy of wastewater treatment plants to guarantee the health of effluent receiving water bodies. Discharge of settleable solids to water a body increases sedimentation rates and often destroy and alter habitats for aquatic organisms. Release of sewage effluents with excess suspended solids to rivers and other aquatic systems can increase levels of pathogens and contaminants including bacteria, protozoa, nitrates and phosphorus, pesticides, mercury, lead and other metals [7]. Pathogens can attach to suspended materials and increase the risk of disease outbreaks. High levels of TSS prevents penetration of light to the submerged plants leading to reduced rates of physiological processes such as photosynthesis and respiration in aquatic organisms. The inability of plants to photosynthesise will eventually lead to death and the decaying process that demands more oxygen use from water.

Biodegradable organic materials such as proteins, carbohydrates and fats can cause pollution if discharged into aquatic environments. Large amounts of biodegradable materials are hazardous to aquatic

***Corresponding author:** Ali Kipchumba Rono, Department of Biological Sciences, Moi University, Kenya
E-mail: alikipchumba@gmail.com

bodies because microorganisms will initiate biochemical reactions by utilising dissolved oxygen in the water to breakdown the wastes. The amount of oxygen organisms need to breakdown wastes in wastewater is referred to as the biochemical oxygen demand (BOD). The BOD test measures the biodegradable fraction of the wastewater by monitoring the assimilation of organic material by aerobic microorganisms and therefore a suitable indicator of treatment efficiency. If effluent with high levels of BOD is discharged into a stream or river, it will accelerate bacterial growth in the river and consume the oxygen levels in the river.

Phosphorus exists in sewage water in forms that include soluble organically-bound phosphate and Inorganic phosphates. The organically-bound phosphate in wastewater is from human excreta and urine. Effluent with excess phosphorus is the main cause of eutrophication in surface water bodies such as lakes and rivers [8]. In wastewater treatment phosphorus remains attached to sediments. Consequently, phosphorus removal in wastewater treatment plants must be enhanced to prevent its release to the environment [9]. In wastewater treatment plants phosphorus in organic materials is finally oxidized to phosphates. This study investigated changes in the quality of effluent received from sewage treatment stabilization ponds in Moi University by direct measurement of TSS, BOD_5 and TN and comparison with NEMA standards. The results of the study are expected to be useful to Sambul community and Moi University by offering information on the effectiveness of sewage treatment plant in removing pollutants from its effluent.

Materials and Methods

Study area

The study was conducted at Sambul river and Moi University, situated in Uasin Gishu County, Kenya (Figure 1). The sewage treatment plant at the University performs biological treatment of sewage through a series of ponds before discharging effluents to Sambul river.

Experimental design and sampling

Moi University sewage treatment plant (STP) (Figure 2) provides biological treatment of wastewater. The waste stabilization ponds constructed are simple to operate and maintain. Routine tasks comprises of cutting the embankment grass, removing scum, and any other floating vegetation from pond surface, keeping the inlet and outlet channels clear and repairing any damage to the embankments.

The STP at Moi University receives and treats domestic wastewater from the entire premises. The mean inflow of influent into the treatment plant is 3,200 m^3/day. STP Inlet (Figure 2) is the point where wastewater is received at the treatment plant. Screening of wastewater occurs at this point. Wastewater flows through a coarse screen (size opening 50 mm), where all the floatable solids are trapped for removal. A fine screen (size opening 25 mm) is also fitted downstream of the coarse screen to trap materials that passes through the coarse screen. If the coarse screen in the channel is blocked due to excessive solids in the wastewater, the level of wastewater rises in the channel and wastewater starts flowing through the emergency bypass channel, which has another coarse screen (size opening 75 mm).

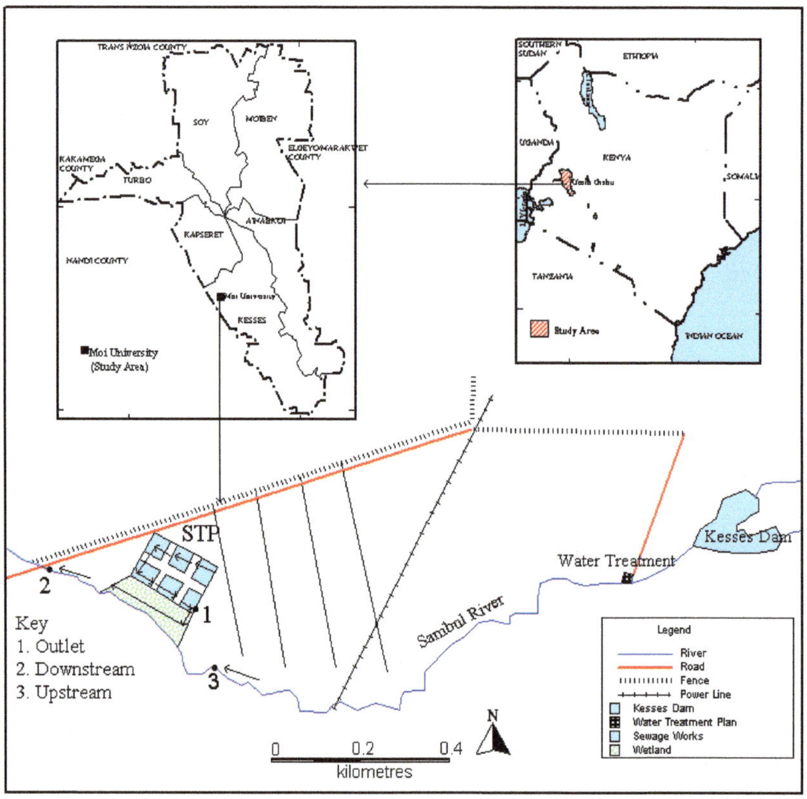

Figure 1: Map showing study sites at river Sambul (by Kanda, Geographic Information Systems Laboratory, MOI University, 2015).

From the inlet channel, wastewater flow into stabilization ponds which are six in number. There are two anaerobic ponds, primary and secondary facultative ponds, and two maturation ponds (Figure 2). An anaerobic pond (width 51 m, length 51 m and depth 3 m) allows for sewage settling, some BOD removal and anaerobic digestion of solids. Anaerobic breakdown occurs when solids settle at the bottom and are eliminated as bacteria decompose organic matter. In addition, organic nitrogen is hydrolysed to ammonia in anaerobic ponds. From the anaerobic pond, wastewater flows into the primary facultative pond (width 80 m, length 220 m, and depth 1.7 m). Phosphorus and nitrogen removal occur in facultative ponds. The primary facultative pond provides some anaerobic digestion in the benthic zone, and aerobic respiration at the water surface. The primary pond discharges wastewater into the secondary facultative pond (width 70 m, length 220 m, and depth 1.7 m) where wastewater is aerobically treated.

Maturation ponds (each width 70 m, length 70 m, and depth 0.5 m) receive effluent from the secondary facultative pond to polish the wastewater. In this ponds pathogens die as the levels of suspended solids is low and pathogens are exposed to ultra violet radiations and perish and remaining BOD is removed by heterotrophic bacteria. Maturation ponds discharge wastewater into the outlet channel where effluent is disinfected by the charcoal and gravel deposits along the channel. From the outlet channel, wastewater effluent is discharged into the wetland. The effluent released into the ecosystem has undergone biological treatment in the stabilization ponds within the sewage treatment plant. At the wetland more nutrients are absorbed by the plants through bio filtration thus reducing the nutrient load. From the wetland wastewater effluent is then discharged to Sambul river. The quality of effluent discharged into river Sambul must be within the standards outlined by environmental agencies such as National Environment Management Authority (NEMA).

Sampling points for this study were selected using a purposive design. This generated water samples that were used to evaluate changes in TSS, BOD_5 and TN at Sambul river. Three sampling points (Figure 1) were selected as follows; sampling point 1 (SP1), which is the outlet where effluent is released from the sewage treatment plant into the Sambul river. At this point, effluent has undergone biological treatment. Sampling point 2 (SP2), which is at downstream of Sambul river where biologically treated effluent is mixed with waters from the Sambul river. The last point was sampling point 3 (SP3), which is upstream of Sambul river a point before the Sambul river water mixes with effluent from the sewage treatment plant. This point served as a control, reflecting the most naturally preserved conditions of river ecosystem without the influence of the sewage effluent. Wastewater samples at each of the three (SP1-SP3) sampling points were collected twice every month from May to August 2015. Wastewater samples were collected in 1 L sterilized plastic bottles. During each sampling period, sample water was used to rinse sample bottles before collection. Triplicate samples were collected from each sampling site. All samples were the labelled, iced and taken to the laboratory for analyses.

Determination of total suspended solids, biochemical oxygen demand (BOD_5) and total phosphorus (TP)

Gravimetric method was used to determine TSS in water samples [10], and the following equation was then used to calculate TSS (mg/L) in a sample.

$$TSS(mg/L) = \frac{W1 - W2}{V}$$

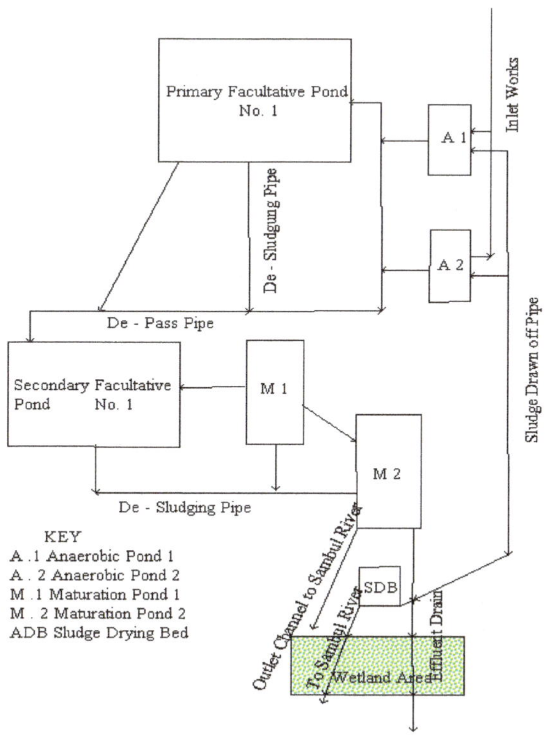

Figure 2: MOI University sewage treatment plan schematic flow diagram (by Kanda, Geographic Information Systems Laboratory, Moi University, 2015).

Where,

W1=mass of dried residue and filter paper in grams.

W2=initial weight of filter paper

V=volume of sample water taken

The BOD_5 (mg/L) of water samples was determined by getting the variation between dissolved oxygen (DO) on day one and day five. Calculation for BOD_5 (mg/L) was then determined by the following formula:

$$BOD(mg/L) = \frac{D1 - D2}{P}$$

Where,

D1=DO (mg/L) value in initial sample

D2=DO (mg/L) value in final sample

P=decimal volumetric fraction of sample used (ml of sample/300 ml).

Wastewater Total Phosphorus (TP) (mg/L) concentrations were determined using potassium per sulphate digestion followed by ascorbic acid procedure [11].

Data analysis

Data from this study was analysed using the Statistical Package for Social Science (SPSS version 22 then one way Analysis of Variance (ANOVA) was used to determine significant differences. Contaminants removal efficiencies from the outlet to downstream sampled points were determined for the tested parameters in percentages.

Results

Samples data for the study period are shown in Tables 1-3 for the outlet, downstream and upstream studied points respectively. Total Suspended Solids (TSS) mean values for the different sampled points ranged from 2.63 ± 0.95 to 22.94 ± 1.98 mg/L (Table 4). The mean TSS for upstream (control) was not significantly different from that of downstream ($F_{4,115}$=708.50; p=0.9999) but was significantly different from that of the outlet ($F_{4,115}$ = 708.50; p=0.5068). There was 79.56% TSS removal at the downstream sampled point (Table 4). The mean value for BOD_5 at upstream (control) was not significantly different from that of downstream ($F_{4,115}$=38965.46; p=0.9734) but was significantly lower than that of outlet ($F_{4,115}$=38965.46; p<0.0001) (Table 4). At downstream, BOD_5 removal efficiency was 87.00%. Total Phosphorus values for the different sampled points ranged from 0.04 ± 0.01 to 2.02 ± 0.04 mg/L (Table 4). Mean total phosphorus level for upstream (control) was significantly lower than that of outlet ($F_{4,115}$=2107.17; p<0.0001) but showed no significant differences from that of downstream ($F_{4,115}$=2107.17; p=0.9972) (Table 4). TP removal efficiency was 95.05% at the downstream sampled point (Table 4).

Discussion

Sewage effluent discharges into aquatic bodies have continued to be a major concern for governments and environmentalist around the world. This is agitated by the physicochemical and microbial content in sewage that are used as indicators of environmental degradation in the effluent receiving environment [6]. Environmental regulations on effluent discharge have emerged currently to ensure that the impact of sewage effluents is greatly reduced. In Kenya the NEMA formulates and issues environmental standards to be adhered to in sewage treatment plants across the country. The findings of this study were within the allowable limits by NEMA, subsequently ensuring environmental health protection [12,13]. There were high levels of BOD_5 at the outlet (17.99 mg/L) was higher than at downstream (2.34 mg/L). The BOD_5 mean values at outlet were within the NEMA allowable standard of 30 mg/L. This implies that the observed concentration of BOD at outlet sampled point would not affect water quality at the effluent receiving Sambul river. Based on the reduced BOD outlet value, the sewage treatment plant at Moi University is relatively efficient in treating

Date/parameter	14/5/15	29/5/15	11/6/2015	25/6/15	9/7/2015	23/7/15	7/8/2015	21/8/15
TSS	22.45	23.98	22.74	22.87	22.67	22.96	22.95	22.89
BOD_5	16.98	17.89	18.02	19.04	17.67	17.88	17.76	17.9
TP	1.98	2.06	2	2.02	2.03	2.01	2.04	2.01

Table 1: Data for outlet sampled point May to August 2015.

Date/parameter	14/5/15	29/5/15	11/6/15	25/6/15	9/7/2015	23/7/15	7/8/2015	21/8/15
TSS	3.89	4.56	4.67	4.78	4.98	4.84	4.65	4.12
BOD_5	2.42	2.35	2.44	2.36	2.56	2.22	2.32	2.46
TP	0.04	0.03	0.02	0.06	0.03	0.05	0.04	0.01

Table 2: Data for downstream sampled point May to August 2015.

Date/parameter	14/5/15	29/5/15	11/6/15	25/6/15	9/7/2015	23/7/15	7/8/2015	21/8/15
TSS	2.56	2.67	2.87	2.42	2.65	2.61	2.66	2.63
BOD_5	1.02	1.06	1.05	1.03	1.04	1.05	1.04	1.05
TP	0.02	0.03	0.01	0.01	0.04	0.05	0.04	0.05

Table 3: Data for upstream sampled point May to August 2015.

Parameter	Outlet (O)	Downstream (D)	Upstream (U)	Removal % increase or decrease (O/D)	NEMA allowable limits
TSS	22.94 ± 1.98a	4.69 ± 0.86b	2.63 ± 0.95b	79.56	30
BOD	17.99 ± 1.02 a	2.34 ± 0.07 b	1.06 ± 0.04 b	87	30
TP	2.02 ± 0.04a	0.05 ± 0.01b	0.04 ± 0.01b	95.05	2 Guideline value
Mean (± SEM) in the same row followed by the same alphabet letter are not significantly different at p=0.05					

Table 4: levels of selected physicochemical parameters and NEMA standards.

sewage. In addition, the removal efficiency of BOD at the downstream point was 87.00% implying that the treatment plant does not increase the pollution of the river. The reduced levels of BOD_5 at outlet also explain the functioning of the sewage stabilization pond [5].

The total suspended solids (TSS) for downstream (4.69 mg/L) was significantly lower than that of inlet (22.94 mg/L). The outlet value was within the NEMA standard of 30 mg/L. The reduced TSS levels at the outlet could be attributed to the sewage treatment plant TSS removal efficiency through the sedimentation of settleable solids. TSS removal of 79.56% was observed in the study implying that the treatment plant is efficient in ensuring that impact of effluents at the receiving Sambul river is reduced and that the riparian community as users of river Sambul is safeguarded [3]. Discharge of effluents with high levels of suspended solids into aquatic bodies lowers water quality and depletes dissolved oxygen available for aquatic life.

The removal of nutrients in sewage effluents is necessary to reduce the effects on the receiving water bodies [14]. The study findings showed Total phosphorus (TP) mean value at the outlet (2.02 mg/L) was higher than that of downstream (0.05 mg/L). The removal efficiency of 95.05% of TP at downstream indicates the relative efficiency of Moi university sewage treatment plant and the study is similar to that by Waiser et al. [15]. The reduced levels of TP at downstream imply that no eutrophication could occur in Sambul river that would significantly deteriorate water quality and hinder aquatic life metabolic and physiological processes.

Conclusion

Treatment of sewage effluents before discharge into aquatic bodies is vital to prevent pollution and protect the entire environment and public health. From this study, it was concluded that effluent discharged to Sambul river from Moi University sewage treatment plant is within the allowable limits of NEMA. Based on the reduced levels of the TSS, BOD_5 and TN at downstream Sambul river, it showed that the sewage treatment plant is relatively efficient in reduction of organic and nutrient loads from the sewage water. This way, effluents with reduced contaminants if released into the ecosystem are incapable of harming aquatic life, declining water quality and causing disease epidemics to the riparian communities downstream.

Author's Contributions

This work was carried out by the author. The author designed the study, wrote the protocol and interpreted the data, anchored the field study, gathered the initial data, performed preliminary data analysis, managed the literature searches and produced the initial draft. The author read and approved the final manuscript.

Acknowledgements

Our gratitude goes to Moi University for financial support and for permitting this study and for provision of laboratory facilities.

References

1. Nzuki S (2008) Initial environmental audit report for Kipevu/west mainland sewage treatment works and sewer system. pp: 170-180.
2. World Health Organization (WHO) and United Nations Children Fund (UNICEF) (2012) Progress on drinking water and sanitation: 2012 Update.
3. Kaluli WJ, Githuku C, Home P, Mwangi BM (2011) Towards a national policy on wastewater reuse in Kenya. J Agricult Sci Technol 13: 116-125.
4. Environmental Management and Co-ordination Act (1999) Laws of Kenya.
5. Navaraj PS (2005) Anaerobic waste stabilization ponds: A low cost contribution to a sustainable wastewater Reuse Cycle.
6. Malik OA, Hsu A, Johnson LA, De Sherbinin A (2015) A global indicator of wastewater treatment to inform the Sustainable Development Goals (SDGs). Environ Sci Policy 48:172-185.
7. Giller PS, Malmqvist B (1998) The Biology of Streams and Rivers. Biology of Habitats Series, Oxford University Press, pp: 231-235.
8. McHale MR, McChesney D (2007) Phosphorus concentrations in stream-water and reference samples-An assessment of laboratory comparability. US Geological Survey Open-File Report 1267.
9. Dates G (1994) Monitoring for phosphorus. Volunte Monit 6: 39.
10. American Public Health Association (APHA) (1998) Standard methods for the examination of water and wastewater. (20thedn), American Water Works Association/ Water Environmental Federation, Washington DC, pp: 1287.
11. Andrew DE, Lenore SC, Arnold EG (1995) Standard methods for the examination of water and wastewater. (19thedn).
12. Jagai JS, Wang QSL, Messier KP, Wade TJ, Hilborn ED (2015) Environmental Health Perspectives 110: 174.
13. Tsuzuki Y (2012) Linking sanitation and wastewater treatment: from evaluation on the basis of effluent pollutant concentrations to evaluation on the basis of pollutant removal efficiencies. Water Sci Technol 65: 368-379.
14. Carey RO, Migliaccio KW (2009) Contribution of wastewater treatment plant effluents to nutrient dynamics in aquatic systems: a review. Environ Manage 44: 205-217.
15. Waiser MJ, Tumber V, Holm J (2011) Effluent-dominated streams. Part 1: Presence and effects of excess nitrogen and phosphorus in Wascana Creek, Saskatchewan, Canada. Environ Toxicol Chem 30: 496-507.

Comparison of Dietary Cadmium Exposure among the General Population from Two Cadmium-polluted Regions in China

WenLi Zhang[1]*, Jiangang Zheng[2], Xuemei Han[1], Dancheng Yao[3], Jingxiu Han[1] and Qi Shang[1]

[1]*Department of Environmental Epidemiology, Institute for Environmental Health and Related Product Safety, Chinese Center for Disease Control and Prevention, Beijing, 100050, China*

[2]*Jiangxi Provincial Center for Disease Control and Prevention, Nanchang, 330046, China*

[3]*Guizhou Provincial Center for Disease Control and Prevention, Guiyang, 550004, China*

Abstract

Background: In China, Dayu County is famous for tungsten, lead–zinc mines that are abundant in Hezhang County. Cadmium (Cd) is a concomitant pollutant with tungsten and lead–zinc mining. The most severe Cd pollution occurred from the 1970s until 2006. Large-scale mining stopped after 2006 in the two counties; however, the long-term mining led to soil Cd pollution in these areas.

Aim: The study aimed to investigate Cd concentration in crops and dietary exposure among the general population in two Cd-polluted regions.

Methods: Local self-seeded grain and vegetable samples were collected and analyzed for Cd content with atomic absorption spectrometry. A three-day dietary interview was carried out among local residents to estimate diet Cd intake.

Results: In total, 957 and 281 crop samples were collected. Three-day diet surveys were conducted among 546 residents and 262 residents, respectively from Hezhang and Dayu Counties. Higher Cd content of rice (0.536 mg/kg) and a ratio of rice to Cd content (71.69%) over the standard limit were found in Dayu, whereas a higher Cd content was found in leafy vegetables (0.386 mg/kg) and ratio of leafy vegetables to Cd content (51.40%) over the standard limit in Hezhang. Rice is the main food consumed and it is a major source of dietary Cd intake for local residents in Dayu; however, maize is the main food consumed and leafy vegetables are the major source of dietary Cd intake for local residents in Hezhang. For the two counties, both average daily Cd intake and cumulative Cd intake of residents in the polluted areas exceeded the limitation of Cd from food, and the consumption is higher in Dayu.

Conclusion: Cd exposure through food is very high among the general population in the two regions. Different measures should be taken in different regions. Also, close attention should be paid to the health of the local residents.

Keywords: Cadmium pollution; Dietary intake; Cadmium intake; Food chain

Introduction

Cadmium (Cd) does not break down in the environment; it can only transfer between its various forms. It can accumulate in soils. Excess Cd in soil can be taken up by plants or crops, and it accumulates in human body via the food chain. Further, because it has a long biological half-life ranging from 10 to 30 years in the human body, it can be highly dangerous to human health [1-3]. The issue of Cd in food has been a worldwide concern since "Itai-itai" disease occurred in Japan in the 1950s. In addition, the Joint FAO/WHO Expert Committee on Food Additives has confirmed that Cd is one of the major food contaminants, and they have formulated maximum levels for Cd to assure food safety [3]. Cd levels in plants are affected by soil properties and crop type; soil pH is the major factor influencing plant uptake of Cd from soils; soil type also affects uptake of Cd by plants. The Cd content of agricultural crops varies according to species and plants grown in the same soil. Accumulation of Cd decreases in the following order: leafy vegetables > root vegetables > grain crops [3-5].

Tungsten resources are abundant in Dayu County, Jiangxi Province, China, and Cd is a contaminant with tungsten ore mining. Industrialized tungsten mining since the 1960s has released wastewater with Cd from the ore dressing directly into the river. Irrigation with Cd-polluted river water has resulted in Cd pollution in soils. Dayu is a rice-growing area, where rice fields account for more than 95% of the total cultivated land area [6]. Lead–zinc mines that are abundant in Hezhang County, Guizhou Province, China, and Cd is concomitant with lead–zinc mining. Mineral waste from artisanal zinc smelting was piled in the farmland, and thus the soil was polluted with Cd [7]. The soil of Hezhang is suited for dryland agriculture, and the main crop is maize. Maize production accounts for about 55% of the county's total grain, and rice cultivation is rare. The two counties share some commonalities. Both have had a similar mining history for almost a hundred years, and for both, the most severe pollution occurred from the 1970s until 2006. Large-scale mining stopped after 2006 in the two counties, but the long-term mining exploitation led to significantly higher Cd content in regions' soil [7]. Because of the different means of pollution, soil properties, and crop species, the Cd enrichment of crops will differ between the two counties. To compare the characteristics of Cd accumulation in crops and assess residents' exposure risks between the two different Cd-contaminated environments, two field surveys were carried out separately on Cd concentration of crops and diet

***Corresponding author:** Wenli Zhang, Department of Environmental Epidemiology, Institute for Environmental Health and Related Product Safety, Chinese Center for Disease Control and Prevention, 29 Nanweilu, Xicheng District, Beijing, 100050, China, E-mail: zhangwlcdc@126.com

exposure. The survey in Dayu was done in April 2007. The survey in Hezhang was done in April 2009. The Cd intake through ingestion was evaluated among the general population in the two polluted regions, and the results provide data for the assessment of human health hazards of Cd pollution for different environments.

Materials and Methods

Study population

There are 11 townships in Dayu County. According to the natural environmental conditions and lifestyle, townships irrigated by waste water that had rice Cd concentration ≥ 0.2 mg/kg were selected as the Cd-exposed area, whereas areas not irrigated by waste water and with a rice Cd concentration < 0.2 mg/kg were regarded as the control area. There are 27 townships in Hezhang County, and based on monitoring data from soil, the soil Cd content is in a range from 1.09 to 6.71 mg/kg in the county. The townships with highest Cd content in the soil were selected as the Cd-exposed area. The townships with the lowest Cd content in the soil were selected as the control area. We selected one exposure and one Control Township from each of the two counties, Dayu and Hezhang.

Cluster random sampling was used to choose subjects from the selected townships, and age range spanned from childhood to old age to represent different physiological stages, except for the youngest age group, which is sorted by the interval of 5 years; the interval between the other age groups is 10 years. Males and females were evenly distributed in each age group.

Dietary data collection

A dietary recall survey was conducted only in contaminated areas. A three-day inquiry method and a food frequency questionnaire were used in the survey. In total 262 and 515 residents were followed for three days in Dayu and Hezhang, respectively, and food consumption data were collected.

Sample collection and test

Locally produced foods grown in the fields surrounding the village were collected. Samples were collected from the subjects' households and stored in sealed polyethylene bags (colorless and transparent), and then sent to the lab quickly to detect Cd content. Foodstuff was sorted into two main categories: cereals and vegetables. Rice, wheat, and maize are the main cereals in China. Vegetables are divided into three subgroups: stem and root vegetables, leafy vegetables, and fruit vegetables, based on the China National Nutrition and Health Survey Workbook [8]. In total, 117 samples from one village in the exposure area and 94 samples from three villages in the control area were collected in Dayu County, whereas 735 samples from ten villages in the exposure area and 86 samples from three villages in the control area were collected in Hezhang County. Sampling site is showed in Figure 1. The Cd content in food was determined by atomic absorption spectrometry (AAS), and the results of Cd content in food were evaluated according to Chinese National Standards [9,10].

Quality control

Standard substances were used throughout the detection procedure for quality control. The Cd standard for rice (rice flour, GBW (E) 080684) and vegetables (tomato leaf, ESP-1) was obtained from the Center of National Standard Research of China. Relative error of Cd content in a standard substance between true value and measured value was in the range of 0.5–5.6%. The detection limit of AAS was 0.1 μg/l. No subject had responses below the detection limit.

Statistic method

Statistical analyses were performed using SPSS 15.0. Measurement data were described using means and standard deviation and were analyzed using the *t*-test to compare the means. Categorical data were described as the proportion and analyzed with the chi-square test. For all analyses, a P value < 0.05 was regarded as statistically significant.

Results

Cadmium content in food

Results of Cd content in food are displayed in Table 1. All vegetable results are given in fresh weight except for rice and maize, which are in dried weight. According to FAO/WHO, maximum levels (MLs) for rice, leafy vegetables, stem and root vegetables, and fruit vegetables are 0.4 mg/kg, 0.2 mg/kg, 0.1 mg/kg, and 0.05 mg/kg, respectively (FAO/WHO, 2014). Chinese National Standards on foods contaminants (GB2762-2005) are similar with the MLs except for rice (0.2 mg/kg), so we used the Chinese National Standards to assess the result of Cd content in food. The mean of Cd content in food is over the MLs of the national standard in the Dayu exposure area. Further, Cd content decreased in the following order: rice > root vegetables > leafy vegetables > fruit vegetables, whereas there are a few samples of fruit vegetables in which the Cd content is over the MLs in the Dayu control area, and there is significant difference for both mean and exceeding ratio comparison between exposure and control in the Dayu region. As far as the Hezhang exposure area is concerned, food Cd contents are also over the MLs; however, the decreasing order is different from Dayu: leafy vegetables > root vegetables > maize > fruit vegetables, whereas Cd content in fruit vegetables is a little higher in the Hezhang control area, and the means and exceeding ratios are significantly different between exposure and control.

General information on study subjects

The basic characteristics of study subjects are listed in Table 2. Age of subjects ranged from 2 to 75 years old, and gender was evenly distributed among age groups. We tested by ANOVA and chi square

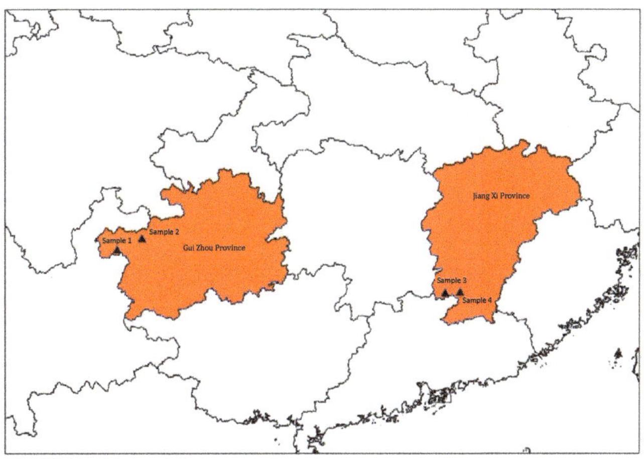

Note: sample 1 and 2 stand for pollution and control respectively at Hezhang County, Gui Zhou Province sample 4 and 3 stand for pollution and control respectively at Dayu County, Jiang Xi Province.

Figure 1: Sampling site in the study.

Region	Type	Exposure area			Control area		
		N	mean ± SD	Exceeding ratio (%)	N	mean ± SD	Exceeding ratio (%)
Dayu	rice	53	0.536 ± 0.507*	71.69a	44	0.072 ± 0.040*	0a
	Stem and root vegetable	30	0.127 ± 0.171*	36.67a	16	0.011 ± 0.011*	0a
	leafy vegetable	24	0.119 ± 0.176*	20.83a	24	0.021 ± 0.023*	0a
	fruit vegetable	10	0.027 ± 0.016*	10	10	0.012 ± 0.015*	10
Hezhang	maize	158	0.137 ± 0.133*	55.06a	21	0.025 ± 0.023*	0a
	Stem and root vegetable	249	0.211 ± 0.184*	82.33a	20	0.039 ± 0.027*	0a
	leafy vegetable	286	0.386 ± 0.698*	51.40a	10	0.050 ± 0.025*	0a
	fruit vegetable	42	0.130 ± 0.083	59.52	35	0.111 ± 0.084	45.71

N: number of samples collected
*: mean comparison between exposure and control group within same region P<0.05.
a: Ratio comparison between exposure and control group within same region P<0.05.

Table 1: Cadmium content of locally produced food in the two regions.

Age group	Dayu region				Hezhang region			
	Male		Female		Male		Female	
	N	%	N	%	N	%	N	%
2～	11	8.4	5	3.8	20	8	14	5.3
6～	22	16.8	19	14.5	45	18.1	50	18.8
15～	43	32.8	44	33.6	39	15.7	30	11.3
25～	9	6.9	17	13	22	8.8	31	11.7
35～	14	10.7	12	9.2	39	15.7	44	16.5
45～	10	7.6	11	8.4	28	11.2	37	13.9
55～	12	9.2	12	9.2	28	11.2	31	11.7
65～	10	7.6	11	8.4	28	11.2	29	10.9
Total	131	100.0	131	100.0	249	100.0	266	100.0

N: number of subjects.

Table 2: Age and gender distribution of the study population.

test and found no significant difference in mean age and age ratio between males and females within the same region (Dayu is F=0.172 and X^2=5.192; Hezhang is F=0.568 and X^2=5.186). We also compared the age ratio between Dayu and Hezhang, and the results show a difference in age structure between Dayu and Hezhang (male, X^2=16.333, P=0.022, F=3.730, P=0.054; female, X^2=31.526, P=0.000, F=4.814, P=0.029). As shown in Table 2, the 15-year-old group had a larger proportion in the Dayu region, which is the reason for the difference in age structure between Dayu and Hezhang. When we compared age structure between Dayu and Hezhang without the 15-year-old group, the result for males was X^2=1.512, P=0.959, F=0.469, P=0.494, and X^2=3.055, P=0.802, F=0.154, P=0.695 for females, so the difference existed only in the 15-year-old group. Overall, there was no difference in age distribution either between males and females or between Dayu and Hezhang, except in the 15-year-old age group.

Dietary consumption of the study population

Average total daily dietary consumption between genders differs among age groups. Total daily dietary consumption differs significantly for overall comparison of all ages in Dayu, however, no significant difference was found in Hezhang. t-tests show no statistically significant difference between males and females within any age group, except in the 6-year-old age group and 15-year-old age group in Dayu. Overall, the main feature of average total daily dietary consumption is the difference among different age groups, and there is little difference based on gender.

Figure 2 displays average total daily dietary consumption in a graph. The 2-year-old age group had the lowest consumption amount, and there was an upward trend with increasing age, with consumption amount reaching the highest levels in the 25- and 35-year-old age groups. Further, the 45-year-old age group showed a slow downward trend. The amount of average total daily dietary consumption was higher for males than that for females, with each age group in each region, whereas it was lower among people in Dayu compared with those in Hezhang. The average total daily dietary consumption rates are 450.1 g, 340.6 g, 677.6 g, and 598.5 g for Dayu males, Dayu females, Hezhang males, and Hezhang females, respectively.

Results of dietary cadmium intake

The approach used for estimating the daily intake of Cd is based on consumption and Cd concentration of each type of food. First, the Cd concentrations were multiplied by the amount of intake of each foodstuff to give the Cd intake for each food item that is one day's data for an individual resident. We then used these data to calculate the average Cd intakes of the individual for each food item during the interviewing date. Finally, the average daily Cd intake of each foodstuff in each age group was attained from the average Cd intakes of the individual. The average daily Cd intake multiplied by the survivor days of each age group equals the cumulative Cd intake of each age group. The cumulative Cd intake of a population is the summary of all age

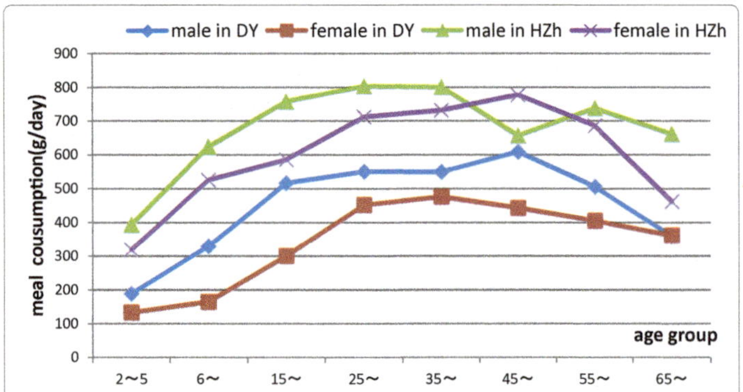

Figure 2: Average total consumption of daily meal for local residents.

groups. The average life expectancy is estimated as 75 years in China. For individuals, mean Cd content of each food type from the same village was used to calculate the Cd intake of each person. Crop Cd is a major source of exposure for most persons living under environmental Cd pollution, and because there are almost no data showing that Cd content of meat is over the standard in these region, meat Cd content is not taken into account in this study.

As seen in Table 3, the 2-year-old age group had the lowest daily Cd intake. Cd intake in the 15-year-old age group in Dayu was somewhat lower than expected. Moreover, excluding the 2- and 15-year-old groups, the average daily Cd intake was above 70 μg/day in all age groups. Average daily dietary Cd intakes of males and females differ from each other at all ages in each region. ANOVA results show that average daily Cd intake is significantly different for overall comparison of all ages in Dayu (F=6.453, P=0.000 for males and F=8.705, P=0.000 for females); however, the differences are not significant for overall comparison of all ages of males and slight for females in Hezhang (F=1.143, P=0.337 for males and F=2.381, P=0.022 for females).

To determine if there were gender differences for average daily dietary Cd, the ANOVA test was used to compare average daily dietary Cd between males and females in each region. Results showed no differences between males and females in either Dayu (F=0.212, P=0.646) or Hezhang (F=0.272, P=0.602). Then we use the t-test to compare average daily dietary Cd between males and females in each age group; there were no statistic differences between males and females within any age group. Overall, the main feature of average daily dietary Cd is the difference among different age groups, and there is little difference for gender.

Figure 3 displays the trend of average daily dietary Cd in different age groups; the features of trend variation are consistent for males and females in Dayu, except for those aged 6 years who had a slightly higher level. We found an upward trend with increasing age, until age 55, when a slow downward trend was shown. The trend variation for average daily dietary Cd in males coincide closely with that in females in Hezhang, and it shows a very slow upward trend with increasing age. Further, the variation between the age groups is small, and it shows a slow downward trend after age 55.

Figure 4 displays the trend of cumulative dietary Cd in different age groups. The trend variation is consistent for males and females in both regions, an upward trend with increasing age. Before age 25, the lines of cumulative dietary Cd for Dayu and Hezhang were very close with each other, and after age 25, it shows a steep upward trend

Age group	Dayu				Hezhang			
	Male		Female		Male		Female	
	A(μg)	C(mg)	A(μg)	C(mg)	A(μg)	C(mg)	A(μg)	C(mg)
2～	78.59	118.14	39.6	79.34	50.46	73.67	29.87	43.61
6～	163.41	594.17	134.13	519.96	87.05	359.63	72.01	280.16
15～	52.17	784.59	35.23	648.53	94.82	705.72	82.62	581.73
25～	76.67	1064.44	113.65	1063.34	114.19	1122.52	91.98	917.45
35～	178	1714.15	195.79	1777.99	108.08	1517.01	97.89	1274.75
45～	286.73	2760.73	214.41	2560.6	91.91	1852.48	112.09	1683.88
55～	196.61	3478.36	215.25	3346.25	106.27	2240.37	97.06	2038.15
65～	139.41	3987.21	182.06	4010.79	103.31	2617.45	70.64	2295.99

Table 3: Average daily Cd intake and cumulative Cd intake from food among residents in the two regions.

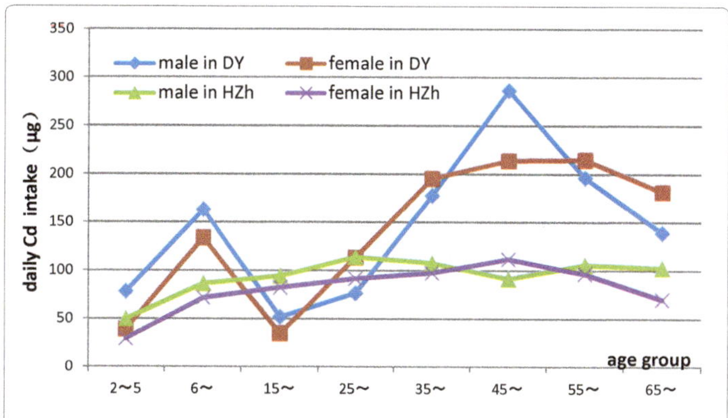

Figure 3: Average daily Cd intake from food among residents.

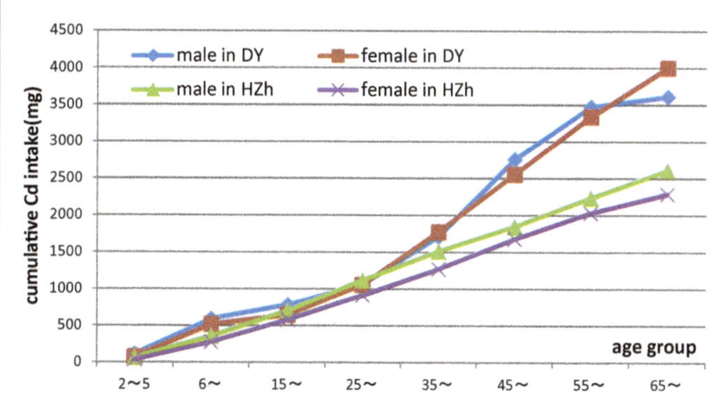

Figure 4: Cumulative Cd intake from food among residents.

in Dayu, whereas it shows a steady upward trend in Hezhang. The line of cumulative dietary Cd for males is almost above that for females, but it is reversed in the 65-year-old group in Dayu. Cumulative lifelong Cd is estimated to be 3617.77 mg and 4010.79 mg for males and females, respectively, in Dayu, and 2617.45 mg and 2295.99 mg for males and females, respectively, in Hezhang.

To find the main factors affecting dietary Cd intake among the populations, the proportions of dietary consumption, as well as the proportion of the dietary Cd intake for each type of foodstuff, were calculated in each age group. Average specific dietary consumption (for specific foods, such as rice or maize) was divided by average total daily dietary consumption equal to the specific dietary proportion in total dietary, and average specific dietary Cd intake (such as rice or maize) was divided by average total daily dietary Cd intake equal to the specific dietary Cd intake proportion in total dietary Cd intake. The overall proportion of dietary consumption, as well as the proportion of the dietary Cd intake for males or females was averaged according to proportion in each age group. Results are shown in Figures 5A and 5B. As seen in the figure, rice has a large proportion of dietary consumption: the proportion is 80% and 83% for males and females, respectively, whereas in Dayu, the contribution of rice Cd to total dietary Cd intake was 96% and 97% for males and females. The proportion of rice Cd intake to dietary Cd intake is higher than the proportion of rice consumption to dietary consumption, which means that rice consumption has made a key contribution to Cd intake, and the high Cd content in rice further promotes the increase of this proportion. Maize is the largest proportion of dietary consumption: the proportion is 26% and 23% for males and females, respectively, in

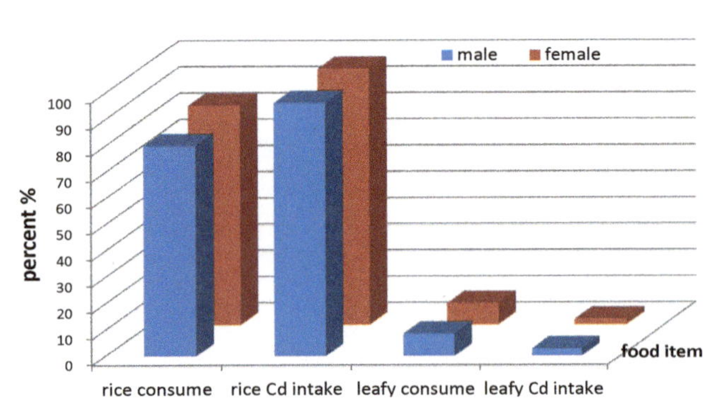

Figure 5A: In Dayu: proprotion of main food in diet and its contribution to total diet Cd intake.

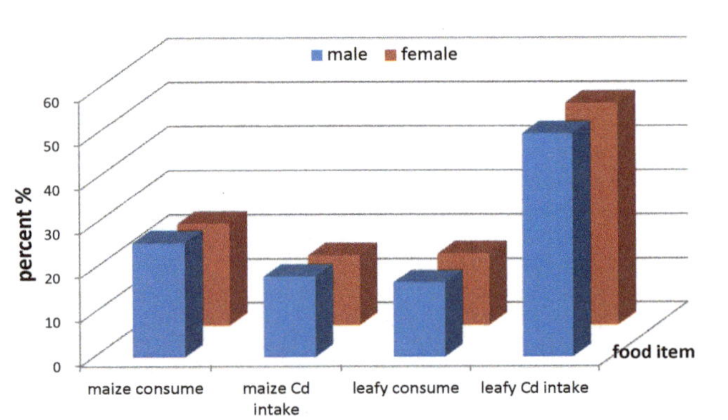

Figure 5B : In Hezhang: proprotion of main food in diet and its contribution to total diet Cd intake.

Hezhang, whereas contribution of maize Cd to total dietary Cd intake is 18% and 16%, for males and females, respectively. The proportion of maize Cd intake to dietary Cd intake is lower than the proportion of maize consumption to dietary consumption, which means that maize consumption has played a key role for dietary consumption contribution but not for the contribution to Cd intake because the low Cd content in maize decreases the proportion. For this reason, leafy vegetables are listed here, but it is obvious that leafy vegetables are not the main dietary consumption; the proportion is 17% and 16% for males and females, respectively, in Hezhang, whereas contribution of leafy vegetable Cd to total dietary Cd intake was 51% and 50% for males and females, respectively. The proportion of leafy vegetable Cd intake to dietary Cd intake is higher than the proportion of leafy vegetable consumption to dietary consumption, which means that although leafy vegetable consumption has not played a key role as a dietary contribution, the high Cd content in leafy vegetables increases its contribution to Cd intake. Overall, the main food for the contribution to Cd intake is rice in Dayu and it is leafy vegetables in Hezhang.

Discussion

Cd exposure estimation is important in accessing the effect of exposure on human health for the population exposed; it has important implications for the prediction of human health hazards and in the prevention of environmental Cd pollution. Food is the most common routes of non-occupational exposure to Cd [11-13]. Our study found cereal and vegetables to have relatively high Cd content. The Cd contents of rice and root vegetables are outstanding in Dayu, while leafy and root vegetables contain relatively high Cd content in Hezhang; The variation feature of average daily dietary Cd intake with age group is similar to that of average total daily dietary consumption, which means that average daily dietary Cd intake and average total daily dietary consumption are correlated with each other closely, more dietary consumption results in more Cd intake. However, the variation of average daily dietary Cd intake with age group in Dayu differs from that in Hezhang. The variation of Cd intake is notable with increasing age in Dayu, but it shows a slowly increasing trend with increasing age in Hezhang. Dietary Cd intakes of residents in different regions of China are quite different; it ranges from 12.0 to 70 μg/day in a no-pollution area, and it is still higher than that in developed countries (10-30 μg/day) [14]. Our study shows that residents in polluted regions take in much more Cd via food in comparison with those living in non-polluted regions in China, and his intake is much higher than that found in European countries. However, our study is similar with the result from a study in a contaminated region of Thailand, where estimated dietary Cd exposures of 188-224 μg/day for males and 99-113 μg/day for females 20-59 years of age who never smoked were reported [15]. A survey in the Jinzu River Basin, in Japan, demonstrated that dietary Cd exposures of most people are 100–200 μg/day [16]. These studies show that higher Cd polluted areas are found mostly in Asia, and that the exposure situations are similar to each other.

The changing trend of cumulative dietary Cd intake is consistent in the two studied regions: cumulative dietary Cd intake increases with age; the changing trend of cumulative dietary Cd intake is similar for males and females within the same region, but the amount of cumulative dietary Cd intake is higher among males than females; cumulative lifelong Cd intakes are estimated to be 3617.77 mg and 4010.79 mg for males and females, respectively, in Dayu, and are 2617.45 mg and 2295.99 mg for males and females in Hezhang. A study from Europe showed that in the non-smoking general population, food accounts for approximately 90% of Cd intake, with the main food commodities contributing to Cd exposure being cereals and vegetables [3,17,18]. The 2002 China National Nutrition and Health Survey showed that Cd contributions from different food groups varied between different regions in China. For the total population, vegetables contributed 27.8% to dietary Cd exposure, followed by rice and flour. For the population in the north, vegetables, flour, and aquatic products, accounting for 34.6%, 27.4% and 13.4%, respectively; In contrast, vegetables contributed 34.4% of the total Cd for the south [14]. Our study agrees with previous studies in China, because Cd polluted regions usually are far from urban areas, so there are further remarkable regional characteristics for dietary Cd intake. Rice intake accounts for 80% and 83% of total dietary consumption, respectively, for males and females, in Dayu, but it is the absolute majority contribution to dietary Cd intake, and it accounted for 96% and 97% among males and females in Dayu. Actually, the food-item contribution to dietary Cd intake we found is very similar to that in the Kakehashi River Basin, which showed that Cd levels in rice were representative of Cd intake, because over 70% of the total Cd intake came from rice (ATSDR) [19]. Although vegetables only make up 17% and 16% of the total dietary consumption for males and females in Hezhang, leafy vegetables are the major contribution to dietary Cd intake, accounting for 51% and 50% among males and females. Therefore, we can draw the conclusion that two major factors affect dietary Cd intake: one is food consumption, and the other is the Cd content in food. Food consumed in larger quantities made the greatest contribution to the dietary exposure. High or low Cd content may change the contribution to the dietary Cd exposure for different Cd-polluted regions, but it is necessary to identify major dietary patterns and consumption, to detect the Cd content that people consume. A combination of dietary consumption and Cd content assessment can lead to comprehensive evaluation of dietary Cd exposure in the general population.

Three approaches are used for estimating the daily intake of Cd in food [1]. Among them is the total-diet collection method, in which the total Cd intake is calculated as the product of the concentration and the estimated amount of food eaten, which is similar to the method used in our study. Estimation of daily Cd intake only reflects the degree of contamination and exposure to Cd at the point of investigation; it is cross-sectional and statically, it reflects little correlation between Cd intake and a gradual process of harm to the body; therefore, cumulative Cd intake was used to reflect the relationship between total Cd intake and injury. Some studies use mean daily intake levels in all people and multiply these with the people's ages to assess the total cumulative lifetime Cd intake [16]; this method does not consider the different of diet consumption at different ages. Therefore, the estimated amount may be greater than the actual intake. Based on previously used methods, we calculate the cumulative Cd intake by multiplying average daily Cd intake and days of length of life at each age group, then the cumulative Cd intake of all age groups were summed up to get the total lifetime Cd intake. This method is better than the conventional method because the influence of age on diet consumption is taken into account. It was found that a non-smoking individual's intake would at least be 4 grams of Cd in Dayu and 2 grams of Cd in Hezhang after lifetime exposure with the current pollution and lifestyle. A health-based guidance value for Cd of 7 μg/kg body weight (b.w.) per week (Provisional Tolerable Weekly Intake (PTWI)) was established previously by the Joint FAO/WHO. This PTWI value corresponds to a daily tolerable intake level of 70 μg of Cd for the average 70-kg man and 60 μg of Cd per day for the average 60-kg woman. Further, an association between Cd exposure and increased urinary excretion of low molecular weight proteins has been noted in humans with a cumulative intake of about 2000 mg or more (for both men and women) [1]. While in regions we studied, no matter the levels of average daily intake of Cd or cumulative Cd intake has exceeded the threshold for health recommended by WHO. Therefore, measurement should be taken to prevent the harmful effects of Cd on human health through the food chain. Because of the different environmental Cd pollution sources and their different characteristics, the manner and extent of its influence on population health are not the same. Therefore, regions should adopt individual measures based on their own characteristics. Thus, more precautions should be noted in local seeded rice in Dayu. However, both regions should focus attention on leafy and root vegetables. Estimation of Cd exposure is very high in the two regions, and close attention should be paid to the health of the local residents.

Conflicts of Interest

The authors declare that there are no conflicts of interest.

Acknowledgment

We give special thanks to both Guizhou Province Center for Disease Control and Prevention and Jiangxi Province Center for Disease Control and Prevention for their great organization, management, and cooperation for the study. The authors also thank the administration and staff of the Center for Disease Control and Prevention of Dayu and Hezhang Counties, for their effective and efficient teamwork and continuous support in collecting information.

References

1. WHO (1992a) Cadmium. Environmental Health Criteria 134. World Health Organisation, International Programme on Chemical Safety (IPCS), Geneva, Switzerland.
2. FAO/WHO (Food and Agriculture Organization/World Health Organization) (2014) Working Document for Information and Use in Discussions Related to Contaminants and Toxins in the GSCTFF (General Standard for Contaminants and Toxins in Food and Feed). 8th session Geneva 31 March-4 April.
3. EFSA (European Food Safety Authority) (2009) Scientific Opinion of the Panel on Contaminants in the Food Chain on a Request from the European Commission on Cadmium in Food. EFSA J 980: 1-139.
4. Williams PN, Villada A, Deacon C, Raab A, Fiquerola J, et al. (2007) Greatly enhanced arsenic shoot assimilation in rice leads to elevated grain levels compared to wheat and barley. Environ Sci Technol 41: 6854-6859.
5. Zeng F, Mao Y, Cheng W, Wu F, Zhang G, et al. (2008) Genotypic and environmental variation in chromium cadmium and lead concentrations in rice. Environ pollut153: 309-314.
6. Cai S, Yue L, Shang Q, Nordberg G (1995) Cadmium exposure among residents in an area contaminated by irrigation water in China. Bull World Health Organ 73: 359-367.
7. Liang S, Liang Z, Li-jun Z (2013) Cadmium pollution caused by artisanal zinc-smelting in Hezhang County. Chin J Public Health 29: 541-543.
8. Yuexin Y, Guangya W (2002) Chinese food composition table [M]. Beijing: Peking University Medical Press (in Chinese).
9. The National Standard of People's Republic of China. Determination of cadmium in foods. National Standard Bureau, China (in Chinese).
10. The National Standard of People's Republic of China. Maximum levels of contaminants in foods, Ministry of Health, China (in Chinese).
11. Vacchi-Suzzi C, Eriksen KT, Levine K, McElroy J, Tjønneland A, et al. (2015) Dietary Intake Estimates and Urinary Cadmium Levels in Danish Postmenopausal Women. PLoS One 10: e0138784.
12. Quraishi SM, Adams SV, Shafer M, Meliker JR, Li W, et al. (2015) Urinary cadmium and estimated dietary cadmium in the Women's Health Initiative. J Expo Sci Environ Epidemiol 26: 303-308.
13. Islam Eu, Yang XE, He ZL, Mahmood Q (2007) Assessing potential dietary toxicity of heavy metals in selected vegetables and food crops. J Zhejiang Univ Sci B 8: 1-13.
14. Zhong MS, Jiang L, Han D, Xia TX, Yao JJ, et al. (2015) Cadmium exposure via diet and its implication on the derivation of health-based soil screening values in China. J Expo Sci Environ Epidemiol 25: 433-442.
15. Satarug S, Swaddiwudhipong W, Ruangyuttikarn W, Muneko N, Ruiz P (2013) Modeling Cadmium Exposures in Low- and High-Exposure Areas in Thailand. Environ Health Perspect 121: 531-536.
16. Kobayashi E, Okubo Y, Suwazono Y, Kido T, Nogawa K (2002) Dose-response relationship between total cadmium intake calculated from the cadmium concentration in rice collected from each household of farmers and renal dysfunction in inhabitants of the Jinzu River basin, Japan. J Appl Toxicol 22: 431-436.
17. Vromman V, Waegeneers N, Maizeelis C, De Boosere I, Van Holderbeke M, et al. (2010) Dietary cadmium intake by the Belgian adult population. Food Addit Contam Part A Chem Anal Control Expo Risk Assess 27: 1665-1673.
18. Akesson A, Julin B, Wolk A (2008) Long-term dietary cadmium intake and postmenopausal endometrial cancer incidence: a population-based prospective cohort study Cancer Res 68: 6435-6441.
19. Agency for Toxic Substances and Disease Registry (ATSDR) (2012) Toxicological profile for cadmium, Atlanta, Georgia.

Effect of Dietary Selenium on Chosen Physiological and Biochemical Parameters in *Labeo rohita* (Hamilton, 1822) Reared in Copper Contaminated Water

Ibrahim SM[1], Theivandran G[2*] and Thamilmaraiselvi B[2]

[1]*Department of Zoology, VOC College, Millepuram, Tuticorin-628 008, Tamil Nadu, India*
[2]*Department of Microbiology, Cauvery College for Women, Tiruchirappalli-620018, Tamilnadu, India*

Abstract

Copper occurs naturally in the aquatic environment as a result of weathering of soils and rocks and land drainage. Copper is a micronutrient at low level and highly toxic to fish at high concentrations above 0.4 ppm can kill some species of fishes in few hours. The reduction of toxic elements in aquatic system by some acceptable methods is needed urgently. The present work was designed to study the effect of selenium on the reduction of copper toxicity in tissues and improvement of growth haematology and proximate composition in the freshwater fish, *Labeo rohita*.

Keywords: *Labeo rohita*; Water; Copper sulphate $CuSO_4$ 7 H_2O; Selenium (sodium selinite)

Introduction

Metals are introduced in aquatic systems as a result of the weathering of soils and rocks, from volcanic eruptions and from a variety of human activities involving mining, processing, or use of metals and/or substances that contain metal pollutants. Copper occurs naturally in the aquatic environment as a result of weathering and land drainage. But pesticides and fungicides and copper in water pipes and other human devices have introduced additional quantities of copper into the environment. Copper can be actually toxic to freshwater fish, the toxicity depending on species, age, size, water, temperature and ionic composition. The reduction of toxic elements in aquatic system by some acceptable methods is needed urgently. In the present study, selenium reduced the toxicity of copper and thereby enhanced the tested physiological and biochemical parameters. The present study concludes that, 100 mg sel kg^{-1} diet was required to reduce copper toxicity when copper exist in the environment.

Materials and Methods

An experimental animal (*Labeo rohita*) is the most important among the three Indian major carp species used in carp polyculture systems. its high growth potential, coupled with high consumer preference, have established *L. rohita* as the most important freshwater species cultured in India, *L. rohita* were collected from the Manimuthar Dam, Triunelveli district, and immediately transported to the laboratory in plastic buckets with the same Dam water. Fishes were acclimated to the laboratory conditions for a month. During acclimation, the animals fed with dried pellets of 35% protein diet. The water in the rearing tanks was changed daily. Well acclimated *L. rohita* (2.15± 0.75 g) were selected from the stock and divided into 6 groups of 10 individuals of each. Both male and female sexes were used. They were not fed for 24 h before commencement of the experiment.

Feed preparation

The ingredients of dried fish meal groundnut oil cake, cod liver oil, egg yolk, tapioca flour, vitamins and minerals mixture were used to prepare the 35% protein diet with appropriate proportions by square method [1]. Five diets were prepared with different levels of selenium (sodium selinite). The experimental diets (0, 25, 50, 100 and 200 mgselkg^{-1} diet) were prepared by adding the appropriate levels of selenium with chosen ingredients and an aliquot of boiled water, mixed well and steam cooked for 15-20 minutes. After moderate cooling pellets (2 mm) were prepared with a hand operated pelletizer and dried in sunlight. After drying, diets were separately stocked in refrigerator.

Test chemical preparation

A stock solution of copper (ppt) was prepared by dissolving exactly 3.93 g of copper sulphate $CuSO_4$ 7 H_2O in 1 lit of distilled water. From this stock solution, the desired concentration (ppm) was prepared by mixing appropriate quantity of fresh water. The static renewable bioassay method was adopted to estimate the LC_{50} [2]. The 96 hr LC_{50} value, its 95% confidence limit and slope function were determined following the method of Litchfied and Wilcoxon. 50% of LC50 was taken as the maximum sub lethal concentration for the present study.

Oxygen consumption was estimated by Winkler's idometric method described by Welsh and Smith [3]. A method derived by Yokayama [4] and later modified by Christensen et al. [5] was followed to estimate the RBC quantitatively using the Haemocytometer. The carbohydrate contents, proteins and lipids of the feed samples were estimated by Anthrone method and Lowry et al., and Bradgon [6,7] respectively.

Statistical analysis

The collected data were subjected in to the Standard deviation and Student's 't' test.

Results

Copper exposed animals fed on selenium free diet elicited the significant ($p<0.05$) reduction in the oxygen consumption (Figure 1) and

***Corresponding author:** Theivandran G, Department of Microbiology, Cauvery College for Women, Tiruchirappalli-620 018, Tamilnadu, India
E-mail: deva.voc.net@gmail.com

specific growth rates and the parameters were improved significantly ($p<0.01$) in copper exposed fish fed on selenium supplemented diets (Table 1, Figures 1 and 2).

The red blood corpuscles of control fish was high and it significantly ($p<0.05$) declined in fish exposed to sub lethal level of copper and fed on selenium free diet. However RBC's count was improved in the copper exposed fish received the dietary supplementation of selenium diets (Table 2 and Figure 3).

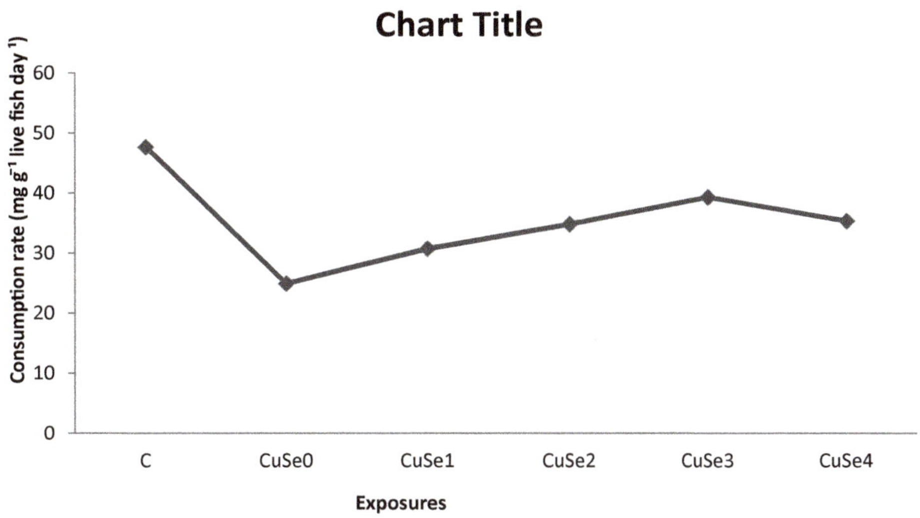

Figure 1: Effect of feeding different levels of dietary selenium on consumption rate in copper exposed *labeo rohita.*

Concentrations of copper(ppm)	Dead / tested	Percent mortality	Lethalconcentration (ppm)			95% confidence limit (ppm)		Slope function (S) ppm	Regression equation(R=a+bx)
			16	50	84	Upper	Lower		
0.100	0/6	0							
0.150	1/6	16.67							
0.200	3/6	50.00	0.09	0.23	0.56	1.82	-1.10	8.77	R=1.84 ± 8.77
0.300	4/6	66.67							
0.400	5/6	84.33							
0.500	6/6	100							

Table 1: Effect of copper levels on *Labeo rohita* exposed for 96 hr.

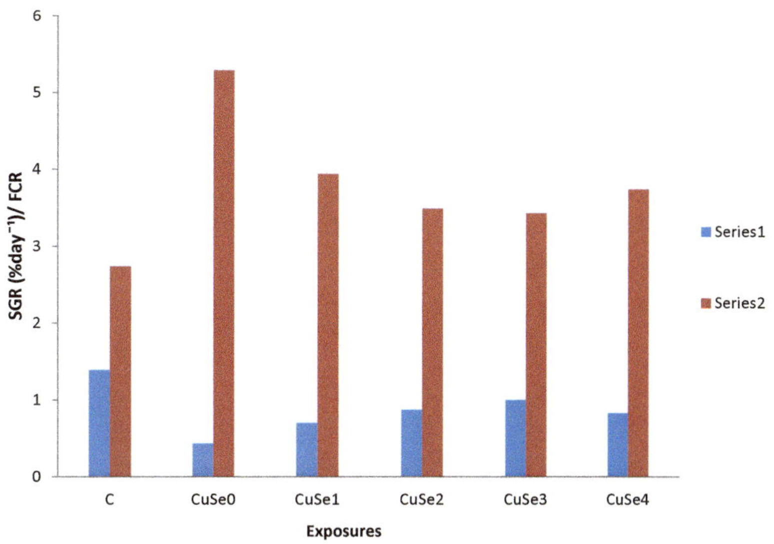

Figure 2: Effect of feeding different levels of selenium on specific growth rate (% day⁻¹) and feed conversion ratio in copper exposed *Labeo rohita.*

Parameters	Exposures					
	Control	CuSe0 0 mg	CuSe1 25 mg	CuSe2 50 mg	CuSe3 100 mg	CuSe4 200 mg
Initial wet weight (gm)	21.06 ± 1.45	21.90 ± 2.13	22.35 ± 2.30	21.15 ± 2.07	22.03 ± 1.77	20.61 ± 2.56
Initial mean wet weight (gm)	2.106 ± 0.20	2.190 ± 0.19	2.235 ± 0.17	2.115 ± 0.22	2.203 ± 0.20	2.061 ± 0.16
Final wet weight (gm)	32.06 ± 3.15	25.00 ± 1.96	27.58 ± 2.36	27.48 ± 2.60	29.61 ± 2.15	26.46 ± 2.19
Final mean wet weight (gm)	3.206 ± 0.27	2.500 ± 0.26	2.758 ± 0.26	2.748 ± 0.25	2.961 ± 0.28	2.646 ± 0.27
Total weight gain (gm)	11.00 ± 1.06	3.10 ± 0.35	5.23 ± 0.47	6.33 ± 0.38	7.58 ± 0.57	5.85 ± 0.49
Feed intake(g dry matter)	30.12 ± 3.86	16.4 ± 1.50	20.61 ± 2.18	22.10 ± 1.85	25.97 ± 2.61	21.86 ± 2.35
Consumption rate(mg live fish g^{-1} day^{-1})	47.67 ± 3.96	24.96 ± 1.57	30.74 ± 2.65	34.83 ± 2.94	39.30 ± 3.13	35.36 ± 3.70
Gain in weight(g wet weight)	11.00 ± 1.06	3.10 ± 0.35	5.23 ± 0.47	6.33 ± 0.38	7.58 ± 0.57	5.85 ± 0.49
Specific growth rate (%days)	1.39 ± 0.04	0.43 ± 0.02	0.70 ± 0.03	0.87 ± 0.06	1.00 ± 0.08	0.83 ± 0.07
Feed conversion rate	2.74 ± 0.18	5.29 ± 0.41	3.94 ± 0.27	3.49 ± 0.36	3.43 ± 0.45	3.74 ± 0.29
Red blood corpuscles count($\times 10^6 mm^{-3}$)	2.57 ± 0.22	1.557 ± 0.12	1.85 ± 0.11	2.02 ± 0.02	2.25 ± 0.05	2.12 ± 0.12
Heamoglobin	8.69 ± 0.45	5.10 ± 0.43	5.39 ± 0.51	5.89 ± 0.55	6.95 ± 0.62	6.04 ± 0.57
Oxygen carrying capacity of blood (mg O_2 g^{-1} Hb)	10.86 ± 0.94	6.38 ± 0.57	6.74 ± 0.66	7.36 ± 0.42	8.69 ± 0.67	7.55 ± 0.72
Rate of oxygen consumption (mg O_2 g^{-1} hr^{-1})	0.33 ± 0.01	0.56 ± 0.04	0.49 ± 0.05	0.44 ± 0.05	0.39 ± 0.04	0.43 ± 0.05

Table 2: Effect of sub lethal exposure of copper and supplementation of selenium (mg kg $^{-1}$ diet) levels on weight gain, selected food utilization parameters, hematological parameters and oxygen consumption in labeo rohita. Each value is the mean ($\bar{x}$ ± SD) of three estimations.

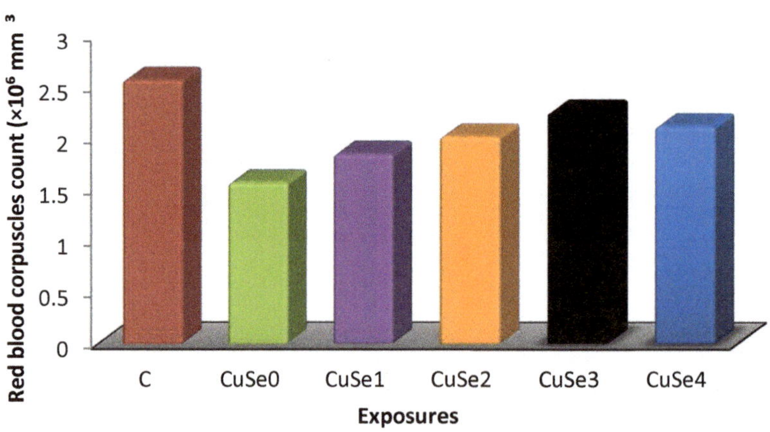

Figure 3: Effect of feeding levels of selenium on red blood corpuscles count in copper exposed *Labeo rohita*.

Haemoglobin content of copper exposed *Labeo rohita* ($CuSO_4$) at the end of the exposure period was 5.10 g% as against 8.69 g% in control. There was two times decline of haemoglobin content in copper exposed fish as compared to control. The exposures of Labeo rohita sublethal level of copper resulted the significant ($p<0.05$) decrease in the RBC count and haemoglobin content leading to anaemia. The anaemia condition may be due to inhibition of erythropoiesis and haemosynthesis and increase in the rate of erythrocyte destruction [8]. Oxygen carrying capacity of blood was also declined in metal exposed *Labeo rohita* may be due to the reduction of RBC count and Hb content. James and Sampath [9] found that the oxygen carrying capacity of blood of *Heteropneustes fossils* declined due to the reduction of RBC count and Hb content which reflected on tissue respiration. Ahmed et al., [10] found that the organic selenium supplementation to fry Nile tilapia enhanced the growth, feed utilization and reduced the susceptibility to *Aeromonas hydrophila* infection. The proximate compositions and energy contents were high in control fish. Copper exposed L. rohita fed on se free diet reduced the proximate compositions and energy and they were improved in copper exposed fish fed on se supplemented diets. Among these supplemented diets, fish fed on 100 mg Se k-1 dirt, significantly ($p<0.05$) improved the selected physiological and biochemical parameters then other levels. Selenium (Se) is an essential trace element required in the diet for normal growth and physiological function of fish [11,12]. Hence supplementation of 100 mgSekg^{-1} diet is considered as optimum dose and selenium protects L. rohita from copper toxicity.

References

1. Hardy R (1980) Fish feed formulation. In: Fish feed technology, ADCP / REP/80/11, FAO of the UN Rome, pp. 233-239.
2. Sprague JB (1973) The ABC's of pollutant bioassay using fish. In: Biological methods for the assessment of Water Quality. ASTM. STP. 528. Amer Soc Testing Materials, pp. 6-30.
3. Welsh JH, Smith (1953) Laboratory exercise in invertebrate physiology. Burgers Publishing Co., Minneapolis.
4. Yokoyama HO, Hasler AD, Bilstad NM (1947) Studies on the origin, development, and seasonal variations in the blood cells of the perch, Perca flavescens. Anat Rec 99: 680.

5. Christenson GM, James TF, Barbura AP (1978) Cell properties, protein and certain physical, chemical properties of brook trout (Salvelinus fontinalis) blood. J Fish Biol, 12: 51-60.

6. Lowry OH, Rosebrough NJ, Farr AL, Randall RJ (1951) Protein measurement with folin-phenol reagent. J Biol Chem, 193: 265-275.

7. BRAGDON JH (1951) Colorimetric determination of blood lipides. J Biol Chem 190: 513-517.

8. Wintrobe MM (1978) Clinical Haematology, London, Kimpton, H., 448, pp.

9. James R, Sampath K (1995) Sublethal effects of mixtures of copper and ammonia on selected biochemical and physiological parameters in the catfish Heteropneustes fossilis. Bull Environ Contam Toxicol 55: 187-194.

10. Ahmed MH, El-Marakby HI, Seden MEA, Tawwab MA, Abou-El-Atta ME (2006) The use of organic selenium (sel-plexo) in practical diets of Nile tilapia, Oreochromis niloticus (L.): effect on growth performance, feed utilization, whole body composition and entropathogenic Aeromonas hydrophilla challenge. In : W. Contreras and K. Fitzsimmons, Editors, 7th International symposium on Tilapia in Aquaculture, 6 – 8 September 2006, Boca del Rio, Veracruz, Mexico, pp. 95-107.

11. Hilton JW, Hodson PV, Slinger SJ (1980) The requirement and toxicity of selenium in rainbow trout (Salmo gairdneri). J Nutr 110: 2527-2535.

12. Bell JG, Cowey JB, Andron JW, Shanks AM (1985) Some effects of vitamin E and selenium deprivation on tissue enzyme levels and indices of tissue peroxidation in rainbow trout (Salmo gairdneri). British Journal of Nutrition 53: 49-57.

Remediation of Some Selected Heavy Metals from Water Using Modified and Unmodified Mushrooms

Thomas Arap Bii[1], Isaac W Mwangi[1]*, Ruth N Wanjau[1], Sauda Swaleh[1], Manohar Ram[1] and Ngila JC[2]

[1]*Department of Chemistry, Kenyatta University, P. O. Box 43844-00100, Nairobi, Kenya*
[2]*Department of Chemical Technology, University of Johannesburg, Doornfontein Campus, P. O. Box 17011, Doornfontein 2028, Corner Beit and Nind Street, South Africa*

Abstract

The dispersal of toxic heavy metals by water from natural and anthropogenic is a worldwide environmental concern due to pollution. Despite some metals playing an important role in body, they are toxic when the level exceeds the tolerance limits while others such as lead have no known physiological value to human beings. Since heavy metals cannot be degraded, then their removal from drinking water is necessary. Mushrooms are readily available in Bomet County and their metal removal ability was investigated. The study aimed at removing heavy metals from water by adsorption using mushroom, as a cost-effective and sustainable method. The raw mushroom was modified with sodium hydroxide and characterization of both the parent material and its modified form was done using Fourier Transform Infrared spectrometry (FTIR). Sorption experiments were carried out using the batch adsorption method and sorption parameters including pH, contact time, adsorbent dose and initial metal ion concentration investigated. The results found out that the sorption capacity for cadmium ions ranged from 1.826-25.285 mg/g by the unmodified edible mushroom (UEM), the modified edible mushroom (EM), unmodified toxic mushroom (UTM) and modified toxic mushroom (TM). For copper ions, sorption capacity ranged from 0.002-4.097 mg/g, while that of the lead ions ranged from 1.345-2.593 mg/g by the UEM, EM, UTM and TM respectively. The sorption capacity showed improvement on modification as sorption of cadmium increased from 1.826-25.285 mg/g by the UEM, EM, UTM and TM. At a pH range of 4-6, the sorbent material was found to remove up to 90% of the metals. The sorbent material had a removal efficiency of 95% of the metals in less than 20 minutes. The UEM and UTM fitted well in Langmuir adsorption isotherm model for cadmium and lead ions. For copper ions, UEM, EM, UTM and TM fitted in the Freundlich model. TM for lead ions best fitted in the Freundlich model. The bio-sorption kinetics was determined by fitting first-order-Lagergreg and Pseudo-second-order kinetics models to the experimental data. It was found that the data for lead was better described by the pseudo-second-order model. For copper ion, the data was best described by Ho's pseudo second order for UEM and UTM, cadmium ions for all sorbents was best described by Lagergreg's first-order kinetics. The FTIR analysis suggested the possibility of the participation of carboxyl groups in metal uptake. The levels of dissolved organic carbon (DOC) were found to be 19.0 mg/L in the raw material and 2.19 mg/L after modification. It was confirmed that modification minimized secondary pollution. This indicated that mushrooms have a potential application for the remediation of metal polluted waters.

Keywords: Mushrooms; Alkali modification; Sorption; Dissolved organic carbon

Introduction

Water has a special property due to its ability to split ionic compounds it interacts with making it a universal solvent [1]. As such water in natural water bodies will have dissolved material of both organic and inorganic origin dispersed in this solvent. The presence of elevated concentrations of the dispersed solutes may render this commodity polluted and thus unsafe for human consumption. This implies that when water interacts with industrial, domestic, or agricultural wastes, it is a serious concern to human populations as well as the flora and fauna. This will result with the water being contaminated with a myriad of different components such as organic compounds, synthetic chemicals, nutrients, organic matter and heavy metals to the environment [2]. Interaction of such waters with the environment leads to loss of biodiversity due to pollution [2]. Anthropogenic activities such as metal processing industries and tanneries introduce heavy metals to water [3]. Unlike other organic waste heavy metals cannot be degraded biologically to products that are harmless and as such the only option is their removal from water. Irrespective of the fact that most heavy metals in the environment are in trace levels, continuous consumption of such water result to bioaccumulation of the same in vital organs and may pose a health risk [4-6]. Their toxicity depends on the type of metal, speciation and the type of organisms that are exposed to it. Examples of heavy metals include lead, zinc, copper, mercury, iron, manganese, cadmium, vanadium, antimony, selenium, arsenic and cobalt. Some metals such as lead interfere with normal body functions such as causing impaired brain development thus making consumers of lead contaminated waters to have challenges in life [7]. This makes heavy metals a concern to environmentalist, heath workers as well as industrial quality control managers [8-10]. Usually, these metals are leached into water from the land by rainwater and accumulated into ground water sources and other water bodies [11,12]. This renders the quality of the blue gold compromised and hence not suitable for consumption [13-16].

To make such water safe for domestic use, dissolved metal ions have to be removed. Removal has previously been achieved by the use of methods such as ion exchange, reverse osmosis and precipitation [17]. However such methods are expensive and not effective when the

***Corresponding author:** Isaac W Mwangi, Department of Chemistry, Kenyatta University, P. O. Box 43844-00100, Nairobi, Kenya
E-mail: isaacwaweru2000@yahoo.co.uk

concentrations are in trace levels ranging from 1-20 μg.l^{-1} [18]. The use of biomass materials such as agricultural products which are abundant has been reported [19]. They, Abia et al. [19] reported that, waste cassava peels had a potential for the removal of trace metal ions from aqueous solutions over a wide range of concentrations. However, it has observed that all biomaterial had a challenge of leaching of organic matter during removal processes thus causing a "secondary pollution" [20]. Such biomaterials were not regeneratable and are susceptible to bacterial action [21,22].

This study therefore aimed to overcome such limitations by developing a cost-effective method by synthesizing an ion exchange biomaterial resin from both edible and toxic mushrooms [23-27]. Mushrooms are very nutritious as they have essential minerals and functional groups very important for life supporting enzymes and medicinal value [28]. This has given the mushrooms, despite growing very rapidly, an ability to contain some trace metals from the soil [29]. This ability was exploited in this study for the metal removal. The process was achieved by anchoring suitable functional groups capable of complexing metals on the biomaterial [30-36]. This would enable it to attract metals on its surface ions and remove them from water. The process was achieved by chemical modification of the mushrooms with a sodium hydroxide, a method exploited to prepare snuff from tobacco leaves to enable them is safe to chew, guard the tobacco leaves against degradation as well as pass the test of time [37]. The resulting solid material was then employed in its raw and modified forms for the removal of heavy metals in synthetic and contaminated waters. Its potential to minimize secondary pollution was also investigated [38-46].

Materials and Methods

Materials and reagents

All the solutions were prepared in double-distilled water, and the reagents were of analytical grade. Sodium hydroxide, sucrose, nitric acid and nitrates of lead, cadmium, and copper, were all supplied by Sigma Aldrich (Kobian, Nairobi Kenya). Mushrooms were obtained from a farm in Bomet County, Kenya.

Instrumentation

The FTIR spectrophotometer with an attenuated total reflectance (ATR) facility (Perkin Elmer 100 with sampling accessory-Waltham, MA, USA) was used to characterize the parent and the modified biomaterials [47-50]. The leached organic matter was determined using UV-Vis spectrophotometer (Specord 200-Germany). A flame atomic absorption spectrometry (FAAS; Buck Scientific 210 VGP) was used to determine the metal content in solution [51-54].

Experimental procedures

The sorbent used in this study was from both the edible and toxic mushroom collected from Sotik, Bomet County. They were cleaned on site to remove dirt, then with double distilled water. They then were transferred to the laboratory and dried in an air oven at 25°C until constant weight was attained [55-60]. The dried material was ground and sieved through a fine-mesh (150-250 mm). Some of the resulting material was modified, characterized and applied in the sorption process while the other was used its raw form. The sorption mechanism for both raw sorbent and its modification form were then investigated.

Modification of both edible and toxic mushroom

The modification process involved treating a solid sample of the dried mushroom (10.0 g) previously ground material with 0.1 M sodium hydroxide [61-64]. The resulting mixture was refluxed at a temperature of 80°C for three hours. The treated biomaterial was then washed with 100 ml of distilled water to bring the pH to neutral with dilute nitric acid. The solution was filtered through a sintered glass crucible and dried in a vacuum [64]. The resultant solid was dried in an oven at temperature of 80°C for two hours and then extent of modification was confirmed with FTIR and the product used for sorption experiments.

Characterization using FTIR

Analysis of the parent and the modified forms was done using FTIR spectrophotometer (FTIR-8400). The FTIR spectra were analyzed, and the functional groups in each were confirmed.

Batch sorption experiments

Sorption studies were carried out on a Lab-line mechanical reciprocating shaker model SSL2 (Harrogate, UK) using 100 ml screw cap plastic bottles. Known weights (0.3 g) of the adsorbent were in each bottle containing a known (40 ml) concentration of the metal ion of interest and the pH of the mixture adjusted to values between pH 3 and pH 7 (desired values) using 0.1 M nitric (V) acid and 0.1 M sodium hydroxide drop wise. The mixtures were then left to equilibrate allowing sufficient time for adsorption. Each of the resulting mixture was then filtered through Whatman No. 42 filter paper and the metal ions in the filtrate were determined by FAAS.

Optimization of pH

The effect of pH on the equilibrium was investigated by mixing 0.02g of sorbents material with 40ml of the test solution of 20 mg.l^{-1} concentration buffered with 0.1 mol.l^{-1} sodium acetate, and at pH values (3-7) by the addition of 0.1 M nitric acid drop wise. The mixture was equilibrated in a plastic screw cap bottles (100 ml) for 1 hour. The mixture was then filtered and the final concentration of the metal ions determined to establish the optimum pH for sorption of each respective metal.

Effect of the contact time on sorption of the metal ions

The effect of time on adsorption of metals by both the edible and toxic (modified and unmodified) mushroom was determined by mixing 0.3 g sorbent material with 40 ml of the metal solution whose concentration was 20 mg.l^{-1}. The initial pH of the model solutions was adjusted to optimal value for each metal [65-68]. The respective mixtures were then allowed to equilibrate at predetermined time intervals of 2-100 minutes in a water bath shaker at 25°C. The sorbent was filtered using Whatman No. 42 and the metal ion concentration in the filtrate determined using FAAS.

Effect of metal ion concentration on sorption

The adsorption capacity was determined by mixing 0.3 g with 40 ml model metal solution (concentration 2-200 mg.l^{-1}) at their optimum pH values with modified and unmodified sorbents. The mixtures were agitated for 1 hour, filtered and the concentration of the metal ions in the filtrate determined by FAAS.

Effect of adsorbent dose on percent recovery of metals ions

In the determination of adsorption capacity of the adsorbent for lead ions, 40 ml of the working solution of ions containing 20 mg.l^{-1} of metal ions into different plastic screw cap bottles and known masses (0.1 g, 0.5 g, 1.0 g, 1.5 g, 2.0 g and 2.5 g) of the adsorbent was added to the solution (specifically at the same and controlled initial pH of solution fixed at 7.0). The respective mixtures were agitated for 1 hour. The same procedure was followed in the determination of the adsorbent for cadmium ions and for copper ions. Content of the metal ions in the solution was then analyzed using FAAS [69-72].

Calculation of metal ion adsorption of modified and unmodified mushrooms

The amounts of metal ions adsorbed per unit mass by the modified and unmodified mushrooms during the batch investigation were calculated as amount adsorbed as given by equation 1 below

$$q_e = \frac{(C_o - C_e)}{W} V \qquad (1)$$

Where C_o and C_e are the initial and equilibrium concentration (mg.l^{-1}) of metal ions in solution, respectively, V is the volume of the solution in (ml) while W is the mass of the adsorbent in (g).

Analysis of environmental water samples

River water samples were collected from Nyangores River (Bomet County, Kenya). Known concentration of copper, cadmium and lead ions were spiked separately in to each of the samples (10 ml). The samples were loaded on to the solid-phase extraction (SPE) column containing 0.3 g of each adsorbent. The retained metal ions were then stripped with 5 mL of 0.5 M nitric acid and their concentration determined by FAAS.

Determination of secondary pollution in the treated water

The determination of leached organic matter was performed using dichromate method according to the procedure proposed by Baumann [73] and Gonzalez [74] by UV-Vis spectrophotometric method. Carbon standard solutions were prepared from sucrose in the range of 10–100 mg.l^{-1}. Separate samples were prepared by placing a known weight of the respective biosorbents material (0.1 g) into a 250 ml round-bottomed flask containing 100 ml of water. Each of the mixture was placed on a magnetic stirrer and agitated for a period of 1 hour to extract soluble organic compounds. The solid was filtered off and the dissolved organic carbon content determined [75-82].

Results and Discussions

Characterization and optimization studies

FTIR characterization: The modified and unmodified adsorbents were characterized with FTIR and the resulting spectra are presented in Figures 1-4. The spectrum shows a band at 1225 cm^{-1} that may be attributed to the presence of organo sulphate groups [83]. This indicates the possibility of the presence of sulphur groups in the adsorbent. The band at approximately 930 cm^{-1} could be attributed to the C-H stretching [83]. There was also a broad band between 3921.0 cm^{-1} and 3710.8 cm^{-1} which had a very low sensitivity. The edible mushroom was modified and the results are presented in Figure 2.

The results show an enhanced broad band at 1878.5 cm^{-1} [84-86]. This could be a shift in the band at 1750.0 cm^{-1} in the unmodified material which was pronounced after modification. This band could be contributed by a keto group within the cellulistic structure [87]. Thus the main structure of the material could be a five membered ring [87]. The spectra also show a band at 1230.5 cm^{-1} that can be attributed to the presence of the amide functional group which is characteristic of charged NH_4^+ functional group [88]. It can also be due to stretching vibration of organic sulphates within the material [83,88]. This indicates the presence of sulphur in the adsorbent. The broad band at 1750 cm^{-1} could be as a result of a carbonyl, C=O, functional group [89-91]. The pattern in the peaks of both the raw and modified form is the same. The only difference is the enhancement of the broad band

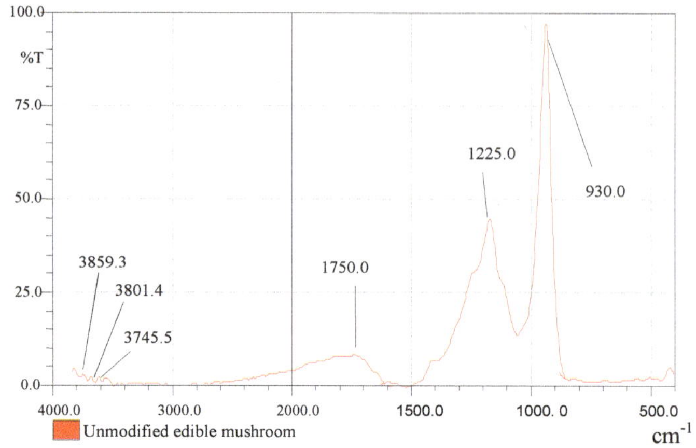

Figure 1: FTIR spectra of the unmodified edible mushroom.

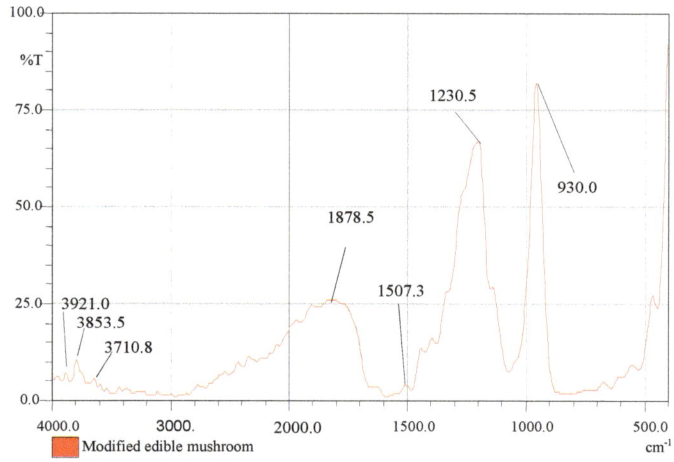

Figure 2: FTIR spectra of the modified edible mushroom.

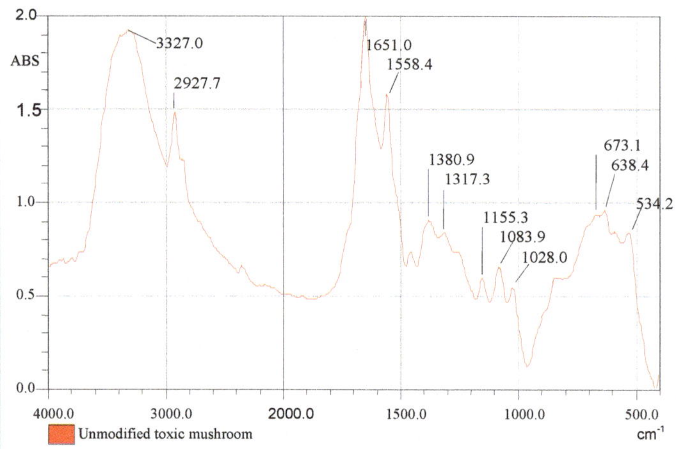

Figure 3: FTIR spectra of the unmodified toxic mushroom.

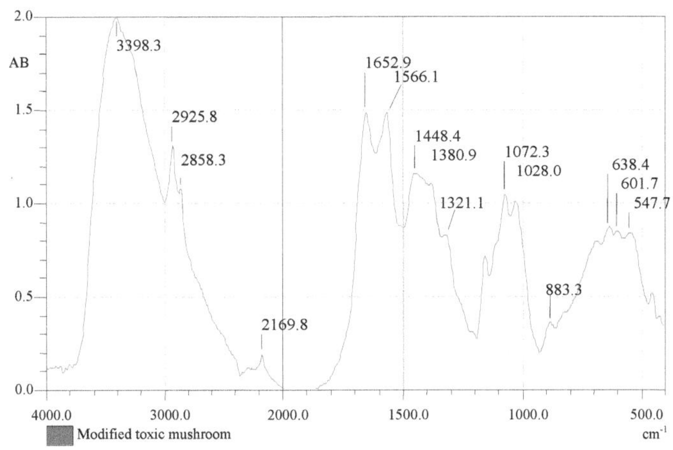

Figure 4: FTIR spectra of the modified toxic mushroom.

in the modified form which could be due to anchoring a keto group within the structure of the material. Both the raw and modified toxic mushroom materials were analysed and the results obtained were as presented in Figures 3 and 4 respectively. Figure 3 shows results upon characterization of the unmodified toxic mushroom species.

The spectra show a strong band at 3327.0 cm^{-1} attributed to either -OH or -NH groups [92,93]. But the band appears as two peaks, thus an indicator that it is most likely due to -NH groups [92,93]. The band at 2927.7 cm^{-1} was assigned to C-H stretching. The band at 1645.2 cm^{-1} may be due to the presence of the carbonyl group [81]. The band at 1317.3 cm^{-1} could be as a result of an amide or sulphamide group [89,90]. The band at 1155.3 cm^{-1} was assigned to the organic sulphates [83,94]. This indicates the presence of sulphur in the adsorbent. The material was modified and the results obtained presented in Figure 4.

The results show a strong band at 3398.3 cm^{-1} attributed to either -OH or -NH groups [92,93]. But the band appears as one peak, thus an indicator that it is most likely due to -NH group [92,93]. Treatment of raw biomass of mushroom with base improved the content of reactive centers such as carbonyl group of amide as shown by a sharp peak at 1652.9 cm^{-1} [95]. The transmitted peak at 2927.7 cm^{-1} present in unmodified toxic mushroom was shifted to 2925.8 cm^{-1} in modified toxic mushroom and the sharp peaks at 1651.0 cm^{-1} and 1558.4 cm^{-1} were broadened after modification [95]. The appearance of new peaks in modified toxic mushrooms at 1072.3 cm^{-1} and 1028.0 cm^{-1} may be due to C-O stretch or sulphonates [83].

The general observation from FTIR-characterization of the mushroom shown the presence of many functional groups such as -NH, C=O, -COOH, S=O, -OH and C-H which are capable of binding to metal ions [96-98].

Optimization of sorption parameters: Adsorption of metal ions is dependent on a number of factors. This study had to establish the conditions for removal of targeted metal ions from water. The parameters studied were effects of contact time, concentration of the initial metal ion, pH and dosage of the adsorbent.

Effect of pH on adsorption of metal ions: Adsorption of metal ions by sorbents is pH dependent as it influences the charge of the binding sites of the sorbent and the chemistry of the metal ion [99]. In this study, the optimum pH for metal ions uptake was determined through batch adsorption experiments. The optimum values were then applied in the subsequent experiments. The effect is due to the fact that, sorbents have sorption sites due to the presence of functional groups, whose charges are influenced by the pH [100]. At low pH, the adsorbent is positively charged because the pH is lower than the isoelectric point or point of zero charge (PZC thus pH < PZC). At such low pH range, adsorption is poor due to the charge on the adsorbent [101]. At high pH (PZC) the adsorbent is negatively charged contributing to a high adsorption [102]. This arises from the fact that when the metal is in solution, it is positively charged and will be attracted to the surface of the negatively charged adsorbent at that pH (pH > PZC), favouring adsorption. At pH > 6, there is metal hydrolysis, leading to precipitation due to formation of hydroxyl metal ion [103]. Results for the adsorption experiments at different pH values are presented in the Figure 5A.

Results show that maximum sorption of Cadmium ions by both unmodified and modified (edible and toxic) mushroom was observed at pH 4.0. This similar to what was reported by other researchers [21] who investigated sorption of metal ions by maize tassels. Mushroom both toxic and edible presents a high content of ionizable groups such as carboxyl, organic sulphates, and amino groups. Presence of these groups was confirmed by FTIR spectroscopic analysis (discussed in Section 4.1). At values lower than pH 4.0, cadmium ions removal was partially inhibited, due to the competition between hydrogen and cadmium ions on the sorption sites. At low pH, high hydrogen ion which restricts the metal cations to reach the sorption site, as in consequence of the repulsive force [104].

The highest bio-sorption efficiency of mushroom (edible and toxic) was observed at pH 4.2 with unmodified toxic mushroom (UTM) having high profile possibly due to the presence of carbonyl groups. Increase in cadmium ions removal, as the pH increased, can be explained in terms of point of zero charge (pzc) of the adsorbent and metal speciation taking place in the solution. The pH value, at which the charge of the solid surface is zero, is referred to as the point of zero charge (pzc). At pH value above pzc, there is a net negative charge on the biomass surface that promotes the uptake of metal ions. As the pH lowered, however, the overall surface charge on the biomass become positive which inhibit the approach of positively charge metal cations. It is likely that protons will then compete with metal ions thereby decreased interaction of metal ions with the sorbent as reported by Sag et al. [105]. The findings also compared well with by Padmavathy et al. [106]. According to Leyva, et al. [107] cadmium ions are the predominant ionic species at pH values below than 7.0 but precipitates above this pH value. As a result, pH optimization was carried at pH values below 7.0.

The results for the sorption of copper, obtained were presented. Figure 5B shows that the best sorption of copper ions by both the modified and unmodified (edible and toxic) mushrooms as a function of pH was observed to have a range between 4.5 and 5.0 for the modified toxic (TM) and unmodified edible (UEM) respectively and 5.8 for both unmodified toxic (UTM) and modified edible (UEM).

The effect of pH can be described in terms of pH_{pzc} of the adsorbent and species of copper ions formed in the solution. When pH is equal to pH_{pzc}, the surface charge of adsorbents is neutral and electrostatic attraction existing between the adsorbent surface and metal cations allows sorption. When pH is lower than pH_{pzc}, the surface charge of adsorbents is positive, which inhibits the approach of positively charged cations. Rivera et al. [108] stated that an increase in the negative surface charge of the adsorbent that augments it to adsorb positively charged species. At pH greater than pH_{pzc}, the surface charge of adsorbents is negative. Thus, cations adsorption on adsorbent is favorable at pH

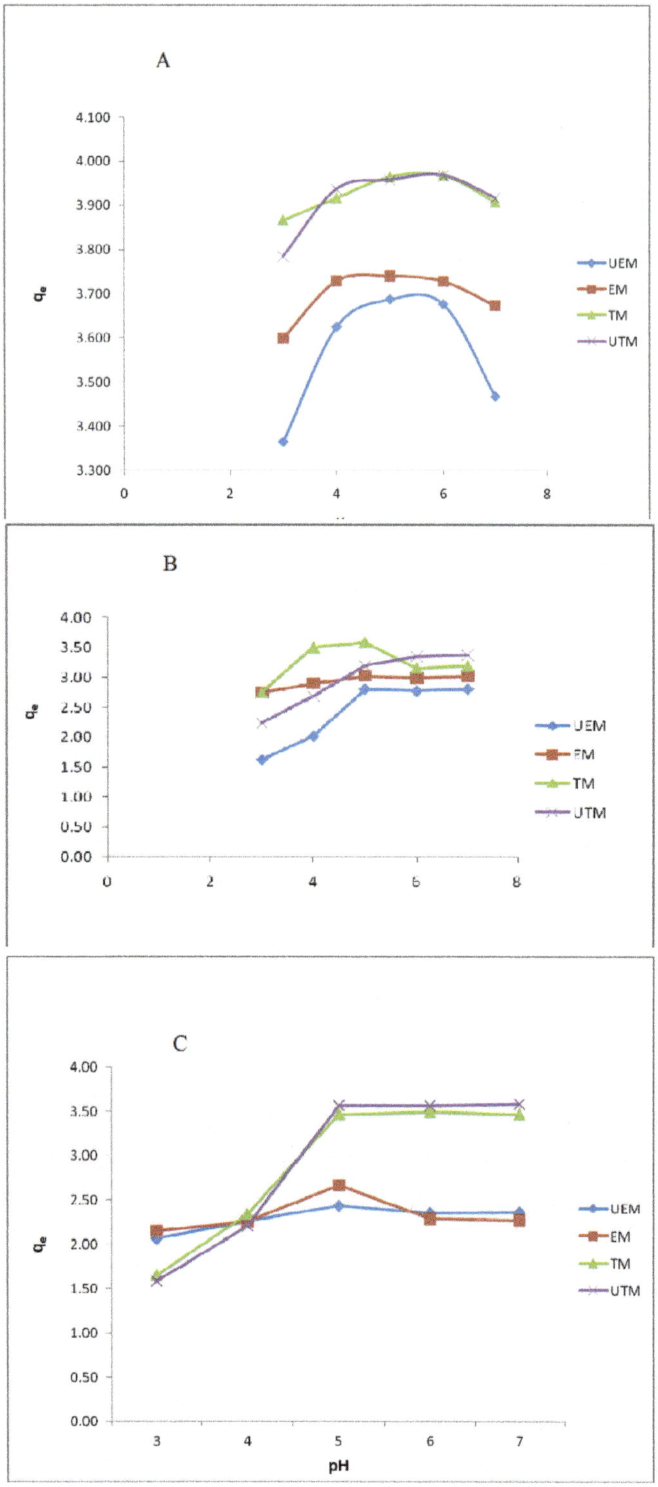

Figure 5: Effect of pH on sorption of cadmium (A) copper (B) and Lead (C).

values greater than pH_{pzc} while anion adsorption is favorable at pH value lower than pH_{pzc} [109].

For sorption of lead ions (Figure 5C), was found to be strongly dependent on the pH of the solution. The optimum pH for the adsorption lead ions was about 5.0 for both modified and unmodified adsorbent. From the experimental data, sorption of lead by the unmodified sorbents (edible and toxic mushroom) had a wider sorption pH region (4.5–5.8) compared to the modified (pH 5.0), as shown in Figure 5C. This was attributed to the fact that modifications lowered the point of zero charge (pH_{pzc}) which resulted in more negatively charged surfaces contributing to the observations made by Zhu, Yang, and Deng [110]. In addition, lead forms polynuclear complexes containing two or more lead atoms, in the presence of ligands at different pH conditions, favoring attachment of the metal ion onto the ligand [111]. At low pH (below 3), there was excessive protonation of the active sites at mushroom surface and this often refuses the formation of links between Pb^{2+} ions and the active site [103]. At moderate pH values (3-6), linked hydrogen ions being released from the active sites and adsorbed amount of metal ions is generally found to increase. At higher pH values (above 6), the precipitation is dominant or both ion exchange and aqueous metal hydroxide formation may become significant mechanisms in the metal removal process [104]. This condition is often not desirable as the metal precipitation could lead to a misunderstanding for the adsorption capacity. The pH value for maximum sorption for lead was observed as 5.1 closer to that reported by other researchers [112,113] who investigated the sorption of metals by marine algae. The pH value of 5.1 was adopted as the optimum value for lead sorption in our study.

Generally, observation from the experiments was that uptake capacities of the three heavy metals was pH dependent [90,91]. Study on the performances of the two species of mushrooms revealed that sorbent modification had a significant effect on the pH of maximum sorption of the metals analyzed [21]. The pH for the maximum adsorption was found to range between 4-6 for all the three metals. This implies that the adsorbent material is suitable for removal of metals from acidic water. Further experiments were carried out at optimal pH values for each respective metal ion.

Effect of contact time: The rate of metal ion adsorption is related to the efficiency of the adsorbent and activity of the metal and therefore controls the residence time of adsorbate at the solid–solution interface [114]. It is also related to the availability of the sorbents binding sites to hold the metal ions [114]. The results on the rate of adsorption are presented in Figure 6A.

The results show time dependent sorption of the selected metals. The rate of cadmium ions uptake in unmodified and modified edible mushroom was fast. The results show that 90% of the total uptake occurred within 20 minutes. The adsorption per gram of adsorbent was between 1.41 $mg.g^{-1}$ and 1.86 $mg.g^{-1}$. This was then followed by a slower uptake rate which resulted into a plateau after the sorption sites were occupied. The initial rapid rate may be due to a physical sorption or ion exchange at cell surface and the subsequent slower phase may be due to other mechanisms such as complexation, micro-precipitation, or saturation of the binding sites [114].

The sorption profile of copper (Figure 6B) show a rate of copper ions uptake of copper by all the adsorbent materials except UTM was 90% within the fast first 40 minutes. The latter removed 90% of the metal after 60 minutes. This was later followed by a slower uptake rate leading to a steady state where no significant adsorption occurred. Sorption by UTM attributed to the fact that modification lowered pH_{pzc} which resulted in more negatively charged surfaces contributing to the observations made by Zhu, Yang, and Deng [110].

The rate of sorption of lead ions (Figure 6C) show that the rate of lead ions uptake by all the adsorbent materials was that 90% of the metal

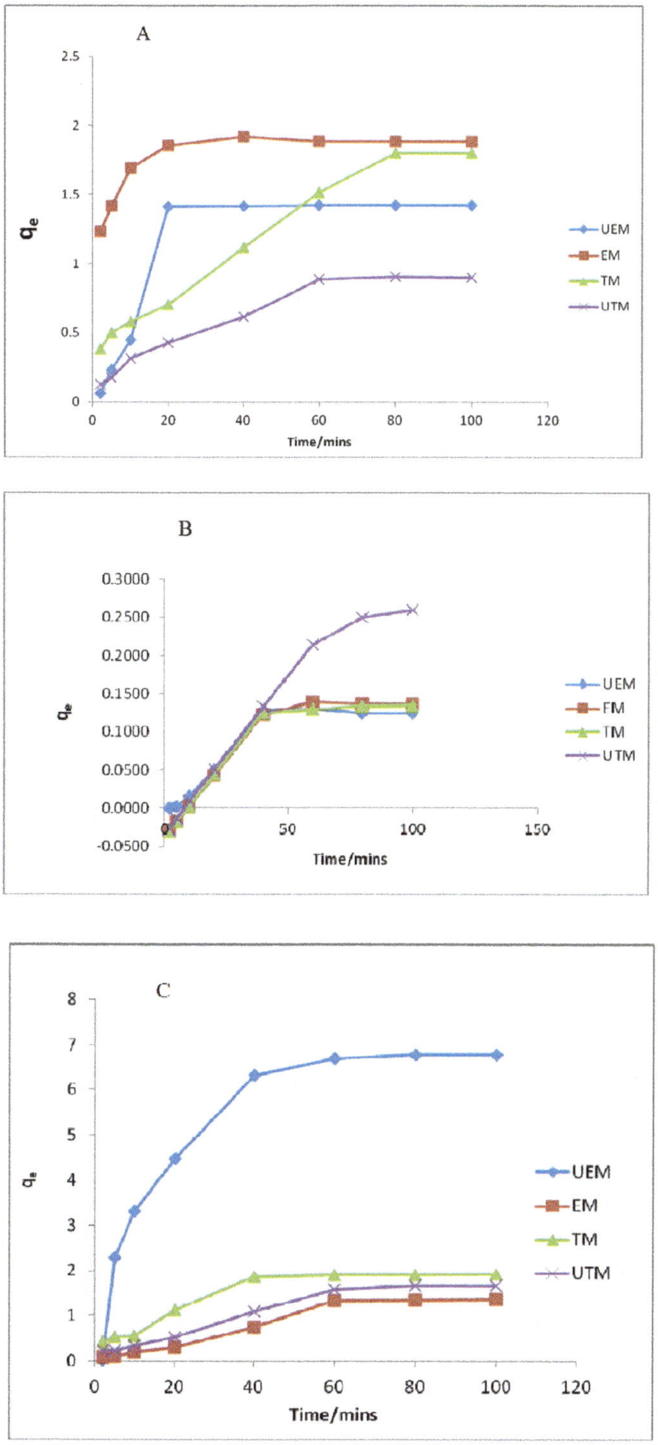

Figure 6: Effect of time on adsorption of cadmium (A) copper (B) and lead (C) ions.

ions were removed from solution within the first 50-60 minutes. After that, the sorption remained constant after equilibrium was attained. This could be because the binding sites were limited and the remaining vacant surface sites were difficulty to be occupied by metal ions due to the formation of repulsive forces between metal ions on solid phase and the liquid phase [2]. The general observation was that the high rate of sorption was due to both chemisorptions and complexation while the slower phase may be due to saturation of the binding site and the subsequent attainment of equilibrium [115]. Mwangi et al. [21] investigated sorption of metals ions by maize tassels and reported that the sorption capacity of cadmium by the unmodified material was relatively lower (≈4 g.kg^{-1}) than that for the modified (≈12.5 g.kg^{-1}).

A similar profile was observed as reported by Mwangi et al. [21], of a general metal uptake were rather fast such that 90% of the total uptake occurred within 10 minutes, after which a slower uptake rate approached a steady state (where no significant sorption occurred). The initial rapid rate (within the first 10 min) may be due to a physical sorption or ion exchange at cell surface and the subsequent slower phase may be due to other mechanisms such as complexation, micro-precipitation, or saturation of the binding sites. It was also observed that the modified sorbent had a better efficiency on the metal uptake.

Fazal and Rafique [116] investigated on mechanistic understanding of cadmium sorption by sulphonated and esterified spent black tea observed that for the fixed cadmium ions concentration and sorbent mass in the interface mixture, the sorption of metal increased as the contact time was increased. The sorption was, however, noted to occur in two phases of fast and slow rates. The biosorption rate was fast initially; about 50% of total cadmium was removed within 16 minutes. In the initial stages the high removal efficiency is due to the abundant availability of active binding sites on biomass, and with gradual occupancy of these sites, the sorption become less efficient in the later stages. The biosorption of cadmium reaches plateau value within 18 minutes. In this study beyond 20 min the cadmium ions desorption from porous sorbent shown increase. So the optimum agitating time for biosorption of cadmium was accepted as 20 minutes.

The results obtained from the study in rate of sorption of the three metals implied that 20, 40 and 60 minutes were sufficient time enough to achieve maximum adsorption for cadmium, copper and lead respectively.

Effect of sorbent dose on % removal of the metals ions: This was investigated by weighing varying masses (0.05-0.30g) of modified and unmodified mushroom (edible and toxic) and packed them in separate columns. Model solutions 40 ml, with a concentration of 20 mg.l^{-1} with their pH values adjusted to each metal's optimum value were loaded onto the column at a flow rate of 3 ml minute^{-1}. The content of the metal ions in the respective solutions were as presented in Figure 7A.

The results show that the percentage removal with Toxic mushrooms (TM) was highest with cadmium ions. This could have been attributed to the fact that the binding sites available in the modified toxic mushroom had high affinity for cadmium. The maximum percentage removal was 97.92% when the sorbent dose was 0.2 g. This could be because modification of raw sorbent with a base improves the content of reactive centers such as carbonyl group of amides hence relatively higher affinity for these metal ions in modified adsorbent [96]. Treating raw sorbent with sodium hydroxide increases the net negative charges on the surface of the sorbent thus possibly leading to the stronger electrostatic force of attraction between negatively charged surface of the adsorbent and positively charged sorbate ions hence efficiency towards improved removal [96]. Sorption of cadmium ions is dependent on the stability constant between the functional groups and the metal upon complexation. The higher that value, the higher the removal. Modification of the edible mushroom lowers the removal because of derivatization of the prior functional groups with a new one with a lower stability constant. This results to a lower adsorption of 95.8% from 96.5%. The unmodified toxic had a

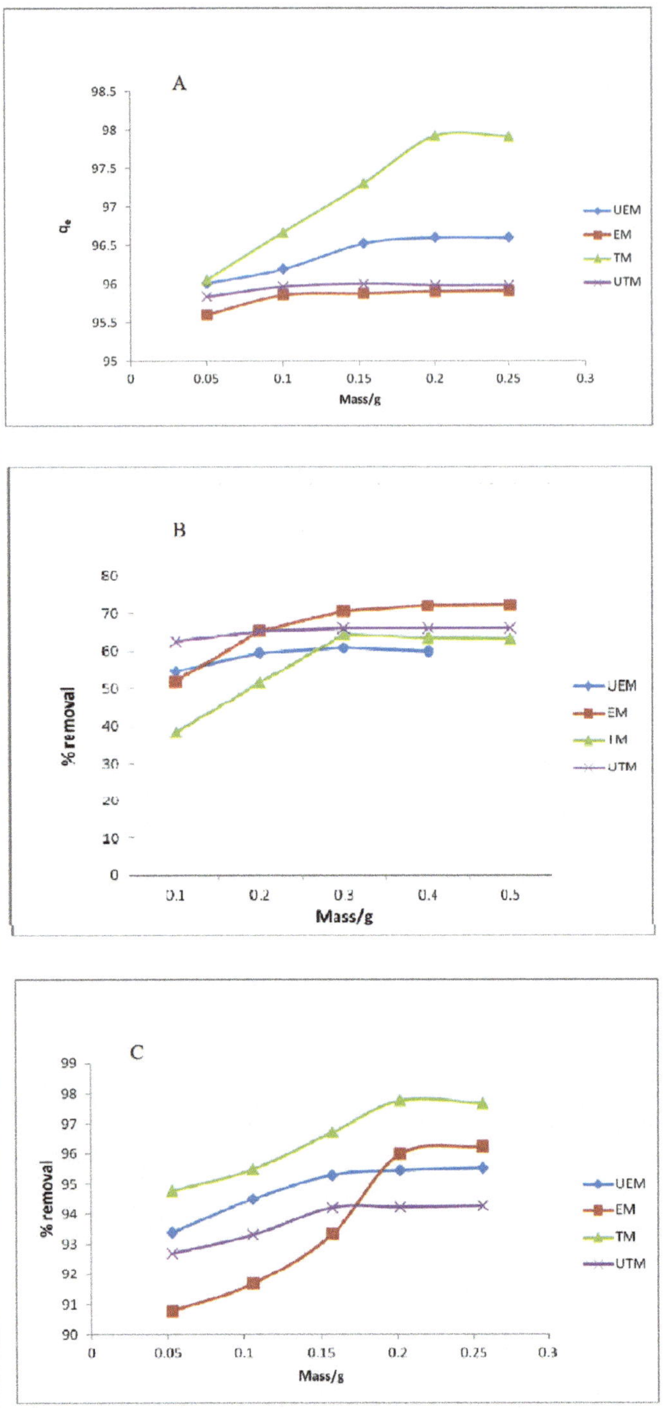

Figure 7: Effect of sorbent dose on adsorption of cadmium (A) copper (B) and lead (C) ions.

higher removal ability which is reduced upon modification due to the same reason of derivatization of the functional group.

The removal efficiency of copper ions as a function of sorbent dose was also investigated (Figure 7B). The results show that the adsorption of copper ions was that the UTM had the highest removal ability (74.7%) when the mass was above 0.3 g. The UEM had the least ability (65.7%) considering the same mass of adsorbent. This was due to the functional groups which bind metals during the sorption process. For the TM, adsorption was above 70% when the adsorbent dose was 0.3 g and 66.26% with adsorbent dose of 0.3 g for UTM and EM.

The percentage removal of lead with TM and EM (modified sorbents) was highest with 97.79% at 0.2 g and 96.24% at 0.25 g respectively for Pb^{2+} ions. This could have been attributed to the fact that the binding sites available in the modified (edible and toxic) mushroom had high affinity for lead ions. Also possibly because modification of raw sorbent with a base improves the content of reactive centers such as carbonyl group of amides hence relatively higher affinity for these metal ions in modified adsorbent [95]. However, maximum adsorption for UEM and UTM (unmodified sorbents) was lower with 95.3% for 0.15 g and 94.2% for 0.15g respectively (Figure 7C).

The general observation made for all the three metals was that, adsorption increased with increase in sorbent dose. This is because the more the sorbent, the more the site of adsorption hence more metal ions will occupy the available adsorption sites [117]. However, the increase of adsorbent against constant metal ions concentration leads to depletion of metal ions in solution hence a plateau. However, unmodified mushroom leached brown colour in the solution during the adsorption process unlike modified sorbent. This is a secondary pollution of water while being treated for the removal of metals. There is need to modify the raw mushroom before their use in the water treatment process, in order to prevent secondary pollution [118].

Effect of initial metal ion concentration (Determination of adsorption capacity): Metal adsorption is significantly influenced by the initial concentration of metal ions in aqueous solutions. In this study, the initial metal ion concentration was varied from 2-200 $mg.l^{-1}$ maintaining the adsorbent dosage at 0.02 g. To obtain the sorption capacity of the bio sorbents, the data collected for each of the three metals was plotted as a function of concentration of the metal ions. The adsorbed metal ions were in $mg.g^{-1}$ of the sorbent (q_e) against the initial metal concentration in solution (C_o). Figures 8A-8C shows results on the effect of initial metal ion concentration on the adsorption.

The Results show that, the maximum adsorption capacity for cadmium ions was found to be 0.47851 $mg.g^{-1}$ on the UEM, 0.6804 $mg.g^{-1}$ on EM, 0.3615 $mg.g^{-1}$, for the UTM and 0.4755 $mg.g^{-1}$ for the TM. The unmodified sorbent had a very low uptake as observed from the sorption plots. This confirms that the modification has a positive effect on sorption of the metals. The profile of sorption of cadmium was observed to increase with the increase in the initial metal concentration up to 150 $mg.l^{-1}$ after which it levels off. An overall observation for the sorption of cadmium show a continuous metal uptake followed by a plateau as the concentration of the metal in the solution was increased.

Sorption of copper showed that the sorption of copper sorption on all the adsorbent increased with increase in concentration up to a maximum at initial metal ion after which a plateau was obtained. At that level, the adsorbent is said to be saturated hence the adsorption capacity [119]. The adsorption capacities of copper by the adsorbents were calculated using the Langmuir and Freundlich models.

The results for lead shows that, the rate of sorption by the adsorbent materials increased with increase in concentration up to a maxima at initial metal ion after which a plateau is attained due to be saturation of the active adsorbent sites [119]. From graph, it was found out that a plateau was leveling at a value of 0.36198 $mg.g^{-1}$ for the UEM, 0.56966 $mg.g^{-1}$ g for the EM, 0.58338 $mg.g^{-1}$ for the UTM and 0.49292 $mg.g^{1}$ for the TM at maximum initial concentration of 150 $mg.l^{-1}$. The

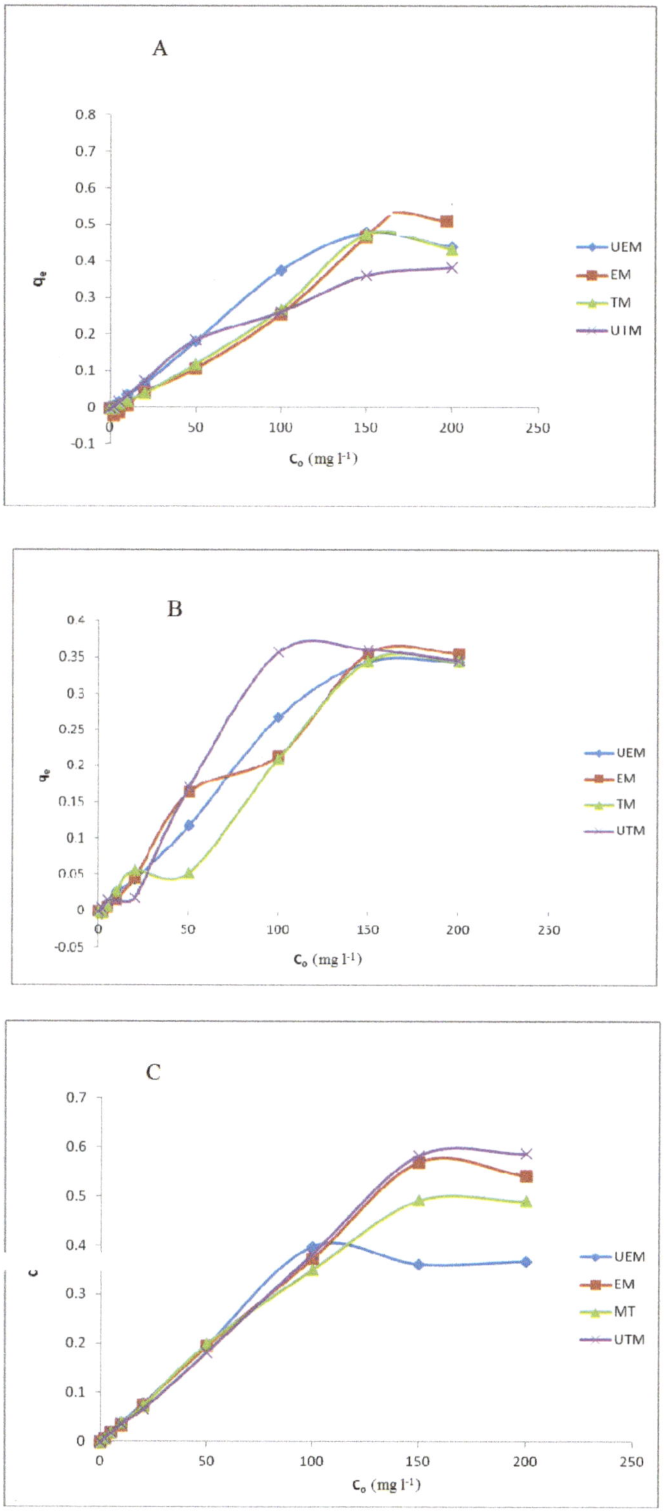

Figure 8: Effect of initial metal ion concentration on adsorption cadmium (A) copper (B) and lead (C).

enhancement of bio sorption capacity of lead ions with increasing initial concentration could be partially explained by higher availability of metal ions for bio-sorption.

Generally the profile of sorption of metal ions was observed to increase with an increase in the initial metal concentration in all the three metals up to 150 mg.l^{-1} after which a plateau was observed for all except for modified edible mushroom. The results also indicate that the metal ions uptake tends to saturate as the initial concentration is increased. This may be explained by a progressive increase in electrostatic mutual interaction between sites that have lower affinity for metal ions as the population of occupied sites increases [120]. A similar trend has been observed in removal of divalent metal ions copper, cadmium, zinc and lead ions by other bio [121-123].

It is also observed that sorption at low metal concentration was high as the sorbent to possess many sites with high affinity for the sorbate species [124]. An overall observation for all the three metals ions shows a continuous metal uptake followed by plateau as the concentration of the metal ion in the solution was increased. This may be due to an increase in metal ions for the fixed amount of the adsorbent material and due to saturation of available sites on the sorbent resulting in a saturation effect showing a steady state on the regression plot [119]. At this stage, the sorbent is said to have attained its operational maximum adsorption capacity.

The transmittance of certain peaks like those of sorption for cadmium, copper and lead ions using TM and sorption for lead ions using EM was observed substantially lower than unmodified. This may be an indication of a lesser degree of bond stretching due to change in chemical environment [125].

Adsorption isotherms: Adsorption isotherm models commonly describe the adsorption phenomena at the solid-liquid interface. They are essential data source for practical design of adsorption systems by determining factors such as adsorption capacity and understanding of the relation between adsorbent and adsorbate as well as the sorption mechanism [126]. Data from the Langmuir and Freundlich models were used to determine sorption capacity and mechanism of sorbent-sorbate interaction in this study. This analytical method was important in the comparison of different adsorbents. The rate of sorption was established using the Lagergreg and Ho's kinetic models.

Langmuir isotherms: Langmuir isotherm has been applied extensively for the sorption of many heavy metal ions. Langmuir assumes adsorption of an ideal gas on an ideal surface, that sorption occurs on a homogeneous surface; no interaction between adsorbed molecules and that adsorption can only occur at fixed sites and can only hold one adsorbate molecule at a time (monolayer). A further assumption is that no transmigration of adsorbed molecules on the adsorption surface [127]. Equation 2 shows the Langmuir equation.

$$\frac{C_e}{q_e} = \frac{1}{bQ_{max}} + \frac{C_e}{Q_{ma}X} \qquad (2)$$

Where; q_e is the amount of solute adsorbed (mg/g), C_e is the equilibrium concentration of the solute in the bulk solution (mg/L), Q_{max} is the monolayer adsorption capacity mg.g^{-1} and **b** is the constant related to the energy of adsorption (L/g). The experimental data were applied to the equation above and a plot of $\frac{C_e}{q_e}$ against C_e ploted. A plot with a good linear correlation indicates that the Langmuir model describes the adsorption process [128,129]. Figure 9 below shows an example Langmuir plots for sorption of cadmium by UEM. Figure 10 below shows an example Langmuir plots for sorption of copper by TM.

Figures 9 and 10 shows Langmuir plots for sorption of Cd^{2+} by UEM with R^2 = 0.8829 and that of Cu^{2+} by TM had R^2 = 0.8782. The

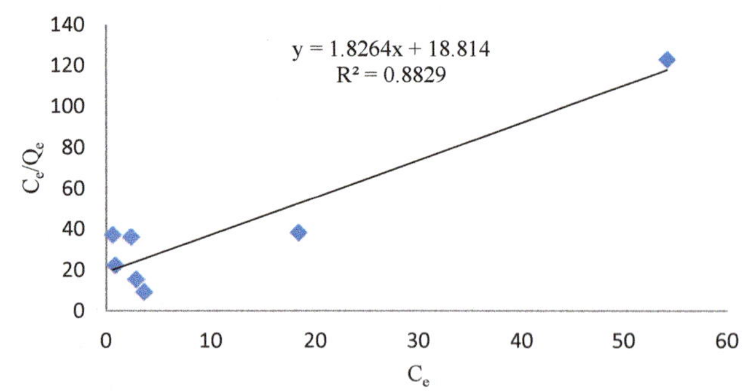

Figure 9: Langmuir adsorption plot for cadmium by unmodified edible mushroom.

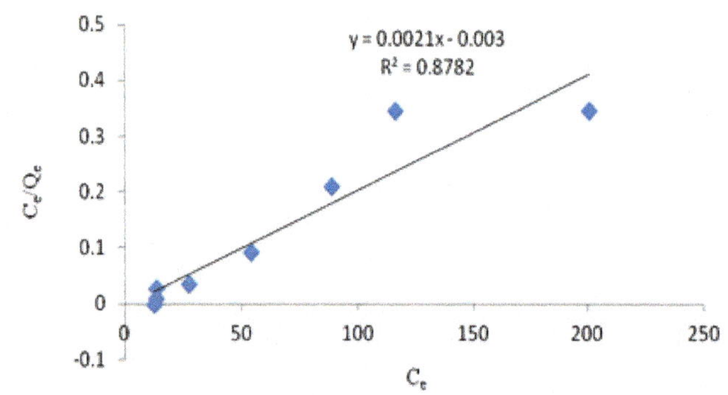

Figure 10: Langmuir adsorption plot for copper by modified toxic mushroom.

sorption of cadmium ions and copper ions with the same biosorbents gave an R^2 value > 0.5 and slightly less than 0.9 indicating that the data has a strong correlation, hence the sorption mechanism can prescribe the Langmuir model. Whenever the R^2 value is lower than 0.5, it can be concluded that the scatter had no correlation and that the adsorption cannot prescribe the Langmuir model and can probably be described by another model.

Freundlich isotherms: The Freundlich isotherm assumes a heterogeneous surface with non-uniform distribution of adsorption heat over the surface (binding sites are not equivalent) and/or a multilayer adsorption [130]. The mono-component Freundlich isotherm equation is given by equation 3 below.

$$\log q_e = \log K_f + \frac{1}{n} \log C_e \qquad (3)$$

Where $\boldsymbol{K_f}$ is the Freundlich isotherm constant related to the sorption capacity, $\boldsymbol{n}$ is the constant related to affinity of metal ions on adsorbent, $\boldsymbol{q_e}$ and $\boldsymbol{C_e}$ has the same meaning as that of Langmuir. By plotting $\log q_e$ versus $\log C_e$, the constant n and K_f can be determined from the slope and intercept, respectively. Good linear correlation implies that the sorption fits the Freundlich model. Figure 11 shows an example of Freundlich plots for sorption of cadmium UEM.

The results shows Freundlich plots for sorption of cadmium by UEM with $R^2 = 0.5337$. This implies that sorption prescribes to a Freundlich model. Figure 12 below shows an example of Freundlich plots for sorption of copper TM.

The results shows Freundlich plots for sorption of copper by TM had $R^2 = 0.916$. This implies that sorption prescribes to a Freundlich model. In both cases of copper and cadmium, it means that the interactions is multi-site adsorption for heterogenous surfaces expecially for highly interactive surfaces. The experimental data from the results extracted on the on the effect of sorption on concentration on sorption obtained upon treatment with both the Langmuir and Freundlich isotherms were recorded in a tabular form. Table 1 below shows the results obtained for adsorption isotherm of cadmium ions.

Table 1 shows that for cadmium ions, EM and TM best fit Freundlich model with $R^2 = 0.8296$ and 0.9211 respectively. UEM and UTM with $R^2 = 0.8829$ and 0.9693 best fit the Langmuir model. The adsorption capacities (Q_{max}) of cadmium ions from Langmuir isotherm model was determined to be 1.826 mg/g for UEM, 25.285 mg/g for EM, 2.3353 mg/g for UTM and 6.5901 for TM. Freudlich constants were calculated for Cd^{2+} where the K_f value was found to be 3.3508 mg/g for UEM, 29.7075 for EM, 4.3947 for UTM and 9.8404 for TM.

Generally for cadmium ions, modified biosorbents had high adsorption capacity on fitting to both Langmuir and Freundlich isotherm models. Modification could possibly have led to stronger electrostatic forces of attraction between negatively charged surface of adsorbent and positively charged sorbate ions, hence efficiency towards removal improved. Table 2 shows the results obtained for adsorption isotherm of copper ions.

Table 2 shows that for copper ions, UEM, EM, UTM and TM

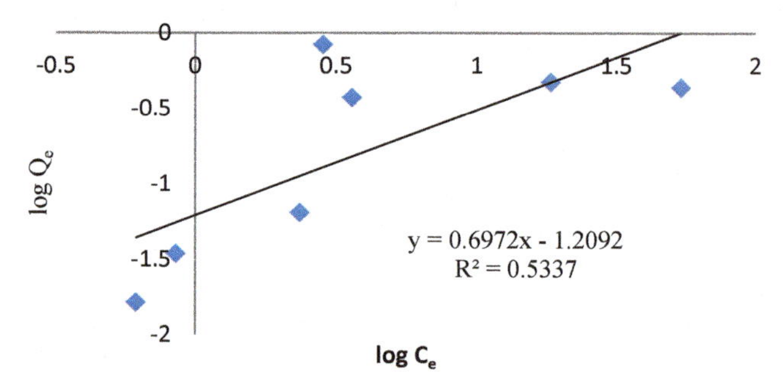

Figure 11: Freundlich adsorption plot for cadmium by unmodified edible mushroom.

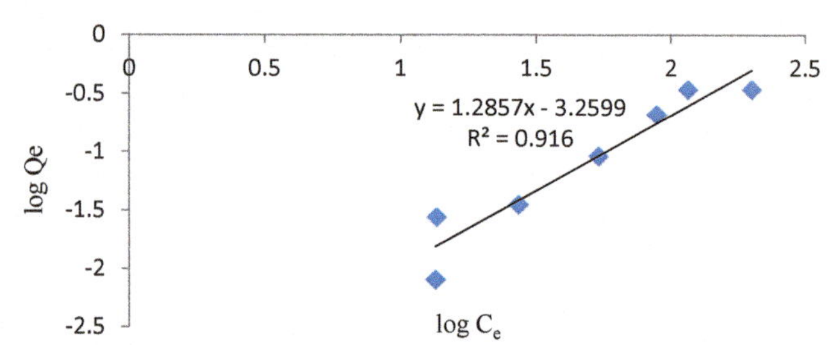

Figure 12: Freundlich adsorption plot for copper by modified toxic mushroom.

	Langmuir		Freundlich		
Sample	Q_{max} mg/g	R^2	K_f mg/g	R^2	Best Model
UEM	1.826	0.8829	3.3508	0.5337	Langmuir
EM	25.285	0.7082	29.7075	0.8296	Freundlich
UTM	2.3353	0.9693	4.3947	0.7117	Langmuir
TM	6.5901	0.0704	9.8404	0.9211	Freundlich

Table 1: Adsorption kinetics of cadmium ions.

best fit Freundlich model with R^2 = 0.8792, 0.8553, 0.635 and 0.916. Adsorption capacities (Q_{max}) was found to be 4.098 for UEM, 2.2136 for EM, 1.7524 for UTM and 0.0021 for TM. UEM was found to have highest adsorption capacity of 4.0978 mg/g. This is possibly because copper species involved are highly interactive with the sorbents [129]. Huang and Vansant also made similar observations and concluded that the electrostatic components, polarization energy and field-dipole energy were important contributors to the phenomenon. The sorption of copper, therefore, could be as a result of more than one mechanism and that a degree of heterogeneity is possible for ionic species involved in the solution and on the surface. Physical adsorption due to Van der Waals forces of attraction could also have contributed to the interaction phenomenon [131]. Table 3 shows the results obtained for adsorption isotherm of lead ions.

(Q_{max}) of lead ions from Langmuir isotherm model was determined to be 2.5936 mg/g for UEM, 1.5536 mg/g for EM, 1.7355 mg/g for UTM and 1.3458 mg/g for TM. K_f from Freundlich model was determined to be 3.6255 mg/g for UEM, 4.3740 mg/g for EM, 0.5568 mg/g and 0.8081 for TM. Adsorption capacities (Q_{max}) for sorption of lead by unmodified biosorbents (both edible and toxic) was high with 2.5936 mg/g for UEM and 1.7355 mg/g for UTM compared with those for modified biosorbents.

Generally for the three metal ions, the adsorption capacity of cadmium ions was higher than that of lead and copper ions in the modified biosorbents. This may be because of the presence of sulfoxide group in the adsorbent which has a selective high affinity for cadmium ions than copper and lead ions [132].

Sorption kinetics

Kinetic studies: The rates of metal adsorption studies were analysed using Lagergreg first-order and Ho's second-order kinetics [133,134]. This gave an insight into the number of molecules taking part in each step of the adsorption. The experimental data were fitted to both the Lagergreg first order and Ho's second-order kinetic expression so as to investigate the molecularity of the adsorption mechanism and the rate-controlling steps [133,134]. The Lagergreg first-order and Ho et al.'s second-order kinetic models are given in equations 4 and 5 respectively.

$$\ln(C_o - C_t) = Kt + A \quad (4)$$

$$\frac{1}{q_e} = Kt + A \quad (5)$$

Where

C_o is the adsorption per unit mass of adsorbent at equilibrium;

K is the adsorption rate constant;

A is the intercept;

C_t is the concentration at time t.

The experimental data for the three metals were then fitted to the Lagergreg pseudo first-order and Ho et al.'s pseudo second-order kinetic models.

The experimental data from the results on the effect of time on sorption obtained were treated with both the kinetics isotherms. The results obtained were as presented in the Tables 4-6. Table 4 below shows kinetics isotherm of cadmium ions with all the adsorbents.

From Table 4, when the experimental data for cadmium was treated with the Lagergreg first-order kinetics and a plot of ln (C_0–C_t) against time obtained, it gave a correlation coefficient value (R^2) of 0.5882, 0.5794, 0.7839 and 0.9578 for UEM, EM, UTM and TM respectively which is high (expected value is 1.0), indicating that there was strong correlation in the scatter. But when treated with Ho's second-order kinetics, the R^2 value was lower than pseudo first order with 0.3585, 0.5503, 0.5955 and 0.8623 for UEM, EM, UTM and TM respectively. This indicated that the sorption of cadmium ions with all the sorbents was likely to be pseudo first order hence suggests that one mechanism controls the adsorption process. Table 5 shows kinetics isotherm of copper ions with UEM, EME, UTM and TM.

From Table 5, when the experimental data for copper was treated with both kinetics isotherms and R^2 values were obtained for all the adsorbents, high value of R^2 were obtained when the data were treated

	Langmuir		Freundlich		
Sample	Q_{max} mg/g	R^2	K_f mg/g	R^2	Best Model
UEM	4.0978	0.0312	24.7667	0.8792	Freundlich
EM	2.2136	0.4773	25.5694	0.8553	Freundlich
UTM	1.7524	0.0386	19.45687	0.635	Freundlich
TM	0.0021	0.8782	26.0469	0.916	Freundlich

Table 2: Adsorption isotherm of copper ions.

	Langmuir		Freundlich		
Sample	Q_{max} mg/g	R^2	K_f mg/g	R^2	Best Model
UEM	2.5936	0.991	3.6255	0.4702	Langmui
M	1.5536	0.8519	4.374	0.7765	Langmuir
UTM	1.7355	0.9175	3.9788	0.5568	Langmuir
TM	1.3458	0.7569	4.5557	0.8081	Freundlich

Table 3: Adsorption isotherm of lead ions.

	Lagergren pseudo first-order		Ho's pseudo second-order		
Sample	R^2	K	R^2	K	Comments
UEM	0.5982	0.0423	0.3585		Sorption, first order
EM	0.5794	0.006	0.5503		Sorption, first order
UTM	0.7839	0.0182	0.5955		Sorption, first order
TM	0.9578	0.0151	0.8623		Sorption, first order

Table 4: Sorption kinetics studies of cadmium ions.

	Lagergreg pseudo first- order		Ho's pseudo second-order		
Sample	R^2	K	R^2	K	Comments
UEM	0.7044		0.7091	0.05	Sorption, second- order
EM	0.6637	0.024	0.6635		Sorption, first- order
UTM	0.8295		0.8306	0.06	Sorption, second- order
TM	0.6281	0.04	0.6274		Sorption, first order

Table 5: Sorption kinetics studies of copper ions.

	Lagergreg pseudo first-Order		Ho's pseudo second-order		
Sample	R^2	K	R^2	K	Comments
UEM	0.6752		0.6809	0.051	Sorption, second- order
EM	0.9159		0.9162	0.06	Sorption, second- order
UTM	0.7719		0.7727	0.056	Sorption, second- order
TM	0.9096		0.9104	0.061	Sorption, second- order

Table 6: Sorption kinetics studies of lead ions.

with Lagergreg pseudo first –order with modified adsorbents compared to unmodified sorbents which exhibited Ho's pseudo second-order. This is possibly because upon modification only one functional group is available in the adsorbent hence demonstrating a single site reaction. For unmodified adsorbents, many functional groups were available hence more than one site available for interactions. The sorption of copper ions on UEM and UTM had higher R^2 values of 0.7091 and 0.8306 respectively hence Ho's pseudo second order, with Lagergreg pseudo first order had a lower value of 0.7044 and 0.6637 for UEM and UTM respectively. For modified sorbents EM and TM, R^2 value was 0.6637 and 0.6281 respectively when treated with Lagergreg pseudo first order. When treated with Ho's pseudo second order R^2 was slightly lower with 0.6635 for EM and 0.6274 for TM. Table 6 shows kinetics isotherm of lead ions with UEM, EM, UTM and TM.

Table 6 above shows experimental data for lead ions where the two chemical reaction rate models used in fitting the experimental data gave an indication of the reaction pathway. From the two models, the pseudo second-order equation fits the experimental data better with a high correlation coefficient R^2 with all the adsorbents. On the basis of this parameter fitting, the biosorption of Pb^{2+} onto the *Agaricus biporus* and *Amanita phalloides* is a second order reaction, and not a first-order one. This is in agreement with the findings of Ghaedi et al. [135], who studied the biosorption of Pb^{2+} ions onto *S. cerevisiae* biomass R^2 for UEM, EM, UTM and TM was 0.6809, 0.9162, 0.7727 and 0.9104 respectively with Ho's pseudo second order. This is slightly higher than Lagergren pseudo second order with R^2 value of 0.6752, 0.9159, 0.7719 and 0.9096 for UEM, EM, UTM and TM respectively.

Analysis of environmental water samples: The results of river samples analysis for the content of copper, lead and cadmium ions are given in the Figures 2-6 [27,28]. Figure 13 shows the profile of percentage removal on the effect of metal ion concentration on adsorption of cadmium using river water.

Figure 14 shows the profile of percentage removal on the effect of metal ion concentration on adsorption of copper using river water.

Generally, the results in Figures 10-12 show that there was high removal of metal ions at low concentration but decreased with an increase in the spiked metal ions concentration. This may be due to an increase of the metal ions against constant adsorbent material (Figure 15).

The sorption of lead by modified adsorbents was found to be the highest followed by copper and then cadmium. This adsorption trend may be explained by the fact that copper ions compared to lead ions, have relatively higher affinity for ligands containing a quaternary N-atom in dissolved organic carbon (DOC) [136]. This would result in the removal of copper by the DOC and hence posing less competition for the lead ions to adsorb on the modified adsorbents [136]. Such ligands, which are smaller than the adsorbent, have a higher affinity for the metal due to their high basicity [137]. The above result shows that the method was effective and reliable as it achieved with recovery values similar to those obtained in model solutions.

Dissolved organic carbon: Absorbance readings to determine the DOC were taken at an optimized wavelength of 606 nm. A calibration curve was prepared whose linear equation was describing was: $y = 0.0052 + 0.006$ and regression $R^2 = 0.983$. This closely agrees to one given by Lewis et al. [138] when he used a 10 mm cell, as they analyzed some Venezuelans and southeastern USA waters. The average DOC content for the modified material was 2.19 mg/L whereas the unmodified material had 19.0 mg/L. The results indicate that modification had a significant effect on minimizing leached organic matter in the treated water.

Conclusions

This study successfully functionalized the mushroom which was confirmed by FTIR analysis. It was confirmed that sorbent modification reduced the problem of leaching organic matter during the treatment process. The average DOC content leached by the modified material was 2.19 mg/L whereas the unmodified material had 19.0 mg/L. This implies that modification was alleviating the secondary pollution phenomena. Kinetic studies confirmed that sorption for lead was of pseudo-second-order but pseudo-first-order for both cadmium and copper. The sorption experiments also revealed that 90% of the metal ions were removed within the first 20 minutes of contact time. The adsorption capacity for cadmium ions was 1.826 mg/g by the unmodified edible mushroom (UEM), 25.285 mg/g by the modified edible mushroom (EM), 2.335 mg/g by the unmodified toxic mushroom (UTM) and 6.5901 mg/g by the modified toxic mushroom (TM). For copper ions, sorption capacity was found to be 4.097 mg/g, 2.213 mg/g, 1.752 mg/g 0.002 mg/g by the UEM, EM, UTM and TM

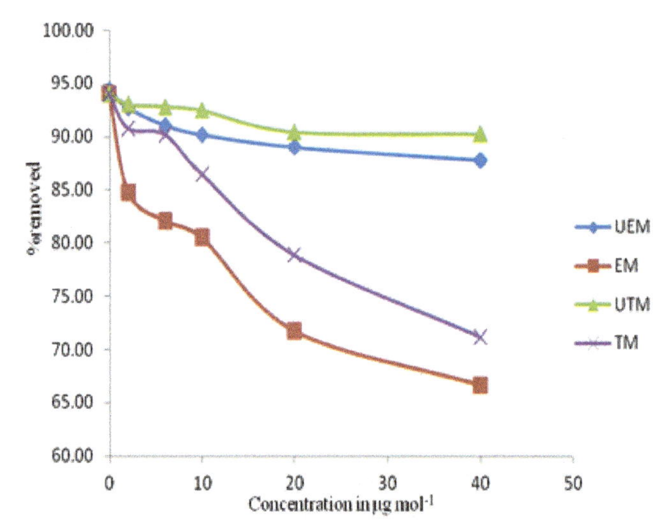

Figure 13: Effect of metal ion concentration on adsorption of cadmium using river water.

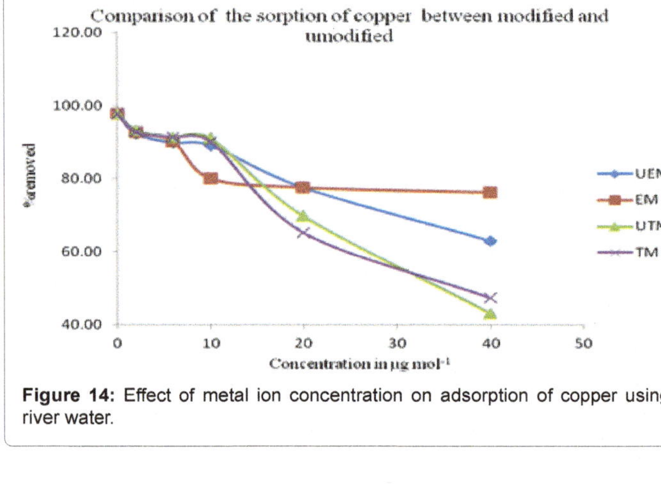

Figure 14: Effect of metal ion concentration on adsorption of copper using river water.

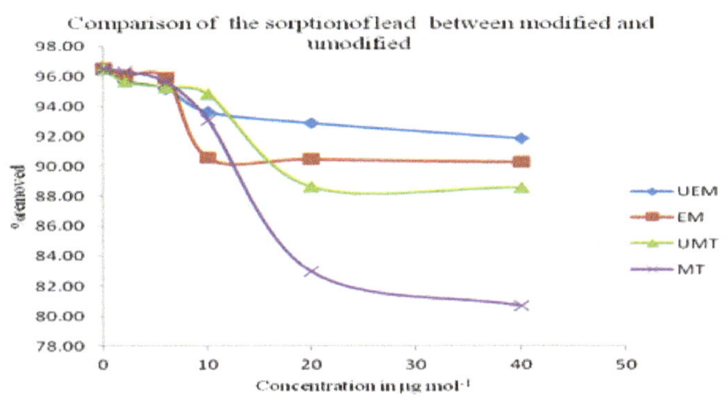

Figure 15: Effect of metal ion concentration on adsorption of lead using river water.

respectively. The sorption for the lead ions was found to be 2.593 mg/g, 1.553 mg/g, 1.733 mg/g and 1.345 mg/g by the UEM, EM, UTM and TM respectively. The results show that the calculated sorption capacities obtained from the modification had a negative effect on sorption of the respective metals but there were numerous advantages such as stability and secondary pollution. It was also observed that the sorbent in the SPE column could be regenerated by stripping the attached metal ions with a dilute acid (1.0 molL^{-1} HNO_3) thus the material could be regenerated when exhausted. In conclusion, our study has shown that the modified mushroom can be used as an effective sorbent for the removal of copper, cadmium, and lead ions in the remediation of contaminated water samples at sub-neutral pH range.

References

1. Soteris A (2005) Seawater desalination using renewable energy sources. Progress in Energy and Combustion Science 31: 242-281.
2. Srivastava N, Majumder CB (2008) Novel bio-filtration methods for the treatment of heavy metals from industrial wastewater. Journal of hazardous materials 151: 1-8.
3. Sarkar B (2005) Heavy metals in the environment. Journal of Environmental Science 23: 33-41.
4. Abu Al-Rub F, El-Naas M, Ashour I, Al-Marzouqi M (2006) Biosorption of copper on Chlorella vulgaris from single, binary and ternary metal aqueous solutions. Biochemistry Processes 41: 457-464.
5. Jarup L (2003) Hazards of heavy metal contamination. Br Med Bull 68: 167-182.
6. Aksu Z, Kutsa T (1991) A Bioseperation process for removing lead (II) ions from wastewater by using vulgaris. Journal of Chemical Technology and Biotechnology 52: 109-118.
7. Ragan P, Turner T (2009) Working to prevent lead poisoning in children: getting the lead out. JAAPA 22: 40-45.
8. Alloway BJ (1995) Heavy Metals in Soils. Blackie Academic and Proffesional 3: 67-115.
9. Alloway BJ (1998) Soil pollution and contamination, causes, effects and control. Royal Society of Chemistry 3: 319-358.
10. Baccar R, Bouzid J, Feki M, Montiel A (2009) Preparation of activated carbon from Tunisia olive–waste cakes and its application of heavy ions. J Hazard Mater 162: 1522-1529.
11. Renault F, Sancey B, Badot P, Crini G (2009) Chitosan for coagulation / flocculation processes –An eco-friendly approach. European Polymer Journal 45: 1337-1348.
12. Pique A, Grandia F, Canals A (2010) Processes releasing arsenic to groundwater in the Caldes de Malavella geothermal area, NE Spain. Water Res 44: 5618-5630.
13. Bae W, Chen W, Mulchandani A, Mehra RK (2000) Enhanced bioaccumulation of heavy metals by bacterial cells displaying synthetic phytochelatins. Biotechnol Bioeng 70: 518-524.
14. Bailey SE, Olin TJ, Bricka RM, Adrian DD (1999) A review of potentially low cost sorbents for heavy metals. Water Research 33: 2469-2479.
15. Brewer GJ (2010) Copper toxicity in the general population. Clin Neurophysiol 121: 459-460.
16. Bushra A, Misbra A, Syeda IZ, Humera (2014) Soils and vegetables enrichment with heavy metals from application sewage. Journal of environment and earth science 4: 68-83.
17. Cha D, Song J, Sarr D (1997) Treatment technologies. Water Environment Resource 69: 676-689.
18. Lodeiro P, Barriada J, Herrero R, de Vicente MES (2006) The marine macroalga cystoseira baccata as biosorbent for cadmium (II) and lead (II) removal: Kinetic and equilibrium studies. Environmental Pollution 142: 264-273.
19. Abia AA, Horsfall M Jr, Didi O (2003) The use of chemically modified and unmodified cassava waste for the removal of Cd, Cu and Zn ions from aqueous solution. Bioresource Technology 90: 345-348.
20. Chen, JP, Yang L (2005) Chemical modification of sargassum sp. for prevention of organic leaching and enhancement of uptake during metal biosorption. Ind Eng Chem Res 44: 9931-9942.
21. Mwangi IW, Ngila JC, Okonkwo JO (2012) A comparative study of modified and unmodified maize tassels for removal of selected trace metals in contaminated water. Toxicological & Environmental Chemistry 94: 20-39.
22. Chen Z, Luo J, Sun Y (2007) Biocidal efficacy, biofilm-controlling function, and controlled release effect of chloromelamine-based bioresponsive fibrous materials. Biomaterials 28: 1597-1609.
23. Cai Y, Shaw WD, Wu X (2008) Risk perception and altruistic averting behavior: removing arsenic in drinking water. Unpublished paper selected for presentation at the American agricultural economics association annual meeting, Orlando F L Biotechnology 41: 826-829.
24. Chang ST, Miles PG (1992) Mushroom biology a new discipline. The Mycologist 6: 64-65.
25. Chaudhari A, Nagpurkar L, Ekhe J (2003) Uptake of heavy metal ions by carbonaceous material obtained from industrial waste lignin using microwave irradiation. Asian journal of chemistry 334: 917-924.
26. Chipasa KB (2003) Accumulation and fate of selected heavy metals in a biological wastewater treatment system. Waste manag 23: 135-143.
27. Crump JA, Okoth GO, Slutsker L, Ogaja DO, Keswick BH, et al. (2004) Effect of point-of-use disinfection, flocculation and combined flocculation-disinfection on drinking water quality in western Kenya. J Appl Microbiol 97: 225-231.
28. Khan MA, Tania M, Liu R, Rahman MM (2013) Hericium erinaceus: An edible mushroom with medicinal values. J Complementary Integr Med 10: 1-6.
29. Allen RO, Steinnes E (1978) Concentrations of some potentially toxic metals and other trace elements in wild mushrooms from Norway. Chemosphere 4: 371 -378.
30. Deans JR, Dixon BG (1992) Uptake of lead and copper ions by novel biopolymer. Water researc 26: 469-472.
31. Donovan M, Hall L, Torek V, Schrock C, Wooldridge M (2002) Demonstration of temperature and -OH mole fraction diagnostic in flames using narrow-line UV- OH absorption spectroscopy. Proceedings of the Combustion Institute 2: 2635-2643.
32. Ediin G, Golantu E, Brown M (2000) Essentials for health and wellness. Health Sciences 4: 368-374.
33. Esser H, Brunnert J (1986) Isolation and partial purification of cadmium-binding components from fruiting bodies of Agaricus bisporus. Environmental Pollution 41: 263-275.
34. Feng N, Guo X (2012) Characteristic of adsorptive capacity and mechanisms on adsorption of copper, lead and zinc by modified orange peel. Journal of Transactions of Nonferrous Metals Society of China 22: 1224-1231.
35. Feng N, Guo, X, Liang S (2009) Adsorption study of copper (II) by chemically modified orange peel. Journal of Hazardous Materials 164: 1286-1292.

36. Friberg L, Nordberg G, Vouk B (1979) Handbook on the toxicology of metals, Elsevier, North-Holland. Biomedical Press 7: 552-555.

37. Groark KP (2010) The Angel in the Gourd: Ritual, Therapeutic, and Protective Uses of Tobacco (Nicotiana tabacum) Among the Tzeltal and Tzotzil Maya of Chiapas, Mexico. Journal of Ethnobiology 30: 5-30.

38. Garg U, Kaur MP, Jawa GK, Sud D, Garg VK (2008) Removal of cadmium (II) from aqueous solutions by adsorption on agricultural waste biomass. Journal of hazardous materials 154: 1149-1157.

39. Giorgi A, Malacalza L (2002) Effect of an industrial discharge on water quality and periphyton structure in a pampeam stream. Environ Monit Assess 75: 107-119.

40. Glassmeyer ST, Furlong ET, Kolpin DW, Cahill JD, Zaugg SD, et al. (2005) Transport of chemical and microbial compounds from known wastewater discharges: potential for use as indicators of human fecal contamination. Environ Sci Technol 39: 5157-5169.

41. Goel PK (2006) Water pollution: causes, effects and control. Newage International 5: 123-130.

42. Graeme K, Pollack C (1998) Heavy metal toxicity; Lead and metal fume fever. Journal of Emergency Medicine16: 171-177.

43. GoK (Government of Kenya) (2004) Kenya country strategy, water and sanitation sector-millennium development goals: Needs assessment report prepared by Dr PMA Odira. Nairobi: Ministry of planning and national development.

44. GoK (Government of Kenya) (2007) Kenya Vision 2030: A Globally competitive and prosperous Kenya. Nairobi: Ministry of planning, national development and vision 2030.

45. GoK (Government of Kenya) (2008) Vision 2030: Sector plan for environment, water and sanitation 2008-2012. Nairobi: GoK.

46. Herrero R, Lodeiro P, Rojo R, Ciorba A, Rodriguez P, et al. (2008) The efficiency of the red alga Mastocarpus stellatus for remediation of cadmium pollution. Bioresour Technol 99: 4138-4156.

47. Pollution (2013). Bioresource Technology. 99: 4138-4156.

48. Hong J, Kim J, Choi J (2002) Hapten synthesis and influence of coating ligands on enzyme-linked immunoreaction. Bulletin-Korean Chemical Society 10: 1413-1419.

49. Jalan J, Somanathan E (2008) The importance of being informed: experimental evidence on demand for environmental quality. Journal of development economics 87: 14-28.

50. Kadirvelu K, Namasivayam C (2000) Agricultural by-products as metal adsorbents: Sorption of lead (II) from aqueous solutions onto coir-pith carbon. Journal of Environmental Science and Technology 21: 1091-1097.

51. Kandasamy J, Vignesswaran T, Chaudhary D (2008) Waste water treatment and technologies. Adsorption and Biological Fitration and water treatment. Encyclopedia of Life Support Systems 30: 173-187.

52. Katana C, Murungi J, Harun, M (2013) Speciation of chromium and nickel in open-air automobile mechanic workshop soils in Ngara, Nairobi, Kenya. World environment 3: 143-154.

53. Kithiia SM (2010) Water Quality Degradation Trends in Kenya over the Last Decade. University of Nairobi, Nairobi, Kenya.

54. Kratochvil D, Volesky B (1998) Advances in the bio-sorption of heavy metals. Trends in biotechnology 16: 291-300.

55. Kremer MJ, Leino E, Miguel AP (2011) Spring cleaning: A Randomized evaluation of source water quality improvement. The quarterly journal of economics 126: 145-205.

56. Kumar U (2006) Agricultural products and by-products as a low cost adsorbent for heavy metal removal from water and wastewater. Journal of Scientific Research and Essay 1: 33-37.

57. Lo W, Chua H, Lam K, Bi S (1999) A comparative investigation on the biosorption of lead by filamentous fungal biomass. Chemosphere 39: 2723-2736.

58. Michael K, Albert QK, Antwi CK (2013) Role of maize tassels in Amelioration of heavy metals from contaminated soils and its effects on vegetables. Journal of environment and earth science 3: 192-198.

59. Michelot D, Siobud E, Dore JC, Viel C, Poirier F (1998) Update of metal content profiles in mushrooms - toxicological implications and tentative approach to the mechanisms of bioaccumulation. Toxicon 36: 1997-2012.

60. Abasi C, Abia A, Igwe J (2011) Adsorption of iron (III), lead (II) and cadmium (II) ionsby unmodified raphia palm (Raphiahookeri) fruit endocarp. Environmental Research Journal 3: 104-113.

61. Miles PG, Chang ST (1997) Mushroom biology: concise basics and current developments. World scientific publishing company Pvt. Ltd., Singapore 32: 112-152.

62. Morán J, Alvarez V, Cyras V, Vázquez A (2008) Extraction of cellulose and preparation of nanocellulose from sisal fibers. Cellulose 15: 149-159.

63. Pereira J, Sousa A, Coelho H, Amado AM, Ribeiro-Claro PJA (2003) Use of FTIR, FT-Raman and 13C–NMR spectroscopy for identification of some seaweed phycocolloids. Biochemical engineering 20: 223-228.

64. Tashiro T, Shimura Y (1982) Removal of mercuric ions by systems based on cellulose derivatives. Journal of applied polymer science 27: 747-756.

65. Perkin-Elmer (1957) The infracord double-beam spectrophotometer. Clinical Science 2: 16.

66. Poe GL, Bishop RC (1999) Valuing the incremental benefits of groundwater protection when exposure levels are known. Environmental and resource economics. 13: 341-367.

67. Richardson JT (1967) Crystal field effects in ion-exchanged faujasites. Journal of Catalysis 9: 178-181.

68. Sadhna G (2013) Study of toxic effects of heavy metals in soil and water. Journal of environment and health science 3: 68-75.

69. Saracoglu S, Soylak M, Elei L (2001) Preconcentration of Cu (II), Fe (III), Ni (II), co (II) and Pb (II) ions in some manganese salts with solid phase extraction method using chromosorb-102 resin. Trace Element Electrolytes 18: 129-133.

70. Satheeshkumar P, Senthilkumar D (2011) Identification of heavy metals contamination by multivariate statistical analysis methods in Poudichevy Mangroves. Journal of Environment and Earth Science 1: 32-35.

71. Schmitt JA, Meisch HU (1985) Cadmium in mushrooms, distribution, growth effects and binding. Trace Elements in Medicine 2: 163-166.

72. Sigel A, Sigel H, Sigel R (2013) Cadmium from Toxicity To Essentiality 5: 14-22.

73. Baumann FJ (1974) Dichromate reflux chemical oxygen demand-Proposed method for chloride correction in highly saline wastes. Anal Chem 46: 1336-1338.

74. Gonzalez JF (1986) Semi-micro determination of COD on fish filleting waste-water. Environmental Technology Letters 7: 269-272.

75. Simeonov L, Kolhubovski M, Simeonov B (2010) Environmental heavy metal pollution and effects on child mental development. Springer 2: 114-115.

76. Skoog DA, Holler F, West D (2007) Principles of instrumental Analysis. USA: Thomson Brooks 6: 169-173.

77. Sanayei Y, Ismail N, Talebi S (2009) Determination of heavy metals in zayandeh road river, Isfahan-Iran. World Applied Sciences Journal 6: 1204-1214.

78. Sridhara CN, Kamala CT, Samuel S, Raj D (2008) Assessing risk of heavy metals from consuming food grown on sewage irrigated soils and food chain transfer. Ecotoxicol Environmental saf 69: 513-524.

79. Taghreed H, Noor AL, Layla K, Laith JAA (2013) Study effects of industrial waste on Tigris water in Al-Grea't city –Baghdad-Iraq. Journal of advance in physics theories and applications 25: 56-93.

80. Thevenod F, Lee WK (2013) Toxicology of cadmium and its damage to mammalian organs. Journal of Metal Ions Life Science 11: 415-490.

81. Torres R, Torres W, Paul P, Pulgarin C (2003) Electrochemical degradation of P- substituted phenols of industrial interest on platinum electrodes.: Attempt of a structure-reactivity relationship assessment. Chemosphere 50: 97-104.

82. Tsezos M, Volesky B (1981) Biosorption of uranium and thorium. Biotechnology 23: 583-604.

83. Khazaei A, Soudbar D, Sadri M, Mohaghegh SMS (2007) Synthesis and characterization of polymer (8-hydroxyquinoline-p-styrene sulphonate) and study of its controlled release behaviour. Iran Polymer 16: 309-317.

84. UNICEF (2010) Equity, safety and sustainability.

85. Van E, Ryan A, Tomasso J, Klaine S (2005) Evaluation of acute copper toxicity to larval fathead minnows in soft surface. Environ Toxicol Chem 24: 408-414.

86. Vetter J (1993) Toxic elements in certain higher fungi. Food chemistry 48: 207-208.

87. Ogawa T, Matsui M (1967) Total Synthesis of Polycyclic Terpenes. Agricultural and Biological Chemistry 31: 1401-1403.

88. Kacurakova M, Ebringerova A, Hirsh J, Hromadkova Z (1994) Infrared studies of arabinoxylans. Journal of Science Food and Agriculture 66: 423-427.

89. Matheickal JT, Yu Q (1996) Biosorption of lead from aqueous solutions by marine alga Ecklonia Radiata. Water Science Technology 34: 1-7.

90. Kapoor A, Viraraghavan T (1998) Bio-sorption of heavy metals on aspergillus niger. Effect of pre-treatment. Biores technology 63: 109-113.

91. Kapoor A, Viraraghavan T (1995) Fungal biosorption an alternative treatment Option for heavy metal bearing wastewaters: a review. Bioresource Technology 53: 195-206.

92. Volesk B, Schiewer S (1999) Encyclopedia of Bioprocess Technology. Fermentation, Biocatalysis, and Bioseparation 39: 351-379.

93. Wasewar KL, Atif M, Prasad B, Mishra IM (2008) Adsorption of zinc using tea factory waste: Kinetics, Equilibrium and Thermodynamics. Environmental Chemistry 36: 320-329.

94. Peak D, Ford RG, Sparks DL (1999) An *in Situ* ATR-FTIR Investigation of Sulfate Bonding Mechanisms on Goethite. Journal of Colloid and Interface Science 218: 289-299.

95. Marshall WE, John MM (1996) Agricultural by-products as metal adsorbents sorption properties and resistance to mechanical abrasion. J Chem Tech Biot 66: 192-198.

96. Welz B, Sperling M (1999) Atomic absorption spectrometry. AAS Spectrometry 71: 353-359.

97. World Health Organisation (2003) Copper in drinking-water. Background Document for Preparation of WHO Guidelines for Drinking-Water Quality. Geneva, World Health Organization.

98. WHO (World Health Organization) (2011) Cadmium in Drinking Water.

99. Unlu N, Ersoz M (2006) Adsorption characteristic of heavy metal ions onto a low cost biopolymeric sorbent from aqueous solution. Journal of Hazardous Materials 136: 272-280.

100. Fourest E, Serre A (1996) Contribution of carboxyl groups to heavy metal binding sites in fungal wall. Toxicological and Environmental Chemistry 54: 1-10.

101. Tripathy T, Ranjan DB (2006) Flocculation: a new way to treat the waste water. Journal of Physical Science 10: 93-127.

102. Singh SP, Ma LQ, Hendry MJ (2006) Characterization of aqueous lead removal by phosphatic clay: Equilibrium and kinetic studies. J Hazard Mater 136: 654-662.

103. Li X, Tang Y, Cao X, Lu D, Luo F, et al. (2008) Preparation and evaluation of orange peel cellulose adsorbents for effective removal of cadmium, zinc, cobalt and nickel, colloids and surfaces. Physicochemical Engineering Aspects 317: 512-521.

104. Hameed BH, Ahmad AA (2009) Batch adsorption of methylene blue from aqueous solution by garlic peel, an agricultural waste biomass. Journal of Hazardous Materials 164: 870-875.

105. Sag Y, Ozer D, Kutsal T (1995) A comparative study of the biosorption of lead (II) ions to Z. ramigera and R. arrhizus. Process Biochemistry 30: 169-174.

106. Padmavathy V, Vasudevan P, Dhingra SC (2003) "Biosorption of Ni (II) ions on Baker's yeast". Process Biochem 38: 1389-1395.

107. Leyva-Ramos R, Rangel-Mendez JR, Mendoza-Barron J, Fuentes-Rubio L, Guerrero-Coronado RM (1997) Adsorption of cadmium (II) from aqueous solution onto activated carbon. Water Science Technology 35: 205-211.

108. Rivera-Utrilla J, Bautista-Toledo I, Ferro-Garcia MA, Moreno-Castilla C (2001) Activated carbon surface modifaications by adsorption of bacteria and their effect on aqueous lead adsorption. Journal of Chemical Technology and Biotechnology 76: 1209-1215.

109. Adedirin O, Adamu U, Nnabuk EO (2011) Removal of Cd (II) from aqueous solution using Bacillus sublitis and Escherichia coli immobilized in agarose gel: equilibrium, kinetics and thermodynamic study. Scholars Research Library 3: 59-76.

110. Zhu J, Yang J, Deng B (2008) Ethylenediamine-modified activated carbon for aqueous lead adsorption. Environmental Chemistry 3: 277-282.

111. Wang Y, Chai L, Chang H, Peng X, Shu Y (2009) Equilibrium of hydroxyl complex ions in Pb^{2+}–H_2O system. Transactions of Nonferrous Metals Society of China 2: 458-462.

112. Jorge L, Michelle GTK, Becker-Hapak K, James M, Dennis WD (1990) Effect of chemical modification of algal carboxyl groups on metal ion binding. Environ Sci Technol 9: 1372-1378.

113. Sheng PX, Ting YP, Chen JP, Hong L (2004) Sorption of lead, copper, cadmium, zinc, and nickel by marine algal biomass. Characterization of biosorptive capacity and investigation of mechanisms. Colloidal and Interface Science 1: 131-141.

114. Demirbasa E, Kobyab M, Senturkb E, Ozkan T (2004) Adsorption kinetics for the removal of chromium (VI) from aqueous solutions on the activated carbons prepared from agricultural wastes. Water SA 30: 533-539.

115. Khani M (2006) Biosorption of uranium from aqueous solutions by nonliving biomass of marine algae. Cystoseira indica. Journal of Biotechnology 9: 101-108.

116. Fazal A, Rafique U (2012) Mechanistic understanding of cadmium sorption by sulfonated and esterified spent black tea. International Journal of Chemical and Environmental Engineering 3: 68-78.

117. Ilhan S, Kilicarslan S, Ozdag H (2004) Removal of chromium, lead and copper from industrial waste by Staphylococcus saprophyticus. Turk Electron Journal of Biotechnology 2: 50-57.

118. Chen J, Lie D, Wang L, Wu S, Zhang B (2002) Dried waste activated sludge as biosorbents for metal removal: adsorptive characterization and prevention of organic leaching. Journal of Chemical Technology and Biotechnology 77: 657-662.

119. Benhima H, Chiban M, Sinan F, Seta P, Persin M (2008) Removal of lead and cadmium ions from aqueous solution by adsorption onto micro-particles of dry plants. Colloids and Surfaces B: Biointerfaces 61: 10-61.

120. Al-Asheh S, Banat F, Mohai F (1997) Sorption of copper and nickel by spent animal bones. Chemosphere 39: 2087-2096.

121. Brown PA, Brown JM, Allen SJ (2001) The application of Kudzu as a medium for adsorption of heavy metals from dilute waste streams. Bioresour Technol 78: 189-195.

122. Al-Subu MM (2002) The Interaction effects of cypress (Cupressu sempervirens), Cinchona (Eucalyptus longifolia) and pine (Pinus halepensis) leaves on their efficiencies for lead removal from aqueous solutions. Advances in Environmental Research 6: 569-576.

123. Salim R, Al- Subu M, Qashow S (1994) Removal of lead for polluted water using decaying leaves. Journal of Environmental Science 29: 2087-2114.

124. Volesky B, May H, Holan ZR (1993) Cadmium adsorption by Saccharomyces cervisiae. Biotechnology 41: 826-829.

125. Horsfall M, Ogban FE, Akporhonor EE (2006) Recovery of lead and cadmium ions from metal-loaded biomass of wild cocoyam (Caladium bicolor) using acidic, basic and neutral eluent solutions. J Biotechnol 9: 52-157.

126. Foo KY, Hameed BH (2010) Insights into the modeling of adsorption isotherm systems. Chemical Engineering Journal 156: 2-10.

127. de Luna M, Flores E, Genuino D, Futalan C (2013) Adsorption of EriochromeBlack T (EBT) dye using activated carbon prepared from waste rice hulls-Optimization, isotherm and kinetic studies. Journal of the Taiwan Institute of Chemical Engineer 16: 646-653.

128. Langmuir I (1918) The adsorption of gases on plane surfaces of glass, mica and platinum. Journal of American chemical society 40: 1362-1403.

129. Nge J, Cheung W, McKay G (2003) Equilibrium studies for the sorption of lead from effluents using chitosan. Chemosphere 52: 1021-1030.

130. Freundlich HMF (1906) Over the adsorption in solution. Journal of Structural

Chemistry 57: 385-470.

131. Bayrak Y (2003) Adsorption isotherms in bleaching hazelnut oil. Journal of American Oil Chemical Society 80: 1143-1146.

132. Krest I, Glodex J, Keusgen M (2000) Cysteine sulfoxides and alliinase activity of some Allium species. Journal of Agricultural and food chemistry 48: 3753-3760.

133. Lagergreg S (1898) The theory of so-called adsorption of soluble substances. Hanlingar 24: 1-39.

134. Ho YS, McKay G, Wase DJ, Foster CF (2000) Study of the sorption of divalent metal ions on to peat. Adsorption science and technology 18: 639-650.

135. Ghaedi M, Ghezelbash GR, MarahelF, Shiva Ehsanipour S, Najibi A, et al. (2010) Equilibrium, thermodynamic and kinetic studies on lead (II) biosorption from aqueous solution by saccharomyces cerevisiae biomass. Clean-Soil, Air and Water 38: 877-885.

136. Topperwien S, Xue H, Asigg A (2007) Cadmium accumulation in Scenedesmus vacuolatus under freshwater conditions. Journal of Environmental Science and Technology 41: 5383-5388.

137. Comuzzi C, Grespan M, Melchior A, Portanova R, Tolazzi M (2001) Thermodynamics of complexation of cadmium (II) by open-chain N-donor ligandsin dimethyl sulfoxide solution. European Journal of Inorganic Chemistry 12: 3087-3094.

138. Lewis WD, Canfield D (1977) DOC in some dark Venezuelan waters and revised equation for spectrophotometric determination of DOC. Arch fur Hydrobiologie 79: 441-445.

The Effect of Fish Size, Age and Condition Factor on the Contents of Seven Essential Elements in *Anguilla anguilla* from Tersakan Stream Mugla (Turkey)

Cahit Kasimoglu*

Institute of Science Biology, Mugla Sitki Kocman University, Mugla-48000, Turkey

Abstract

The correlation coefficients between age, weight, total length and condition factor (K) and trace metals (Co, Cr, Cu, Fe, Mn, Ni and Zn) in *Anguilla anguilla* were determined. Totally 160 fish samples were collected from Tersakan Stream (Mugla) between September 2011 and August 2012. Highly significant ($P<0.01$) relationship between trace metal concentrations in fish muscle and age, weight, total length and condition factor (K) was found. Significantly positive relationship among trace metals (Co, Cr, Cu, Fe, Mn, Ni and Zn) and fish age, weight and total length was observed, while all trace metal concentrations were negatively correlated with fish condition factor (K). The comparison of trace metal concentrations in the muscle of *Anguilla anguilla* with the tolerable values indicated that the trace metal concentrations in muscle tissue did not exceed the WHO/FAO, EPA and IAEA–407 guidelines.

Keywords: *Anguilla anguilla*; Trace metals; Food safety; Mugla

Introduction

Trace elements may exert benefical or harmful effects on plant, animal and human life depending upon the concentration [1]. These elements are introduced into the environment through various routes such as smelting processes, fuel combustion and industrialization [1,2]. They find their way into rivers, lakes or oceans through atmospheric fallout, dumping wastes, accidental leaks, runoff of terrestrial systems (industrial and domestic effluents) and geological weathering [1,3,4].

In aquatic environment, fishes reflect the effects of numerous interacting biotic and abiotic factors. In many cases the variation in elemental concentration in fish tissue has been attributed to variation in size and age [5]. Therefore, many studies have focused attention on the dependence between the contents of metals and fish biometry (weight and length) and age [6-10]. Variation of elemental concentration in fish tissue due to size has been reported [11-15].

As a consequence, pollutants discharged in the aquatic environment are likely to accumulate in fish and represent a potential risk not only to the fish, but also to other fish consumers, particularly humans. Catadromous eels, widely distributed throughout the world, are one of the top predator in freshwater ecosystem [16]. Due to the long-life cycle and the spesific biological and ecological features of anguillid eels, they are vulnerable to adverse impacts from nature and human activities [17,18]. Accordingly, the eel populations have declined dramatically in recent years and causes are attributed to over-fishing, construction, climate change, other enviromental factors, especially enviromental pollution [19,20]. Hence, anguillid eels has been widely used as bioindicator for environmental monitoring to assess the aquatic system quality in many countries [21,22].

This study investigates relationship of fish size and nutritional quality, occurrence of trace metals and further identifies the optimum size, i.e. total length and body weight, of this species suitable for human consumption.

Material and Method

Study area

Dalaman- Tersakan Stream (36°45’51”N, 28°49’20”E) is located in province Mugla in the southwest of Turkey (Figure 1). The sampling site (Tersakan Stream) is a temperate stream which is impacted by unpredictable environmental conditions associated with a Mediterranean climate. Its length is 30 km and this stream has temporal and spatial water flow variations throughout the water course (48–780 m^3/s). The lower section of the stream was channelized by local authorities to prevent seasonal floods. The stream flows into the Mediterranean Sea [23] (Figure 1). There are eight known species inhabiting the stream of which the most abundant are *Mugil cephalus* (Linnaeus), *Leuciscus cephalus* (Linnaeus), *Gambusia affinis* (Baird and Girard) and *Anguilla anguilla* (Linnaeus) [23].

Sampling and analysis

Water and fish samples were collected 4 times (from September 2011 to August 2012) for every 3 months at four stations in the Tersakan Stream. The sampling bottles were pre-conditioned with 5% nitric acid and later rinsed thoroughly with distilled de-ionized water. At each sampling site, the polyethylene sampling bottles were rinsed at least three times before sampling was done. Pre-cleaned polyethylene sampling bottles were immersed about 10 cm below the water surface. About 0.5 L of the water samples were taken at each sampling site. Samples were acidified with 10% HNO_3, placed in an ice bath and brought to the laboratory. The samples were filtered through a 0.45 µm micropore membrane filter and kept at 4°C until analysis. A total of 160 samples of eel (Anguilla anguilla) were captured by backpack electrofishing with a battery-powered unit (550 V,5-100 Hz) at four different stations along the tersakan stream in the period of Semptember 2011 to August 2012). Fish samples were immidiately transported to the laboratory in a thermos flask with ice. The fish species selected were based on their

***Corresponding author:** Cahit Kasimoglu, Ph. D, Institute of Science Biology, Mugla Sitki Kocman University, Mugla-48000, Turkey
E-mail: cahitkasimoglu@hotmail.com

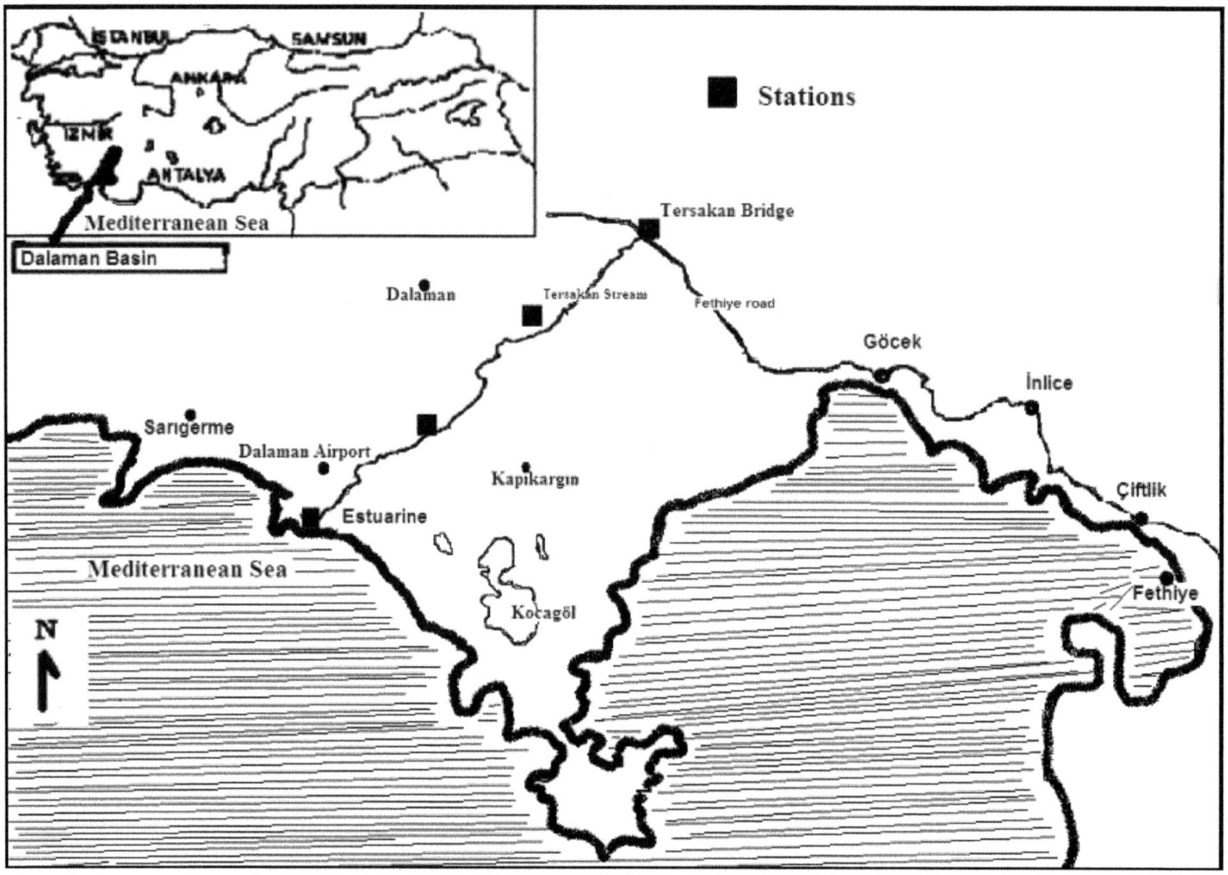

Figure 1: Sampling area of *Anguilla anguilla* specimens in Mugla province (Turkey).

Element	Wavelength (nm)	Detection limit (mg/L)
Co	228.616	0.0070
Cr	267.716	0.0071
Cu	327.393	0.0097
Fe	238.204	0.0046
Mn	257.610	0.0014
Ni	231.604	0.0150
Zn	202.548	0.0040

Table 1: Spectral lines used in emission measurements and detection limit for the elements measured by ICP-AES.

abundance along the river watercourse and frequency in use as food by the inhabitants of these areas.

The samples were carefully cut opened using a plastic knife in order to remove the muscle, and freeze dried and pulverized into a uniform particle size prior to analysis. The small sized particles were subjected to acid digestion using nitric acid. Approximately 5 g of muscle from each sample were dissected, washed with deionized water, weighed, packed in polyethylene bags, and stored at -20°C prior to analysis.

The muscle tissue samples were digested with concentrated nitric acid. Dissected samples were transferred to a 100 mL teflon beaker. Thereafter, 10 mL ultrapure concentrated HNO_3 (Merck) was added, and the sample heated at (100, 150, 210 and 280°C on a hot plate for 0.5, 0.5, 0.5 and 2 hours) with a DK-20 Heating Digester respectively. Two mL of 1 N HNO_3 was added to the residue, and the solution continuously evaporated on the hot plate, until it was digested in every sample. After cooling, a further 10 mL of 1 N HNO_3 was added. The solution was transferred, diluted and filtered through 0.45 μm nitrocellulose membrane fitler [24].

Two mL of 1 N HNO_3 was added to the residue, and the solution continuously evaporated on the hot plate, until it was digested in every sample. After cooling, a further 10 mL of 1 N HNO_3 was added. The solution was transferred, diluted and filtered through 0.45 μm nitrocellulose membrane fitler [24].

All samples were analysed in duplicates for Co, Cr, Cu, Fe, Mn, Ni and Zn by ICP/AES (Optima 2000-Perkin Elmer), which is a fast multi-element technique with a dynamic linear range and moderate-low detection limits [25]. Detection limits are given in Table 1. The detection limit is defined as the lowest analytical signal to be distinguished qualitatively at a specified confidence level from the background signal [26]. The accuracy of analytical procedure was checked by analyzing the standard reference materials (Water: SRM-143d, National Institute of Standards and Technology; Fish: DORM–2, National Research Council). Recovery rates ranged from 79 to 96% for all the elements analysed.

Statistically analysis

Statistical analysis of data was carried out using SPSS 14.0 statistical package program. One-way analysis of variance (ANOVA) was used to assess whether metal concentrations varied as significantly among tissues. Relationship of metal levels in muscle tissue with metal levels

Age (years)	Fish Number (N)	Total length (mm)	Body weight (g)	K
3	61	216,70 ± 39,23 A	42,02 ± 8,65 A	0,42 ± 0,12 A
4	53	279,45 ± 56,37 B	49,12 ± 4,46 B	0,22 ± 0,02 B
5	26	352,65 ± 51,57 C	54,53 ± 7,45 C	0,12 ± 0,01 C
6	12	443,58 ± 77,77 D	58,20 ± 9,52 D	0,06 ± 0,01 CD
7	8	571,37 ± 64,272 E	63,93 ± 9,77 E	0,03 ± 0,007 D
Total	160	372,75 ± 89,54	53,56 ± 8,454	0,26 ± 0,11

[A,B,C,D,E]: Different letters in each column indicate statistically significant differences ($p<0.05$).

Table 2: Total length, weight, age and condition factor (K) (mean ± SE) of *Anguilla anguilla.*

Trace metals in water		Co	Cr	Cu	Fe	Mn	Ni	Zn
		LOD	0.3 ± 0.34	1.15 ± 0.49	27.62 ± 6.1	0.63 ± 0.24	0.48 ± 0.14	1.78 ± 0.53
FAO [30]		0.05	0.1	0.2	5	0.2	0.2	2
TWPCR [16]	I	10	20	20	300	100	20	200
	II	20	50	50	1000	500	50	500
	III	200	200	200	5000	3000	200	2000
	IV	>200	>200	>200	>5000	>3000	>200	>2000

I: High quality water, II: weakly polluted water, III: Polluted water, IV: High polluted water
LOD: Limit of Detection

Table 3: Concentrations of trace metal in water of Tersakan Stream (mean±standard error (SE)).

Trace metals in muscle	Co	Cr	Cu	Fe	Mn	Ni	Zn
	0.07 ± 0.02	1.73 ± 0.27	1.66 ± 0.28	31.21 ± 6.75	2.04 ± 0.65	1.29 ± 0.25	22.11 ± 4.67
FAO/WHO limits [32]	-	-	5	5	100	-	40
EPA [33]	27	4,1	54	410	190	27	410
IAEA-407 [31]	-	0,73	3,28	146	3,52	0,60	67,10
TFC [25]	-	-	20	-	20	-	50

Table 4: Trace metal concentrations (µg/g wet weight, mean±standart errors (SE)) in muscle tissue of *Anguilla anguilla.*

Parameters	*Anguilla anguilla*
Cr muscles/ Cr water	.73**
Cu muscles/ Cu water	.82**
Fe muscles/ Fe water	.94**
Mn muscles/ Mn water	.89**
Ni muscles/ Ni water	.90**
Zn muscles/ Zn water	.96**

Table 5: Correlation coeffitients between metal concentration in water and muscle tissue.

in water samples and fish biometric parmeters was performed using the Pearson's correlation analysis.

Results and Discussions

Fish biometric parameters, total length, weight and condition factor (K), are shown in Table 2. The range of mean values of eel's was (216.70-571,37 mm), body-weight 42.02-63.93 g and condition factor 0.03-0.42.

Sampled fish individuals belonged to 5 age groups (3, 4, 5, 6, 7 years old). Fish total length, body weight and condition factor among different age groups were found statistically significant different ($p<0.05$) (Table 2).

The average concentrations of trace metals in water samples from the Tersakan Stream are presented in Table 3. According to the findings, pH value in water varied between 7.55–8.06. Metal concentrations in water followed the order: Fe > Zn > Cu >Mn > Ni > Cr, while Co levels were below limit of detection. İron is one of the most abundant elements in the earth's crust [27,28]. In water Fe may occur as complex and diverse mixture of soluble and insoluble forms such as inorganic and organic complex and or associated with colloids and suspended particulate matter. Similar finding was observed by Bamishaiye et al [29]. High metals concentrations in water can retard the growth and development in fish, particularly the developmental stages, resulting in possible changes in fish size.

Metal concentrations in Tersakan Stream's water samples were compared with international standards. The obtained results showed that the concentrations of Cr, Cu, Fe, Mn and Ni in water exceeded the values of FAO [30] (Table 3). Accordingly, the water taken from Tersakan Stream is not proper for irrigation.

Metal concentrations in Tersakan Stream's water samples were compared with international standards. The obtained results showed that the concentrations of Cr, Cu, Fe, Mn and Ni in water exceeded the values of FAO [30] (Table 3). Accordingly, the water taken from Tersakan Stream is not proper for irrigation. Trace metal concentrations in water samples were determined to be class I according to Turkish Water Pollution Control Regulations (TWPCR) [31]. According to our results, Tersakan Stream might be classified as a stream of good water quality but, periodically impacted by pollution exposure. Especially the use of fertilizers and pesticides in agriculture, domestic and urban sewage and industrial waste is believed to would increase the metals concentrations. Regular environmental monitoring in the stream is also very important.

All measured metal concentrations in fish muscle tissue were higher than total dissolved metal concentrations in water (Table 4). The trace metal concentrations in muscle of *Anguilla anguilla* were lower than the concentrations issued by WHO/FAO [32], EPA [33], IAEA-407 [34] and TFC [35] guidelines. Accordingly, there is a little risk for the human consumption of these fish species. Thus, these fish can be considered for now to be fit and safe for consumption. Because the levels of heavy metals are below the permissible limits (Table 4). On the other hand, this study shows that a potential danger may occur in the future depending on the agricultural development.

Age Total length (mm) weight (g)	Co	Cr	Cu	Fe	Mn	Ni	Zn
3 (175–265 mm) (34,2–49,5 gr)	0,01 ± 0,14 B	1,11 ± 0,23 E	1,05 ± 0,24 C	20,3 ± 2,14 C	1,11 ± 0,15 C	0,89 ± 0,05 E	15,64 ± 2,48 C
4 (248–326 mm) (45–53,5 gr)	0,03 ± 0,18 B	1,65 ± 0,25 D	1,67 ± 0,36 B	29,04 ± 3,63 C	1,91 ± 0,39 C	1,29 ± 0,09 D	20,56 ± 3,67 C
5 (304–400 mm) (51,5–59 gr)	0,06 ± 0,82 B	2,37 ± 0,36 C	2,4 ± 0,28 A	45,01 ± 6,8 B	2,97 ± 0,46 B	1,59 ± 0,06 C	30,29 ± 6,22 B
6 (400–490 mm) (54,5–61,5 gr)	0,16 ± 0,07 A	2,83 ± 0,45 B	2,65 ± 0,62 A	47,51 ± 5,89 B	3,54 ± 0,91 B	2,03 ± 0,16 B	32,02 ± 4,77 B
7 (518–635 mm) (60–67,5 gr)	0,18 ± 0,09 A	3,26 ± 0,13 A	2,43 ± 0,38 A	59,42 ± 4,76 A	4,72 ± 0,98 A	2,29 ± 0,23 A	40,31 ± 7,48 A

[A,B,C,D,E]: Different letters in each column indicate statically significant difference in metal levels of different age groups (p<0.05).

Table 6: Trace metal concentrations in muscle of *Anguilla anguilla* (µg/g wet weight) sorted according to fish age groups.

	Age	Length	Weight	K
Co	.67**	.68**	.66**	-.65*
Cr	.83**	.81**	.84**	-.68*
Cu	.61**	.55**	.68**	-.73**
Fe	.79**	.78**	.81**	-.92**
Mn	.63**	.61**	.68**	-.96**
Ni	.76**	.75**	.80**	-.87**
Zn	.61**	.57**	.63**	-.76**

*; P < 0.05 **; P < 0.01 ***; P < 0.001

Table 7: Correlation matrix between total lengths, weight, age and condition factor and Heavy metal concentrations in muscle of *Anguilla anguilla*.

In order to understand interrelationship between the metal concentration in water and fish muscle tissues, Pearson's correlation analysis was carried out (Table 5). It is evident that metal concentration in water and in eel's muscle tissue were highly significantly positively correlated, which may indicate that the increased concentration of heavy metals in water is reflected in increased metal concentration in the fish muscle. In consistence to the present findings, Bird [36] reported increased heavy metal concentrations in water and accordingly in fish. Higher contamination of trace metals in fish tissues were correlated with agricultural runoff entering the water bodies. Chaisermartin [37] also reported that agricultural runoff with fertilizers contains heavy metals and cause pollution in the receiving water bodies, contaminating their fish. Bioconcentration factor, expressed as the ratio of trace metals Cr, Cu, Fe, Mn, Ni and Zn in the fish muscle tissue and water samples were high. This led to the conclusion that metal concentration (Cr, Cu, Fe, Mn, Ni and Zn) in the muscle tissues of fish accumulated from the river water.

The distribution of trace metal concentrations in muscle tissues of *Anguilla anguilla* according to age groups (3-7 years old), total length and body weight are given in Table 6.

An increasing trend in Co, Cr, Cu, Fe, Mn, Ni and Zn values in muscle tissue was observed with increasing fish length, weight and age (Table 6). The effect of fish age on heavy metal accumulation showed a similar pattern to those of fish weight and length. Generally, the growth-dependent variations of trace element levels are known to be influenced by various factors such as metabolic rate and growth dilution of the elements [5,22]. The concentrations of all analysed trace metals in the muscle tissues were statistically significant with eel's age (p<0.05), i.e. trace metal concentrations were generally higher in tissue of older and bigger compared to younger and smaller fish. This was more conspicuous when the correlations coeffitients were considered, which showed a highly significant (p<0.01) positive correlations between concentrations of Co, Cr, Cu, Fe, Mn, Ni, Zn and body length and weight (Table 7). Ziyadah [38] reported that the tissues tend to accumulate high concentrations of heavy metals with the increase of fish size. Many studies have also demonstrated a positive relationship between body size (length, weight and age) and trace metal concentration [9,13,39,40].

Correlation analysis resulted in negative correlation of all trace metal concentrations and fish condition factor (K). On the other hand, it was found that the correlations between the trace metal concentrations of muscles and the condition factor (K) of fish samples varied with characteristic opposite trends compared to those related to, length, weight and age (Table 6). The negative relationship between the trace metal concentration and the condition factor of fish suggests the relative dilution effect of the lipid content of tissues [41].

İt is generally accepted that trace element accumulation in living organisms, which is controlled by spesific uptake, detoxification and elemination mechanisms, depends significally also on the size-specific metabolic rate of organisms [42]. Some metals do not increase in concentrations with age or size because they are thought to be under homeostatic control [43]. So, in present study, the positive correlation between some metals and fish age and sizes may be due to loss of homeostasis capacity of *Anguilla anguilla* under chronic metal exposure leading to bioaccumulation. This assumption is supported well also by the fact that lipid as a percent of body weight is usually lower in younger fish, decreases during spawning and reaches its peak at the end of the main feeding period [44]. These variation may explain also the opposite correlations observed in present study between heavy metals-size and heavy metals-condition factors of fish [20].

It could be concluded that, *Anguilla anguilla* is a good species for biomonitoring pollution as it could withstand the adverse conditions in the ecosystem. The concentrations of heavy metals in fish muscle vary significantly not only as function of fish sizes and age, but is influenced in a remarkable degree by the fish condition [41].

Conclusions

Fishes are suitable indicators of heavy metal contamination in aquatic ecosystem because they occupy different trophic levels and are of different sizes and ages. Since fishes are located at the end of the aquatic food chain they reflect the water quality status and represent indicators of water pollution, particularly heavy metals.

The trace metal concentrations in muscle of *Anguilla anguilla* were compared with the tolerable values in fish. The result obtained showed that the trace metal concentrations in muscle tissue did not exceed the WHO/FAO, EPA, IAEA-407 and TFC guidelines. Therefore, according to the results of this study, the consumption of *Anguilla anguilla* from Tersakan Stream can be safe for human health in spite of possible contamination with heavy metals. The knowledge on heavy metal concentrations in fish is important and necessary for the management water plans involving polluted aquatic systems and for the evaluation of potential risk to human health.

Acknowledgment

I am grateful to Dr. İbrahim ÖRÜN for his help in the statistical analysis. This research is a part of Ph.D. thesis and supported financially by Mugla Sıtkı Kocman University Scientific Research Project Coordination Department. (Project No: 2011/14).

References

1. Forstner U, Wıttman GTW (1983) Metal Pollution in the Aquatic Environments. Springer-Verlag, Berlin, Germany.
2. Adeyeye EI, Akinyugha NJ, Fesobi ME, Tenabe VO (1996) Determination Of Some Metals İn Clarias Gariepinus (Cuvier And Vallenciennes), Cyprinus Carpio (L.) And Oreocromis Niloticus (L.) Fishes in a Polyculture Freshwater Pond And Their Encironmentes. Aquaculture 147: 205-214.
3. Perry J, Vanderklein EL (1996) Water Quality Management of a Natural Resource. Blackwell Science.
4. Al-Yousuf MH, El-Shahawi MS, Al-Ghais SM (2000) Trace Metals in Liver, Skin And Muscle Of Lethrinus Lentjan Fish Species in Relation To Body Length And Sex. The Science Of Total Environment 256: 87-94.
5. Phillips DJH (1980) Quantitative Aquatic Biological İndicators. Applied Science Publishers, London.
6. Kroupa M, Hartvıch P (1990) Concentrations of Some Heavy Metals in Fish. Zivocisna Vyroba 35: 937-943.
7. Falandysz J (1994) Metal Content İn The Muscular Tissue Of Turbot Psetta Maxima From The Gdansk Bay, Bromat. Chem Toksykol 27: 37-39.
8. Kostecki M (2000) Heavy Metals in Flesh and Liver Of Some Fish Species in Dzierzno Duze Dam-Reservoir. Arch Ochr Srod 26: 109-125.
9. Filipović Marijić V, Raspor B (2006) Age and tissue dependet Metallothionein and Cytosolic Metal Distribution in a Native Mediterranean Fish, Mullus Barbatus, From The Eastern Adriatic Sea. Comparative Biochemistry And Physiology Part C.143: 382-387.
10. Filipović Marijić V, Raspor B (2012) Site-Specific Gastrointestinal Metal Variability İn Relation To The Gut Content and Fish Age of İndigenous European Chub from the Sava River. Water, Air & Soil Pollution 223: 4769-4783.
11. Salam A, Ansari TM, Tariq N, Akhtar QA (2002) Effect of Body Size on Metal Concentrations in Farmed Cirrhinus Mrigala. Asian Fish Sci. 15: 329-334.
12. Canli M, Atli G (2002) The Relationships Between Heavy Metal (Cd, Cr, Cu, Fe, Pb, Zn) Levels And The Size of Six Mediterranean Fish Species. Environ Pollut 121: 29-136.
13. Farkas A, Salánki J, Specziár J (2003) Age-And Size-Specific Patterns Of Heavy Metals İn The Organs Of Freshwater Fish Abramis Brama L. Populating A Low-Contaminated Site. Water Res. 37: 959-964.
14. Luczynska J, Tonska E (2006) The Effect of Fish Size on the Content of Zinc, Iron, Copper And Manganese in the Muscles of Perch (Perca Fluviatilis L.) And Pike (Esox Lucisus L.). Arch Pol Fish 14: 5-13.
15. Promya J, Chitmanat C (2011) The Effects of Spirulina Platensis And Cladophora Algae on The Growth Performance, Meat Quality and İmmunity Stimulating Capacity Of The African Sharptooth Catfish (Clarias Gariepinus). Int J Agric Biol 13: 77-82.
16. Tesch FW (2003) The Eel. Blackwell Publishing Company. UK. The Fisheries Of British Isles.
17. Robinet T, Feunteun E (2002) Sublethal Effects of Exposure to Chemical Compounds: An Cause For The Decline İn Atlantic Eels. Ecotoxicology 11: 265-277.
18. Pierron F, Baudrımont M, Bossy A, Bourdıneaud JP, Brethes D, et al. (2007) Impairment of Lipid Storage by Cadmium in European Eel (Anguilla Anguilla). Aquatic Toxicology 81: 304-311.
19. Brusle J (1990) Effects of Heavy Metals on Eels, Anguilla Sp. Aquatic Living Resources 3: 131-141.
20. Castonguay M, Hodson PV, Couıllard CM, Eckersley MJ, Dut JD, et al. (1994) Why is Recruitement of American Eel, Anguilla Rostrata, Declining in the ST Lawrence River And Gulf. Canadian Journal Of Fisheries And Aquatic Science 51: 479-488.
21. Edwards SC, Cecılıa L, Leod M, John NL (1999) Mercury Contamination of the Eel (Anguilla Anguilla) and Roach (Rutilus Rutilus) in East Anglia, Uk. Enviromental Monitoring And Assessment 55: 371-387.
22. Langston WJ, Chesman BS, Burt GR, Pope ND, Mcevoy J (2002) Metallothionein in Liver of Eels Anguilla Anguilla from the Thames Estuary: An Indicator Of Environmental Quality. Marine Environmental Research 53: 263-293.
23. Barlas M, Yilmaz F, Dirican S, Yorulmaz B, Yuvarlak C (Koycegiz Muğla) 's Investigation of Fish Fauna, IV. Symposium on Fisheries, 423-436, 28-30 June 2000, Erzurum - 2012.
24. Alam MGM, Tanaka A, Allınson G, Laurenson LJB, Stagnitti F, et al. (2002) A Comparison of Trace Element Concentrations in Cultured and Wild Carp (Cyprinus Carpio) Of Lake Kasumigaura, Japan. Ecotoxicol. Environ. Saf 53: 348-354.
25. Sturgeon RE (2000) Current Practice and Recent Developments in Analytical Methodology for Trace Metal Analysis of Soils, Plants and Water. Commun. Soil. Sci. Plant 31: 1512-1530.
26. Kackstaetter UR, Heınrıchs G (1997) Validity of lowcost Laboratory Geochemistry for Environmental Applications. Waterair Soil Pollut. 95: 119-131.
27. Atsdr (2003) Agency For Toxic Substances And Disease Registry.
28. Turkmen A (2003) İskenderun Körfezi' Nde Deniz Suyu, Askıdaki Katı Madde, Sediment Ve Dikenli Taş İstiridyesi' Nde (Spondylus Spinosus Schreibers, 1793) Oluşan Ağır Metal Birikimi Üzerine Araştırma. Atatürk Üniversitesi, Fen Bilimleri Enstitüsü, Doktora Tezi, Erzurum.
29. Bamishaiye EI, Ogbonna O, Jimoh WL, Amagn E (2011) Advances in Applied Science Research 2: 62-68.
30. FAO (1985) Water Quality for Agriculture. Irrigation and Drainage Paper No. 29, Rev. 1. Food And Agriculture Organization Of The United Nations, Rome.
31. TWPCR (2004) Turkish Water Pollution Control Regulations-25684 (İn Turkish). Çevre Ve Orman Bakanlığı, 25684 Sayılı Resmi Gazete, Ankara.
32. FAO/WHO (1989) Evaluation of Certain Food Additives and the Contaminants Mercury, Lead and Cadmium, WHO Technical Report, Series No. 505.
33. Anonymous (2005c) Enviromental Protection Agency (EPA), National Recommended Water Quality Criteria Correction.
34. Wyse EJ, Azermard S, Mora SJ (2003) Report on the World Wide İnter-Comparison Exercise For The Determination of Trace Elements and Methyl Mercury in Fish Homogenate IAEA-407. IAEA/AI/144 (IAEA/Mel/72), IAEA, Monaco.
35. TFC (2002) Turkish Food Codes, Official Gazette, 23 September, No: 24885.
36. Bird SC (1986) The Effect of Hydrological Factors On Trace Metal Contamination in The River Tawe, South Wales, Environmental Pollution.
37. Chaisemartin C (1983) Natural Adaptation to Fertilizers Containing Heavy Metals of Healthy And Contaminated Populations of Austropotamo Biuspalipes (Le). Hydrobiology 17: 229-240.
38. Zyadah MA (1999) Accumulation of Some Heavy Metals İn Tilapia Zilli Organs From Lake Manzalah, Egypt. Tr. J. Of Zoology 23: 365-372.
39. Mastala Z, Balogh KV, Salanskı J (1992) Reliability of Heavy Metal Pollution Monitoring Utilizing Aquatic Animal Versus Statistical Evaluation Methods. Arch. Environ. Cont. Tox. 23: 476-483.
40. Linde AR, Sánchez-Galán S, Izquıerdo JI, Arrıbas P, Marrañón E, et al. (1998) Brown Trout as Biomonitor of Heavy Metal Pollution: Effect of Age on The Reliability of The Assessment. Environ. Res. Section B 40: 120-125.

41. Authman MMN (2008) Oreochromis Niloticus as a Biomonitor of Heavy Metal Pollution with Emphasis on Potential Risk and Relation to Some Biological Aspects. Global Veterinaria 2: 104-109.

42. Newman MC, Doubet DK (1989) Size-Dependence of Mercury (II) Accumulation Kinetics İn The Mosquitofish, Gambusia Affinis (Baird And Girard). Archive of Environmental Contamination and Toxicology 18: 819-825.

43. Evans DW, Dodoo DK, Hanson PJ (1993) Trace Elements Concentrations in Fish Livers İmplications of Variations With Fish Size in Pollution Monitoring. Marine Pollution Bulletin 26: 329-334.

44. Weatherrly AH, Gıll HS (1987) The Biology of Fish Growth. Orlando, FL: Academic Press.

Physico-chemical Characteristics and Levels of Polycyclic Aromatic Hydrocarbons in Untreated Water from Ngong River, Kenya

Mobegi Erick K, Nyambaka Hudson N and Nawiri Mildred P*

Department of Chemistry, School of Pure andApplied Science, Kenyatta University, Nairobi, Kenya

Abstract

River Ngong, one of the major rivers in the Nairobi river basin receives enormous amounts of solid and liquid waste from industrial and domestic discharges. This exposes the river to high pollution. An environmental concern arises especially on its unknown levels of hazardous persistent organic pollutants in particular the polycyclic aromatic hydrocarbons (PAH). Notably, its untreated water from some sections along its course is used for domestic purposes hence poses human health risks that may be attributed to long term exposure and accumulation of PAH. United States Environmental Protection Agency (US EPA) not only identifies sixteen priority PAHs that are carcinogenic and mutagenic but alongside other world bodies including WHO, have recommendations for their allowable threshold in water. Previous studies on this river have focused on pollutants such as heavy metals, microbiological contaminants, polychlorinated biphenyls and many other persistent organic pollutants. However there is no literature on the levels of PAHs in this river hence the need for this study. Data from Nairobi cancer registry indicate that cancer cases are on the rise. This can be attributed to a number of factors ranging from nutrition, physical life style to fears of exposure to possible sources of carcinogenic compounds such as PAHs. The aim of the study was to determine levels of sixteen (16) US EPA priority PAHs in water in the Ngong River and compare the levels to the WHO/US EPA standard maximum level. Physico-chemical parameters and levels of the sixteen US EPA priority PAHs in water sampled from sections along the Ngong River are reported, the latter using GC-MS. In some sampling points, turbidity, dissolved oxygen and electrical conductivity were found to divert from the recommended thresholds and overall, the total concentrations of the PAHs ranged between 2.69 ng/L and 14.22 ng/L. Although PAH levels were found to be within acceptable levels, precaution need to be adhered to in the use of untreated water from River Ngong in fear of body accumulation of PAHs and their dangers.

Keywords: Physico-chemical parameters; Polycyclic aromatic hydrocarbons; Ngong river; Nairobi river

Introduction

Environmental degradation has become a major concern globally and in particular water pollution which ranks second to air pollution in the order of important environmental issues. Persistent organic pollutants (POPs) such as polycyclic aromatic hydrocarbons (PAHs), polychlorinated biphenyls (PCBs), phenols and pesticides of different chemical, physical and toxicological properties enter water bodies as a result of human activities [1]. The PAHs have a number of sources enlisted to include emissions from automobiles, industrial and domestic sources such as combustion procedures and inadvertent runoffs among others [2-4].

The existence of over 100 PAHs in the environment is increasingly becoming a concern worldwide with world water bodies across continents documenting presence of PAHs ranging from 0.001 ng/L to magnitudes of 29325 ng/L [5-8]. Irrespective of the levels, it is worth noting that these organic compounds are resistant to degradation, bioaccumulate through the food chain and are known to be carcinogenic [9-11]. Sixteen (16) PAHs are infact considered as priority by the European Union environmental protection agency, US EPA and WHO these being naphthalene, acenaphthylene, acenaphthene, fluorine, phenanthrene, anthracene, fluoranthene, benzo(a)anthracene, chrysene, benzo(b)fluoranthene, benzo(k)fluoranthene, benzo(a) pyrene, indeno(1,2,3-cd)pyrene, dibenz(ghi)anthracene, benzo(ghi) perylene and pyrene [12,13]. The concern therefore is that PAHs are detrimental to human beings (can impair survival and growth by causing abnormal reproduction and development), plants and animals if their levels of toxicity exceed recommended threshold [14,15].

Ngong River is one of the major rivers in the Nairobi river basin and flows through the heart of Nairobi city which is one of the major industrial cities in Kenya. The city continues to experience exponential growth in industrialization and population which has led to environmental pollution ranging from uncollected garbage, human waste from informal settlements, industrial wastes in the form of gaseous emissions, liquid effluents, agro-chemicals, and other wastes especially petro-chemicals and metals from micro- enterprises [16,17]. Consequently, the aforementioned are a direct indication to pollution of water bodies among them the Ngong River. Notably, the untreated water from some sections along the river course way is used for domestic purposes hence poses human health risks that may be attributed to long term exposure and accumulation of PAH. It is in this regard that physico-chemical parameters such as dissolved oxygen (DO), pH, turbidity, and conductivity which also provide an indication of water pollution and the levels of the 16 US EPA priority PAHs were investigated in water from Ngong River.

Materials and Methods

Chemicals and reagents

All chemicals and reagents were of analytical grade.

***Corresponding author:** Nawiri Mildred P, Department of Chemistry, School of Pure and Applied Science, Kenyatta University, Nairobi, Kenya
E-mail: nawiri.mildred@ku.ac.ke

Dichloromethane (DCM) hexane, high purity nitrogen, anhydrous sodium sulphate, florisil, phosphate buffer (pH 6.8), Silica gel and aluminium oxide, were obtained from Sigma Aldrich, UK. A PAH standard mix containing the 16 priority PAHs was obtained from Dr. Ehrenstofer Augsburg, Germany.

Instruments and instrumental conditions

LabQuest vernier multi-parameter meter was used for measurement of physiochemical parameters. An Agilent 7890A Gas Chromatography and 5975C Mass Selective Detector was used for the analysis of PAHs. Column oven temperature was programmed from 35 to 280°C with initial temperature maintained for 5 min then 10°C/min to 280°C for 10.5 min and then for 29.9 min at 50°C/min to 285°C. The GC was fitted with a HP-5 MS low bleed capillary Column (30 m × 0.25 mm i.d., 0.25 µm film thickness). The carrier gas was helium.

Sampling, physiochemical measurements and pretreatment

Sampling sites Mbagathi, Marter, Donholm and Kangundo along the Ngong River were selected based on the physical appearance of the water, land use patterns and surrounding economic activities. Water samples for analysis were collected directly into clean sterilized amber colored 500 mL sample bottles which had been pre-washed with distilled water and dried. Three replicate samples were collected at the same sampling point. The bottles were sealed with aluminium foil and then put in ice-box containing wet ice and then transported to the laboratory where they were kept in the fridge at -4°C prior to PAH extraction. The physico-chemical parameters, pH, temperature, conductivity, turbidity and dissolved oxygen were measured on site.

Extraction and clean-up

Extraction was carried out by liquid-liquid extraction following the procedure outlined by Anyakora and Herbert [5]. The samples were passed through glass wool to remove any sediments and coagulants. 100 ml of water was measured using a glass measuring cylinder, transferred into a beaker. Phosphate buffer was added to the samples. This was followed by addition of 20 ml dichloromethane (DCM) in a separator funnel. The mixture was shaken for 2 minutes and allowed to separate for 5 minutes. The lower organic layer was then collected into a clean and dry 250 ml conical flask and the extraction process repeated twice each time with 20 ml dichloromethane.

Sample cleanup was done using silica gel. A chromatographic column 25 cm × 1.5 cm diameter was packed with 5 g of silica gel followed by freshly baked 5 g of anhydrous sodium sulphate. The column was conditioned using 15 ml hexane and discarded. The sample extract was transferred into the column and eluted with 100 ml hexane and 100 ml dichloromethane. The eluate was then concentrated to about 3 ml using a rotary evaporator which was further reduced to about 0.5 mL using a stream of gaseous nitrogen. The concentrated extracts were then put in vials and stored in a fridge at -4°C ready for analysis.

Identification and quantification

Identity of PAHs in the samples was confirmed by the retention time and abundance of quantification/confirmation ions and mass spectra match against the calibration standards in the authentic PAHs standards. Confirmation of identity of the analytes was done using NIST/EPA/NIH MASS SPECTRAL LIBRARY (NIST 05) and NIST MASS SPECTRAL SEARCH PROGRAM Version 2.0d. Quantification of individual PAH compounds was performed by the method of internal standardization. The quantitation was based on the ratio of the peak height of the quant ion to that of the internal standard.

Results and Discussion

Physico-chemical parameters

The result obtained from physico-chemical analysis of water samples collected from different sampling points are compared with WHO and US EPA water quality standards. Table 1 shows the mean values of the physicochemical parameters at different sampling sites along the Ngong River.

pH influences the biological activity of the water microflora. During this study pH value ranged from 6.42 ± 0.00 to 6.96 ± 0.01. Except for Donholm Bridge water sample, these values are within maximum permissible limit (6.5-8.5) prescribed by WHO. However, these pH values were consistent with those for any kind of open natural water system, which vary from 5.5 to 10.5 [18]. The observed pH value indicates that the present water samples are slightly acidic. This can be attributed to deposition of acid forming substances and high organic content which can decrease the pH because of the carbonate chemistry [19]. In most cases where the water is neither very acidic nor very alkaline, the pH is regulated by the CO_2-HCO_3 system [20].

The maximum value of dissolved oxygen was noted to be 6.27 ± 0.0 mg/L at Donholm bridge water samples while the minimum value of 2.57 ± 0.09 mg/L was at Mbagathi road bridge water. The values of DO contents of water samples were above the standard limit (5 mg/L) as recommended by WHO [12] except for Mbagathi and Marter sampling points did not meet this threshold. This might be due to decay and decomposition of organic matter [20] responsible for the depletion of oxygen. The low level of DO might be due to turbidity effect which minimized the dissolve oxygen content or due to inflow of sewage with less oxygen content into the Ngong River from informal settlements such as Kibera slums.

Turbidity values fluctuated between 68.00 ± 0.64 to 85.27 ± 0.55 NTU, the minimum being in Donholm Bridge water and maximum in water from Marter bridge. High turbidity recorded in this river could be associated with sediments resulting from soil erosion from the observable farming activities and urban runoff pollution [21]. Additionally, the variations of turbidity depend on the inflow of rain water carrying suspended particles [22].

Physicochemical Parameters (n=3)					
Sampling points	pH	Dissolved Oxygen (mg/L)	Turbidity (NTU)	Temperature (°C)	Conductivity (µScm^{-1})
Mbagathi	6.96 ± 0.01^d	2.57 ± 0.09^a	69.57 ± 0.19^d	22.73 ± 0.07^d	865.00 ± 1.53^d
Marter	6.61 ± 0.00^c	2.83 ± 0.07^b	85.27 ± 0.55^a	22.37 ± 0.03^a	425.00 ± 1.53^a
Donholm bridge	6.42 ± 0.00^a	6.27 ± 0.03^d	68.00 ± 0.64^c	22.70 ± 0.06^c	646.00 ± 1.53^b
Kangundo	6.53 ± 0.01^b	5.73 ± 0.07^c	73.77 ± 0.23^b	22.67 ± 0.15^b	677.00 ± 1.53^c

Mean values followed by the same small letter within the same column are not significantly different (One-way ANOVA, Duncan's test, α=0.05).

Table 1: Mean Values of Physicochemical Parameters of water from Ngong River.

Water temperature plays an important function in the solubility of salts and gases. Additionally, chemical, biochemical and biological characteristic of an aquatic system is greatly influenced by temperature. The temperature of Ngong river water is largely influence by local climatic conditions. Values of water temperature ranged from 22.37 ± 0.03°C to 22.73 ± 0.07°C, the minimum values was recorded in winter at the Marter bridge and maximum at Mbagathi Road. The temperature difference might be due to differences between the collection times.

The conductivity of Ngong river water ranged from 425.00 to 865.00 µS cm^{-1}. The conductivity levels were found to be within values that have been reported in some water bodies from other parts of the world. For example, Sekabira et al., [23] reported conductivity levels ranging between 208.0 and 1884.0 µs/cm determined in Nakivubo Stream water in Kampala, Uganda. On the other hand the levels were lower compared to those reported in Chemu, Korle and Kpeshie lagoons, Ghana which recorded an average high conductivity of 5.677, 40.83 and 40.50 S/m respectively [7]. According to the WHO, the limit for conductivity for drinking and portable water is 700 µScm^{-1} [24]. This threshold was met in all sampling sites for both rivers except at Mbagathi Road Bridge (865.00 ± 1.53 µScm^{-1}) and Outer Ring road bridge (749.33 ± 4.26 µScm^{-1}) sampling sites where this limit was exceeded.

Levels of polycyclic aromatic hydrocarbons

The concentration (ng/L) and distribution of PAHs in water along sites on the Ngong River are presented in Table 2. Except for benzo(a)pyrene, benzo(ghi)perylene, acenaphthylene, dibenz(a,h) anthracene and acenaphthene, which were below the detection limits of the analytical instrument, eleven (11) of the priority PAHs were detected at various points along the river. Worth noting is that chrysene and benzo(a)anthracene were detected at all the sampling sites. The concentrations of the PAHs ranged from 2.32 ng/L to 14.22 ng/L, with an overall mean concentration of 11.53 ng/L. There were significant differences ($p<0.05$) in the mean concentration of PAHs at different sampling sites for all the detected PAHs except for benzo(k) flouranthene, flourene and pyrene. Although these levels are found to be below the WHO's limit of 50 ng/L [13], they fall within ranges reported in studies across the globe although in the USA and China values as high as 29325 ng/L have been reported [5-8].

The variation in the distribution of PAHs of various ring sizes along the sampling sites can be explained one, based on the surrounding activities and two, the dilution effect as the river flows. An example is the Mbagathi site which had the highest total PAHs concentration of 14.22 ng/L, is close to the expansive Kibera slums where domestic sewage outfalls into the river, while at Marter Bridge a number of industries are located around this sampling site. Furthermore, at both sites, heavy traffic is an ordinary activity. Industrial waste water and atmospheric deposition as a result of air emission from industries in addition explains the presence of the PAHs in the sampling sites. In terms of number of rings, two and three ring PAHs concentration were the lowest in all water samples although larger ring sized PAHs were detected too. Such a wide range of PAHs at different concentrations indicates that there are potentially many different sources of PAHs in the Ngong River, possibly including combustion, atmospheric fallout, oil residues, sewage discharges, outfalls and industrial wastewater [3].

Data analysis

One-Way ANOVA at 95% was employed using SPSS version 20 for windows. Values of $p<0.05$ were considered significantly different and SNKtest at $\alpha=0.05$ indicated differences among means. Student's T-test was applied for mean comparisons between means.

Conclusion

The results show an indication of some pollution in the river. However, some values of the physico-chemical parameters were found to be within acceptable range. pH levels were within the 6.5-8.5, dissolved oxygen 5 mg/L, conductivity 700 µScm^{-1} as provided by WHO. The results indicated that Ngong River is contaminated with a number of the US EPA priority PAHs. However, these levels were however below the threshold recommended by WHO and US EPA (50 ng/L). The type of PAHs detected varied from one location to another. The commonly found PAHs compounds in the water samples were fluorene, phenanthrene, fluoranthene, pyrene, benzo(b)fluoranthene,

PAHs	No.of Rings	Study sites Mbagathi	Marter	Donholm	Kangundo	p-value (0.01)	Overall Mean
		Mean ± SE (ng/L; n=12)					
Naphthalene	2	ND	ND	0.52 ± 0.06	0.14 ± 0.00	<0.001	0.33 ± 0.06
Acenaphthylene	3	ND	ND	ND	ND	ND	ND
Acenaphthene	3	ND	ND	ND	ND	ND	ND
Fluorene	3	3.14 ± 0.51	2.62 ± 0.03	ND	ND	0.33	2.88 ± 0.25
Phenanthrene	3	1.87 ± 0.02^{b}	1.82 ± 0.04^{b}	ND	0.27 ± 0.00^{a}	<0.001	1.32 ± 0.18
Anthracene	3	0.68 ± 0.00	0.69 ± 0.01	ND	1.28 ± 0.00	<0.001	0.88 ± 0.07
Fluoranthene	4	1.86 ± 0.04^{b}	1.88 ± 0.05^{b}	ND	0.12 ± 0.00^{a}	<0.001	1.28 ± 0.20
Pyrene	4	1.66 ± 0.00	1.62 ± 0.08	ND	<LOD	0.623	1.64 ± 0.04
Benzo(a)anthracene	4	2.06 ± 0.06^{c}	2.43 ± 0.32^{c}	1.18 ± 0.04^{b}	0.19 ± 0.00^{a}	<0.001	1.47 ± 0.02
Chrysene	4	2.95 ± 0.22^{c}	2.48 ± 0.00^{b}	0.24 ± 0.01^{a}	0.07 ± 0.00^{a}	<0.001	1.43 ± 0.27
Benzo(b)fluoranthene	5	ND	ND	0.07 ± 0.00	0.06 ± 0.00	0.002	0.07 ± 0.00
Benzo(k)fluoranthene	5	ND	ND	0.08 ± 0.00	0.08 ± 0.00	0.092	0.08 ± 0.00
Benzo(a)pyrene	5	ND	ND	ND	ND	ND	ND
Dibenz(a,h)anthracene	5	ND	ND	ND	ND	ND	ND
Indeno(1,2,3-cd)pyrene	6	ND	ND	0.08 ± 0.01	0.48 ± 0.00	<0.001	0.15 ± 0.01
Benzo(ghi)perylene	6	ND	ND	ND	ND	ND	ND
∑ PAH measured		14.22	13.53	2.32	2.69		11.53

Mean values followed by the same small letter within the same row are not significantly different (One-way ANOVA, Duncan's test, $\alpha=0.05$); ND: Not detected.

Table 2: Mean ± SE Concentrations (ng/L) and Distribution of PAHs in the Ngong River Water.

naphthalene, dibenz(a,h)anthracene, dibenz(a,h)anthracene and chrysene. The predominance of low and medium molecular weight PAHs (PAHs with 2-3 rings) in water from both rivers reflects petrogenic origin, related to petroleum including crude oil and its refined products. Additionally, the presence of PAHs with 4-6 rings such as fluoranthene, pyrene, benzo(a)anthracene and chrysene indicates pyrogenic origin and combustion of fossil fuels and of recent organic material. Although PAH levels were found to be within acceptable levels, precaution need to be adhered to in the use of untreated water from River Ngong in fear of body accumulation of PAHs and their dangers.

References

1. Tarek S, Mohamed H (2006) Determination of persistent organic pollutants in water of new Damietta Harbour, Egypt. Egyptian Journal of Aquatic Research 32: 235-245.
2. Oren A, Aizenshtatb Z, Chefetz B (2005) Persistent organic pollutants and sedimentary organic matter properties: A case study in the Kishon River, Israel. Environ Pollut 141: 265-274.
3. Okoro D (2007) Source determination of polynuclear aromatic hydrocarbons in water and sediment of a creek in the Niger Delta region. African Journal of Biotechnology 7: 282-285.
4. Liu Y, Chen L, Jianfua Z, Qinghuib H, Zhilianga Z, et al. (2008) Distribution and sources of polycyclic aromatic hydrocarbons in surface sediments of rivers and an estuary in Shanghai, China. Environmental Pollution 154: 298-305.
5. Anyakora C, Herbert C (2006) Determination of polynuclear aromatic hydrocarbons (PAHs) in selected water bodies in the Niger Delta. African Journal of Biotechnology 21: 2024-2031.
6. Zhang S, Qiang Z, Shameka D, Odi E, Guangdi W (2007) Simultaneous quantification of polycyclic aromatic hydrocarbons , polychlorinated biphenyls, and pharmaceuticals and personal care products in Mississippi river water, in New Orleans, Louisiana, USA. Chemosphere 66: 1057-1069.
7. Essumang D, Adokoh C, Afriyie J, Mensah E (2009) Source assessment and analysis of polycyclic aromatic hydrocarbon in the Oblogo waste disposal sites and some water bodies in and around the Accra metropolis of Ghana. Journal of Water Resource and Protection 1: 456-468.
8. Zhang W, Chaohai W, Chunhua F, Zhe Y, Man R, et al. (2011) Distribution and health-risk of polycyclic aromatic hydrocarbons in soils at a coking plant. Journal of Environmental Monitoring 13: 3426-3436.
9. Palma-Fleming H, Carlos C, Myriam G, Viviana P, Marta G, et al. (2008) Polycyclic aromatic hydrocarbons and polychlorinated biphenyls in coastal environments of Valdivia and Valparaíso, Chile. Journal of the Chilean Chemical Society 53: 1533-1538.
10. Li A, Uchimura T, Tsukatani H, Imasaka T (2010) Trace analysis of polycyclic aromatic hydrocarbons using Gas Chromatography–Mass Spectrometry based on nanosecond multiphoton ionization. Japan Society for Analytical Chemistry 26: 842-846.
11. Wang J, Kai Z, Bo L, Eddy Y (2011) Occurrence, source apportionment and toxicity assessment of polycyclic aromatic hydrocarbons in surface sediments of Chaohu. Journal of Environmental Monitoring 13: 3336-3342.
12. US Environmental Protection Agency (1984) Federal Register, Rules and Regulations, Polynuclear Aromatic Hydrocarbons 49.
13. WHO (1998) Polynuclear aromatic hydrocarbons. In: Guidelines for drinking-water quality, 2nd ed. Health criteria and other supporting information, Geneva 2: 123-152.
14. WHO (1997) Water Pollution Control - A Guide to the Use of Water Quality Management Principles. Great Britain: WHO/UNEP.
15. Lajide L, Ololade I (2010) Exposure level and bioaccumulation of polycyclic aromatic hydrocarbons (PAHs) in edible marine organisms. Journal of Environmental Indicators 5: 69-88.
16. Okoth F, Otieno P (2001) Pollution Assessment Report of the Nairobi River Basin, UNEP. AWN, Nairobi pp: 9-20.
17. Budambula N, Mwachiro E (2006) Metal status of Nairobi river waters and the bioaccumulation in Labeo Cylindricus. Water, Air, and Soil Pollution 169: 275-291.
18. Njenga N, Oyake L, Kamau G (2009) Effect of Nairobi industrial area's effluents on levels of Ngong River pollution. International Journal of BioChemiPhysics 17: 1-12.
19. Fella H, Mohamed B, Fatouma B, Mohand SH (2013) Preliminary Study on Physico-Chemical Parameters and Phytoplankton of Chiffa River. Journal of Ecosystems 9.
20. Leena S, Choudhary S (2013) Physico-chemical characteristics of river water of Ganga in middle Ganga plains. International Journal of Innovative Research in Science, Engineering and Technology 2: 4349- 4357.
21. Akoto O, Bruce T, Darko G (2008) Heavy metals pollution profiles in streams serving the Owabi reservoir. African Journal of Environmental Science and Technology 11: 354-359.
22. Begam N, Purushothama R, Narayana J, Kumar KPR (2006) Water quality studies of TV stations Reservoir at Davangere city, Karnataka (India). J Environ Sci Eng 48: 281-284.
23. Sekabira K, Oryem H, Basamba T, Mutumba G, Kakudidi E (2010) Heavy metal assessment and water quality values in urban stream and rain water. International Journal of Environment, Science and Technology 7: 759-770.
24. WHO (2003) Background document for preparation of WHO Guidelines for drinking-water quality. Geneva, world Health Organization.

Detection of Extended Spectrum Beta-Lactamase Producing *Escherichia Coli* on Water at Hafr Al Batin, Saudi Arabia

Sulaiman Ali Al Yousef, Eman Saleh Farrag, Ahmed Mohamed Ali, Sabry Younis Mahmoud*

Department of Medical LaboratoryTechnology, College ofApplied Medical Science, University of HafrAl Batin, Kingdom of SaudiArabia

Abstract

In this study antibiotic resistant *Escherichia coli* from household water collected in Hafr AlBatin, Saudi Arabia, were characterized by prevalence of extended spectrum betalactamase (ESBL). Samples were collected from drinking and washing water at 12 locations. From the 144 samples obtained, Millipore membrane filters incubated on trypton soya agar plates produced colonies that yielded 34 isolates of *E. coli* as verified by biochemical tests. Isolates suspected ESBL producing were tested by using MicroScan analysis and disc diffusion test as ESBL phenotypic confirmatory methods. Phenotypically confirmed ESBL isolates were examined for antimicrobial susceptibility against 30 antibiotics and amplification of bla_{VEM}, bla_{CTX}, bla_{TEM}, bla_{GES} and bla_{SHV} genes by polymerase chain reaction. Out of 34 *E. coli* isolates, only 6 (17.6%) were positive for ESBL producing according to MicroScan analysis. Disk diffusion as a confirmatory test indicates sensitivity of MicroScan system. PCR results indicated that the VEB was the most prevalent (83.3%) followed by CTX gene (16.6%) between these isolates. The ESBL-producing *E. coli* isolates was fully susceptible to pip/tazo (100%) and fully resistance to ampicillin, cefazolin and pepracillin (100%). This study showed that ESBL-producing *E. coli* are multidrug-resistant and existent in Hafr Al Batin's water. Also, data indicated that wastewater maybe contributes as a source and reservoir of antibiotic resistance.

Keywords: Antimicrobial resistance; ESBL; *E. coli*; Water

Introduction

One of the major ways of diffusion of pathogenic and/or antibiotic resistant bacteria, through water, soil and air. Multidrug resistant bacteria have been revealed in different water sources including rivers, lakes, groundwater and drinking water [1-4]. Water consumption, can command to habitation of the gastrointestinal tract of humans and animals by bacteria containing resistance genes and barter genes with bacteria previously present in the intestinal tract [5,6].

Some bacteria can make beta-lactamase enzymes that cleave the beta-lactam ring and then disrupt the antimicrobial action [7]. An extended spectrum beta-lactamase (ESBL)-producing bacteria is capable to disrupt the third generation cephalosporins and monobactum [8]. Also, shown to be able to combat quinolones [9]. ESBL-producing bacteria especially of Enterobacteriaceae, foremost *E. coli* and *Klebsiella pneumonia* [10]. ESBL-producing *E. coli* have 3 various resistance mechanisms. The most widespread is production of beta-lactamase which hydrolyzes the beta-lactam ring in penicillins and cephalosporins. The second is mutation which decrease beta-lactam uptake [11]. The third resistance mechanism is existence of efflux pumps, which exports antibiotics outside of the cell.

Beta-lactam antibiotics work by prevents cell wall biosynthesis in the bacteria. The most common reason of resistance to beta-lactam antibiotics like penicillin is production of enzymes named beta-lactamases. Beta-lactamases are a family of enzymes produced by numerous of Gram positive and negative bacteria that disrupt beta-lactam antibiotics by slit the beta-lactam ring [12].

Penicillins and concerned antibiotics have been utilized extensively for the control and treatment of bacterial infections. Efficiency of this group of antimicrobial agents has been amended, because of the evolution of multidrug-resistant strains of bacteria [13]. Through the years, incalculable penicillin derivatives [14] have been designed and tested, and a variety of new β-lactam ring systems have been introduced such as cephalosporins, cephamycins, oxacepems, clavulanic acid and carbapenems.

Saudi Arabia with an area of 2.15 million km^2 is a desert and water deficit country, with limited fresh water-supplies. It is also recognized by low annual rainfall and privation perennial rivers. The water resources in the Saudi Arabia are surface and underground deposits. Water collected through rainfall (surface water).

In Hafr Al Batin city about 100% of water supply comes from groundwater. Water transported by car to the house where stored in the wells dedicated to it, which may be located next to the sewage tanks. No available information about extended-spectrum beta-lactamases (ESBLs) producing bacteria in water Hafr Al Batin. The present study was to gain insight into the prevalence of AMR *E. coli* in washing and drinking water, and to check the possible role of sewage tanks as AMR contamination source of water.

Materials and Methods

Water sampling

Water samples were taken from 12 locations in the Hafr Al-Batin city, Saudi Arabia (Figure 1). Three sampling sites were located on each location. Two of which were from washing water that taken from house tanks in most of residential neighborhoods and one from drinking water from charity refrigerators. Sampling was done twice, the first on January (winter) and the second on July (summer). Two samples were collected each time at each collection site at two different times, and we

***Corresponding author:** Sabry Younis Mahmoud, Department of Medical Laboratory Technology, College of Applied Medical Science, University of Hafr Al Batin, Kingdom of Saudi Arabia, E-mail: symahmud@uod.edu.sa

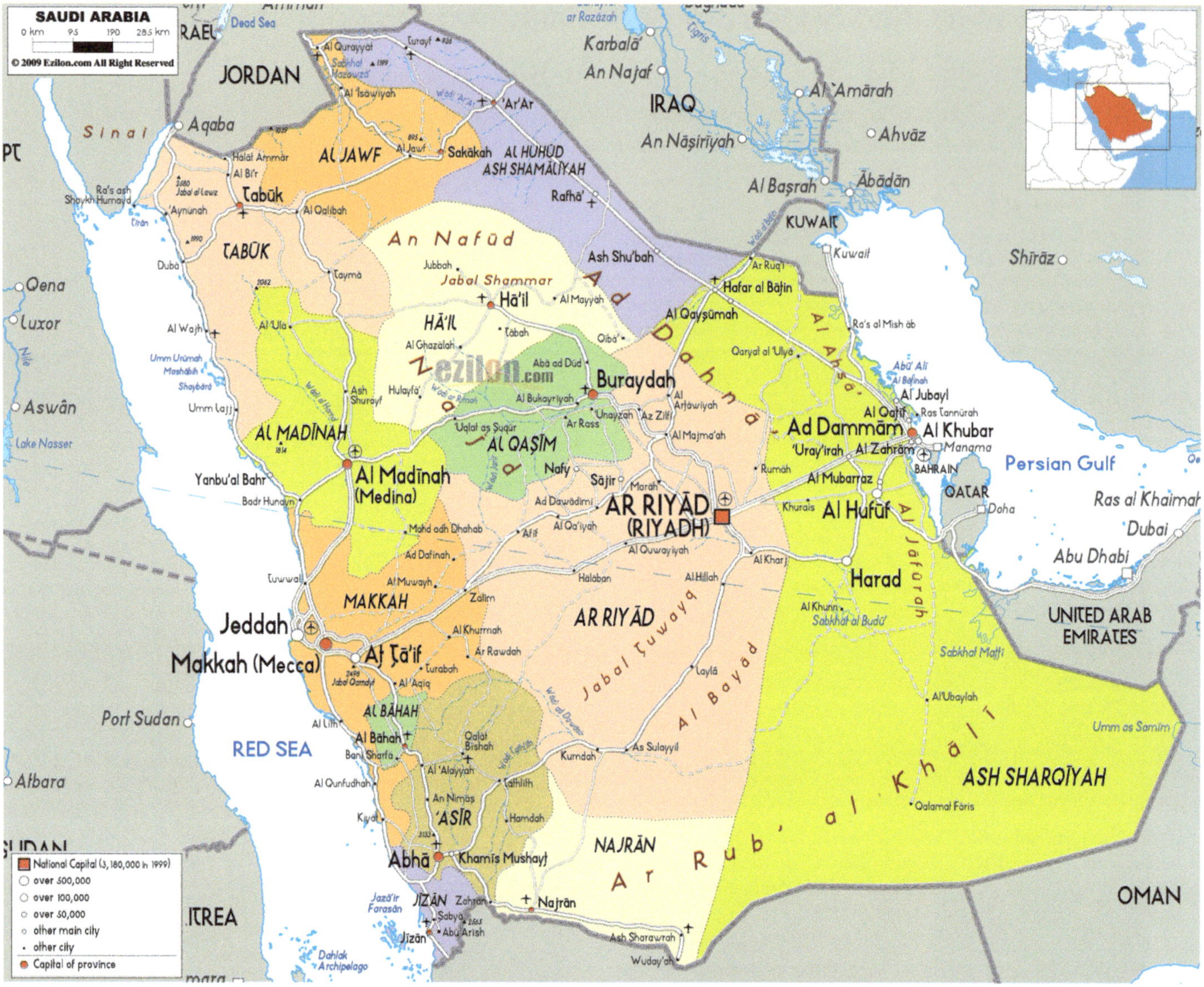

Figure 1: Location map of the study area of Hafr Al-Batin city, Saudi Arabia.

used three replicates from each sample for tests. Collected samples were transported by standard methods as mentioned in APHA, 1989 [15].

Isolation of *E. coli*

100 mL water sample is filtered through 0.45 µm pore size membrane filters (Millipore, the Netherlands). Filters were incubated on tryptone soya agar (TSA) at 36 ± 2°C during 4-5 hours, and subsequently transferred to tryptone bile agar (TBA) then incubated for 19-20 hours at 44 ± 0.5°C as described by Sing et al. [16]. *E. coli* identified using the TSA/TBA method (i.e. indole-positive), were supplementary confirmed as *E. coli* by testing for ß-glucuronidase-activity on Brilliance *E. coli*/coliform agar. Beta-glucuronidase-positive colonies identified using tryptone bile x-glucuronide agar (TBX) was extra confirmed by MicroScan [17] for identification and antibiotic susceptibility.

Analysis of antimicrobial resistance

Generally, 99 *E. coli* isolates (91 from washing water and 8 from drinking water) were obtained. These were screened for confirmation of identification and its susceptibility to a panel of antimicrobials of human and veterinary clinical relevance, using MicroScan. MicroScan® instrumentation (auto SCAN®-4 and WalkAway® System) (Siemens Healthcare Diagnostics Inc, USA) was used. Panels used were MicroScan Dried Gram Positive MIC/Combo, Dried Gram Positive Breakpoint Combo and Dried Gram Positive ID Type 2 or 3. Also, MicroScan Dried Gram Negative MIC/Combo panels and Dried Gram Negative Breakpoint Combo Panels were used. MicroScan panels were designed for use in determining agent susceptibility and/or identification to the species level of rapidly growing aerobic and facultative Gram positive cocci or aerobic and facultatively anaerobic Gram negative bacilli. MICs obtained for ceftazidime and cefotaxime with CA were compared with those obtained with the same drugs without CA. Subsequently, strains were considered as ESBL-positive or -negative in accordance with CLSI recommendations (CLSI, 2010). The tests were performed as recommended by supplier guidelines [17]. Susceptible, intermediate and resistant isolates were arranged according to antibiogram results.

Multi drug resistance was defined as resistance to 3 or more different classes of antimicrobials according to CLSI, 2010 [18].

Confirmation of ESBL-producing *E. coli*

Suspected ESBL-*E. coli* isolates (after MicroScan analysis) were confirmed for ESBL-production by double disc synergy test (DDST) according to CLSI guidelines (CLSI, 2010) [18]. DDS test results analysis. Susceptibility testing was performed (McFarland 0.5 standard) on Mueller–Hinton agar (MHA) by placing discs on the agar surface containing 30 mg cefotaxime or ceftazidime, with and without 10 mg CA. Plates were incubated at 37°C for 24 h. According to CLSI guidelines, strains were considered positive for ESBL production whenever zone diameters increased by ¢5 mm for cefotaxime or ceftazidime when tested in combination with CA. This method was considered the gold standard for method comparison.

DNA isolation and genotyping

A single colony from each ESBL-producing isolate was transferred into 100 μL of sterile distilled water and the bacterial DNA was extracted by using boiling method included microwave pre-heating according to Ahmed et al. [19]. PCR screening for presence of different beta-lactamase genes was performed. PCR was carried out and specific primers (Table 1) were used for VEB, CTX, TEM and GES genes. PCR mixtures were prepared by using 5 μL template DNA (about 500 pg of DNA), 12.5μL PCR master mix; 1 × PCR buffer [Tris-Cl, KCl, $(NH_4)2SO_4$, 1.5mM $MgCl_2$] (pH 8.7), 200 μM dNTP, and 1 μL of each 10 pM primer and 0.5U Taq DNA polymerase in a final volume of 25 μL. The amplification reaction was carried out in a Thermal Cycler (Eppendorf master cycler®, MA) with an initial denaturation (94°C for five minutes) followed by 30 cycles of denaturation (94°C for 25 seconds) annealing (52°C for 40 seconds), and extension (72°C for 50 seconds) and a single final extension at 72°C for six minutes [20]. The amplified products were electrophoresed on 2% agarose gel and visualized on a gel document system after staining with ethidium bromide (0.5 mg/mL). A non-ESBL-producing strain (*E. coli* ATCC 25922) was used as a negative control.

Results and Discussion

Water borne infections still ravage the global community and are responsible for millions of deaths per year. Water that looks clear and pure may be sufficiently contaminated with pathogenic microorganisms to be a health hazard.

ESBL-producing *E. coli*

Ninety-nine *E. coli* isolates were detected and isolated from the water samples using the TSA/TBA method. Only 34 isolates were confirmed as *E. coli* by MicroScan. ESBL-producing strains (6⁄34) were found in all the samples analyzed by MicroScan system (Table 2). Five ESBL-producing *E. coli* were detected from the washing water samples at five locations (Al-Khalediyah, Al-Rabwah, Al-Rawdhah, Al-Nayefiyah and Al-Nakheel) at summer season only, no winter was found. On the other hand, ESBL-producing *E. coli* was detected and isolated from drinking water in one location (Al-Baladiyah) out of 12.

Genes	Primer used (5'-3')	PCR product size
VEB	CGACTTCCATTTCCCGATGC GGACTCTGCAACAAATACGC	585 bp
CTX	CAAAGAGAGTGCAACGGATG ATTGGAAAGCGTTCATCACC	205 bp
TEM	TAATCAGTGAGGCACCTATCTC GAGTATTCAACATTTCCGTGTC	800 bp
GES	ATGCGCTTCATTCACGCAC CTATTTGTCCGTGCTCAGG	846 bp
SHV	GGTTATGCGTTATATTCGCC TTAGCGTTGCCAGTGCTC	867 bp

Table 1: The sequences of primers used in PCR amplification of beta-lactamase genes.

Location	Season	Washing water isolates		Drinking water isolates		Total/ confirmed (%)
		Total	+	Total	+	
Abu-Musa al-Asha'ari	Winter	1	0	0	0	1/0 (0.0)
	Summer	4	0	0	0	4/0 (0.0)
Al-Aziziah	Winter	0	0	0	0	0/0 (0.0)
	Summer	2	1	0	0	2/1 (50.0)
Al-Khalediyah	Winter	1	0	0	0	1/0 (0.0)
	Summer	2	0	0	0	2/0 (0.0)
Al-Rabwah	Winter	0	0	0	0	0/0 (0.0)
	Summer	1	1	1	0	2/1 (50.0)
Al-Muhammadiyah	Winter	0	0	0	0	0/0 (0.0)
	Summer	3	0	0	0	3/0 (0.0)
Al-Baladiyah	Winter	1	0	0	0	1/0 (0.0)
	Summer	3	0	2	1	5/1 (20.0)
Al-Rawdhah	Winter	0	0	0	0	0/0 (0.0)
	Summer	4	1	0	0	4/1 (25.0)
Al-Nayefiyah	Winter	0	0	0	0	0/0 (0.0)
	Summer	1	1	0	0	1/1 (100.0)
Al-Sulaimaniyah	Winter	0	0	0	0	0/0 (0.0
	Summer	0	0	0	0	0/0 (0.0)
Al-Faisaliyah	Winter	0	0	0	0	0/0 (0.0)
	Summer	1	0	0	0	1/0 (0.0)
Al-Masyef	Winter	0	0	0	0	0/0 (0.0)
	Summer	4	0	0	0	4/0 (0.0)
Al-Nakheel	Winter	0	0	0	0	0/0 (0.0)
	Summer	2	1	1	0	3/1 (33.3)
Total		30	5	4	1	34/6 (17.6)

Table 2: Total number and ESBL-*E. coli* isolates, which confirmed (+) by MicroScan system from 13 locations.

The prevalence of ESBL-producing *E. coli* varied between waters that differed with regard to region, type of water, and time of the year at sampling. Perhaps, prevalence may vary with the number of faecal dirtiness sources in the neighborhood of sampling sites and factors affecting when and how often releasing takes place (climate or season). Our results agreed with Blaak et al. [21]; Adnan et al. [22].

All ESBL-producing isolates also showed resistance to other antimicrobials: 100% to ampicillin, cefazolin, cefepime, cfuroxime, mezlocillin, piperacillin, trimethoprime, trimethoprime⁄sulfamethoxazole; 83.4% to ciprofloxacin, gentamicin, levofloxacin, norfioxacin, tetracycline, tobramycin; 50% to cefoxitin; 33.3% to ertapenem and 16.6% to fosfomycin, imipenem, meropenem, nitrofurantain, tigecycline (Table 3). The study by Babypadmini and Appalaraju [23] reported 74% resistance to trimethoprim/sulfamethoxazole in ESBL-producing *E. coli* pathogens by disk diffusion method, which is lower than our results. This difference may be due to use of different methods of evaluation for determining the susceptibility. We determined the antimicrobial resistance by the MicroScan method which is more sensitive than disk diffusion method.

Validation of ESBL-producing isolates

ESBL confirmation test was carried out by disk diffusion method. For validity testing of 6 *E. coli* ESBL- producing isolates, no errors were showed between the two test methods, MicroScan and disk diffusion test.

ESBL genes

VEB and CTX-M genes were detected in 5 and 1 ESBL-producing

Antibiotics	Washing water-*E. coli* isolates										Drinking water-*E. coli* isolate	
	1		2		3		4		5			
	MIC	Interps	MIC	Interps	MIC	Interps	MIC	Interps	MIC	Interps	MIC	Interps
Amikacin	≤16	S	>32	R	≤16	S	≤16	S	≤16	S	>32	R
Amox/K Clav	>16/8	R	>16/8	R	≤8/4	S	16/8	I	16/8	I	>16/8	R
Ampicillin	>16	R	>16	R	>16	R	>16	R	>16	R	>16	R
Cefazolin	>16	R	>16	R	>16	R	>16	R	>16	R	>16	R
Cefepime	>16	R	>16	R	>16	R	>16	R	>16	R	>16	R
Cefotaxime	>32	ESBL	>32	ESBL	>32	ESBL	>32	ESBL	>32	ESBL	>32	ESBL
Cefotaxime/K Clav.	≤0.5		≤0.5		≤0.5		≤0.5		≤0.5		≤0.5	
Cefoxitin	≤8	S	>8	R	≤8	S	≤8	S	>8	R	>8	R
Ceftazidime	>16	ESBL	>16	ESBL	>16	ESBL	>16	ESBL	>16	ESBL	>16	ESBL
Ceftazidime/k clav.	2		2				≤0.25		≤0.25		2	
Cefuroxime	>16	R	>16	R	>16	R	>16	R	>16	R	>16	R
Ciprofloxacin	>2	R	>2	R	≤1	S	>2	R	>2	R	>2	R
Colistin	≤2	S	≤2	S	≤2	S	≤2	S	≤2	S	>4	R
Ertapenem	≤2	S	>4	R	≤2	S	≤2	S	≤2	S	>4	R
Fosfomycin	≤32	S	≤32	S	≤32	S	≤32	S	≤32	S	>32	R
Gentamicin	>8	R	>8	R	≤4	S	>8	R	>8	R	>8	R
Imipenem	≤4	S	≤4	S	≤4	S	≤4	S	≤4	S	>8	R
Levofloxacin	>4	R	>4	R	≤2	S	>4	R	>4	R	>4	R
Meropenem	≤1	S	≤1	S	≤1	S	≤1	S	≤1	S	>8	R
Mezlocillin	>64	R	>64	R	>64	R	>64	R	>64	R	>64	R
Moxifloxacin	>1	R	>1	R	1	I	>1	R	>1	R	>1	R
Nitrofurantoin	≤32	S	≤32	S	≤32	S	≤32	S	≤32	S	>64	R
Norfloxacin	>8	R	>8	R	>4	S	>8	R	>8	R	>8	R
Pip/tazo	64	I	≤16	S	≤16	S	≤16	S	≤16	S	≤16	S
Piperacillin	>64	R	>64	R	>64	R	>64	R	>64	R	>64	R
Tetracycline	>8	R	>8	R	≤	S	>8	R	>8	R	>8	R
Tigecycline	≤1	S	≤1	S	≤1	S	≤1	S	≤1	S	>2	R
Tobramycin	>8	R	>8	R	≤4	S	>8	R	>8	R	>8	R
Trimeth/sulfa	>2/38	R	>2/38	R	>2/38	R	>2/38	R	>2/38	R	>2/38	R
Trimethoprim	>8	R	>8	R	>8	R	>8	R	>8	R	>8	R

S = Susceptible, I = Intermediate, R = Resistant, MIC = Minimum Inhibitory Concentration (mcg/m), Interps = Interpretation,
ESBL = Extended spectrum beta-lactamase. Suspected ESBL (confirmatory test needed to differentiate ESBL from other beta-lactamase)

Table 3: Antibiotic susceptibility pattern among ESBL-*E. coli* resistant isolates.

E. coli isolates from washing and drinking water samples, respectively. The results of ESBL genotyping showed that VEB gene was the most prevalent (83.3%) followed by CTX gene (16.6%), TEM (0.0%), GES (0.0%) and SHV (0.0%). ESBL-producing *E. coli* isolate from drinking water was carried CTX gene (Figures 2 and 3). In washing water, only VEB gene was predominant. VEB was the most frequent resistant gene in ESBL-producing *E. coli* isolates in this study. The study by Rezai et al. [24] reported that TEM resistant gene was most prevalent. TEM,

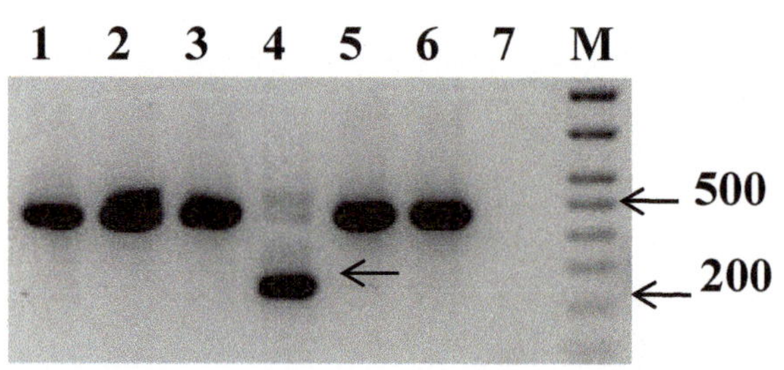

Figure 2: Agarose gel showing the 585 and 205bp PCR fragments band for VEB and CTX genes from ESBL-producing *E. coli* isolates, respectivily. Lanes: M: molecular weight marker; 1, 2, 3, 5 and 6: VEM positive washing water isolates; 4: CTX positive drinking water isolate and 7: negative control.

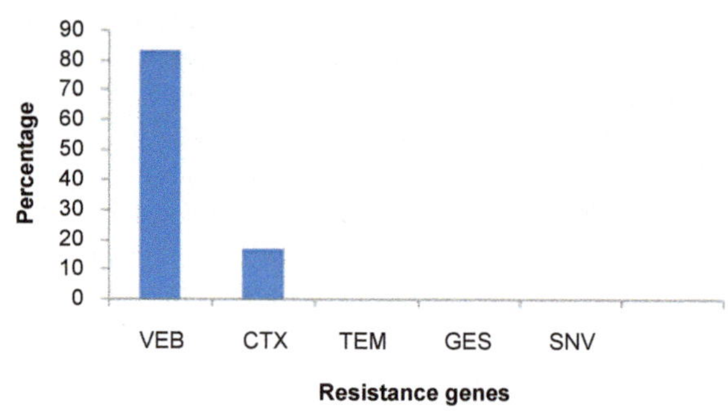

Figure 3: Distribution of resistance genes in ESBL-producing *E. coli* isolates.

GES and SHV resistant genes were not found in ESBL-producing *E. coli* isolates which is in line with the low frequency of this gene in ESBL-producing *E. coli* strains [25].

Conclusion

Fecal contamination of tank water in Hafr Al Batin city by sewage may be occurs through discharge of untreated sewage through sewage leaking or during heavy rainfall. Water distribution and storage

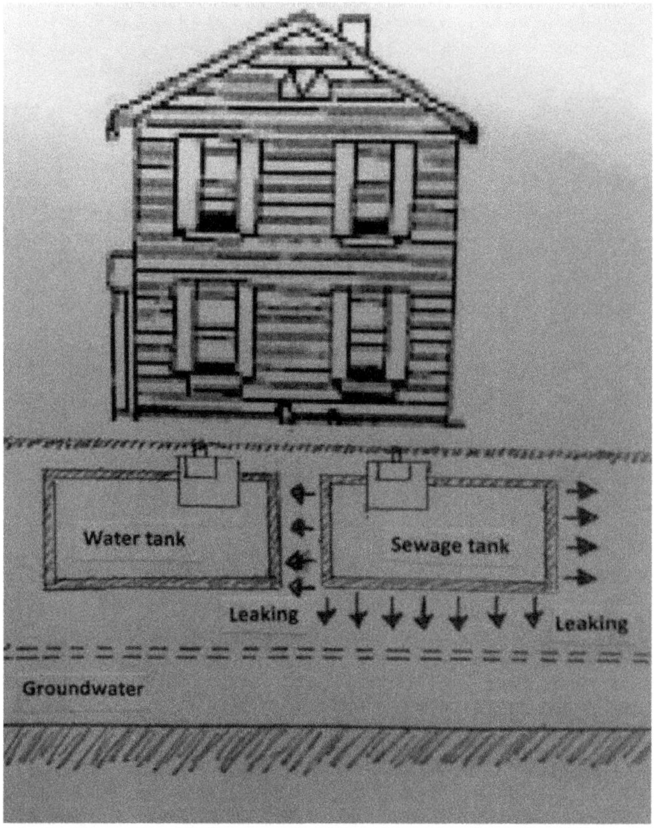

Figure 4: Diagram showing pathway of contaminants emission from sewage tank to adjacent water tank and groundwater.

systems in Hafr Al-Batin city could serve as an incubator for growth of certain ARB populations and as an important reservoir for the spread of antibiotic resistance to opportunistic pathogens. Out of six ESBL-producing *E. coli* isolates, five were obtained from washing water tanks adjacent to sewage tanks as shown in proposed pathway of contamination transport (Figure 4). Leaks happen when a pipe isn't sealed in a specific spot. Most old septic tanks didn't have any sealants applied to the mating joints and troubleshooting difficult. Also, heavy vehicles can crush septic system drain lines causing leakage and toxic smells and odors. The septic cover can cause leaks because they may not be well-secured. From here, we recommend that the system of quantitative and qualitative microbial risk estimate must be applied on houses water tanks at Hafr Al-batin city. This process was successfully implemented for estimate exposure and infection hazard of bacterial and other pathogens from consumption of drinking water and washing water. Water-borne transmission has been demonstrated to be a relevant route of transmission for faecal bacterial species, including *Salmonella* and *E. coli* [26,27]. To evaluate risks of human exposure, the system of quantitative microbial risk assessment could be used [28]. This process was successfully implemented for estimate exposure and infection hazard of bacterial and other pathogens from consumption of drinking water and recreational water [29,30].

Our results suggested the possible emissions of the ESBL-producing *E. coli* from sewage tanks to the environment. Water distribution systems in Hafr Al-Batin city could serve as an incubator for growth of certain ARB populations and as an important reservoir for the spread of antibiotic resistance to opportunistic pathogens.

Acknowledgment

The authors extend their appreciation to the Deanship of Scientific Research at University of Dammam for funding this work through research project no. 2014027.

References

1. Hamelin K, Bruant G, El-Shaarawi A, Hill S, Edge TA, et al. (2007) Occurrence of virulence and antimicrobial resistance genes in Escherichia coli isolates from different aquatic ecosystems within the St. Clair River and Detroit River areas. Appl Environ Microbiol 73: 477-484.
2. Baquero F, Martínez JL, Cantón R (2008) Antibiotics and antibiotic resistance in water environments. Curr Opin Biotechnol 19: 260-265.
3. Marti E, Jofre J, Balcazar JL (2013) Prevalence of Antibiotic Resistance Genes and Bacterial Community Composition in a River Influenced by a Wastewater Treatment Plant. PLoS ONE 8: e78906.
4. Ramírez-Castillo FY, González FJA, Garneau P, Díaz FM, Guerrero-Barrera AL, et al. (2013) Presence of multi-drug resistant pathogenic Escherichia coli in the San Pedro River located in the State of Aguascalientes, Mexico. Front Microbiol 4: 147.
5. Coleman BL, Salvadori MI, McGeer AJ, Sibley KA, Neumann NF, et al. (2012) The role of drinking water in the transmission of antimicrobial-resistant E. coli. Epidemiol Infect 140: 633-642.
6. Finley RL, Collignon P, Joakim-Larsson DG, McEwen SA, Li XZ, et al. The sourge of antibiotic resistance: the important role of the environment. Clin Infect Dis 57: 704-710.
7. Tenover CF (2006) Mechanism of Antimicrobial resistance in Bacteria. Am J Infect Control 34: S3-S10.
8. Paterson DL, Bonomo RA (2005) Extended-spectrum beta-lactamases: a clinical update. Clin Microbiol Rev 18: 657-686.
9. Karah N (2010) Plasmid-mediated quinolone resistance determinants qnr and aac(6')-Ib-cr in Escherichia coli and Klebsiella spp. from Norway and Sweden. Diagn Microbiol Infect Dis 66: 425-431.
10. Katsanis GP, Spargo J, Ferraro MJ, Sutton L, Jacoby GA (1994) Detection of Klebsiella pneumoniae and Escherichia coli strains producing extended spectrum beta lactamases. J Clin Microbial 32: 691-696.
11. Jacoby GA, Sutton L (1991) Properties of plasmids responsible for production of extended-spectrum b-lactamases. Antimicrob Agents Chemother 35: 164-169.
12. Kong K F, Schneper L, Mathee K (2010) Beta-lactam antibiotics: From antibiotic to resistance and bacteriology. APMIs 118: 1-36.
13. Fisher JF, Meroueh SO, Mobashery S (2005) Bacterial resistance to β-lactam antibiotics: compelling opportunism, compelling opportunity. Chem Rev 105: 395-424.
14. Sassiver ML, Lewis A (1977) In: Perlman D (ed.) Structure-Activity Relationships among the Semisynthetic Antibiotics; Academic Press: New York, USA 87-160.
15. APHA (1989) Standard Methods for the examination of water and waste water, American Public Health Association, 17th Ed., Washington, DC.
16. Singh M, Taneja N, Bhatti D, Sharma M (2002) Comparison of ReadyCult® Coliform 50 with conventional methods for detection of total coliforms and faecal Escherichia coli in drinking water samples. Eco Env Cons 8: 15-20.
17. Snyder JW, Munier GK, Johnson CL (2008) Direct Comparison of the BD Phoenix System with the MicroScan WalkAway System for Identification and Antimicrobial Susceptibility Testing of Enterobacteriaceae and Nonfermentative Gram-Negative Organisms. J Clin Microbiol 46: 2327-2333.
18. Clinical and Laboratory Standards Institute (2010) Performance standards for antimicrobial susceptibility testing; twentieth informational supplement M100-S20. CLSI, Wayne, PA, USA.
19. Ahmed OB, Asghar AH, Elhassan MM (2014) Comparison of three DNA extraction methods for polymerase chain reaction (PCR) analysis of bacterial genomic DNA. Afr J Microbiol Res 8: 598-602.
20. Edelstein M, Pimkin M, Palagin I, Edelstein I, Stratchounski L (2003) Prevalence and Molecular Epidemiology of CTX-M Extended-Spectrum -Lactamase-Producing Escherichia coli and Klebsiella pneumoniae in Russian Hospitals. Antimicrobial Agents and Chemotherapy 47: 3724-3732.

21. Blaak H, Lynch G, Italiaander R, Hamidjaja A, Schets M, et al. (2015) Multidrug-resistant and extended spectrum Beta-lactamase-producing Escherichia coli in Dutch surface water and wastewater. PLoS ONE 10: e0127752.

22. Adnan N, Sultana M, Islam OK, Nandi SP, Hossain MA (2013) Characterization of ciprofloxacin resistant extended spectrum β-Lactamase (ESBL) producing Escherichia spp. from clinical waste water in Bangladesh. Advances in Bioscience and Biotechnology 4: 15-23.

23. Babypadmini S, Appalaraju B (2004) Extended spectrum-lactamases in urinary isolates of Escherichia coli and Klebsiella pneumoniae-prevalence and susceptibility pattern in a tertiary care hospital. Indian J Med Microbiol 22: 172-174.

24. Rezai MS, Salehifar E, Rafiei A, Langaee T, Rafati M, et al. (2015) Characterization of Multidrug Resistant Extended-Spectrum Beta-Lactamase-Producing Escherichia coli among Uropathogens of Pediatrics in North of Iran. BioMed Research International 1-7.

25. Hawkey PM (2008) Prevalence and clonality of extended-spectrum β-lactamases in Asia. Clin Microbiol Infect 14: 159-165.

26. Søraas A, Sundsfjord A, Jørgensen SB, Liestøl K, Jenum PA (2014) Zhou D (ed.)High Rate of Per Oral Mecillinam Treatment Failure in Community-Acquired Urinary Tract Infections Caused by ESBL-Producing Escherichia coli. PLoS ONE. Public Library of Science (PLoS) 9: e85889.

27. Rangel JM, Sparling PH, Crowe C, Griffin PM, Swerdlow DL (2005) Epidemiology of Escherichia coli O157:H7 Outbreaks, United States, 1982–2002. Emerging Infectious Diseases 11: 603-609.

28. Haas CN, Rose JB, Gerba C, Regli S (1993) Risk Assessment of Virus in Drinking Water. Risk Analysis 13: 545-552.

29. Schets FM, Schijven JF, de Roda Husman AM (2011) Exposure assessment for swimmers in bathing waters and swimming pools. Water Research 45: 2392-2400.

30. Schijven JF, Teunis PFM, Rutjes SA, Bouwknegt M, de Roda Husman AM (2011) QMRAspot: A tool for Quantitative Microbial Risk Assessment from surface water to potable water. Water Res 45: 5564-5576.

Fish Tissue Bio-concentration and Interspecies Uptake of Heavy Metals from Waste Water Lagoons

Francis Orata* **and Faith Birgen**

Department of Pure andApplied Chemistry, Masinde Muliro University of Science andTechnology, P.O. Box 190-50100, Kakamega, Kenya

Abstract

Fish products may be important vectors of human exposure to heavy metals. However the understanding of how heavy metals bioconcentrate in different fish species (sp.) through contaminated environments such as wastewater treatment lagoons is very limited. In this study bioconcentration and Interspecies Uptake of Heavy Metals by three fish species (*Oreochromis niloticus*, *Clarius gariespinus* and *Protopterus aethiopicus*) receiving naturally contaminated wastewater was investigated. Tissue bioconcentration factors (BCFs) were estimated within a typical municipal wastewater lagoon, in Kenya. It was assumed that the three fish species were able to reach steady state with their dietary intake of heavy metals and through aqueous contact. Heavy metals concentration of up to 11.72, 11.27, 5.29, 4.12 and 4.74 mg/kg for, muscle, skin, liver, scales and gills respectively were obtained through analysis using ICP-MS instrumentation. The results showed variation in the estimated BCFs obtained for *O.niloticus* as compared to those obtained for *P.aethopicus* and *C.gariespinus*. The tissue BCFs ranged from 0.20 for Cd in scales of *C. gariespinus* to 11.27 for Zn in muscles of *P.aethipicus*. Results suggest that heavy metals uptake by the fish sp. and their transfer to various tissue organs do not exclusively depend on concentration levels by aqueous exposure alone but largely depends on their feeding mode, diet and biochemical needs of individual fish sp. Elevated concentrations of Pb and Cd in scales and skin were more than those obtained in muscles and liver which suggested depuration process, considering that Pb and Cd are non-essential metal ions to these fish sp. Fish need to be carefully screened to ensure that unnecessary high level of some toxic trace metals are not transferred to man. The bio concentration and interspecies uptake data estimated here provide a useful tool for predicting human exposure to Heavy metals via fish under different contamination scenarios.

Keywords: Bio-concentration; Heavy metals; Waste water lagoons; Interspecies uptake; Aquatic pollution; Risk assessment

Introduction

Bioconcentration is a situation in which the levels of a toxin in an organism exceed the levels of that toxin in the surrounding environment. This term is often used specifically in reference to aquatic environments and aquatic organisms. Bioconcentration factors (BCFs) are used to express bioconcentration levels in a numeric way. Arithmetically, BCFs are obtained by dividing the levels of a toxin in an organism by the levels in the surrounding water to find a ratio. The higher the ratio, the more severe the bio concentration of toxins, in this study, the heavy metals level in fish. Bioconcentration factor (BCF) information is a requirement for risk assessment of chemicals and is used by regulatory regimes such as the Registration, Evaluation, Authorization and Restriction of Chemicals (REACH), which regulates chemicals in the European Union (EU) [1]. In several regulatory regimes, including REACH in the EU, the criteria for a chemical being categorised as bioaccumulative or very bioaccumulative is based on BCF and not biomagnification (BMF). In addition, risk assessment requires a BCF to estimate concentrations in prey for the investigation of risks from secondary poisoning. Unlike BMF which involves exposing organisms to the test chemical via the diet (in a dietary study) BCF is derived from the aqueous exposure method [1].

African aquatic systems are experiencing increasingly heavy metal pollution as a result of industrial growth and development [2]. Urbanization and industrialization cause increase in release of industrial and domestic effluents which constitute the largest sources of heavy metal pollution and therefore contribute to a steady increase of metallic contaminant in aquatic and terrestrial environment in most parts of the world [3]. Over 60% of industries are located in the three cities namely; Nairobi, Mombasa and Kisumu in Kenya. This results to industrial discharge of effluents containing heavy metals into the terrestrial and aquatic ecosystems within the cities. Typically, Kisumu city has one lagoon system (Nyalenda wastewater) which acts as a sink or reservoir receiving effluents of over 11,000 M^3 daily. Figure 1 shows the location of the study site (Nyalenda Waste Stabilization lagoons).

With contamination of aquatic ecosystems, aquatic animals are constantly exposed to heavy metals, hence results in bio transference and biomagnification of hazardous heavy metals to humans, as a result of consumption of fish [4]. Because of their chemical stability, heavy metals tend to accumulate into the different tissues of aquatic organisms [5]. Studies have shown that sediments form a major sink of heavy metals [6]. Therefore the sediments contamination is also an exposure route for heavy metals in human, through the consumption of the Benthoic organisms and bottom feeder fish such as *Clarius* sp. and *Protopterus* sp.

Both essential and non-essential elements are persistent and bioaccumulative and therefore they are potentially significant in ecotoxicology [7]. Contamination by heavy metals causes a long lasting effect on ecological balance of the environment and destroys the diversity of different aquatic species living in it [8,9]. Tissue accumulation of heavy metals via polluted waters depends on concentration, time of exposure,

***Corresponding author:** Francis Orata, Department of Pure and Applied Chemistry, Masinde Muliro University of Science and Technology, P.O. Box 190-50100, Kakamega, Kenya, E-mail: fraora@yahoo.com

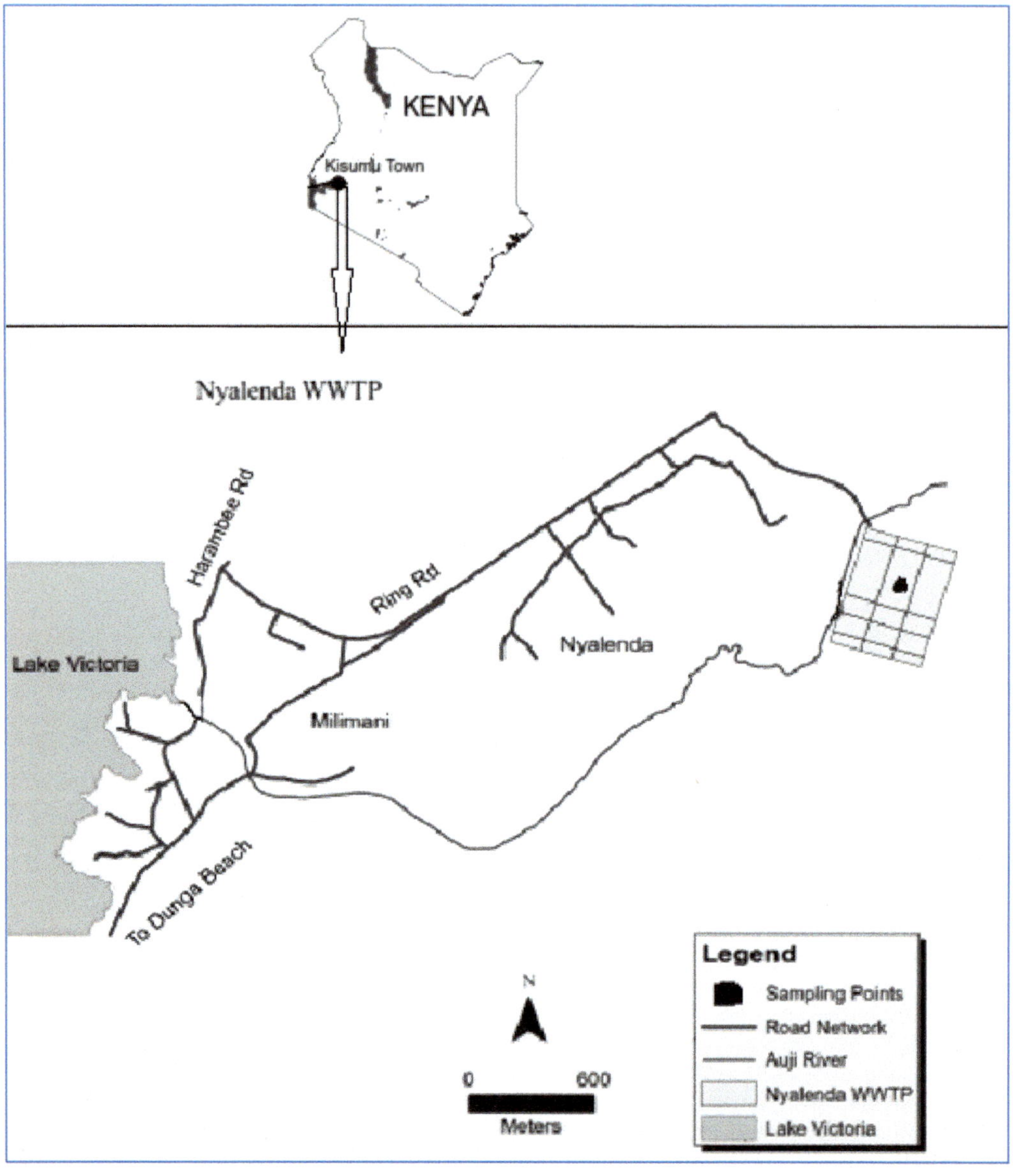

Figure 1: The map showing the location of Nyalenda waste water lagoons at Kisumu City: Adopted from Lake Victoria North Water & Sewerage Company.

way of metal uptake, environmental conditions (water temperature, pH, hardness, salinity), and intrinsic factors (fish age, feeding habits) [10]. Various metals show different affinity to fish tissues. Most of them accumulate mainly in liver, kidney and gills. Fish muscles, comparing to the other tissues, usually contain the lowest levels of metals [10]. Fish liver is one of organs which is most susceptible to storage of heavy metals and therefore commonly used in analysis [11]. The results of many field studies of metal accumulation in fish living in polluted waters show that considerable amounts of various metals may be deposited in fish tissues without causing mortality [10]. Biomonitoring of trace elements is essesntial to access ecosystem health [12].

Nyalenda Waste Stabilization lagoons (NWSL) treat wastewater before discharge into Winam Gulf (Lake Victoria). During and before the time of study, there lacked information on their efficiency of the lagoons in waste water treatment [13]. Sewage effluent and surface runoffs supply the water bodies and sediment with huge quantities of inorganic anions and heavy metals [14]. The NWSL is a source of Fishery especially the *Oreochromis niloticus*, *Clarius gariespinus* and *Protopterus aethiopicus*. Fish from NWSL are commercially available and is part of the diet of the local population in the Lake Victoria region of East Africa. Therefore there is need to monitor the levels of heavy metals in fish, sediments and water with a view of assessing the current status of pollution in waste water lagoons. In addition, Surface water is the main source of drinking water in riparian states of East Africa that share Lake Victoria water. Going by the limited number of study publications on environmental pollutant indicators such as heavy metals BCFs for fish within the region, there is need for more studies and data report for global readership.

Materials and Methods

Sample collection

Samples of fish, water and sediments were collected in Nyalenda

Wastewater Lagoons which is located 4 kilometers north of Kisumu. Sediments sampling was restricted to the lagoons and up to a depth of 100 cm below the water surface. Sediments were collected using stainless steel spoons which were previously washed at intervals with water in sediment environment. Water sample was collected within the ponds where fish was obtained in 1 L polypropylene bottle. Fish Sampling was done at the same site where sediments were collected and samples put in clearly labeled plastic bags. The fish were then placed in an ice box and transported to the laboratory and stored in a deep freezer prior for analysis.

Fish identification, maturity, diet and mode of feeding

The identification of the fish species were carried out at the Biological department of Masinde Muliro University of Science and technology. The species of the fish collected were identified as; *Oreochromis niloticus, Clarias gariepinus* and *Proptopterus Aethipicus.* In addition, all the fish were confirmed as mature. The maturity length of *O. niloticus* (Nile tilapia) is 20 cm [15]. Nile tilapia are known to feed on phytoplankton, periphyton, aquatic plants, invertebrates, benthic fauna, detritus, bacterial films [16] and even other fish and fish eggs. Depending on the food source, they feed either via suspension filtering or surface grazing, trapping plankton in plankton rich bolus using mucus excreted from their gills [17]. *O. niloticus* have been observed to exhibit trophic plasticity according to the environment and the other species they coexist with [18]. *C. gariespinus* reaches maturity towards the end of the second year (350 mm Total length (TL), rarely towards the end of the first year, 240 mm TL). *Clarius gariespinus* and *Protopterus aethiopicus* respectively reaches maturity at around 24 cm TL and 70 - 76 cm TL [9,19] towards the end of the second year (350 mm TL, rarely towards the end of the first year, 240 mm TL).

Laboratory apparatus and chemicals

The laboratory apparatus comprised the following: glass wool, propylene bottles, 50 ml volumetric flask (Pyrex borosilicate glass, Germany) stainless steel raddle with elongated handle, samples drying papers, pestle and motors, dissection kit, unit sieve, weighing balance, digestion block, electric hot block. All the chemicals were purchased from Sigma-Aldrich through Kobian limited, Kenya. Reagents used were all of analytic grade, concentrated nitric acid, hydrochloric acid, sulphuric acid and hydrogen peroxide purchased from Sigma-Aldrich through Kobian limited, Kenya.

Pretreatment and digestion of fish tissues

Upon identification weight and length of the fish were measured, and recorded as follows; 1650 g and 70 cm, 1762 g and 72 cm, 152.9 g and 25.2 cm) for *C. gariespinus*, *P.aethopicus* and *O. niloticus* respectively. Samples were allowed to thaw for at least 6 hours, and then rinsed with deionized water. Fish muscles were dissected from dorsal lateral part ventrally to the dorsal fin. Liver, gills, scales (from *O. niloticus*) and skin (from *C. gariespinus* and *P.aethopicus*) were then separated from respective fish sp. An aliquot of 5 g of weighed fish tissues and whole liver and gills for *O. niloticus* samples were put in digestion tubes and placed on aluminum digestion block. 4.0 ml of nitric acid (14N), a drop of (30%) silicon antifoaming reagent, and 4 ml of concentrated sulphuric acid (18.3) were then added to the samples tubes. The resultant mixture was heated at 60°C for 25 minutes. Further 12 ml concentrated nitric acids were added and digested as follows: 90 - 100°C for one hour, 100 -110°C for one hour, 110 -120°C for one hour. The digest were cooled to 80C then 2.0 ml of 30% (V/V) hydrogen peroxide was added to oxidize the organic matter present. The digests were then returned to hot block and heating continued at 130˚C for 30 minutes. This was then followed by filtration through small piece of glass wool (soaked in (10% V/V)) nitric acid and rinsed with dionized water). The digests were transferred into 50 ml volumetric flask and made to the mark using deionized water and stored in propylene bottles ready for ICP analysis. The *O. niloticus* scales obtained were placed on a clean petri dish and oven dried to a constant weight at 100°C. An aliquot of 2 grams of dry scale were weighed into digestion tubes and 5 ml of concentrated nitric acid (14 N) added. The resultant mixture was digested at 100°C for 3 hours (until brown fumes ceased). After cooling to 80°C, 1 ml of 30% w/v H_2O_2 was added. The digests were further returned to the hot plate and further heated at 100°C for one hour. The digests were finally cooled and filtered through glass wool and made to the mark with deionized water. The digest were finally transferred to 50 ml flask added to the mark and stored in polypropylene bottles ready for analysis.

Sediment and water analysis

The sediments were oven dried at 10°C for 10 hours, ground and sieved through 60 µm sieve size. Coarse particles were discarded. For analysis 2.0 g of sediments were weighed and put in digestion tubes and 20 ml mixture of concentrated hydrochloric (36%) and nitric acid (1:4) were added and digested at 70°C for 25 minutes, 100°C for 3 hours, followed by addition of 1 ml hydrogen peroxide and further heated for 30 minutes at 100°C on aluminum heating block. The digest were then allowed to cool and filtered into 50 ml volumetric flask, through a piece of cotton that was previously soaked in acid (10% nitric acid) and rinsed with deionized water. Water sample were collected using 500 ml polypropylene bottles that had been previously rinsed with deionized water and further rinsed with the sample water and acidified with 10% nitric acid. Water samples were then digested using procedure highlighted [20].

Instrumental analysis

Determination of the elements in all samples was carried out by inductively coupled plasma mass spectrophotometer ICP-MS SPECTRO ACROS (Germany) with ICP-OES analyzer. The detection wavelengths of the analytes were set at 228.8 nm, 324.8 nm, 213.8 nm, 267.72 nm and 283.3 nm respectively for Cd, Cu, Zn, Cr and Pb respectively. The quantification limits for individual analytes were as follows; Zn 0.01 mg/kg, Cu 0.14 mg/kg, Pb 0.02 mg/kg, Cd 0.1 mg/kg, and Cr 0.1 mg/kg.

Bioconcentration factors (BCFs) estimations

The calculated estimates of BCF in this study were as reported by Gobas, et al. [21] where bioconcentration factors (BCF) are defined as the ratio of the steady-state metal ions concentrations in the three fish sp. vs the concentration in water/sediments. Using the same equation 1, BCFs were tested based on sediment concentration for *P.aethiopicus* and *C. gariespinus* which are bottom feeders.

The assumption made in this estimation was (i) that the surrounding conditions, feeding, excretion by fish and chemical concentrations in the lagoon had attained steady-state; (ii) that the ratio of k1/k2 which is known as the kinetic BCF is equal to the steady state, and (iii), that BCF obtained by the ratio of the steady-state concentration in fish to that in water followed the First-order kinetics.

This assumption is shown by the following mathematical expressions;

Rate of uptake = $K_1 \times$ [Cwater]................................**Expression 1**

Rate of depuration = $K_2 \times$ [Cfish]........................ .**Expression 2**

Where K_1 = First-order rate constant for uptake into fish (day^{-1}).

K_2 = First-order rate constant for depuration/elimination from fish (day^{-1}).

[Cwater] = Concentration in water (mg /L).

[Cfish] = Concentration in fish (mg kg^{-1} wet weight).

At steady state the rate of uptake = the rate of depuration and so combining the two expressions gives the following relationship (expression 3) which is:

$$BCF = \frac{\text{Concentration in fish at steady state}\left(\text{mg / kg wet fish}\right)}{\text{Concentration in water at steady state}\left(\text{mg / L}\right)} = \frac{K1}{K2}$$

....**Expression 3**

The BCFs were calculated using the following equation;

$$BCF = \frac{\text{Concentration in fish at steady state}\left(\text{mg / kg wet fish}\right)}{\text{Concentration in water}\left(\text{or sediment}\right)\text{at steady state}\left(\text{mg / L}\right)}$$

....**Equation 1**

Results and Discussion

Concentration of heavy metals in fish tissue

All metals analyzed in this study were present in all the three fish sp. Zn was found to be the most abundant metal ion in all the tissue organs of the fish analyzed and it ranged from 0.68-10.2 mg/kg as shown in Figure 2. The highest concentration of Zn was in *P.aethiopicus* muscles 10.2 mg/kg, followed by the gills, skin and liver with respective concentrations of 3.44, 2.78 and 1.92 mg/kg. for *P. gariepinus*, the Zn concentration in organ tissues was in the following order; muscles, liver, gills and skin. In contrast to *P.aethiopicus* and *C. gariespinus*, Zn concentration in *O. niloticus* had higher Zn levels (4.12 mg/kg) in the scales, followed by the gills, muscles, and then liver. Cd was only detected in *P.aethiopicus* skin and gills with concentrations of 6.20 mg/kg and 1.60 mg/kg respectively. Cd is a non-essential heavy metal in the biochemical role in the life processes of fish and its obtained concentration in skin was of interest. The Cd present in the skin could indicate Cd ions depuration process through the skin. Cu was higher in muscles of *C. gariespinus* and *P.aethiopicus* compared to the other tissue organs but higher in scales of *O. niloticus*. The highest concentration of Cu was in *P.aethiopicus* muscles (4.63 mg/kg). Pb was detected in all fish sp tissue organs. Scales and skin were observed to contain more Pb than muscles and liver, in both fish sp. (Figure 2). Cr was detected in all the fish species tissue organs.

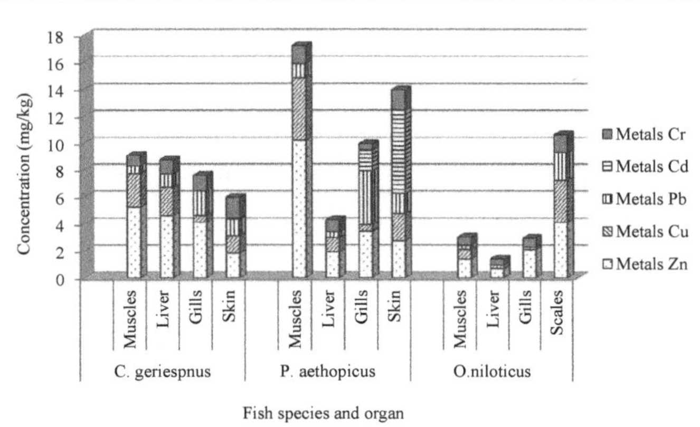

Figure 2: Tissue concentrations and distribution of Heavy metals in *Oreochromis niloticus*, *Clarius gariespinus* and *Protopterus aethiopicus*.

There was no positive correlation of metal ions concentration for the various organ tissues studied. Previous studies indicated that the variability observed in the metal levels of different species depended on ecological needs, metabolism [22], age, size and length of the fish [23] feeding habits [24] and their habitats [25].

Concentration in water and sediments

Table 1 above shows results of the heavy metals concentration in wastewater collected from several sampling points in Nyalenda lagoons. The mean concentration range of Heavy metals in the wastewater were from 0.51 ± 0.2 mg/kg (Cu) to 1.02 ± 0.5 mg/kg (Pb) as shown in Table 1. Sediments had higher heavy metals concentration in orders of magnitude of more than 4 to about 30 more than the concentrations in wastewater except for Cr. This study recommends Cr speciation study as it is of interest to know the form (oxidation state) of Cr ion in fish organ tissues, wastewater and sediments.

Bioconcentration factors

The BCFs were calculated using equation 1 above and results shown in Table 2. Zn had the highest BCF for all the fish sp. and ranged from 0.78 for *O.niloticus* liver to 11.72 for *P.aethiopicus* muscles. The BCFs for Cd were only calculated for the gills and skin of *P.aethiopicus* where the concentrations were above the levels of detection. Estimation of BCFs for Zn, Cu, Pb, Cd, and Cr in *C. gariespinus* and *P.aethiopicus* tissues based on sediments concentration are tabulated in Table 3. The two fish sp. is bottom feeders and ingests heavy metals that are adsorbed on sediment materials. The concentrations of heavy metals in sediments were several orders of magnitude higher than the concentrations in wastewater, for

Sample	Zn	Cu	Pb	Cd	Cr
Wastewater	0.87 ± 3.0	0.51 ± 0.2	1.02 ± 0.5	0.55 ± 0.20	1.10 ± 0.4
Sediments	28.38 ± 3.7	4.19 ± 0.8	13.15 ± 2.0	32.88 ± 2.5	1.38 ± 0.3

Table 1: Concentration of Heavy metals in wastewater (mg/L) and sediments (mg/kg).

Fish sp.	Metal	Fish Tissue BCF			
		Liver	Muscle	Gills	Scales
Orechromis niloticus	Zn	0.78	1.61	2.39	4.74
	Cu	0.55	1.39	0.35	6.06
	Pb	0.24	0.3	ND	2.03
	Cd	ND	ND	ND	ND
	Cr	0.41	0.54	0.62	1.23
		Liver	Muscle	Gills	Skin
Clarius gariespinus	Zn	5.29	6.02	4.74	2.11
	Cu	4.1	4.84	0.96	2.49
	Pb	1.01	0.57	1.8	1.19
	Cd	ND	ND	ND	0.2
	Cr	0.93	0.71	1.02	1.39
		Liver	Muscle	Gills	Skin
Protopteru aethipicus	Zn	2.21	11.72	3.95	3.2
	Cu	2.06	9.08	0.96	3.88
	Pb	0.44	1.01	3.88	1.47
	Cd	ND	ND	2.91	11.27
	Cr	0.93	0.71	1.02	1.36

ND: Below detection limit.

Table 2: BCFs in L/kg for Zn, Cu, Pb, Cd, Cr in Fish sp. tissues based on aqueous contact.

Fish sp.	Metal	Fish Tissue BCF			
		Liver	Muscle	Gills	Scales
Clarius gariespinus	Zn	1.1	1.25	0.98	0.44
	Cu	1.51	1.79	0.36	0.92
	Pb	0.04	0.02	0.06	0.04
	Cd	ND	ND	ND	0.01
	Cr	0.03	0.02	0.03	0.05
Protopterus aethiopicus		Liver	Muscle	Gills	Skin
	Zn	0.46	2.43	0.82	0.66
	Cu	0.76	3.36	0.36	1.43
	Pb	0.02	0.04	0.14	0.05
	Cd	ND	ND	0.12	0.47
	Cr	0.03	0.02	0.03	0.05

ND: Below detection limit.

Table 3: BCFs for Zn, Cu, Pb, Cd, Cr in *Clarius gariespinus* and *Protopterus aethiopicus* tissues based on sediments concentration.

Zn, Cu, Cd, and Pb. Cr had highest concentration in sediments than in wastewater (Table 1). Therefore it was necessary to compare the BCFs that were based on sediment heavy metals concentration and those based on aqueous contact (wastewater concentration). Based on sediment concentration, the BCFs indicated bioaccumulation of heavy metals in Liver and Muscles of *C.gariespinus* for Zn and Cu which had factors>1.0. For *P. aethopicus,* bioaccumulation was observed in Muscles for Zn and Cu and Skin for Cu.

Bioconcentration of Heavy metals in *O.niloticus* were observed as follows, Zn in muscles (BCF 1.61), gills (BCF, 2.39) and scales (BCF, 4.74). Cu bioconcentration was observed in muscles and scales with BCFs of 1.39 and 6.06 respectively, Pb and Cr bio concentration were observed in scales with BCFs of 2.03 and 1.23 respectively. Bioconcentration in *C. gariespinus* was observed for Zn in the entire organ tissues analyzed, Cu in liver, muscles and skin, Pb in gills and Cr in skin and the values were tabulated (Table 2). For *P.aethiopicus,* bioconcentration was observed as follows; Zn with BCFs of 2.21, 11.72, 3.95, and 3.20 for liver, muscles gills and skin respectively. Cu for liver, muscles and skin, Pb in gills and skin, Cd in gills and skin and Cr in skin (Figure 3). This variability in BCFs results show that bioconcentration of heavy metals in the fish sp. studied does not exclusively dependent on the concentration levels by aqueous exposure alone. As observed in Figure 3, the concentration of heavy metals in wastewater sample was lower than the concentrations obtained in various fish organs. These observations were an indication of bio-concentration in some of the fish organs analyzed. The bioconcentration observed in fish in this study as be explained as follows; *C. gariespinus* and *P.aethiopicus* are bottom feeders and a close similarly in the observed bioconcentration pattern of Zn, Cu, Pb and Cr points to the mode of feeding and diet as a major factor that determine heavy metals accumulation in these two fish sp. The difference in the amounts of various metal ions accumulation in fish body result from different affinity of metals to fish tissues, different uptake, deposition, and excretion rates [10]. Generally, metal levels in live fish usually follow the ranking: Fe > Zn > Pb > Cu > Cd > Hg. The levels of Zn may be very high, up to over 300 µg/g d. w., while the maximum concentrations of lead and copper are lower and usually do not exceed 10 µg/g d. w. Cadmium is accumulated by the fish in very low amounts, below 1 µg/g d.w. [10]. Previous studies in both, field and laboratory established a relationship between metal concentrations in fish and in the water [26,27]. It should however, be emphasized that body metal level is related to its waterborne concentration only if metal is taken up by the fish from water. If diet and feeding mode is the main source of metal, such a relationship does not necessarily occur.

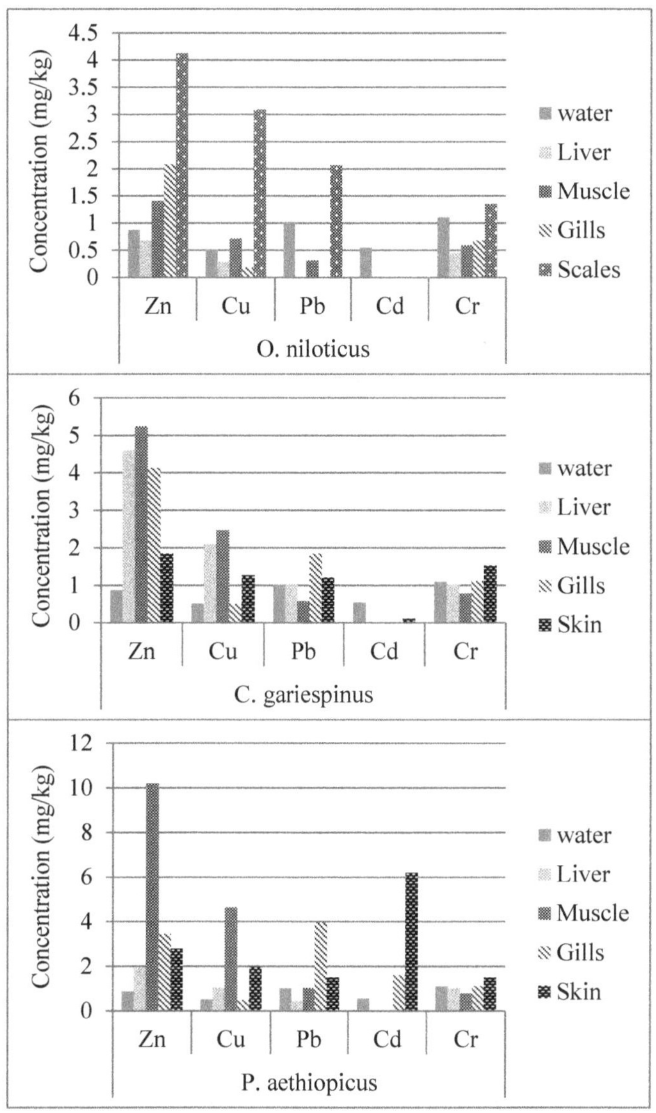

Figure 3: Heavy metals levels in organ tissues in *O. niloticus, C. gariespinus* and *P.aethiopicus* compared to levels in water.

Conclusion

Bioaoncentration transfer factors (BCFs) were estimated for heavy metals uptake in *O. niloticus, C. gariespinus* and *P.aethiopicus* for all the metals analysed except Cd. The BCFs ranged from 0.20 for Cd in scales of *C. gariespinus* to 11.72 for Zn in muscles of *P.aethiopicus*. The BCFs obtained based on wastewater heavy metals concentration and those obtained based on sediment concentration showed variations and therefore no organ tissue metal ions concentrations correlation among the fish species studied. The results suggested that heavy metals uptake by the fish in Nyalenda lagoons and their BCFs were largely depended on their feeding mode and diet. Scales and skin were observed to contain more Pb than muscles and liver, in the three fish sp. Cd present in the skin (BCF of 11.27 for *P.aethiopicus*) could indicate Cd ions depuration process through the skin, considering that Pb and Cd are non-essential metal ions to these fish sp. Furthermore, results from this study discourage consumption of fish catch from Nyalenda lagoons under the present levels of heavy metals.

Acknowledgement

We also thank the Lake Victoria North Water Services Board and Lake Victoria South Water Services Board for making sampling exercise possible. We also acknowledge the Department of Biological Sciences, Masinde Muliro University of Science and Technology for Fish taxonomy Studies.

References

1. Crookes M, Brooke D (2011) Introduction. In Estimation of fish bioconcentration factor (BCF) from depuration data. Environment Agency. Environment Agency, Evidence. Horizon House, Deanery Road, Bristol, BS1 5AH. 11-12.
2. Lwanga MS, Kansiime F, Denny P, Scullion J (2003) Heavy metals in Lake George, Uganda, with relation to metal concentrations in tissues of common fish species. Hydrobiologia 499: 83-93.
3. Jibiri NN, Adewuyi GO (2008) Radionuclide contents and physico-chemical characterization of solid waste and effluent samples of some selected industries in the city of Lagos, Nigeria. Radioprotection 43: 203-212.
4. Mansour S A, Sidky MM (2003) Ecotoxicological studies. 6. The first comparative study between Lake Qarun and Wadi El-Rayan wetland (Egypt), with respect to contamination of their major components. Food Chemistry 82: 181-189.
5. Uysal K, Emre Y, Kose E (2008) The determination of heavy metal accumulation ratios in muscle, skin and gills of some migratory fish species by inductively coupled plasma-optical emission spectrometry (ICP-OES) in Beymelek Lagoon (Antalya/Turkey). Microchem J 90: 67-70.
6. Adeniyi AA,Yusuf KA (2007) Determination of heavy metals in fish tissues, water and bottom sediments from Epe and Badagry Lagoons, Lagos, Nigeria. Environmental Monitoring and Assessment 37: 451-458.
7. Storelli MM, D'ddaabbo A, Morano R, Bruno C, Marcotrigiano GO (2005) Trace Elements in loggerhead turtles (Caretta caretta) from the Eastern Mediterranean Overview and Evaluation. Environ Pollut 135: 163-170.
8. Kamaruzzaman B, Razarudin Y, Shazili I, MohdLokman H, Sulong I (2001) The geochemistry of Ba, U, Cd and Mn in the surface sediments of Matang Mangrove, Malaysia. Oriental Journal of Chemistry 17: 377-382.
9. Farombi EO, Adelowo OA, Ajimoko YR (2007) Biomarkers of oxidative stress and heavy metal levels as indicators of environmental pollution in African cat fish (Clarias gariepinus) from Nigeria Ogun River. Int J Environ Res Public Health 4: 158-165.
10. Jezierska B, Witeska M (2006) The metal uptake and accumulation in fish living in polluted waters. Soil and Water Pollution Monitoring, Protection and Remediation 69: 107-114.
11. Yilmaz AB (2003) Levels of heavy metals (Fe, Cu, Ni, Cr, Pb, and Zn) in tissue of Mugil cephalus and Trachurus mediterraneus from Iskenderun Bay, Turkey. Environ Res 92: 277-281.
12. Mico C, Peris M, Sanchez J, Recatala L (2006) Heavy metal content of agricultural soils in a Mediterranean semiarid area: the Segura River Valley (Alicante, Spain). Spanish Journal of Agricultural Research 4: 363-372.
13. Musungu P, Jondiko J, Lalah J ,Ongeri D, Chepkui R, et al. (2013) The extent of nutrient removal by Wastewater Treatment Plants along the Nyalenda wigwa stream and the River Kisat (Kenya). Ecohydrology & Hydrobiology 13: 236-240.
14. ECDG (2002) "European Commission DG ENV. E3 Project ENV.E.3/ ETU/2000/0058. Heavy metals in waste," Final report.
15. Bwanika GN, Makanga B, Kizito Y, Chapman LJ, Balirwa J (2004) Observations on the biology of Nile tilapia, Oreochromis niloticus, L., in two Ugandan Crater lakes. African Journal of Ecology 42: 93-101.
16. FAO (2005-2012) In: Rakocy J.E (ed.) Cultured Aquatic Species Information Programme. Oreochromis niloticus. Cultured Aquatic Species Information Programme. FAO Fisheries and Aquaculture Department.
17. Fryer G, Iles TD (1972) The Cichlid Fishes of the Great Lakes of Africa: Their biology and Evolution. T.F.H. Publications, Hong Kong.
18. Bwanika G, Murie D, Chapman L (2007) Comparative Age and Growth of Nile tilapia (Oreochromis niloticus L.) in Lakes Nabugabo and Wamala, Uganda. Hydrobiologia 589: 287-301.
19. Mlewa CM, Green JM (2005) Biology of the marbled lungfish, Protopterus aethiopicus Heckl, in Lake Baringo, Kenya. African Journal of Ecology 42: 338-345.
20. Lalah JO, Ochieng EZ, Wandiga SO (2008) Sources of heavy metal input Into Winam Gulf, Kenya. Bull Environ Contam Toxicol 81: 277-284.
21. Gobas FAPC, Wolf D, Burkhard W, Verbruggen LP, Plotzke K, et al. (2009) Revisiting bioaccumulation criteria for POPs and PBT assessments. Integrated Environmental Assessment and Management 5: 624-637.
22. Canlı M, Kalay M (1998) Levels of heavy metals (Cd, Pb, Cu, Cr and Ni) in tissue of Cyprinus carpio, Barbus capito and Chondrostoma regium from the Seyhan River, Turkey. Turkish Journal Zooogy 22: 149-157.
23. Al-Yousuf M, El-Shahawi MS, Al-Ghais SM (2000) Trace metals in liver, skin and muscle of Lethrinus lentjan fish species in relation to body length and sex. Sci Total Environ 256: 87-94.
24. Watanabe KH, Fesimone FW, Thiyagarajah A, Hartley WR, Hindrichs AE (2003) Fish tissue quality in the lower Mississippi River and health risks from fish consumption. Science of the Total Environment 302: 109-126.
25. Canli M, Atli G (2003) The relationships between heavy metal (Cd, Cr, Cu, Fe, Pb, Zn) levels and the size of six Mediterranean fish species. Environ Pollut 121: 129-136.
26. Linde AR, Arribas P, Sanchez-Galan S, Garcia-Vazquez E (1996) Eel (Anguilla anguilla) and brown trout (Salmo trutta) target species to assess the biological impact of trace metal pollution in freshwater ecosystems. Arch Environ Contam Toxicol 31: 297-302.
27. Zhou T, Weis P, Weis JS (1998) Mercury burden in two populations of Fundulus heteroclitus after sublethal methylmercury exposure. Aquat Toxicol 42: 37-47.

Permissions

All chapters in this book were first published in JPEC, by OMICS International; hereby published with permission under the Creative Commons Attribution License or equivalent. Every chapter published in this book has been scrutinized by our experts. Their significance has been extensively debated. The topics covered herein carry significant findings which will fuel the growth of the discipline. They may even be implemented as practical applications or may be referred to as a beginning point for another development.

The contributors of this book come from diverse backgrounds, making this book a truly international effort. This book will bring forth new frontiers with its revolutionizing research information and detailed analysis of the nascent developments around the world.

We would like to thank all the contributing authors for lending their expertise to make the book truly unique. They have played a crucial role in the development of this book. Without their invaluable contributions this book wouldn't have been possible. They have made vital efforts to compile up to date information on the varied aspects of this subject to make this book a valuable addition to the collection of many professionals and students.

This book was conceptualized with the vision of imparting up-to-date information and advanced data in this field. To ensure the same, a matchless editorial board was set up. Every individual on the board went through rigorous rounds of assessment to prove their worth. After which they invested a large part of their time researching and compiling the most relevant data for our readers.

The editorial board has been involved in producing this book since its inception. They have spent rigorous hours researching and exploring the diverse topics which have resulted in the successful publishing of this book. They have passed on their knowledge of decades through this book. To expedite this challenging task, the publisher supported the team at every step. A small team of assistant editors was also appointed to further simplify the editing procedure and attain best results for the readers.

Apart from the editorial board, the designing team has also invested a significant amount of their time in understanding the subject and creating the most relevant covers. They scrutinized every image to scout for the most suitable representation of the subject and create an appropriate cover for the book.

The publishing team has been an ardent support to the editorial, designing and production team. Their endless efforts to recruit the best for this project, has resulted in the accomplishment of this book. They are a veteran in the field of academics and their pool of knowledge is as vast as their experience in printing. Their expertise and guidance has proved useful at every step. Their uncompromising quality standards have made this book an exceptional effort. Their encouragement from time to time has been an inspiration for everyone.

The publisher and the editorial board hope that this book will prove to be a valuable piece of knowledge for researchers, students, practitioners and scholars across the globe.

List of Contributors

Anne Priac, Amandine Poupeney, Coline Druart and Grégorio Crini
Chrono-Environment, UMR 6249 usc INRA, University of Franche-Comté/CNRS, 16 route de Gray, 25000 Besançon, France

Pereira MDG, Sachdev RDL, Santos AVD and Souza LA
State University of Bahia, Department of Exact and Earth Sciences, R. Silveira Martins, Salvador-BA, Brazil

Souza CLM and Pinto PADC
State University of Bahia, Department of Technology and Social Sciences Av. Dr. Chastinet Guimarães, S/N, Juazeiro-BA, Brazil

Ribeiro JN, Oliveira JPD3 and Guimarães MCC
Federal University of Espírito Santo, Center for Health Sciences, Maruípe Avenue, S/N, Vitória-ES, Brazil

Ribeiro AVFN
Federal Institute of Espírito Santo, Undergraduate Coordination, Av. Vitória, Vitória-ES, Brazil

Ali M Abdullah
Holding Company for Water and Wastewater and IGSR, Alexandria University, Cairo, Egypt

Salah El-Dien Hussona
Department of Environmental Studies, IGSR, Alexandria University, Cairo, Egypt

Nadia Morin-Crini, Coline Druart, Caroline Amiot, Frédéric Gimbert, Etienne Chanez and Grégorio Crini
Department of Chrono-Environment, UMR UFC/CNRS 6249 USC INRA, University of Franche-Comté, 16 Route de Gray, 25030 Besançon cedex, France

Jadoon S and and Khalel H
Department of Natural Resources Engineering and Management, University of Kurdistan, Hewlêr, Iraq

Amin AA
Erbil Technology Institute, Erbil Polytechnic University, Kurdistan, Iraq

Malik A
Institute of Molecular Biology and Biotechnology, University of Lahore, Lahore, Pakistan

Ali M Abdullah
Reference Laboratory for Drinking Water, Holding Company for Water and Wastewater, Alexandria University, IGSR, Egypt

Qureshimatva Umerfaruq M and Solanki HA
Department of Botany, University School of Sciences, Gujarat University, Ahmedabad-380 009, Gujarat, India

Archita Bharti
Green Health Foundation, NGO, Mumbai, India

Devansh Jalota
Dhirubhai Ambani International School, Mumbai, India

Ghaly AE and Ramakrishnan VV
Department of Process Engineering and Applied Science, Dalhousie University, Halifax, Nova Scotia, Canada

David Noel S and Rajan MR
Department of Biology, Gandhigram Rural Institute-Deemed University, Gandhigram-624302, Tamilnadu, India

Liu Guo-feng
Freshwater Fisheries Research Center of Chinese Academy of Fishery Sciences, Wuxi, 214081, China
Institute of Agricultural Resource and Environment, Jiangsu Academy of Agricultural Science, Nanjing 210014, China
State Key Laboratory of Lake and Environment, Nanjing Institute of Geography & Limnology, CAS, Nanjing 210008, China

Han Shi-qun
Institute of Agricultural Resource and Environment, Jiangsu Academy of Agricultural
Science, Nanjing 210014, China

Zhang lu and Fan Cheng-xin
State Key Laboratory of Lake and Environment, Nanjing Institute of Geography & Limnology, CAS, Nanjing 210008, China

He Jun
Fisheries Technical Guidance Station of Wuxi Municipal Bureau of Agriculture, Forestry, Wuxi 214023, China

Hans W. Paerl
Institute of Marine Sciences, University of North Carolina, USA

Joachim Ibeziako Ezeji
Rural Africa Water Development Project (RAWDP), Nigeria

Adeyemi Olalekan
Department of Environmental Science, Federal University of Petroleum Resources, PMB 1221, Effurun, Nigeria

Ali M Abdullah
Holding Company for Water and Wastewater, Cairo, Egypt

Abdullahi SA
Faculty of Engineering, Ahmadu Bello University, Zaria, Nigeria

Ali M Abdullah
Holding Company for Water and Wastewater, Cairo, Egypt

Qing Li, Fengxiang Qiao and Lei Yu
Innovative Transportation Research Institute, Texas Southern University, Houston, Texas, USA

Redouane F and Mourad L
LAAR University of Science and Technology, Mohammed Boudiaf , BP 1505, Oran, Algeria

Ogunola Oluniyi Solomon
International Studies in Aquatic Tropical Ecology, University of Bremen, Germany

Thava Palanisami
Global Centre for Environmental Remediation, University of Newcastle, Adelaide, Australia

Sakhuja N
Department of Applied Sciences, JMIT Radaur, District Yamunanagar, India

Sharma V
PG Department of Industrial Chemistry, Guru Nanak Khalsa College, Yamunanagar, India

Aya FC
Department of Soil Science and Environmental Management, faculty of agriculture and natural resources management, Ebonyi State University, P.M.B 053, Abakaliki, Nigeria

Nwite JN
Ebonyi State Ministry of Environment, Abakaliki, Nigeria

Rono AK
Department of Biological Sciences, Moi University, Kenya

WenLi Zhang, Xuemei Han, Jingxiu Han and Qi Shang
Department of Environmental Epidemiology, Institute for Environmental Health and Related Product Safety, Chinese Center for Disease Control and Prevention, Beijing, 100050, China

Jiangang Zheng
Jiangxi Provincial Center for Disease Control and Prevention, Nanchang, 330046, China

Dancheng Yao
Guizhou Provincial Center for Disease Control and Prevention, Guiyang, 550004, China

Ibrahim SM
Department of Zoology, VOC College, Millepuram, Tuticorin-628 008, Tamil Nadu, India

Theivandran G and Thamilmaraiselvi B
Department of Microbiology, Cauvery College for Women, Tiruchirappalli-620018, Tamilnadu, India

Thomas Arap Bii, Isaac W Mwangi, Ruth N Wanjau, Sauda Swaleh and Manohar Ram
Department of Chemistry, Kenyatta University, Nairobi, Kenya

Ngila JC
Department of Chemical Technology, University of Johannesburg, Doornfontein Campus, Doornfontein 2028, Corner Beit and Nind Street, South Africa

Cahit Kasimoglu
Institute of Science Biology, Mugla Sitki Kocman University, Mugla-48000, Turkey

Mobegi Erick K, Nyambaka Hudson N and Nawiri Mildred P
Department of Chemistry, School of Pure and Applied Science, Kenyatta University, Nairobi, Kenya

Sulaiman Ali Al Yousef, Eman Saleh Farrag, Ahmed Mohamed Ali and Sabry Younis Mahmoud
Department of Medical Laboratory Technology, College of Applied Medical Science, University of Hafr Al Batin, Kingdom of Saudi Arabia

Francis Orata and Faith Birgen
Department of Pure and Applied Chemistry, Masinde Muliro University of Science and Technology, Kakamega, Kenya

Index